Y0-AAP-575

CMOS Digital Circuit Technology

MASAKAZU SHOJI

Computing Science Research Center
AT&T Bell Laboratories

PRENTICE HALL, *Englewood Cliffs, New Jersey 07632*

Library of Congress Cataloging-in-Publication Data

Shoji, Masakazu
CMOS digital circuit technology / Masakazu Shoji
p. cm. -- (Computer system series)
Bibliography: p.
Includes index.
ISBN 0-13-138850-9
1. Metal oxide semiconductors, Complementary. I. Title.
II. Series.
TK7871.99.M44S52 1987
621.381'73--dc19 87-28071
CIP

Editorial/production supervision
and interior design: *Carolyn Fellows*
Cover design: *Photo Plus Art*
Manufacturing buyer: *Paula Benevento*

Published by Prentice-Hall, Inc.
A Division of Simon & Schuster
Englewood Cliffs, New Jersey 07632

Prentice Hall Computer System Series, Dr. Amar Mukherjee, Editor.
Also available: Amar Mukherjee, *Introduction to* n*MOS and CMOS VLSI Systems Design*.

Printed in the United States of America
10 9 8 7 6 5 4 3 2

ISBN 0-13-138850-9 025

Prentice-Hall International (UK) Limited, *London*
Prentice-Hall of Australia Pty. Limited, *Sydney*
Prentice-Hall Canada Inc., *Toronto*
Prentice-Hall Hispanoamericana, S.A., *Mexico*
Prentice-Hall of India Private Limited, *New Delhi*
Prentice-Hall of Japan, Inc., *Tokyo*
Simon & Schuster Asia Pte. Ltd., *Singapore*
Editora Prentice-Hall do Brasil, Ltda., *Rio de Janeiro*

Contents

Preface

CMOS (complementary metal-oxide-semiconductor) technology is now the forefront technology for very large scale integration (VLSI) of electronic systems. As more systems are integrated using CMOS and as technology evolves, higher performance is demanded from CMOS VLSI chips. Fast and reliable operation of CMOS circuits is the basic requirement for high-performance CMOS VLSI chips.

The technique of designing high-performance CMOS circuits has long been restricted to the technical know-how of the practicing engineers, which has never been systematized. Furthermore, some authors have thought that there is nothing left to be discovered in MOS digital circuits. This viewpoint is totally unfounded. Because of the complexity of a VLSI chip and because of the nonlinearity of circuit operation at extremely high frequencies, the circuit design problem is often close to basic science, especially physics. This important point has long been overlooked, and because of this neglect some misleading expectations have arisen. A short historical review of the developments in the areas closely related to CMOS circuit technology proves this point.

Scaling down of the feature size of CMOS technology in the last 10 years has been truly remarkable. Ten years ago a CMOS feature size of 5 to 7.5 $ mu $m was state-of-the-art, and there continued to be skepticism as to whether or not CMOS was a feasible VLSI technology. By the end of 1986, 1 $ mu $m feature-size CMOS technology was available and used increasingly to produce commercial VLSI chips. This technology allows integration of several hundred thousand random logic FETs within a single chip about 1 cm square. There is no longer any doubt that CMOS is a feasible VLSI technology. Indeed, it is the technology of choice.

Chip performance improvement by scale-down has certain limits, however. As the feature size of the technology approaches 1 $ mu $m, effects such as carrier drift velocity saturation, trapping of hot carriers in the gate oxide, and source-drain punch-through do not permit sizeable improvements in active-device capability through further scaledown. Sidewall capacitance limits reduction in interconnect capacitance. As speed improvement by scaledown of feature size diminishes, the speed improvement attained using more sophisticated CMOS circuits becomes important to further the advance of the technology.

Techniques using a large number of FETs efficiently have been developed significantly during the last 10 years. The early, simple microprocessor architecture has been upgraded, and several high-performance architectures have emerged. The state-of-the-art high-performance 32-bit microprocessors, such as the Intel 80386, MC68020, and WE32100, are all implemented using CMOS.

Another direction of development in the last decade is in the area of computer-aided design (CAD). The direct motivation was the large number of FETs that a VLSI designer must deal with and the obvious limitations of a human designer in the task of design verification. In the area of design verification, CAD's contribution has been indisputable. The results have been the virtual elimination of chip debugging problems: Modern VLSI microprocessors often work on the first mask set without having a single short or open circuit or logic mistake.

CAD directed toward eliminating human designers entirely from the VLSI design process has produced mixed results. Techniques directed toward "silicon compilation" have not yet produced state-of-the-art microprocessors. The objective set by early advocates of silicon compilers may have been too ambitious. Further, early CADs often did not satisfy the real needs of IC designers. They were often not user-friendly: Designers had to spend large amounts of time just to make the software work and to interpret confusing results. These experiences, however, created a healthy criticism of the future direction of CMOS VLSI design methods. The present trend of the VLSI industry is to train many experienced designers who can select and use the most efficient methods of design—some automated, some manual. Such a selection and integration of various design methods is essential in the design of compact high-performance chip.

The logic and circuit simulators developed in recent years are powerful tools to optimize design. FET sizes and the parasitics can be optimized to attain the highest performance from a given circuit configuration. The problem has emerged, however, that a designer who is proficient in using powerful software tools is often unable to conceive a circuit structure that is worthwhile optimizing, or finds it difficult to choose the best circuit from many different alternatives. It is obviously impossible to include all the possible circuit configurations in a silicon compiler and let the software make a choice. This type of design decision depends on the experience of the designer and on the understanding of the working mechanisms of the circuits.

As shown in this book, new inventions in CMOS circuits are still being developed. Several examples from the author's own work presented in Chapters 5 and

8 prove this point. It is an unbiased observation of an experienced designer that state-of-the-art performance is expected only from a chip that contains such new inventions in some key components, not from an assembly of conventional circuits that were optimized by software. It is not possible to expect creative intelligence from software at the present time, even from sophisticated software such as a silicon compiler.

This observation does not lead to the conclusion that silicon compilation is unnecessary. On the contrary, ever-increasing numbers of FETs make high-level design software, like the silicon compiler, ever more necessary. Before we attempt to write a truly useful silicon compiler, however, we must upgrade CMOS circuit technology to a systematic science and improve the knowledge of CMOS circuits to the degree of sophistication wherein there are few inventions in circuit configurations left and there are no major new discoveries left in the optimization techniques. To attain this level of circuit technology, we must understand the properties of CMOS circuits in detail. This book is dedicated to that objective.

This explicit objective of improving understanding influenced the selection of topics discussed in this book and the methods of analysis. Since the stress is laid on an understanding of the electrical phenomena in CMOS circuits, CMOS circuits are analyzed using lumped equivalent circuits, rather than layouts. Circuit layout is an important part of design methodology. As for the details of VLSI design methods, many excellent books are available. Therefore, readers are referred to the comprehensive reference books that have been published recently.

ACKNOWLEDGMENTS

This book came to existence through a suggestion of Simon M. Sze, who gave the author continuous encouragement and many valuable comments on the manuscript.

Professor Fred Rosenberger of Washington University, St. Louis, read the entire manuscript, corrected the English, pointed out misleading sentences, and gave valuable technical comments. Professor A. Mukherjee of the University of Central Florida, Professor E. Chenette of the University of Florida, Professor S. M. Kang of the University of Illinois, Professor D. L. Scharfetter of the University of California, Berkeley, Professor S. Soclof of the California State University, and Professor A. J. Strojwas of the Carnegie-Mellon University read all, or part of the manuscript and gave valuable suggestions for improvement.

This book was written in the stimulating atmosphere of the Computing Sciences Research Center of AT&T Bell Laboratories. The Center recognizes that MOS circuit research is an important foundation block of computer science and provided warm support for the project. I owe many members of the Center for stimulating discussions and support: above all T. G. Szymanski, J. H. Condon, P. J. Weinberger, A. G. Fraser, D. Ditzel, R. McLellan, and the members of the CRISP CPU development group of the Center.

Typesetting of this book was carried out using UNIX-based documentation software (GRAP, TBL, EQN and TROFF). The author is grateful for B. W. Kernighan's help in every phase of text development. He provided the newest version of GRAP software, by which all but one of the figures were drawn (the exception is Fig. 7.1, which is a chip photograph). The entire text is processed by a DEC VAX11/750 computer in about 3 hours and 15 minutes. This speed and flexibility is attractive to authors in solid-state physics and electrical engineering.

I learned the art of CMOS VLSI design as a member of the WE32000 series CPU chipset design group. I owe much to the management and members of the group, especially B. T. Murphy, R. H. Krambeck, H. F. S. Law, S. M. Kang, P. W. Diodato, M. S. C. Chung, A. D. Lopez, B. W. Colbry, R. C. Beairsto, D. H. Blahut, and M. J. Killian. A long and continued interaction with Messers Krambeck and Diodato provided stimulation for the circuit research. Mr. Diodato read the first draft and gave valuable comments.

I wish to acknowledge warm support from, and stimulating discussions with, the managers and staff scientists of the Advanced Packaging Technology Project of AT&T Bell Laboratories, especially A. Heller, K. L. Tai, N. Teneketges, M. Jukl, J. Savicki, and D. Schmidt. This new technology provided deep insight into many problems of CMOS circuit design.

During my schooling I was filled with enthusiasm for electronic circuits by Professor A. van der Ziel of the University of Minnesota and Professor Koichi Shimoda of the University of Tokyo. Professors van der Ziel and Shimoda showed and impressed me as a young student with the importance of circuit engineering in basic physics and semiconductor device physics. Further, my immediate family in Japan included several members who showed deep interest in electronic circuits in the early days of electronics. In the 1940s my parents supported my interest in vacuum-tube circuits. The early influence of my immediate family, and my excellent later education convinced me that circuit engineering is not just accumulated knowledge, but a respectable applied science, based directly on the principles of physics. This conviction has been the constant driving force behind my circuit research.

Finally, I am indebted to my wife Elzbieta and son Martin, who gave continuous support in spite of the many difficulties that the family of a high-technology engineer have to live with.

1

CMOS Technology and Devices

1.1 INTRODUCTION

In this chapter the characteristics of CMOS VLSI chip components, field-effect transistors (FETs), and interconnects are summarized.

Properties of CMOS FETs are now well understood and simulation techniques of FET characteristics based on basic semiconductor physics are well established. Several books have been published on MOSFETs [01] [02] [03] [04] [05] [06]. The advanced knowledge of FETs derived from numerical simulation clarified the performance limits of CMOS VLSI chips [07], and the new device physics is reflected in the device models of advanced circuit simulators. At present, practically all the characteristics of a simple CMOS circuit can be studied quantitatively using a circuit simulator such as SPICE [08]. The circuit simulator helps the designer to analyze proposed circuits. Further, the simulation results give the designer clues for optimizing circuit performance.

To conceive a CMOS digital circuit that meets the design requirements and that has many other desirable attributes, such as reliability, is not just a semiconductor device or logic design problem. The desirable properties of digital CMOS circuits on which the designer's attention is focused originate more from the way the devices are interconnected than from the characteristics of the individual device. This book is intended to improve the understanding by CMOS VLSI designers of the working mechanisms of CMOS (FET-level) circuits, and of the electrical phenomena that occur when a number of FETs, diodes, and interconnects are integrated into CMOS digital gates and into a CMOS VLSI chip. To attain this objective, we do not require the latest details of the device physics of FETs. We prefer device theory that is based on

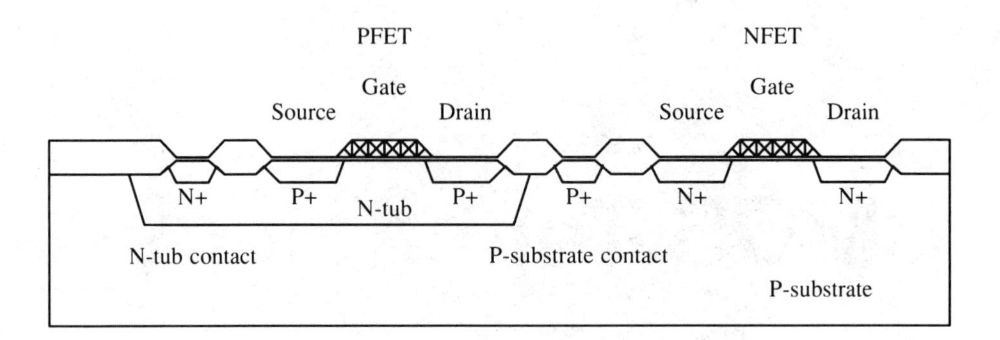

Figure 1.1 Cross section of bulk CMOS integrated circuit.

a simple physical model. In this chapter we summarize the results of such simplified device theories, mostly in closed form, that are used in later chapters to obtain many closed-form results that assist in understanding of CMOS circuits.

1.2 STRUCTURE OF CMOS VLSI

Many varieties of CMOS VLSI device structures have been proposed and fabricated [09]. Figure 1.1 shows the basic structure provided by a typical present-day P-substrate (N-tub) CMOS VLSI processing technology [10]. P-channel FETs (PFETs) and N-channel FETs (NFETs) are fabricated within N-tubs and P-substrate, respectively. The positively biased polysilicon gate of an NFET induces negative electronic charge (an inversion layer) on the surface of the P-substrate. The mobile negative charge carries current from the drain to the source. In a PFET, positive holes induced on the channel by the gate, which is biased negatively, conduct the channel current.

The starting substrate material of a CMOS IC is either P-type or N-type. If the substrate is P-type, N-tubs are mutually isolated structures, each of which is able to take independent positive potential relative to the global P-type substrate. If the substrate is N-type, P-tubs are mutually isolated. Whether or not the more conductive NFETs are on isolated tubs determines the radiation sensitivity of the memory circuit [11]. It is, however, the recent trend that P-type material is used more and more as substrate. Using an independent tub as the base, a bipolar transistor can be made. The collector is connected globally to a power supply potential. A version of CMOS-bipolar technology (abbreviated as "BiMOS" technology) that allows good, independent NPN transistor has been developed [12].

Isolated N-tubs or P-tubs have relatively high sheet resistance (several kilohms per square). The high sheet resistance allows us to design a large resistor using a narrow diffused tub region.

Modern CMOS often uses an epitaxially grown layer on a high-conductivity substrate. The structure has the advantages of reduced latch-up susceptibility and less substrate noise in large-scale dynamic devices such as memory. This is because the highly conductive substrate provides a low-resistance ground or V_{DD} reference that prevents current injected into the substrate to build up a potential difference. This voltage clamping action is very effective in preventing latch-up. The epitaxial material grown on the high-conductivity substrate has high purity to ensure that the subsequent P- and N-tub doping processes create accurately doped P- and N-tubs for FET threshold control. Such a CMOS technology is called a twin-tub CMOS technology.

For cost-sensitive applications, however, traditional bulk CMOS built on moderately doped substrate is still widely used. Both P- and N-type substrates are used. The P-substrate process has the advantage of compatibility with NMOS technology. In this process, NFETs are optimized for performance [10].

Interconnects are, basically, one level of polysilicon or polysilicide and one level of aluminum metal. Silicided polysilicon interconnects are now quite widely available. In the new generation, 1-μm-feature-size CMOS technology, two levels of aluminum interconnects are often available [13]. The structural components of a twin-tub CMOS device are identified in Fig. 1.2 with typical dimensions.

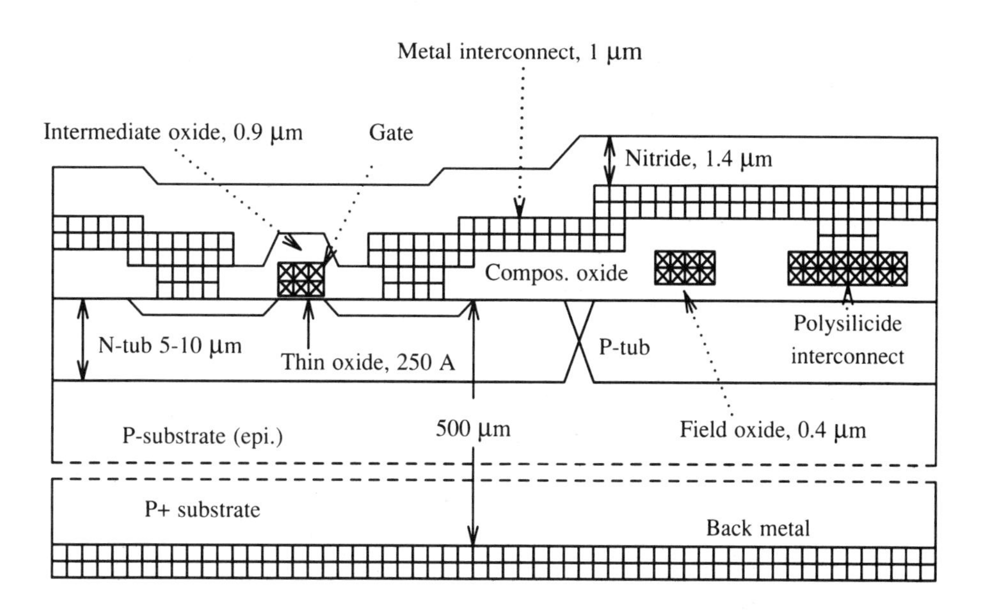

Figure 1.2 Cross-sectional view of twin-tub CMOS integrated circuit.

Polysilicon or polysilicide interconnects are separated from the silicon substrate either by thin oxide (approximately 250 angstroms) or field oxide (approximately 4000 angstroms or 0.4 μm). Metal interconnects are separated from polysilicon by intermediate oxide (approximately 0.9 μm thick). Where polysilicon is absent, field and intermediate oxide layers form a composite oxide (approximately 1.3 μm thick). The metal interconnects are placed on the composite oxide. The entire structure is capped by nitride (1.4 μm thick) to prevent penetration of harmful sodium ions.

The substrate is about 500 μm thick, on the top of which a 15- to 20-μm-thick epitaxial layer is grown. P- and N- tubs are produced by a combination of ion implantation and diffusion, and the junction depth is about 7 to 10 μm. P+ and N+ drain/source islands of modern FETs are very shallow, about 0.3 to 0.5 μm deep, and this small depth reflects the requirement of the self-aligned processing technology, and the requirement that the drain voltage does not influence the threshold voltage of the devices (short-channel effect) [14]. The back side of the substrate is metallized by gold that serves as a solid voltage reference for either ground (P-substrate) or V_{DD} (N-substrate).

Insulating substrate CMOS devices [often called silicon-on-sapphire (SOS) or silicon-on-insulator (SOI) devices] have the structure shown in Fig. 1.3.

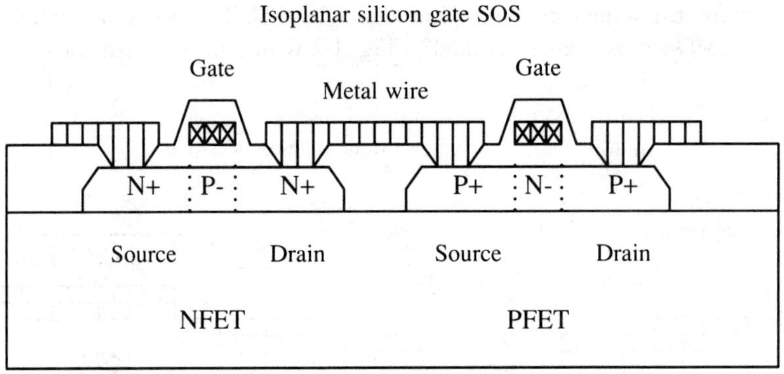

Isoplanar silicon gate SOS

Gate
Gate
Metal wire
N+
P-
N+
P+
N-
P+
Source
Drain
Source
Drain
NFET
PFET

Figure 1.3 Cross-sectional view of SOS CMOS integrated circuit.

Since the parasitic capacitances of wiring and of drain/source islands are smaller than those of bulk CMOS devices, an SOS IC achieves significantly higher frequencies. The chip speed improves by a factor of 1.6 to 2.6, depending on the type of circuit.

A thin (from 0.5 to a few micrometers thick) epitaxial silicon layer is grown on a sapphire substrate [15]. The FETs are isolated by removing the epitaxial silicon layer between the devices. In this structure the area between the FETs is significantly lower, and therefore metal wires have step coverage problems. Various techniques are employed to reduce the step. Sources and drains of the FETs are formed by diffusing impurities all the way through the epitaxial layer to the substrate.

In SOS structures the substrate region under the gate has no connection. This is not an ideal device operating condition: The substrate potential is pulled by the gate and by the drain potential, thereby creating strange phenomena [16]. The most significant is the phenomenon called "kinks." When the drain voltage is increased to a certain point the substrate voltage is pulled up, thereby causing a reduction in the threshold voltage and an increase in the drain current. Generally, SOS CMOS FETs are significantly more leaky than comparable bulk CMOS FETs.

In SOS technology, deep-depletion NFETs are often used instead of enhancement-mode MOS NFETs. A deep-depletion NFET uses a lightly doped N-type epitaxial layer in the channel region [17] [18]. The gate work function, epitaxial silicon layer doping, and thickness are chosen so that the space-charge layer pinches the channel off at zero gate voltage. When positive gate voltage is applied, the conducting channel forms from the bottom of the epitaxial silicon layer. A variation of a deep-depletion FET in bulk CMOS is a buried-channel FET [19] [20]. Buried-channel FETs are used quite widely in standard CMOS technology [21].

1.3 PROCESS STEPS OF BULK CMOS DEVICES

The CMOS process went through many generations of development [22]. A typical bulk CMOS process sequence is as follows. For twin-tub CMOS technology [23] [24] the process sequence is as illustrated in Fig. 1.4 on page 8. The starting material is an N-on-N+ epitaxial wafer. The N+ substrate is highly doped, typically of the order of 0.01 Ω·cm in resistivity. The epitaxial layer is of high-resistivity N-type, typically 50Ω·cm. Using oxide and nitride layers as the masks, phosphorus and boron are implanted in the area of N-tubs and P-tubs, respectively, and diffused into the high-resistivity N-type epitaxial layer. After that, the field oxide is formed on the surface by oxidation in wet oxygen.

Field oxide is removed from the area where a thin gate oxide is formed. The gate oxide layer is grown by oxidization in dry oxygen. Threshold adjustment implantation for PFETs is carried out through the thin oxide. Various techniques are used to attain precise control of FET threshold voltages [25]. A polysilicon layer of about 7000 angstroms is deposited by chemical vapor deposition, and doped by phosphorus. This N+ polysilicon gate provides an NFET threshold voltage of 0.7 V, but the threshold voltage of PFETs would be too high, about 1.9 V. The PFET threshold voltage is reduced by implanting boron in the PFET channel. This processing step precedes the polysilicon deposition, and the process is called threshold adjustment implantation [26].

Polysilicon gate areas are defined by a photomask, and the features are plasma etched. Using polysilicon gate features as a part of masks, P+ and N+ islands are implanted and diffused. This "self-aligned" processing technology overcomes many problems of the older CMOS technologies [27]. Intermediate oxide is then deposited and is re-flowed, to produce a smooth surface. Window areas for contacts are defined and plasma etched. Various techniques are used here to smooth out discontinuities at the window edge. Aluminum is deposited to a thickness of about 1 μm, the metal interconnect area is defined by a photomask, and the unnecessary areas are plasma etched. After sintering the aluminum, a nitride layer is deposited and then bonding pad windows are etched. Deposition of metal on the back side of the silicon substrate completes the process.

The advantage of this twin-tub CMOS technology is that both PFETs and NFETs have optimum substrate doping concentrations. In a CMOS process starting from an N-substrate, however, P-tubs must be doped at least an order of magnitude higher than the N-substrate to maintain stable process control. The high and unbalanced substrate doping creates many undesirable effects on NFETs, such as high drain island parasitic capacitance, strong back-bias effect, and susceptibility to avalanche breakdown.

1.4 PHYSICS OF DEVICE OPERATION

In this section we study the mechanisms of operation of CMOS FETs based on elementary physics, and then we give a crude estimate of device parameters in the next section.

Figure 1.5(a) on page 11 shows the cross section of an NFET. The source and drain are N+ diffused regions. At room temperature donors of the N+ regions are all ionized. The source and drain consist of positively charged donor ions that are incorporated in a crystal lattice, and freely moving negative electrons. The substrate is P-type, consisting of negatively charged acceptor ions and freely moving positive holes.

Across the metallogical boundary of P- and N-type regions, some free electrons diffuse into the P-type region, and some free positive holes diffuse into the N-type region. Because of this exchange the N-type region obtains a positive charge, and the P-type region obtains a negative charge. The positive and negative charges are in the vicinity of the metallogical boundary of the N- and P-type regions, and they make a dipole layer of positive charge on the side of the N-type region and a negative charge on the side of the P-type region of the boundary. The positive charges on the N-type region are ionized donors, and the negative charges on the P-type region are ionized acceptors.

The potential across the dipole layer (which is often called a depletion layer) is such that the N-type region is positive with respect to the P-type region. This potential difference limits the diffusion of free electrons and holes, and a steady-state potential difference is established between the N- and P-type regions. This potential difference is called a diffusion potential.

The diffusion potential exists strictly within the device. When a circuit is closed by metal wire W and two contacts S (source) and Sub (substrate) [Fig. 1.5(a)], the potential differences generated across the metal-semiconductor junction S and Sub by the similar mechanism cancel the diffusion potential of the PN junction, and therefore no current flows around the closed circuit.

The substrate surface between the source and the drain is called a channel, and in this case is P-type. Therefore, the channel is not populated by electrons. If a positive voltage V_G is applied to the gate relative to the source and substrate, however, the positively biased gate exercises a force on the electrons in the source (and in the drain) to pull them out to the channel. If this force is strong enough to compensate the energy barrier of the diffusion potential, the channel becomes populated by electrons.

We wish to study in some detail what happens when the gate voltage is increased (from zero). Since the substrate is P-type and contains mobile positive holes, they must be expelled from the surface of the substrate before the conducting channel, which consists of electrons, is formed. To expel holes from the surface, extra gate voltage must be applied. This mechanism is shown in Fig. 1.5(b). Below the conducting channel of electrons there is a layer depleted of holes that separates the channel from the substrate. This negatively charged layer of ionized acceptors sustains potential ϕ_S, which is called the surface potential. This is an important parameter used in MOSFET theory later, in Section 1.6. The gate voltage necessary to form a conducting channel of electrons is called the threshold voltage, V_{TH}. If substrate doping is higher, more holes must be expelled from the surface before a channel is formed, and therefore, the threshold voltage becomes higher. Furthermore, if the source is biased positively with respect to the substrate, the thickness of the hole-depleted region of Fig. 1.5(b) becomes wider. The gate voltage (measured relative to the source) must be higher to expel more holes, and therefore the threshold voltage (measured relative to the source) becomes higher. This back-bias effect is discussed in Section 1.9.

We saw that the threshold voltage is determined from the number of holes that must be expelled from the substrate surface before a conducting channel is formed. Therefore, it depends on the size and shape of the electrodes of the device [length of channel, width of FET, and depth of source (drain) diffusion]. The effects are significant in scaled-down CMOS and are known as short- and narrow-channel effects. This topic is also discussed in Section 1.9.

In the remainder of this section we consider the essential physics of a FET without including the detailed mechanisms of threshold voltage. This is for simplicity, but we do not lose any essential properties of the device. We assume that the threshold voltage is zero.

When voltage V_D is applied between the drain and source, current I_D flows. Electrons that are drained from the channel to the "drain" are replenished from the "source" and a steady-state current is established (the names of the electrodes, "source" and "drain," originate from this mechanism).

Channel current flows by two distinctly different mechanisms. When the gate voltage is small, the number of electrons pulled to the channel surface is small and the channel retains the properties of P-type silicon. Then, source, substrate, and drain may be regarded as emitter, base, and collector, respectively, of an NPN bipolar transistor. Since the substrate-drain PN junction (collector junction) is reverse biased, most of the applied drain voltage V_D is sustained by the reverse-biased PN junction, and the channel (i.e., the base of the NPN transistor) has no field. Electrons injected from the source to the channel diffuse to the drain and are collected by the reverse-biased PN junction. This mechanism of channel current is called subthreshold conduction, and it is discussed in Section 1.11.

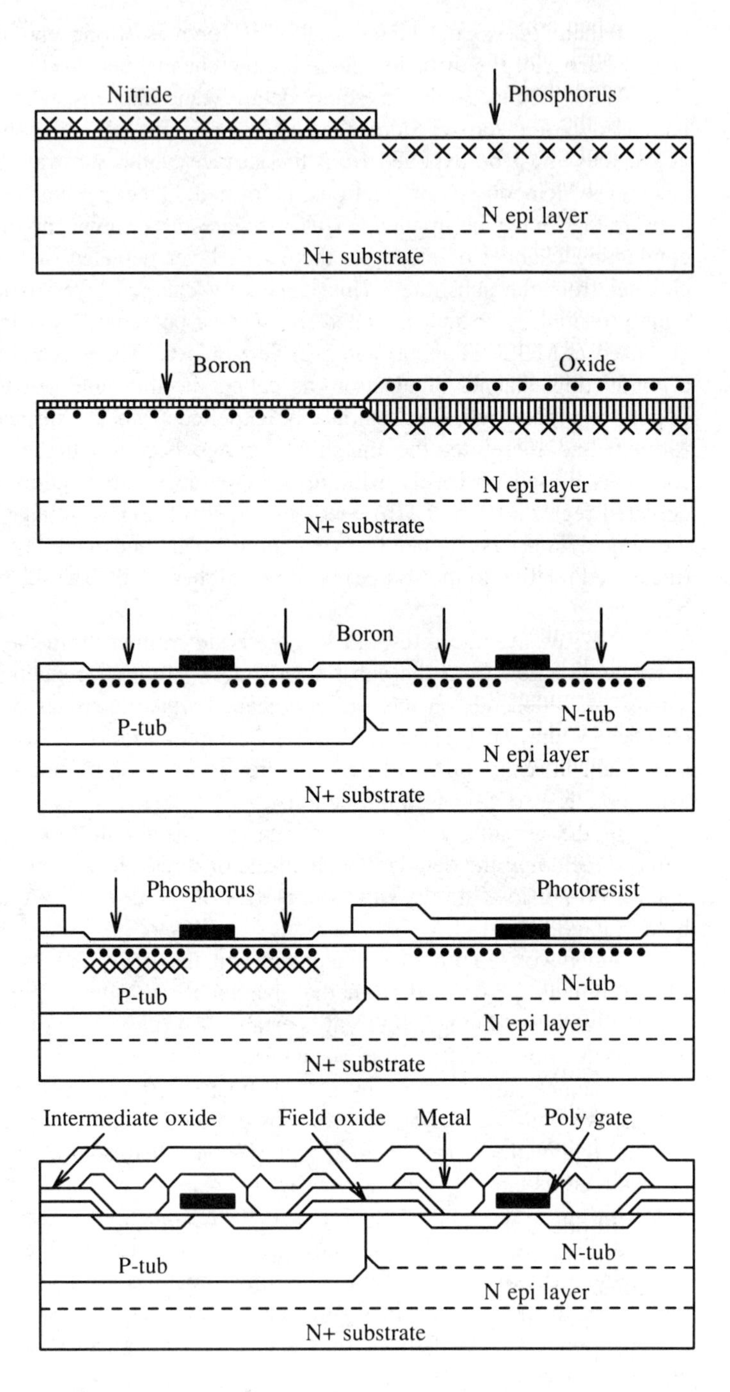

Figure 1.4 Process sequence of twin-tub CMOS technology.

When a high gate voltage is applied, many electrons are pulled to the surface of the channel, and the channel may be regarded as N-type silicon that bridges the N+ source and drain. If higher gate voltage is applied, proportionally more electrons populate the channel, and therefore more current flows. Channel current is a linear function of gate voltage. The FET is a gate-voltage-controlled variable resistor.

Channel current I_D is, however, not a linear function of drain voltage V_D. The channel is a resistor, whose resistivity is determined by the number of electrons pulled to the channel. The resistivity is, however, not uniform along the length of the channel. Resistivity depends on the gate voltage observed from the particular location of the channel. In Fig. 1.5(a), the resistivity of the channel at location A is less than the resistivity of channel at location B, since the potential at point B is higher than that at A due to the ohmic drop along the channel. The number of electrons in the channel decreases as the drain voltage increases. Because of that the channel current I_D increases less than proportional to drain voltage V_D. When the drain voltage increases, the rate of increase of channel current with drain voltage decreases. When the drain voltage equals the gate voltage, the rate of increase becomes zero. The channel current saturates and stays unchanged upon further increase in the drain voltage. The characteristics of a FET based on this physical model are derived in Section 1.6.

When the gate and drain voltages are equal, the induced charge density vanishes at the drain's end of the channel. If the drain voltage is still higher, the conductive channel terminates at point E somewhere between the source and drain, as shown in Fig. 1.5(c). The right side of point E [the area circled in Fig. 1.5(c)] is now depleted P-type material. Therefore, a saturated NFET is analogous to an NPN bipolar transistor as discussed in subthreshold conduction. We study this similarity in some detail.

To the left of point E, the electric field originates from the positive charge of the gate, and terminates at the negative charges, which are electrons on the channel and ionized acceptors in the depletion layer. To the right of point E, however, the electric field originates from the drain (strictly speaking, from the positively charged donor ions on the drain side of the depletion layer of the PN junction), and terminates at any of the following three destinations:

1. Negative charge of ionized acceptors in the depletion layer of the PN junction;
2. Negative charge on the gate electrode;
3. Negatively charged electrons that are injected from the end of the conducting channel E, to the right.

The third one is the most interesting. An electron that flies in the space between point E and the drain induces increasing positive charge on the vertical sidewall of the drain as it approaches the drain (strictly speaking, the positive charge is the ionized donor in the depletion layer of the PN junction), and this increasing positive charge is equivalent to current flowing into the drain. Current between point E and drain D is carried by this static induction mechanism [inset, Fig. 1.5(c)]. This is very much like the conduction mechanism in the reverse-biased base-collector junction. The theory of Section 1.6 shows that the density gradient of electrons at the end of channel E is steep. Then the diffusion current is significant at the end of the conducting channel at E. This is the region that works effectively as the base of an NPN transistor. Electrons diffuse into the region from the left side of point E, and the number of electrons that enter the reverse-biased junction is controlled by the gate voltage. This mechanism shows, furthermore, that to the right of point E a two- or three-dimensional field configuration must be studied to understand the phenomena quantitatively in the vicinity of the drain.

In the scaled-down CMOS devices, another detail must be included in the theory. Electrons in the channel of a scaled-down CMOS FET are subject to a 10^4- to 10^5-V/cm field. At this high field, the drift velocity of electrons is not proportional to the field. The drift velocity of the electrons limits at the saturation velocity v_∞, which is approximately 5×10^6 cm/sec. This effect is significant in 1-μm-feature-size CMOS. This problem is discussed in Section 1.10.

The gate electrode is capacitively coupled to the other electrodes. How this capacitance is accounted for can be understood from Fig. 1.5(d)–(k). In a self-aligned process there is an inevitable overlap of gate to source and drain regions, and the overlapped areas contribute capacitances C_{SO} and C_{DO} of Fig. 1.5(d). When the gate voltage is small, the channel is nonexistent, and the gate is capacitively coupled directly to the substrate. The equivalent circuit is shown in Fig. 1.5(e).

When a channel begins to form [Fig. 1.5(f)], the channel partially shields the gate from the substrate. Since the channel resistance is high, the channel is not at the same potential as the source and the drain. The channel voltage follows the gate voltage, and through the channel the gate is still coupled to the substrate. The equivalent circuit is shown in Fig. 1.5(g). When a channel is strongly formed as shown in Fig. 1.5(h), the substrate is shielded from the gate, and the channel-to-substrate capacitance becomes a part of the source (and drain)- to-substrate capacitance, as shown in Fig. 1.5(i).

When the NFET goes into saturation, the field profile changes from Fig. 1.5(b) to Fig. 1.5(c). The source and the drain are effectively separated, and in the limit of strong saturation there is a high resistance (in the bipolar analogy an equivalent current generator) that represents region E-Drain of Fig. 1.5(c), as shown in Fig. 1.5(k). The gate-to-drain capacitance is then the geometrical capacitance that is determined by the configuration of the gate and drain electrodes (C_{DO}).

Splitting the gate capacitance into various equivalent-circuit components includes arbitrariness. A distributed RC chain model of the FET channel removes the arbitrariness [28] [29].

1.5 ORDER-OF-MAGNITUDE ESTIMATE OF DEVICE PARAMETERS

In this section we estimate the performance of a scaled-down CMOS technology, using the concepts of the basic physics. Estimates of this type are useful in

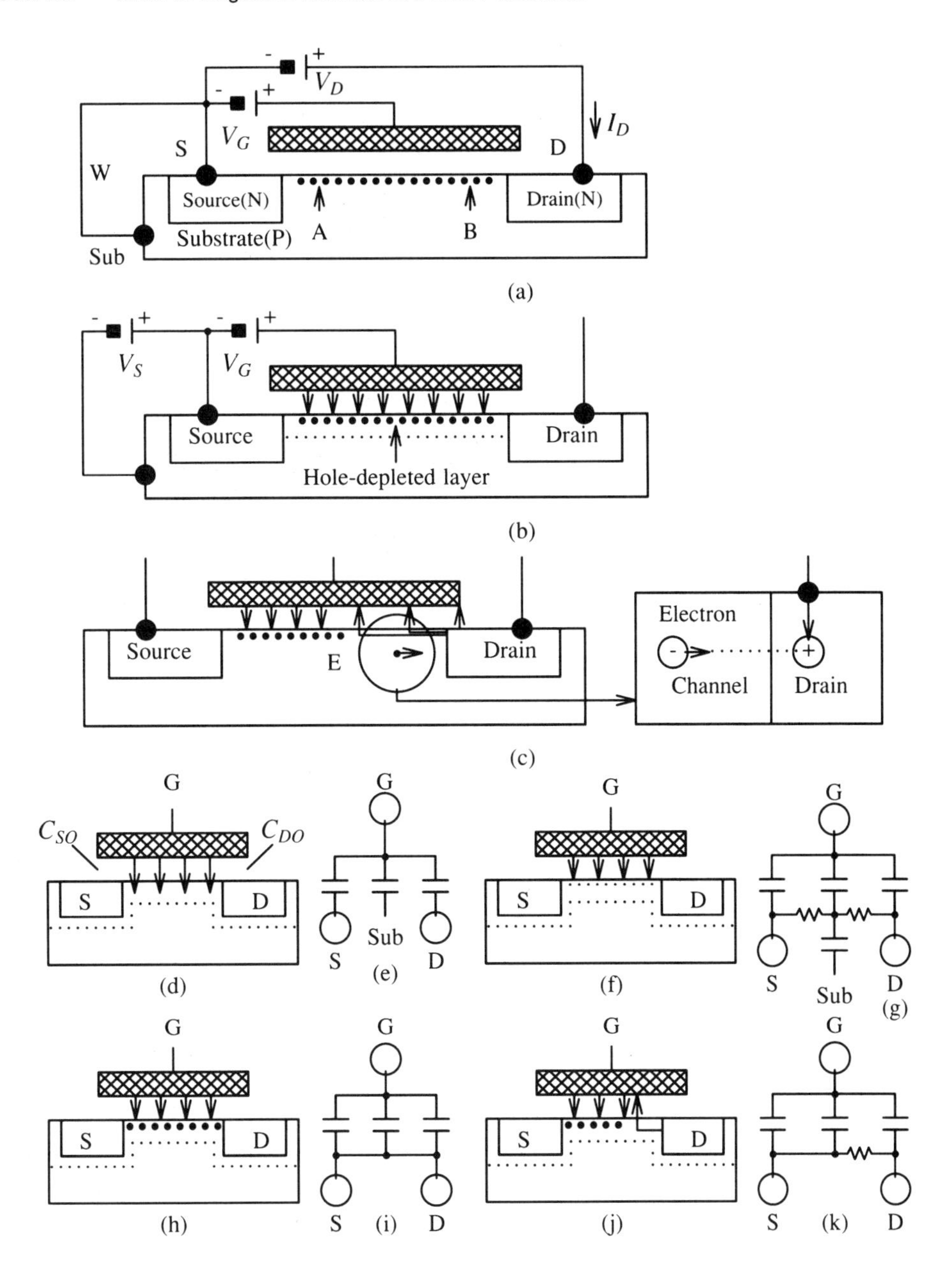

Figure 1.5 The way NFET works.

conceiving new circuits and also in a feasibility study in a VLSI development project. More accurate parameter values must be derived from numerical methods, and therefore no claim for high precision will be made.

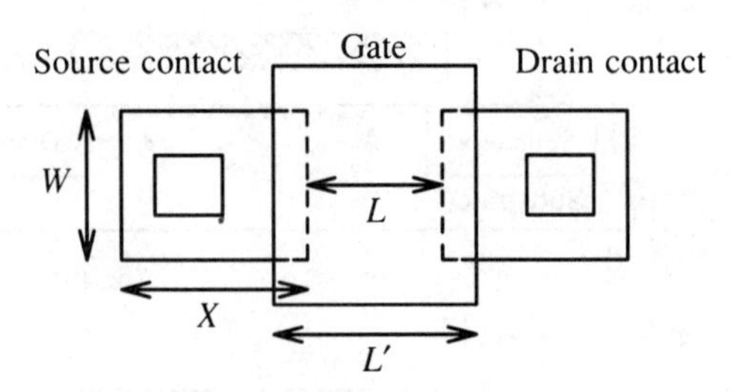

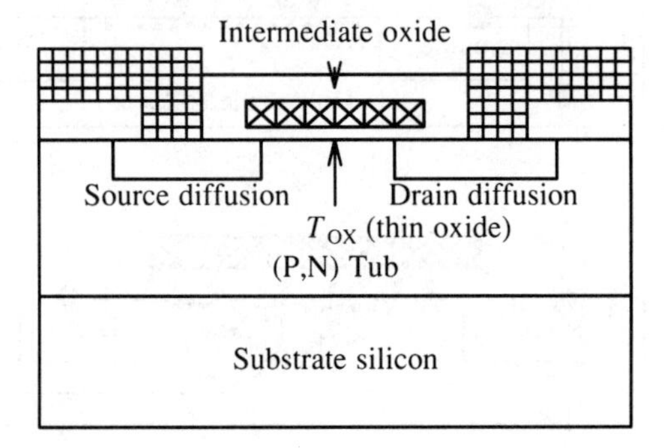

Figure 1.6 Top and cross-sectional views of FET.

(A) Structure of a FET

Figure 1.6 shows a schematic of a FET. The contact size is set by the design rules, and in 1.75-μm-feature-size CMOS, which is used as the example in this section, it is a square hole 1.75 μm by 1.75 μm in size. Minimum FET width (often called size) W is determined by the thin oxide to the contact hole design rule. The minimum-size FET is 4 μm wide. The drawn length of the polysilicon gate, L', is set so that the electrical channel length L, after taking due corrections for the outdiffusion of the source and drain islands, is centered at the specified value. $L = 1.3$ μm is typical. Then L' becomes about 2 μm. Dimension X is determined by the contact hole to thin oxide and by the contact hole to polysilicon gate design rules, and is about 4.5 μm. The thickness of the thin oxide under the gate electrode T_{OX} is about 250 angstroms. Standard parameters used in this book are summarized in Table 1.1.

From these size parameters and physical constants the electrical parameters are estimated in the following.

(B) Gate Capacitance C_G

The gate capacitance of the minimum-size FET is estimated to be

$$C_G = \frac{\varepsilon_{ox} L' W}{T_{ox}} = \frac{0.345\ \text{pF/cm} \times 2 \times 10^{-4}\text{cm} \times 4 \times 10^{-4}\text{cm}}{2.5 \times 10^{-6}\text{cm}} = 1.1 \times 10^{-2}\ \text{pF} \qquad (1.1)$$

The channel of FET has capacitance to the substrate as well as to the gate. This capacitance, C_B, is given by the formula

$$C_B = \left\{ \frac{\varepsilon_S q N_A}{2V} \right\}^{1/2} LW$$

where V is the channel to the substrate potential plus the diffusion potential. This formula is derived in Section 1.16. The ratio of the two capacitances is, assuming that $L \approx L'$,

TABLE 1.1 DEFINITION OF COMMON PARAMETERS

ε_{OX}	Dielectric constant of oxide (0.345 pF/cm)
ε_S	Dielectric constant of silicon (1.06 pF/cm)
q	Electronic charge (1.6×10^{-19} C)
μ	Carrier surface mobility ($cm^2/V \cdot s$)
L	FET electrical channel length (cm)
L'	FET designed channel length (cm)
W	FET width (size) (cm)
W_A, W_D	Depletion layer thickness (cm)
X	FET source/drain length (cm)
T_{OX}	Thin oxide thickness (cm)
T_{FOX}	Field oxide thickness (cm)
T_{COX}	Composite oxide thickness (cm)
N_A	P-substrate doping (cm^{-3})
N_D	N-substrate doping (cm^{-3})
V_{DD}	Power supply voltage (V)
V_G	Gate voltage (V)
V_D	Drain voltage (V)
V_S	Substrate voltage (V)
V_{TH}	FET threshold voltage (V)
V_{THP}	PFET threshold voltage (V)
V_{THN}	NFET threshold voltage (V)
V_{FB}	Flatband voltage (V)
ϕ_S	Surface potential (V)
ϕ_F	Fermi level of substrate (V)
I_D	Drain current (A)
C_G	Gate capacitance (F, pF)
C_B	Substrate capacitance (F, pF)
C_D	Drain capacitance (F, pF)
C_{OX}	ε_{OX}/T_{OX} (F/cm^2)

$$\frac{C_G}{C_B} = \left(\frac{\varepsilon_{OX}}{\varepsilon_S} \right) \left(\frac{V}{V_B} \right)^{1/2}$$

where $V_B = qN_A T_{OX}{}^2/2\varepsilon_S$. In this equation $\varepsilon_{OX}/\varepsilon_S = 1/3$, and since $N_A = 2 \times 10^{16}\,cm^{-3}$,

$$V_B = \frac{1.6 \times 10^{-19} \times 2 \times 10^{16} \times (2.5 \times 10^{-6})^2}{2 \times 10^{-12}} = 0.01$$

we obtain $C_G/C_B = 5.3$ if $V = 2.5\,V$. The capacitance of the channel to the gate dominates over the capacitance to the substrate.

(C) Drain Current

The threshold voltage of a 1.75-μm CMOS FET is about 0.6 to 0.8 V. The mobile charge induced on the channel is, by simple electrostatics,

$$Q = C_G(V_{DD} - V_{TH}) = 1.1 \times 10^{-2}\,\text{pF} \times 4.3\text{V} = 4.7 \times 10^{-2}\,\text{pC}$$

The average field in the channel is

$$E = \frac{V_{DD}}{L} = \frac{5}{1.3 \times 10^{-4}} = 3.8 \times 10^4 \text{ V/cm}$$

At this field the drift velocity of electrons in the channel v is about 6×10^6 cm/s (see Fig. 1.14 on page 30). The drain current (the charge that flows through the FET in 1 second) is

$$I = \frac{Q}{2}\left(\frac{v}{L}\right) = \frac{4.7 \times 10^{-14}}{2}\,\frac{6 \times 10^6}{1.3 \times 10^{-4}} = 1.08\text{mA}$$

where the factor 1/2 takes into account the nonuniform charge of the channel. This is a slight overestimate since not everywhere in the channel is the saturation drift velocity reached.

(D) Drain/Source Capacitance

The capacitance of an abrupt PN junction formed within the semiconductor doped to N_A acceptors per cubic centimeter is given by the formula (derived in Section 1.16)

$$C = \left(\frac{\varepsilon_s q N_A}{2V}\right)^{1/2} \text{F/cm}^2 = \frac{4 \times 10^{-8}}{\sqrt{V}}\text{F/cm}^2 \tag{1.2}$$

if $N_A = 2 \times 10^{16}\text{cm}^{-3}$ and if the N-type side is very heavily doped, where V is the voltage (in volts) developed across the PN junction plus the diffusion potential. Assuming that the diffusion potential is 0.64 V, and the applied voltage V is zero,

$$C = \frac{4 \times 10^{-8}}{0.8} = 5 \times 10^{-8}\text{F/cm}^2$$

and if $V = 5\,\text{V}$, $C = 1.68 \times 10^{-8}\text{F/cm}^2$. Therefore, the junction capacitance of the bottom of the drain is estimated to be the average of the two,

$$3 \times 10^{-8}\text{F/cm}^2 \times 20 \times 10^{-8}\text{cm}^2 = 0.006\text{pF}$$

The capacitance of the vertical wall of the drain can be estimated by assigning the area that is the length of the periphery of the drain times the depth of the diffusion. If the periphery is 18 μm long, and the depth is 0.5 μm, the capacitance is

$$3 \times 10^{-8}\text{F/cm}^2 \times 18 \times 10^{-4}\text{cm} \times 0.5 \times 10^{-4}\text{cm} = 0.0027\text{pF}$$

Because of doping nonuniformity, the capacitance is higher than this estimate. The total drain island capacitance (sum of the bottom and the sidewall capacitances) is of the order of 0.01 pF, comparable to the gate capacitance.

(E) Wiring Capacitance

Metal wiring is on the composite oxide that is from 1 to 1.5 μm thick (T_{COX}). The width is about the same as the designed gate width L'. The capacitance per unit length of the metal wire is

$$C_{MET} = \frac{\varepsilon_{ox} L' f}{T_{COX}} = \frac{0.345\text{pF/cm} \times 2 \times 10^{-4}\text{cm} \times f}{1 \times 10^{-4}\text{cm}} = 0.69f \text{ pF/cm} \qquad (1.3)$$

where f is a nondimensional factor that takes into account fringing field effects. The value $f = 3$ is reasonable. Then

$$C_{MET} = 2.1 \text{ pF/cm}$$

The relative magnitude of wiring capacitance to gate capacitance in a VLSI chip is estimated as follows. In 1.75-μm CMOS, the density of FETs is from 200 μm^2 per FET for a very dense manual layout to 4000 μm^2 per FET for a sparse automated layout. Then the length of an edge of a square that contains a FET ranges from 14 to 67 μm. If the edge of the square is the length of the wiring that connects the FETs, the capacitance falls in the range 0.003 to 0.014 pF. This compares with 0.011 pF of gate capacitance for the minimum-size FET. The capacitance of wiring is in the range 1/4 to 1.5 times the gate capacitance of the FET.

1.6 FIELD-EFFECT TRANSISTORS

Figure 1.6 shows the structure of an N-channel FET. A pure P-type epitaxial layer is grown on the P+ substrate, and later the P-tubs are formed by doping the epitaxial layer to several $\times 10^{16}$ acceptors per cubic centimeter. The P-tub doping is required to produce an enhancement-mode NFET.

The characteristics of enhancement-mode MOSFETs have been studied for 25 years. The theory, based on a simple gradual-channel approximation with negligibly low substrate doping [30] and developed to explain the characteristics of enhancement-mode PFET, has been upgraded to sophisticated theories that include the effects of substrate doping. As the technology scales down, there are many effects that a closed-form theory is unable to handle, such as the short-channel effect and velocity saturation of the carriers in the channel. As a consequence, although the closed-form theory provides correct qualitative description of FET characteristics, quantitative results are not very accurate. For circuit designers, however, the qualitative characteristics with calibrated parameter values are useful as a guide to design good circuits. Therefore, we present here a simplified form of the classical theory.

Suppose that, in Fig. 1.6, the source of the NFET is grounded, then a negative substrate bias of *magnitude* V_S (we choose $V_S > 0$; the substrate is always at the lowest potential), a positive gate voltage V_G that is high enough to create a channel, and a positive drain voltage V_D are applied. A coordinate system x is set up such that $x = 0$ at the source edge and $x = L$ at the drain edge. Along the channel there is a potential profile measured relative to the source as a function of x, $V(x)$ [where $V(0) = 0$]. It is derived from the analysis of the surface space charge [06] that the strongly inverted surface of the semiconductor at the source end of the channel has potential with respect to the deep interior of the substrate,

$$\phi_S(0) = 2\phi_F + V_S$$

where ϕ_F is the Fermi level of the P-type substrate material measured from the center of the energy gap. ϕ_F is positive. At location x of the channel the surface potential relative to the deep inside of the semiconductor is given by

$$\phi_S(x) = V(x) + 2\phi_F + V_S \tag{1.4}$$

The substrate side of the channel is perfectly depleted. The width W_A of a perfectly depleted layer having charge density qN_A per unit volume and sustaining potential $\phi_S(x)$ is given by simple electrostatics (see Section 1.16),

$$W_A = \left\{ \frac{2\varepsilon_S \phi_S(x)}{qN_A} \right\}^{1/2}$$

where ε_S is the dielectric constant of silicon (1.06 pF/cm) and N_A is the acceptor concentration of the substrate. The sheet charge density per unit area of the depletion layer is given by

$$Q_D = qN_A W_A = \sqrt{2\varepsilon_S qN_A \phi_S(x)}$$

If the gate voltage referenced to the source is V_G, the total charge induced on the semiconductor surface is given by

$$C_{OX} \left[V_G - V_{FB} + V_S - \phi_S(x) \right]$$

where $C_{OX} = \varepsilon_{OX}/T_{OX}$ is the gate oxide capacitance per unit area and V_{FB} is the flatband voltage, which depends on the work-function difference and the fixed positive charge at the oxide-semiconductor interface.

The mobile charge in the inversion layer, $Q_I(x)$, is the total charge minus the charge bound in the depletion layer,

$$Q_I(x) = C_{OX} \left\{ V_G - V_{FB} + V_S - \phi_S(x) \right\} - \left\{ 2\,\varepsilon_S qN_A \phi_S(x) \right\}^{1/2} \tag{1.5}$$

We assume that charge $Q_I(x)$ is within a negligibly thin layer of the silicon surface (thickness less than T_{OX}). If this assumption is not satisfied, a correction factor is needed in the following theory [31].

Using the width of the channel W and the carrier mobility μ, the drain current I_D is given by

$$I_D = WQ_I(x)\mu\frac{dV(x)}{dx} \tag{1.6}$$

By integrating the equation by x from $x=0$ to $x=L$ using $dV(x)/dx \cdot dx = dV$, we obtain

$$I_D = \frac{\mu W}{L}\int_0^{V_D}\left\{C_{OX}(V_G - V_{FB} - 2\phi_F - V) - (2\varepsilon_S qN_A)^{1/2}(V + 2\phi_F + V_S)^{1/2}\right\}dV$$

$$= B\left[(V_G - V_{FB} - 2\phi_F - \frac{V_D}{2})V_D - \left(\frac{2}{3}\right)H\left\{(V_S + 2\phi_F + V_D)^{3/2} - (V_S + 2\phi_F)^{3/2}\right\}\right] \tag{1.7}$$

where

$$B = \frac{\mu W}{L}C_{OX} \quad \text{and} \quad H = \frac{\left(2\varepsilon_S qN_A\right)^{1/2}}{C_{OX}}$$

This formula is valid subject to the condition that the channel exists all the way from the source to the drain. The limit of validity is determined by looking for the maximum of I_D as a function of V_D, by solving

$$\frac{dI_D}{dV_D} = 0$$

Solution of this equation is the maximum drain voltage, $V_{D\max}$, and Eq. (1.7) is valid when $V_D < V_{D\max}$. $V_{D\max}$ is given by

$$V_{D\max} = V_G - V_{FB} - 2\phi_F + F(H, V_G, V_S) \tag{1.8}$$

where

$$F(H, V_G, V_S) = \frac{H^2}{2} - H\left\{V_G - V_{FB} + V_S + \frac{H^2}{4}\right\}^{1/2}$$

Using function F, the saturation current $I_{D\max}$ is given by

$$I_{D\max} = \frac{B}{2}\left\{(V_G - V_{FB} - 2\phi_F)^2 - F(H, V_G, V_S)^2\right\} \tag{1.9}$$

$$-\frac{4H}{3}\frac{B}{2}\left\{[V_G - V_{FB} + V_S + F(H, V_G, V_S)]^{3/2} - (V_S + 2\phi_F)^{3/2}\right\}$$

and

$$I_D = I_{D\max} \quad \text{when} \quad V_D > V_{D\max} \tag{1.7a}$$

When $V_D > V_{D\max}$, the FET is in the saturation region. In the saturation region the drain current of the FET is independent of the drain voltage.

Equations (1.7)–(1.9) are derived for NFETs. The equations can be used for PFETs if signs of variables and parameters are properly changed. The mobility of carriers is higher for NFETs than for PFETs, and furthermore, mobility depends on crystallographic orientation [32]. Hole mobility is $145\,\mathrm{cm^2/V \cdot s}$ for <100> crystallographic orientation and $240\,\mathrm{cm^2/V \cdot s}$ for <110> orientation. The electron mobility is $500\,\mathrm{cm^2/V \cdot s}$ for <100> orientation. Therefore, NFETs are about two to three times more conductive than PFETs [33].

It is interesting to examine the limit of $V_D \to 0$,

$$I_D \to B\,(V_G - V_{TH})V_D$$

where

$$V_{TH} = V_{FB} + 2\phi_F + H\,(V_S + 2\phi_F)^{1/2} \tag{1.7b}$$

is the threshold voltage of the FET. When $V_{TH} > 0$, FET carries no current when $V_G = 0$. Such a FET is called an enhancement-mode FET. When $V_{TH} < 0$, FET carries current when $V_G = 0$. Such a FET is called a depletion-mode FET. The threshold voltage depends on V_S, the bias of the substrate relative to the source (the back-bias effect). Since $H \approx N_A^{\ 1/2}$, the threshold voltage increases with increased substrate doping; by increasing the substrate doping, a depletion-mode FET can be made an enhancement-mode FET.

There is positive charge trapped within the thin gate oxide. The positive charge makes a PFET made on a pure N-type substrate an enhancement-mode FET, but an NFET made on a pure P-type substrate, a depletion-mode FET. An established technique to fabricate an enhancement-mode NFET is to dope the substrate. A FET made on a substrate doped to 1 to $3 \times 10^{16}\mathrm{cm}^{-3}$ is an enhancement-mode NFET. Relatively high substrate doping reduces breakdown voltage of the drain to the substrate PN junction, and therefore an enhancement-mode NFET is unable to sustain high drain voltages.

Figures 1.7(a) and 1.7(b) show the FET characteristics derived from the theory of this section. The results are plotted using normalized parameters defined by

$$v_S = \frac{V_S}{V_{DD}}, \quad v_G = \frac{V_G}{V_{DD}}, \quad \text{and} \quad v_D = \frac{V_D}{V_{DD}}$$

where V_{DD} is the power supply voltage. In a digital CMOS circuit an NFET carries the maximum current when $V_G = V_{DD}$ and $V_D = V_{DD}$. Therefore, it is convenient to normalize the drain current using the maximum current.

$$i_D = \frac{I_D}{I_{D\,\mathrm{MAX}}} \quad \text{where} \quad I_{D\,\mathrm{MAX}} = \frac{B}{2}(V_{DD} - V_{TH})^2$$

$I_{D\,\mathrm{MAX}}$ is the saturation current of a FET made on a pure substrate material. In Figs. 1.7(a) and 1.7(b),

$$v_{TH} = \frac{V_{TH}}{V_{DD}} = 0.1, \quad f = \frac{\phi_F}{V_{DD}} = \frac{0.365}{5} = 0.073, \quad \text{and} \quad h = \frac{H}{\sqrt{V_{DD}}} = 0.2668$$

were used. These are the parameter values appropriate for a substrate doped to 2×10^{16} acceptors per cubic centimeter.

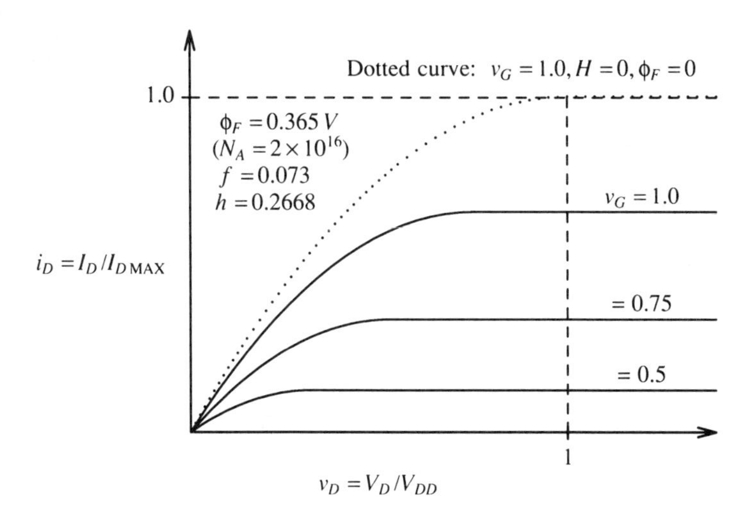

Figure 1.7(a) Normalized current-voltage characteristic of FET.

The standard theory presented here has an advantage over the simple gradual-channel theory [30], of including the effects of substrate charge and the associated back-bias effect. But the sophistication increases the complexity.

The theory approaches the simple theory in the limit as $H \to 0$. Then we have, from Eq. (1.7),

$$I_D = B\left[V_G - V_{TH} - \frac{V_D}{2}\right]V_D \quad \text{when } 0 < V_D < V_G - V_{TH}\ (=V_{D\max}) \qquad (1.10a)$$

$$I_D = \frac{B}{2}(V_G - V_{TH})^2 \quad \text{when } V_D > V_G - V_{TH}\ (=V_{D\max}) \qquad (1.10b)$$

Although less accurate, Eqs. (1.10a) and (1.10b) are quite often used to analyze CMOS circuit problems.

Analysis of FETs in the saturation region requires an elaborate theory. Interested readers are referred to the work by Geurst and Pao [34] [35], but one important subject will be discussed before concluding this section. The FET theory of this section leads to a conclusion that drain current I_D is independent of drain voltage V_D when $V_D > V_{D\max}$, as shown by curve A of Fig. 1.8(a). In a short-channel FET, the predicted current saturation does not occur; drain current increases gradually with increasing drain voltage, as shown by curve B. This gradual increase in drain current can be explained from the basic physics of FETs discussed in Section 1.4.

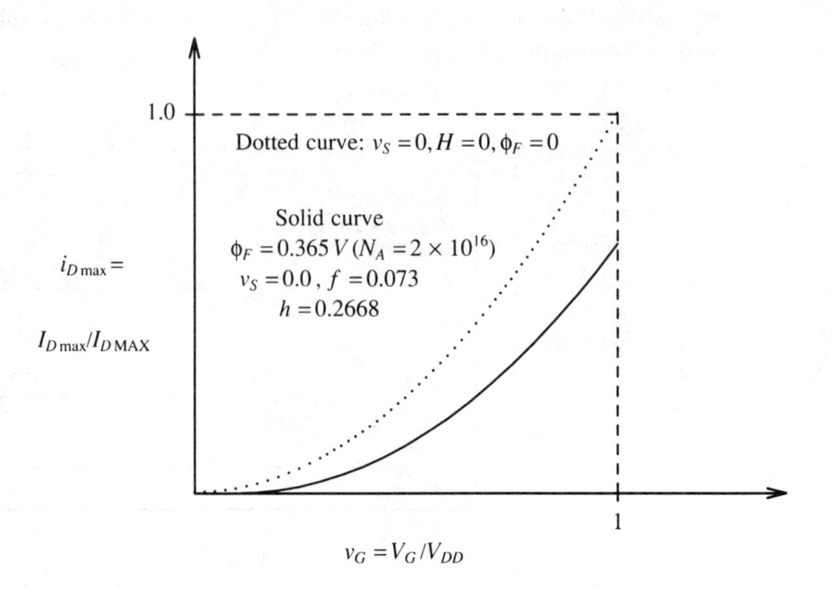

Figure 1.7(b) Normalized saturation current versus normalized gate voltage.

Figure 1.8(b) shows the cross section of an NFET in saturation ($V_D > V_G - V_{TH}$). The conducting channel terminates at point α, and there is an electron depleted region between point α and β, having length ΔL. If the electric field in region $\alpha\beta$ is approximately one-dimensional, its length ΔL is given by

$$\Delta L = \left\{ \frac{2\varepsilon_S (V_D - V_{D\max})}{qN_A} \right\}^{1/2}$$

where $V_{D\max}$ is the saturation voltage of the FET. Length ΔL increases with increasing drain voltage V_D. This equation can be derived analogously to Eq. (1.2) using the fact that the voltage sustained between points α and β is $(V_D - V_{D\max})$. The details are given in Section 1.16. This depletion region can be interpreted as the base-collector space-charge region of an NPN bipolar transistor, whose emitter is the channel and whose base is the edge of the channel. Electrons are injected by diffusion from the terminating channel to the equivalent base region. This current can be found by noting that the voltage at point α equals $(V_G - V_{TH})$ (we assume gradual channel approximation). The channel current is given by Eq. (1.9), using effective channel length $(L - \Delta L)$ instead of L. This is equivalent to replacing parameter B by $B(L/(L - \Delta L))$. Since ΔL increases, the saturation current [which is proportional to $B(L/(L - \Delta L))$] increases with increasing drain voltage.

There is another mechanism of current increase in the saturation region. In Fig. 1.8(b) line γ represents the edge of the depletion layer in the substrate. If substrate doping N_A is higher, curve γ is closer to the channel, and therefore the assumption of uniform field between points α and β is well satisfied. If substrate doping is low, however, the electric field originating from the drain creeps into the space between the channel and the depletion layer edge (field lines 1 and 2), and charge is induced on the channel by the drain. The charge increases with increasing drain voltage, thereby increasing drain current with increasing drain voltage. Electric flux, which influences the area of the channel close to the source (field line 1), has the effect of decreasing the threshold voltage of a FET with increasing drain voltage. In a short-channel FET, the threshold voltage depends on the drain voltage. This effect is known as the short-channel effect. The short-channel effect occurs when the channel length L is less than the lower limit determined from the gate oxide thickness, PN junction depth, and depletion layer widths [36] [37], and will be discussed in Section 1.9.

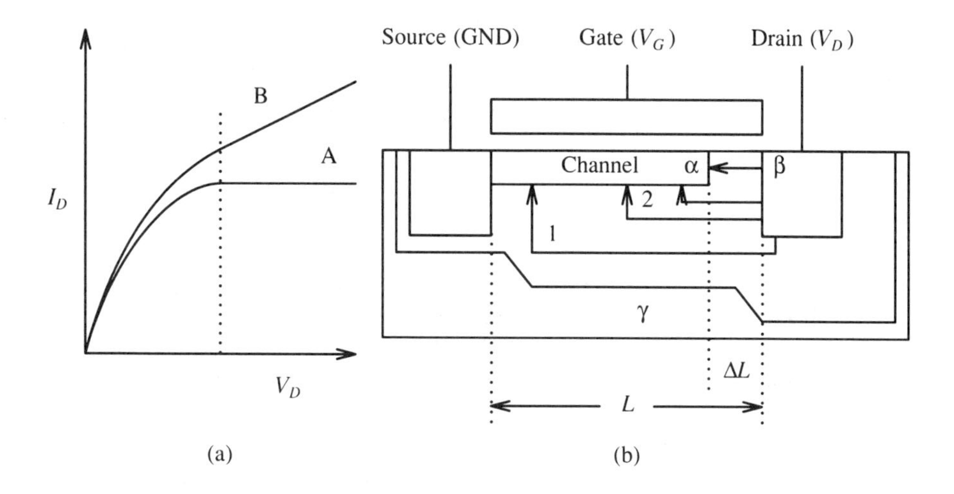

Figure 1.8 Saturation of FET.

1.7 SIMPLIFIED FET THEORY

In the limit as $H \to 0$ the theory of Section 1.6 gives a very simple FET characteristic [30]. The simplicity allows several practical circuit problems solved in closed forms using the theory. Many analog circuits have been analyzed using this approximation [38]. In digital CMOS circuits the effects of nonuniform shape of the FET channel can be analyzed in closed form. Although the closed-form analysis can be carried out using Eq. (1.7), we take advantage of simplicity in this limit. The analysis of this section has application in the gate delay algorithm discussed in Chapter 3.

Let the width of the channel of a FET, $W(x)$, be a function of coordinate x taken along the channel, $x = 0$ at the source and $x = L$ at the drain. Then Eq. (1.6) is

$$I_D = W(x)C_{OX}\left\{V_G - V_{TH} - V(x)\right\}\mu\frac{dV(x)}{dx} \tag{1.11}$$

Equation (1.11) can be integrated by separation of variables as follows:

$$I_D\int_0^x \frac{dx}{W(x)} = C_{OX}\mu\int_0^{V(x)}\left\{V_G - V_{TH} - (1/2)V(x)\right\}dV(x)$$

and therefore

$$I_D = \frac{C_{OX}\mu}{\int_0^L \frac{dx}{W(x)}}\left\{V_G - V_{TH} - (1/2)V_D\right\}V_D \tag{1.12}$$

This equation shows that parameter B of a nonuniform FET is obtained by substituting

$$\int_0^L \frac{dx}{W(x)} \rightarrow \frac{L}{W} \tag{1.13}$$

into parameter B of Eq. (1.10a), where the left-hand side of Eq. (1.13) is the effective number of squares of the channel. The effective number of squares is a concept used to calculate the resistance of a sheet resistor that has nonrectangular shape. Determination of the effective number of squares requires solution of the Laplace equation. When variation of the channel width is gradual, however, Eq. (1.13) gives the effective number of squares accurately. Detailed proof of the existence of the effective number of squares was given by Grignoux and Geiger [39]. Substitution of Eq. (1.13) is possible only when the drift velocity of the carriers is proportional to the field (constant μ), and therefore the channel is in the state of ohmic conduction. This condition is satisfied at the low-field limit or at the long-channel limit.

FETs in a scaled-down CMOS work within the region where the effect of drift velocity saturation is significant. When such FETs are connected in series, and if the effects of the charge stored between FETs is neglected, however, the series-connected FETs may be considered as a single long-channel FET that is in the low-field, long-channel regime. Then the current of the FET is inversely proportional to the number of squares of the FET channel, as shown by Eq. (1.13). This is equivalent to substituting an FET by a linear resistor proportional to the number of squares of the channel. This equivalence is fundamental to the delay algorithm in many gate-level timing simulators.

A trapezoidal NFET (the drain side is wider than the source side) is less susceptible to hot electron effects [40].

1.8 FETS WITH A SPECIAL SHAPE OR STRUCTURE

Most of the FETs used in a CMOS VLSI chip have a straight gate electrode and a rectangular source and drain diffused areas on the two sides of the gate. The source and the drain areas are contacted by metal conductors at many locations along the device width. A designer is able to create FETs that have significant differences from this standard structure. In this section properties of such specialty devices are reviewed. Special devices are never used in automated VLSI design. They have important applications, however, in high-performance custom design.

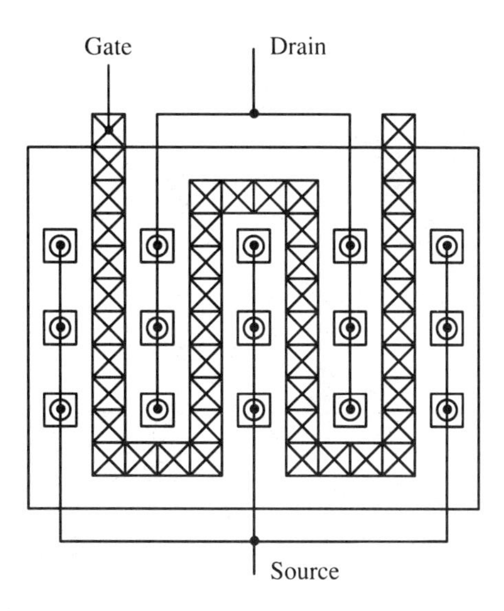

Figure 1.9 Very wide FET, whose gate is driven only from one end.

1. A very wide channel FET, whose extended gate (width of FET is W) is driven from only one end, as shown in Fig. 1.9 has an interesting application [41]. The gate material has a series resistance of about 25 Ω/square if the gate is made of doped polysilicon, and 4 Ω/square if the gate is made of polysilicide. If the gate is polysilicide and is 2 μm wide, the resistance r_G is 2 Ω/μm. The capacitance c_G is, assuming 250 angstroms of thin oxide, $c_G = 2.76 \times 10^{-3}$pF/μm. When one end of the polysilicon gate is pulled up, the signal propagates to the other end of the gate after time

$$T_D = 0.4 c_G r_G W^2 = 2.2 \times 10^{-3} W^2 \text{ps}$$

where W is the width of the FET measured in μm. This formula is derived in Section 6.8. When $W = 1000$ μm, T_D is 2.2 ns. Until time T_D is reached, the far end of the FET is inactive. The active width increases approximately as the square root of time.

A FET of this type does not conduct along its entire width immediately after input switching. Rather, the switching progresses gradually until the time determined by the time constant of the signal propagation along the gate. Many output drivers use this technique to control the discharge current spike. This technique is useful to control noise spike (Sections 7.12 and 9.5) when a very large output driver is required to reduce dependence of the output delay on the load capacitance.

2. In applications such as a databus driver or a PLA pulldown device with large load capacitance, the ratio of the channel current to the drain capacitance determines the figure of merit for the device. There are three different layout structures of a FET, shown in Figs. 1.10(a)–(c). The drain capacitance of an FET, C_D, is given by the formula,

$$C_D = C_A A_D + C_P P_D \tag{1.14}$$

where A_D and P_D are the area and the periphery of the drain, and C_A (pF/μm^2) and C_P (pF/μm) are the coefficients. A rough estimate of C_A and C_P in a scaled-down CMOS is

$$C_A = 3 \times 10^{-4}\text{pF/μm}^2 \text{ and } C_P = 3 \times 10^{-4}\text{pF/μm}$$

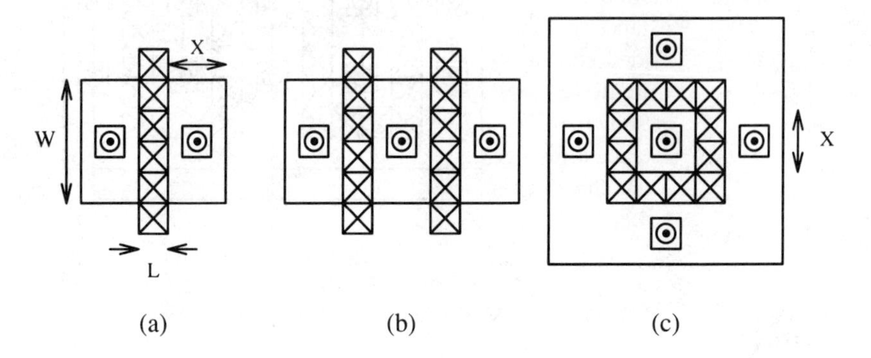

Figure 1.10 Various layouts of FETs.

Let the saturation current per unit width of the device be J_S (A/μm). An approximate value of J_S is

$$J_S = 150\,\mu\text{A/μm}$$

for the typical process of the 1.75-μm CMOS technology at 65 degrees centigrade. The drain current-to-drain capacitance ratios of the three FET structures are as follows:

Structure (a) $$\frac{J_S W}{C_A(XW) + C_P \times 2(X+W)} = \frac{750\,\mu\text{A}}{1.35 \times 10^{-2}\text{pF}}$$

Structure (b) $$\frac{2J_S W}{C_A(XW) + C_P \times 2(X+W)} = \frac{1500\,\mu\text{A}}{1.35 \times 10^{-2}\text{pF}}$$

Structure (c) $$\frac{4J_S X}{C_A X^2 + C_P \times 4X} = \frac{3000\,\mu\text{A}}{1.35 \times 10^{-2}\text{pF}}$$

Exact calculation of current of the square FET structure can be carried out using Eqs. (1.12) and (1.13), but we use approximation here. We assumed that $W = X = 5\,\mu$m.

The results show that the design (b) is twice, and design (c) is four times better than the standard design (a). The square FET was used in the RCA 1802 microprocessor design [42]. For an accurate estimate of performance the gate-to-drain overlap capacitance must be included, and the capacitance makes the difference between structure (a) and (c) somewhat less. Even if the overlap capacitance is included, the square FET is still much better than an FET with structure (a) or (b). A square FET of Fig. 1.10(c) is often used in critical circuits such as an output driver or a clock buffer.

A problem of conventional CMOS technology is that the area and the capacitance of the source/drain islands are significant. The area is substantially more than the area of the contact hole. Several techniques have been developed to reduce the area and the capacitance—for example, by growing oxide around the polysilicon gate [43].

3. A contact hole to silicon and the uncontacted diffused source/drain area of an FET have series resistance. The resistance originates from the resistance of the diffused area, the spreading resistance, as well as from interface imperfections. The series resistance significantly affects the performance of submicron-feature-size FETs [44]. Silicided contact techniques have been developed to alleviate the problem [45]. It is therefore desirable that many contact holes are attached to the source/drain area of a FET, as shown in Fig. 1.11(a). If this is not done, as shown in Fig. 1.11(b), the series resistance affects the switching characteristics.

The *I-V* characteristic of a FET in a gradual-channel low-field regime is

$$I_D = B\left\{V_G - V_{TH} - (1/2)V_D\right\}V_D \tag{1.15}$$

When a series resistance R_S exists between the source and ground, the source voltage V_S is $V_S = I_D R_S$. The following substitutions must be carried out in the equation above:

$$V_G \to V_G - V_S,\ V_D \to V_D - V_S \text{ and } V_{TH} \to V_{TH}(V_S)$$

For simplicity we assume that the back-bias effect is negligible, or $V_{TH}(V_S) \approx V_{TH}$. By substitution we obtain a quadratic equation for I_D that is solved as

$$I_D = \frac{1}{{R_S}^2}\left[\left[R_S(V_G - V_{TH}) + \frac{1}{B}\right] - \left\{\left\{R_S(V_G - V_{TH}) + \frac{1}{B}\right\}^2 - 2{R_S}^2\left[V_G - V_{TH} - \frac{1}{2}V_D\right]V_D\right\}^{1/2}\right]$$

When $R_S B$ is small, the equation simplifies to

$$I_D = \frac{B}{1 + BR_S(V_G - V_{TH})}\left\{V_G - V_{TH} - (1/2)V_D\right\}V_D \tag{1.15a}$$

and the effect of the series resistance is to reduce parameter B by a factor of $1/(1+BR_S(V_G - V_{TH}))$. Since the maximum channel conductance of the FET is defined by

$$G_0 = \lim_{V_D \to 0} \frac{I_D}{V_D} = B(V_G - V_{TH})$$

the reduction factor can be written as $1/(1+G_0 R_S)$. If $G_0 = 200\,\mu\text{S}$ and $R_S = 100\ \Omega$, reduction of B is only 2%, and therefore the problem is not significant for a minimum-size FET. If a FET 10 times the minimum size has only one contact, however, $G_0 = 2\,\text{mS}$ and therefore 20% reduction in the parameter B is expected. In a wider FET, a correspondingly larger number of source/drain contacts should be used.

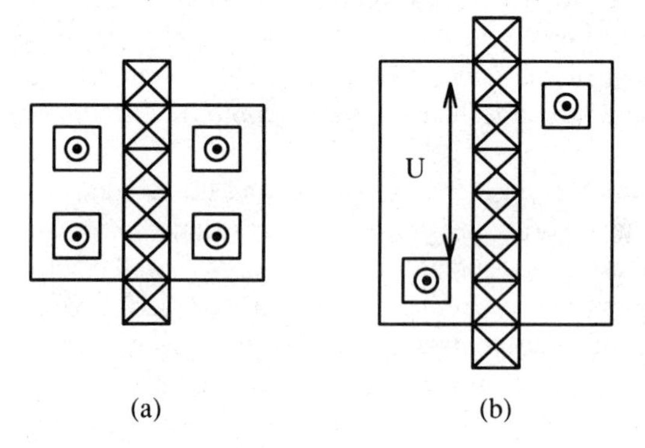

Figure 1.11 Poorly contacted FETs.

When a resistance exists in series with the drain contact, a substitution,

$$V_D \to V_D - I_D R_D$$

in the I-V characteristic of a FET leads to a result which simplifies to

$$I_D = \frac{B}{1+BR_D(V_G - V_{TH} - V_D)} \left\{ V_G - V_{TH} - (1/2)V_D \right\} V_D \qquad (1.15b)$$

in the limit of small $R_D B(V_G - V_{TH} - V_D)$. When $V_G - V_{TH} \approx V_D$ the effect is small, since the device becomes a current generator. When V_D is small, however, the series resistance of the source and of the drain give the same factor of degradation in B.

1.9 THRESHOLD VOLTAGE OF A FET

The most significant conclusion of the theory of Section 1.6 is that the threshold voltage of a FET increases with increasing source-to-substrate bias voltage. The effect is known as a back-bias effect, or a body effect. In many CMOS logic gates FETs are connected in series, and therefore in the process of switching, source to substrate back-bias exists that significantly affects the switching characteristics. Figure 1.12 shows the back-bias voltage versus the threshold voltage, measured for 1.75-μm CMOS FETs. The threshold voltage varies from 0.8 V to 1.25 V in PFETs and from 0.55 V to 1.1 V in NFETs. This effect has one significant consequence in circuit choice: A source-follower is not a preferred circuit in CMOS digital ICs.

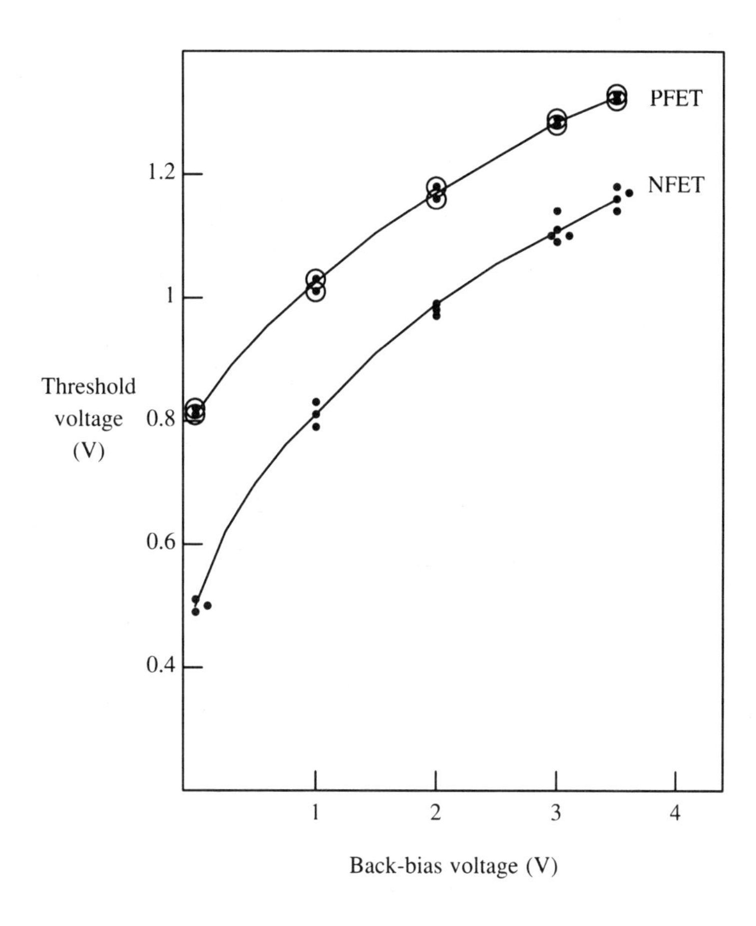

Figure 1.12 Threshold voltage versus source-substrate bias voltage.

Figure 1.12 shows measured values to check the threshold voltage formula, Eq. (1.7b) of Section 1.6. Since

$$H = \frac{\left(2\varepsilon_S q N_A\right)^{1/2}}{C_{OX}} = 0.41\left(\frac{N_A}{10^{16}}\right)^{1/2}$$

where N_A was normalized by $10^{16}/\text{cm}^{-3}$ for convenience. T_{OX} was 250 angstroms.

Then Eq. (1.7b) is

$$V_{TH} = V_{FB} + 2\phi_F + 0.41 \left(\frac{N_A}{10^{16}} \right)^{1/2} (V_S + 2\phi_F)^{1/2}$$

and therefore,

$$\Delta V_{TH}(V_S) = 0.41 \left(\frac{N_A}{10^{16}} \right)^{1/2} \left[(V_S + 2\phi_F)^{1/2} - (2\phi_F)^{1/2} \right] \tag{1.16}$$

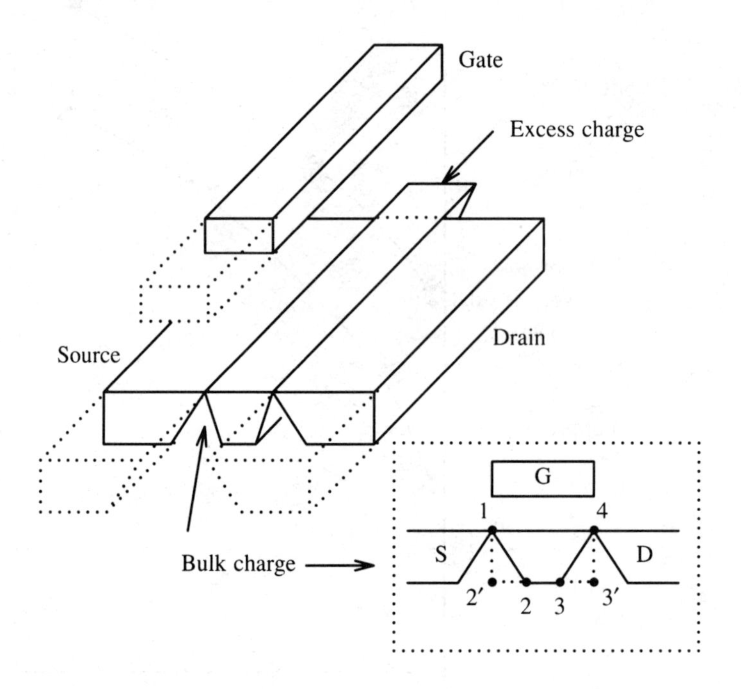

Figure 1.13 Short- and narrow-channel effects of FET.

For $N_A = 1 \times 10^{16}$, $\phi_F \approx 0.35$ V at 300 degrees Kelvin. Therefore,

$$\Delta V_{TH}(\text{at } 3.5\text{ V}) = 0.5 \left(\frac{N_A}{10^{16}} \right)^{1/2}$$

and for $N_A = 1 \times 10^{16}$, the body effect increases the threshold voltage by about 0.5 V when $V_S = 3.5$ V.

The mechanism of back-bias effect is that the gate charge must first compensate the bulk charge that exists on the substrate side of the channel before creating the conducting channel. The charge depends on the structural detail of the FET. The bulk charge exists within the trapezoid column 1234 of Fig. 1.13. The trapezoid shape is because the space-charge layers of the source and the drain diffused islands extend into the space underneath the channel. The bulk charge is less than that computed assuming a rectangular column 12′3′4. This latter assumption is, however, valid in a long-channel FET. Therefore, the threshold voltage of an FET decreases with decreasing channel length. The effect is called a short-channel effect [46].

In a very narrow FET, the bulk charge extends sidewise under the field oxide as shown in Fig. 1.13. The excess bulk charge increases the threshold voltage. The effect is called a narrow-channel effect. The threshold voltage changes by the short- and narrow-channel effects are additive [47]. The short- and narrow-channel effects are significant in scaled-down CMOS of less than 2.5-μm feature size.

1.10 EFFECTS OF DRIFT VELOCITY SATURATION

The FET theory of this chapter assumed that the carriers in the channel move with the velocity that is a linear function of the local electric field (μ is constant). This assumption is reflected to Eqs. (1.6) and (1.11). FETs of 1-μm CMOS technology have a channel electric field of $5\,\text{V}/10^{-4}\text{cm} = 50$ kV/cm. With this field, the carrier velocity is not a linear function of the field. To include this effect accurately in the theory is extremely difficult; here we provide only a very simplified model to derive qualitative results [48] [49] [50].

The drift velocity of the carriers including saturation, $v(E)$, as a function of field E can be approximated by a formula of the form

$$v(E) = \frac{\mu E}{1 + \dfrac{\mu}{v_\infty} E} \tag{1.17a}$$

where μ is the low-field mobility and v_∞ is the saturation velocity at the extremely high fields. Figure 1.14 shows the measured drift velocity of the carriers in the channel versus electric field [51] [52] [53] [54]. This relationship can be approximated by Eq. (1.17a). Equation (1.6) of Section 1.6 becomes

$$I_D = W C_{OX} (V_G - V_{TH} - V) \frac{\mu \dfrac{dV}{dx}}{1 + \dfrac{\mu}{v_\infty} \dfrac{dV}{dx}} \tag{1.17b}$$

The effects of substrate charge are not included. By integrating this equation from $x = 0$ to $x = L$ the current-voltage characteristic is derived as

$$I_D = \left(\frac{\mu C_{OX} W}{L} \right) \frac{(V_G - V_{TH}) V_D - (1/2) V_D^2}{1 + \dfrac{\mu}{L v_\infty} V_D} \tag{1.18}$$

when $0 < V_D < V_{D\max}$. When $V_D > V_{D\max}$ the saturation current $I_{D\max}$ is given by

$$I_D = I_{D\max} = \frac{1}{2}\left(\frac{\mu W C_{OX}}{L}\right) V_{D\max}^{\;2} \tag{1.19a}$$

where

$$V_{D\max} = \left(\frac{L v_\infty}{\mu}\right)\left[\left\{1 + 2\frac{\mu}{L v_\infty}(V_G - V_{TH})\right\}^{1/2} - 1\right] \tag{1.19b}$$

In the limit of large $(\mu / L v_\infty)$,

$$I_{D\max} = W C_{OX} v_\infty (V_G - V_{TH})$$

This equation is in contrast to Eq. (1.10b) of Section 1.6, which gives a square dependence of $I_{D\max}$ on $V_G - V_{TH}$. This linear dependence is universally observed in CMOS FETs of about 1-μm feature size. Figure 1.15 shows a measured $I_{D\max}$ versus V_G, and the dependence is indeed linear when V_G is 1 to 2 V above V_{TH}.

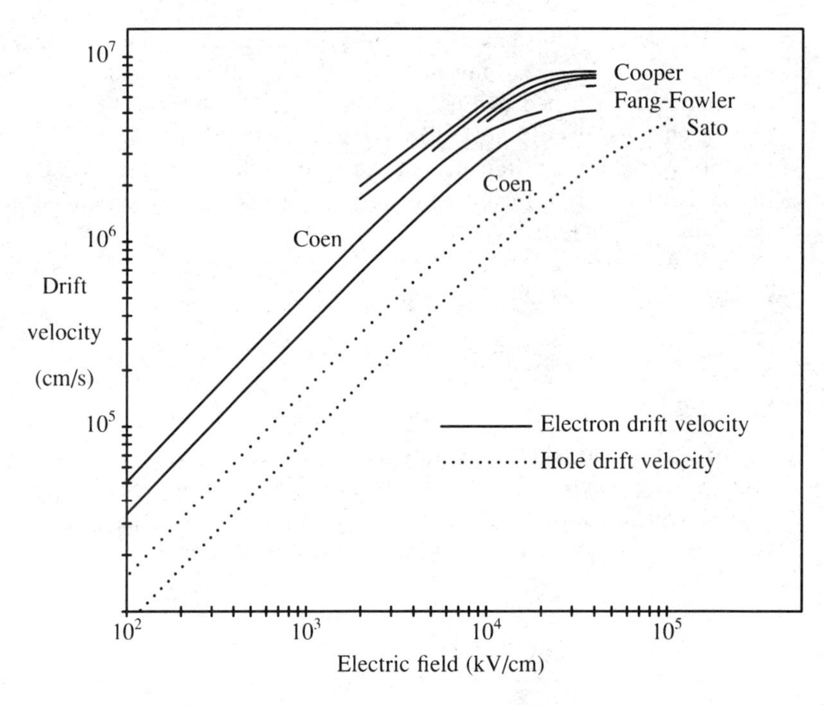

Figure 1.14 Velocity-field characteristics of channel carriers.

The theory of this section should be regarded as only qualitative. Equation (1.17b) does not treat the electron transport problem within the channel accurately, and further the two-dimensional current flow is not included. The theory was used only to show the most essential linear relationship with minimum complexity and to provide a useful interpolation formula.

For convenience of application of this FET characteristic in circuit analysis, a voltage V_0 that specifies the effects of carrier velocity saturation is defined by

$$V_0 = \frac{Lv_\infty}{\mu}$$

The I-V characteristic is then written as

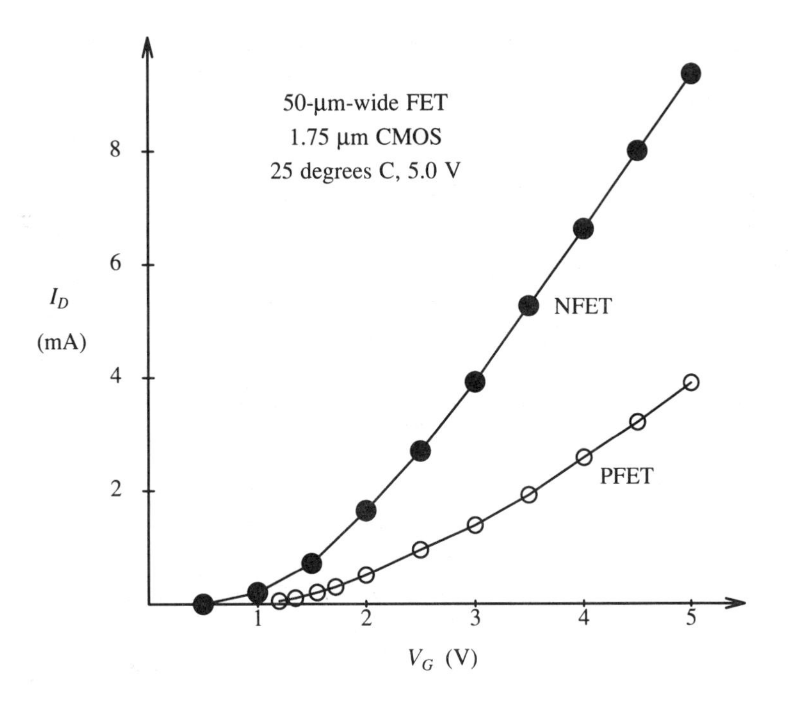

Figure 1.15 Saturation current versus gate voltage of a scaled-down FET.

$$I_D = 0 \quad V_G < V_{TH}$$

$$I_D = \frac{B}{1+\dfrac{V_D}{V_0}}\left\{V_G - V_{TH} - (1/2)V_D\right\}V_D \quad V_G > V_{TH} \text{ and } V_D < V_{D\max} \tag{1.20}$$

$$I_D = \frac{B}{2}V_{D\max}{}^2 = I_{D\max} \quad V_G > V_{TH} \text{ and } V_D > V_{D\max}$$

where

$$B = \frac{\mu C_{OX} W}{L}$$

and

$$V_{D\max} = V_0 \left\{ \left[1 + 2\frac{V_G - V_{TH}}{V_0} \right]^{1/2} - 1 \right\}$$

is the drain voltage where the current saturation begins. In the limit of $V_0 \to \infty$, $V_{D\max} \to V_G - V_{TH}$, in agreement with the gradual channel, low-field theory. Figure 1.16 shows the normalized saturation voltage defined by $v_{D\max}' = V_{D\max}/(V_G - V_{TH})$ and the normalized saturation current, $i_{D\max} = I_D/[0.5B(V_G - V_{TH})^2]$ versus $v_0 = V_0/(V_G - V_{TH})$. Physically, decrease in the saturation current is the direct consequence of decrease of the saturation voltage. When $v_0 < 1$ the effect is quite significant.

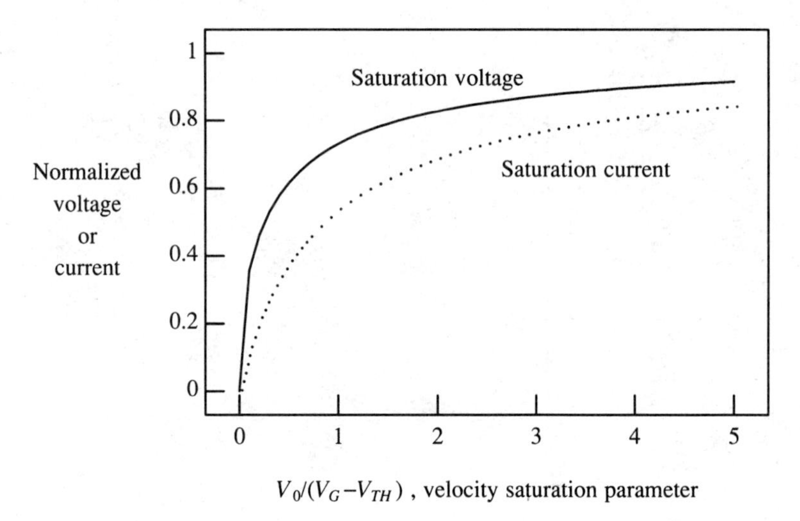

Figure 1.16 Effects of carrier velocity saturation on FET characteristics.

Figure 1.17 shows the saturation current normalized to the saturation current of the low-field FET model versus the normalized gate voltage $v_G = (V_G/V_{DD})$. In the limit of $v_0 \to \infty$ the relationship is quadratic. When v_0 decreases, however, the relationship changes over to a linear dependence [49], in agreement with the data of Fig. 1.15. This linear relationship is the most direct evidence of carrier velocity saturation. Scaled-down CMOS FETs work in the region where this effect is significant. The reduced drain current and the reduced drive capability of FET are reflected in delay times of gate, as discussed in Section 3.6.

Figure 1.14 shows that the saturation drift velocities of electrons and holes at high fields are about the same. At about 0.5-μm channel length the high-field condition completely dominates in the channel, and therefore the current drive capability of NFETs are degraded to be about equal to PFETs [55] [56].

1.11 SUBTHRESHOLD LEAKAGE CURRENT

When the gate voltage of an NFET is below V_{THN}, the simple theory shows that channel current vanishes. In a real NFET a small current called the subthreshold current flows [57], and the current cannot be explained from the simple model. Figure 1.18 shows the cross section of a subthreshold device. By the effect of positive gate bias V_{GB}, which is the sum of the gate-source voltage V_{GS} and the source-substrate voltage V_{SB}, the surface of the semiconductor is depleted to depth W_A.

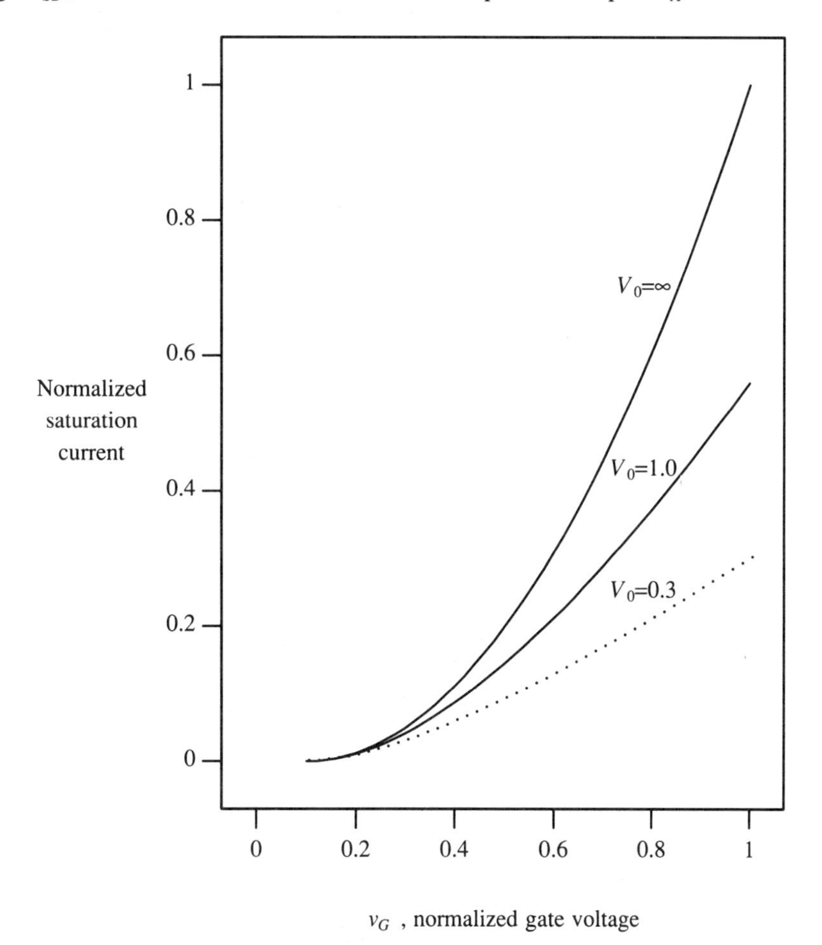

Figure 1.17 FET saturation current versus gate voltage change from square to linear dependence.

Since no channel current flows, the surface charge should not exist. This condition is written as

$$\varepsilon_{OX}\frac{V_{GB}-\phi_S}{T_{OX}}=qN_A W_A \quad \text{and} \quad \phi_S=\frac{qN_A W_A{}^2}{2\varepsilon_S}$$

where ϕ_s is the surface potential relative to the deep interior of the P-tub and where $V_{GB}=V_{GS}+V_{SB}$. By eliminating W_A from the two equations, we obtain

$$(\beta V_{GB}-\beta\phi_S)^2=a^2(\beta\phi_S)$$

where $\beta=q/kT$, $a=\sqrt{2}(\varepsilon_S/\varepsilon_{OX})(T_{OX}/L_B)$ (a parameter introduced by Brews [02]), and L_B is the Debye length. Solving this equation for $\beta\phi_S$ yields

$$\beta\phi_S=\beta V_{GB}+\frac{a^2}{2}-a\left\{\beta V_{GB}+\frac{a^2}{4}\right\}^{1/2} \tag{1.21}$$

If V_{GB} is given, the surface potential satisfying the condition of no surface charge is uniquely determined.

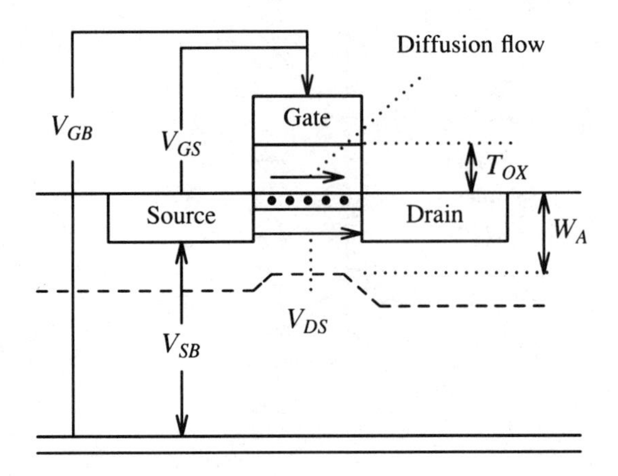

Figure 1.18 Mechanism of subthreshold channel current.

Since the surface of the semiconductor has higher potential than the interior, the minority electron density at the surface is enhanced by a factor of $\exp(\beta\phi_S)$ from that of the interior of the P-tub, where the minority electron density is $n_i{}^2/N_A$ (n_i is the intrinsic carrier density). The subthreshold current is carried, by these minority electrons, by a mechanism similar to a bipolar junction transistor [58].

Existence of the source/drain-diffused island has an effect on the surface minority electron density. Since the source-diffused island is biased to positive V_{SB} volts relative to the P-tub, the island effectively shields the P-tub, thereby maintaining the potential of the minority electrons. This effect contributes a factor $\exp(-\beta V_{SB})$ to the minority carrier density. This factor can be understood easily by noting that if the gate is removed, minority electron density at the surface (in the vicinity of source island) becomes $(n_i{}^2/N_A)\exp(-\beta V_{SB})$. Because of the vertical potential gradient at the surface, the electrons exist only within a layer having a thickness of the Debye length L_B near the surface. If the effects of nonuniform electron density within the layer is included, an additional numerical factor $(2\beta\phi_S)^{-1/2}$ must be multiplied. The sheet density of the minority electrons at the surface is given from these considerations as

$$N_I = N_A L_B \left\{ \frac{n_i}{N_A} \right\}^2 \frac{\exp(\beta\phi_S)}{(2\beta\phi_S)^{1/2}} \exp(-\beta V_{SB}) \tag{1.22}$$

When the drain is biased to positive V_{DS} volts relative to the source, the surface minority electron density at the drain-side edge of the channel is given by multiplying N_I by a factor $\exp(-\beta V_{DS})$. Since there is no horizontal field in the channel, current is carried by diffusion. The drain-to-source current is given by

$$I_D = qD_N W \frac{N_I}{L} \left\{ 1 - \exp(-\beta V_{DS}) \right\}$$

$$= \frac{W}{L} \left\{ \frac{\mu kT}{q} \right\} qN_A L_B \left\{ \frac{n_i}{N_A} \right\}^2 \exp(-\beta V_{SB}) \exp(\beta\phi_S) \left\{ 1 - \exp(-\beta V_{DS}) \right\} \frac{1}{(2\beta\phi_S)^{1/2}} \tag{1.23}$$

where L is the channel length, W the channel width, D_N the diffusion coefficient of electrons, μ the mobility, and $\beta\phi_S$ is given by Eq. (1.21). The Einstein relation $D_N = \mu kT/q$ was used to derive Eq. (1.23). The subthreshold leakage current depends exponentially on the gate voltage through parameter $\beta\phi_S$, but if V_{DS} is larger than several times the thermal voltage (kT/q), the current no longer depends on the drain voltage. A typical measured result is shown in Fig. 1.19 [59].

The subthreshold current must be kept below a certain maximum to guarantee the integrity of a dynamic multi-input NOR circuit, as used in PLAs. If the channel current of an FET is I_D, the maximum subthreshold current is $I_{SS\max}$, and the number of FETs driving the bit line is N,

$$I_D \gg N I_{SS\max}$$

is required. Since N can be as high as one hundred in a dynamic PLA, $I_{SS\max}$ must be maintained three to four orders of magnitude less than the channel current of the FET. Subthreshold current is important, since the current could limit use of dynamic PLAs in submicron CMOS [60].

1.12 DELAY CHARACTERIZATION OF A FET

The figure of merit of a FET as a switching device is determined by the time the FET takes to discharge its own parasitic capacitance. The parameter that represents the on-state resistance of an FET is

$$\frac{V_{DD}}{I_{D\max}(V_G = V_{DD}, V_D = V_{DD})} \tag{1.24}$$

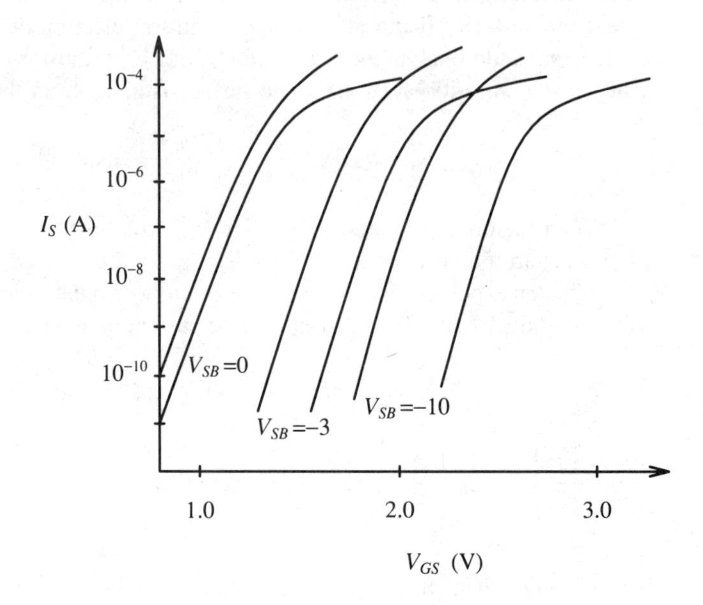

Figure 1.19 Subthreshold current versus gate voltage.

The parasitic capacitance of the FET depends on the way the FET is used, but is typically

$$C = C_G + C_D \tag{1.25}$$

where C_G is the gate capacitance and C_D is the capacitance of the drain diffused island. Time constant of the FET, T_{FET}, is then defined by

$$T_{FET} = \frac{V_{DD}}{I_{D\max}}(C_G + C_D) \tag{1.26}$$

Figure 1.20 shows the saturation current $I_{D\max}$ versus feature size of CMOS technology, Fig. 1.21 the parasitic capacitance versus feature size, and Fig. 1.22 the time constant versus feature size of the technology. The factor $(V_{DD}/I_{D\max})$ of Eq. (1.26) is, essentially, the inverse of the transconductance of the FET.

1.13 FET PERFORMANCE IMPROVEMENT BY SCALING

It is a matter of practical interest to predict how much improvement in switching delay of a circuit is expected by reducing the feature size. Here we wish to consider a practical case of constant voltage scaling. When the feature size of the CMOS technology is in the range from 5 μm to about 1 μm, the power supply voltage has been almost

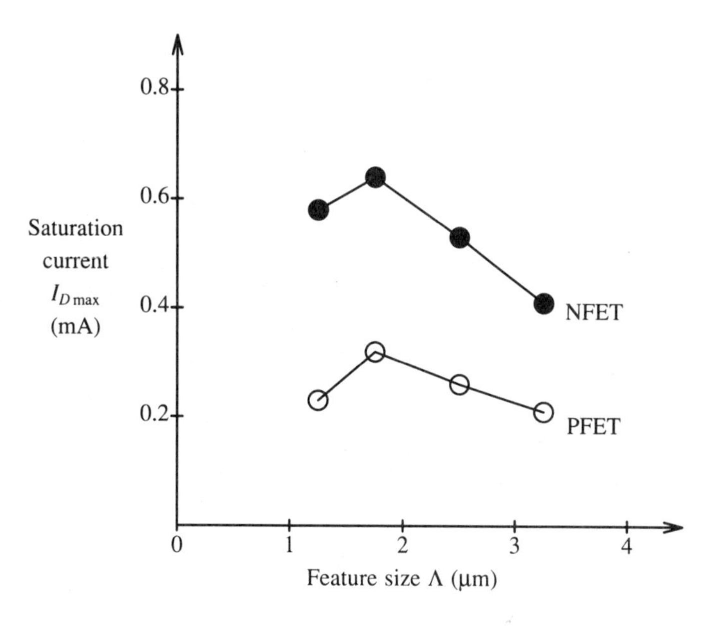

Figure 1.20 Saturation current of the minimum size FET versus feature size.

exclusively chosen at 5 V. This power supply voltage is convenient, above all, for compatibility with bipolar TTL ICs that serve as the system's interface.

The average field in the channel of a FET of from 5-μm to 1-μm CMOS technologies is in the range 10^4 to 5×10^4 V/cm. With reference to the velocity-field characteristic of the surface carriers shown in Fig. 1.14, this is the field range where the channel of the FET is neither in the low-field ohmic transport nor in the high-field saturated velocity transport. Therefore, we need to study two different cases.

The time constant of a FET, T_{FET}, defined in Section 1.12 is written as

$$T_{FET} = \left\{ \frac{V_{DD}}{\varepsilon_{OX} v_{\infty}(V_{DD} - V_{TH})} \right\} \left[\frac{T_{OX}}{W} \right] C(W, L) \tag{1.27a}$$

for the saturated velocity transport and

$$T_{FET} = \left\{ 2 \frac{V_{DD}}{\varepsilon_{OX} \mu (V_{DD} - V_{TH})^2} \right\} \left[\frac{T_{OX} L}{W} \right] C(W, L) \tag{1.27b}$$

for the ohmic transport, where

$$C(W, L) = \varepsilon_{OX} \left[\frac{WL}{T_{OX}} \right] + C_A WL + 2C_P(W + L) \tag{1.28}$$

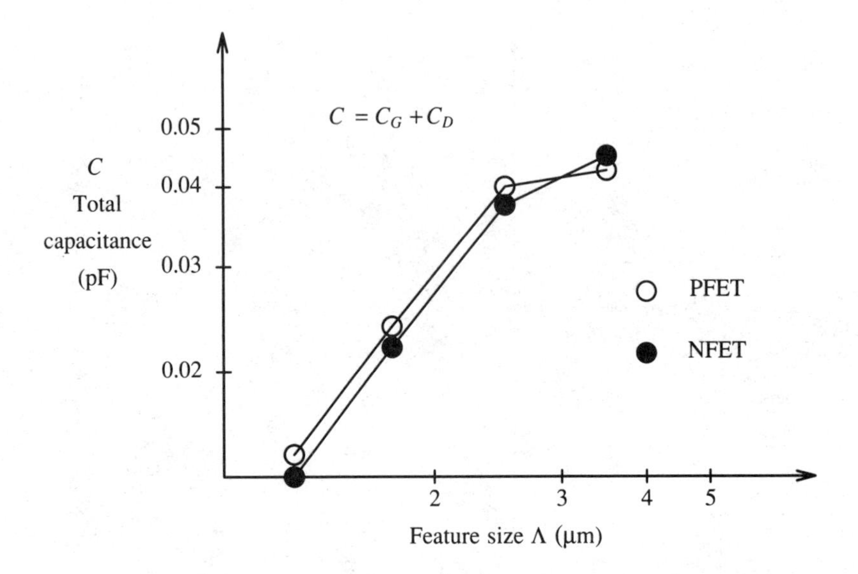

Figure 1.21 Total parasitic capacitance of FET versus feature size.

is the sum of the gate capacitance and the drain-diffused island capacitance. In Eqs. (1.27) and (1.28), the horizontal dimensions of the FET (parameters such as L, L', and X of Fig. 1.6 are represented by a single parameter L, and a vertical dimension such as W in Fig. 1.6 by a single parameter W. This is to make the parameter dependence clear. Correction factors of the order of unity can be applied for better precision, but we do not go into such detail in the following qualitative theory. In the equation C_A is the capacitance of the drain PN junction per unit area. In an abrupt PN junction formed within a material uniformly doped to N_A acceptors/cm^3, C_A is given by (Section 1.16)

$$C_A = \left[\frac{\varepsilon_s q N_A}{2(V + V_{dif})} \right]^{1/2}$$

where the junction voltage V is the average node voltage during switching, say $V_{DD}/2$ and V_{dif} is the diffusion potential. C_P is given by

$$C_P = C_A l_d$$

where l_d is the effective depth of the drain island wall. Parameter l_d includes the effects of doping gradient within the substrate, and therefore l_d may not be equal to the physical depth of the PN junction.

The parameters in Eqs. (1.27a) and (1.27b) have the following dependence on the feature size Λ of the technology. As the technology scales down in the range 5 to 1 μm, N_A stays approximately within the range 1 to 3×10^{16} carriers/cm^3 to maintain

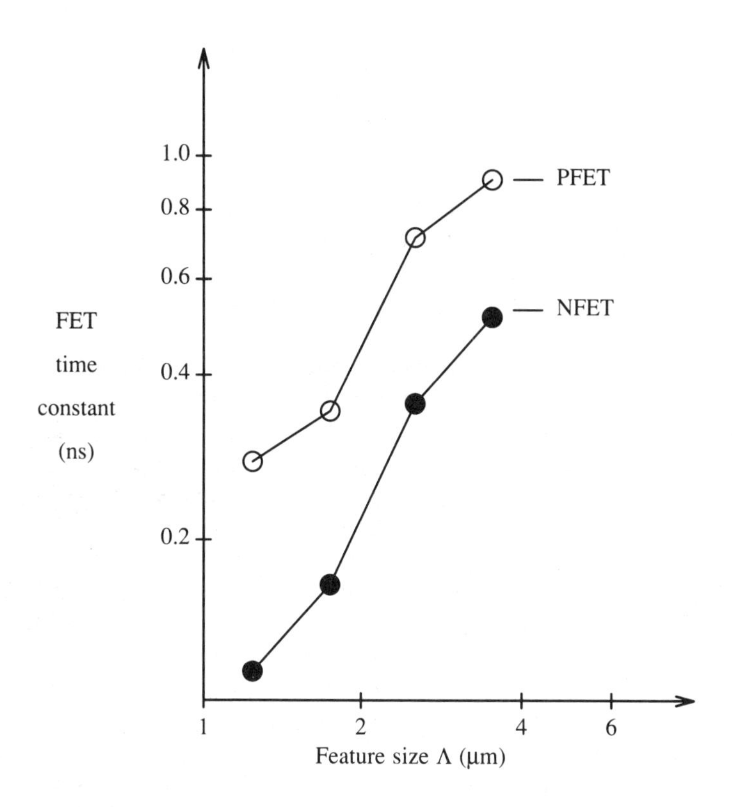

Figure 1.22 FET time constant versus feature size of CMOS technology.

the FET threshold voltage. Junction depth l_d of a self-aligned technology decreases as the channel length decreases with decreasing feature size. T_{OX} decreases but less than proportional to feature size. In a technology of feature size approaching 1 μm at $V_{DD} = 5$ V, T_{OX} cannot be reduced significantly below 250 angstroms, since the hot carrier capture effect by thin oxide [61] becomes significant. Therefore, l_d and T_{OX} scale down weaker than proportional to Λ. Channel length L scales down weaker than Λ, especially when the gate polysilicon is narrower than 2 μm. FET width W scales down about proportional to feature size, from the strong demand for device packing density. As a result, the total parasitic capacitance $C(W, L)$ of Eq. (1.28) scales stronger than feature size Λ, but weaker than Λ^2. $C(W, L)$ versus Λ shown in Fig. 1.21 supports the estimates. The dependence of C in Fig. 1.21 is given approximately by

$$C(W, L) \approx \Lambda^{3/2}$$

Since T_{OX} and L scale down weaker than Λ, typically like $\Lambda^{1/2}$, Eq. (1.27b) shows that T_{FET} of a FET in the ohmic transport regime decreases with feature size Λ like

$$T_{FET} \approx C(W,L) \approx \Lambda^{3/2} \tag{1.29}$$

and T_{FET} of a FET in the saturated velocity transport

$$T_{FET} \approx \frac{C(W,L)}{\Lambda^{1/2}} \approx \Lambda \tag{1.30}$$

Equations (1.29) and (1.30) define the limits of improvement in delay by scaling down FETs while keeping the power supply voltage V_{DD} constant. The delay falls within the limits $T_D \approx \Lambda$ and $T_D \approx \Lambda^{3/2}$. The scaling law discussed here is similar to that reported by Chi and Holmstrom, but the author's argument is not entirely the same [62].

An interesting observation can be made from an examination of the factors contributing to T_{FET}. Saturation current $I_{D\max}$ versus feature size Λ is shown in Fig. 1.20. There is a peak in the saturation current at about 1.75-μm feature size. The peak in the saturation current is an inevitable consequence of constant-voltage scaling; below the feature size where the current attains the maximum, the decreasing FET size overcomes increasing conductance of the FET channel.

The saturation current maximum was predicted by Chatterjee, Hunter, Holloway, and Lin [63]. They studied constant-voltage and quasi-constant-voltage scaling of FETs. They found theoretically that the saturation current of a FET has a maximum at about 1-μm feature size for constant-voltage scaling, and at about 0.5-μm feature size for quasi-constant-voltage scaling. The maximum does not exist in the constant field scaling [64]. In their quasi-constant-voltage scaling, the power supply voltage was reduced as $\Lambda^{1/2}$, and the thin oxide thickness was reduced like Λ. In the constant-voltage scaling T_{OX} was scaled less, like $\Lambda^{1/2}$, because of the hot-carrier trapping effect. By reducing the power supply voltage, however, T_{OX} can be further scaled down and the FET channel current improves further. Fine-line MOS technology (0.5 μm or below) uses power supply voltages that fall in the range 1.5 to 3.5 V. The reduced power supply voltage relieves the problem of hot-carrier capture by the oxide.

Figures 1.21 and 1.22 show that the mechanism of improvement in delay by scaling down a 5-V CMOS technology beyond the 1.75-μm feature size is due to reduction in the gate and drain capacitances. We note that the very high switching speed of SOS devices is due to reduced parasitic capacitance. In a reduced-feature-size technology, an effort to reduce parasitic capacitance is more rewarding than improving FET conductance. Decreasing the parasitic capacitance of a FET, however, has its own limit. The node capacitance is now determined by wiring capacitance. When the capacitance between the nearby wires (sidewall capacitance) becomes comparable to the capacitance to the substrate, further scale-down in wiring is not effective.

Saturation currents of an NFET and a PFET having the same 1.75-μm feature size differ by a factor from 2 to 3. At submicron feature size, the difference will diminish to a factor from 1.3 to 1.6, because velocity saturation of electrons occurs at the lower fields than holes [55].

At submicron feature size, the effects of contact series resistance originating from the resistance of the source-drain diffusion islands and from contact imperfections limit the current-drive capability of FETs. This effect is estimated from Eqs. (1.15a) and (1.15b) as

$$B \rightarrow \frac{B}{1 + R_s B (V_G - V_{TH})}$$

where R_s is the contact series resistance. The term $R_s B (V_G - V_{TH})$ can be about unity in submicron devices.

1.14 WIRING PARASITICS

Wires used to interconnect FETs are one of three kinds: highly conductive metal wires, less conductive polysilicon, or polysilicide wires, and still less conductive diffusion wires. One-micrometer-thick aluminum wires have sheet resistance (per square resistance), of 0.04 Ω worst case, at the room temperature. Silicided wires have about 4 Ω/square, and doped polysilicon wires about 25 Ω/square. Diffused N+ wires have about 20 *to* 30 Ω/square and diffused P+ wires about 130 Ω/square. The aluminum wires are about 100 times more conductive than wires of polysilicide. These are the resistances for one metal, one polysilicon/silicide technology. If more than one level of aluminum metal interconnects is available, the thickness of the first-level metal wires decreases by a factor of 2, and the sheet resistance doubles.

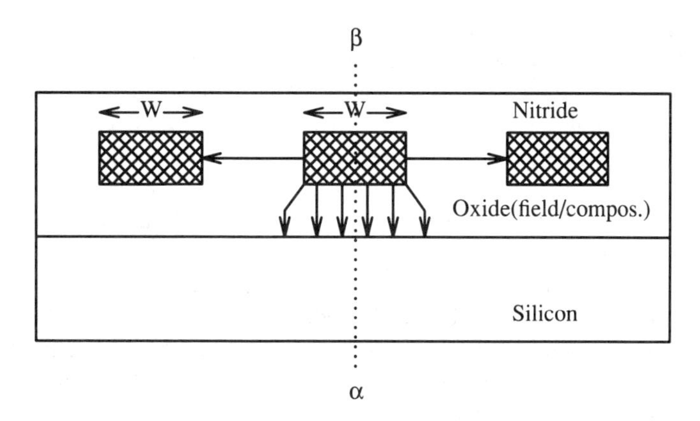

Figure 1.23 Electric field profile of the interconnect wires.

The series resistance of a wire is determined by counting the number of squares and multiplying by the sheet resistance. The series resistance of a metal wire is not always negligible in a CMOS circuit. In 1.75-μm CMOS, the resistance of an inverter

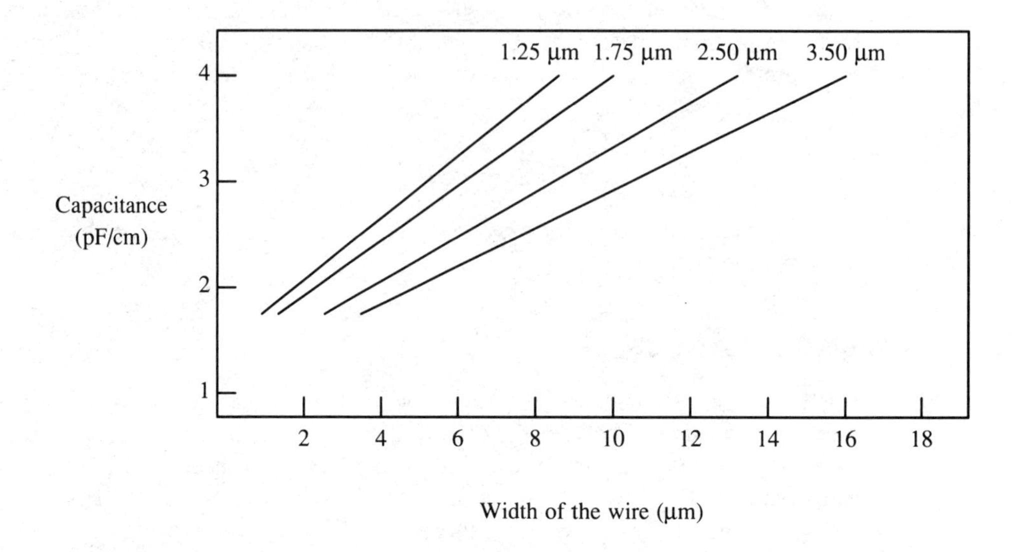

Figure 1.24 Capacitance versus width of metal wire.

10 times the minimum size (at steady state) is 150 to 360 Ω. If a 2-μm-wide wire connects the output of the inverter to a destination 1 cm away, the 5000 square wire has 200 Ω of series resistance. This is comparable to the resistance of the driver itself. In small-feature-size CMOS devices, the wiring delay can be more than the gate delay [65].

The capacitance of wires can be determined from a field analysis using a computer program that solves the Laplace equation. The cross-sectional structure with approximate vertical dimensions of typical CMOS devices is shown in Fig. 1.2. The vertical dimensions (the direction perpendicular to the silicon surface) are determined by the processing technology only, and a designer cannot control them. The horizontal dimensions (the width and the shape of the wires) are subject to the designer's control.

To compute accurate wiring capacitance, the effects of nearby wires must be taken into account. In Fig. 1.23 the capacitance of the metal wire at the center is computed assuming that the metal wires on both sides and the silicon substrate are grounded. Wiring capacitance per centimeter of four recent generations of CMOS technology is shown in Figs. 1.24 and 1.25. Because of the sidewall capacitance, the wiring capacitance does not decrease significantly below 2 pF/cm [66].

The parasitic capacitance of wires having a width different from the default width, to a very good approximation, increases linearly with width. This relationship can be understood from Fig. 1.23. The electric field of a metal wire of width W is shown. If the figure is cut at line $\alpha\beta$ and a metal section of width W' is inserted, such that a uniform field exists below the extra section and no field above, the resulting field is a very close approximation of the electrostatic field of the wire having width

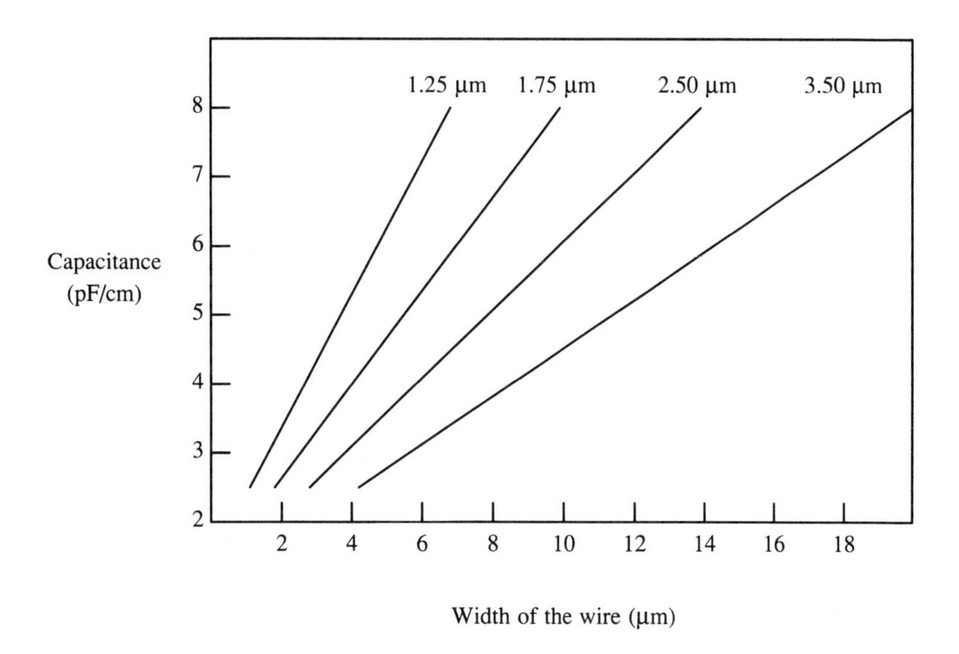

Figure 1.25 Capacitance versus width of polysilicon wire.

$W + W'$. From this construction

$$C(W + W') = C(W) + \frac{\varepsilon_{OX} W'}{T_{COX}} \tag{1.31}$$

where T_{COX} is the thickness of the composite oxide below the metal wire.

The conventional capacitance computation assumes a two-dimensional field configuration. A more precise, three-dimensional capacitance computation is now used, especially in memory design [67].

The capacitance of diffusion wire of width W and length L is calculated using the two capacitance coefficients C_A (capacitance per unit square area of junction) and C_P (capacitance per unit peripheral length) [68] by

$$C_{WIRE} = C_A WL + C_P(2W + 2L) \tag{1.32}$$

A rough estimate can be made using $C_A = 3 \times 10^{-16}$F/μm^2 and $C_P = 3 \times 10^{-16}$F/μm, which are easy numbers to remember. In scaled-down CMOS, the sidewall capacitance becomes quite large [69].

The inductive parasitics are not usually significant except for analysis of power bus noise. Calculation of wiring inductance is not elementary. The silicon substrate is conductive enough to prevent penetration of electric field, but even if doped very heavily, the substrate is not conductive enough to prevent penetration of magnetic field originating from the current of the conductor, as shown in Fig. 1.26. The magnetic

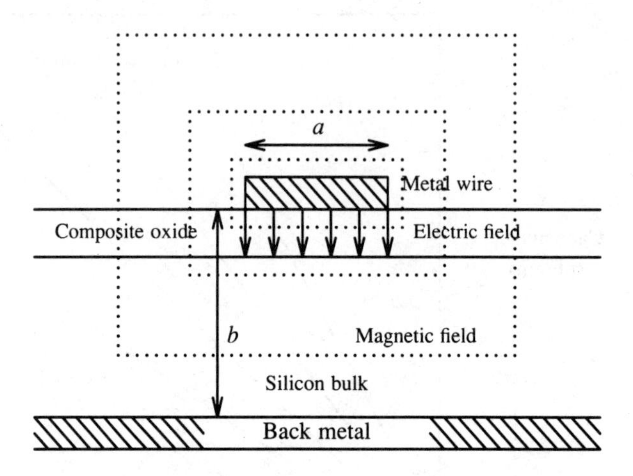

Figure 1.26 Magnetic field profile around metal wire.

field is stopped only by the metal on the back side of the silicon substrate. To a first approximation the magnetic flux is assumed to exist in the entire space above the back metal.

If the width of the metal wire is a and the distance of the metal wire from the back metal is b, the formula

$$L = 2 \times 10^{-9} \times \log\left(8\frac{b}{a} + 4\frac{a}{b}\right) \quad \text{H/cm} \tag{1.33}$$

gives inductance [70]. A typical value of b is 500 μm. Figure 1.27 shows inductance versus a/b. Since the electromagnetic field in the vicinity of the conductor is very different from the field pattern of a TEM-mode wave [71], the velocity of propagation of a signal through a microstrip line on silicon is very slow. Hasegawa et al. called the propagation mode a slow wave [70].

1.15 SILICIDED INTERCONNECTS AND SECOND-LEVEL METAL INTERCONNECTS

Doped polysilicon used as gate material has high resistivity. When polysilicon is used for interconnects, 25 Ω/square sheet resistance creates serious signal delay problems in scaled-down CMOS circuits. As discussed in Chapter 8, this type of signal delay is one of the most serious VLSI design problems. Reduction of the sheet resistance is very desirable.

Refractory metals such as molybdenum (Mo) and tungsten (W) were studied for use as material for gates and interconnects. Although feasibility was established, there are problems originating from weakness in high-temperature oxidizing ambient and to

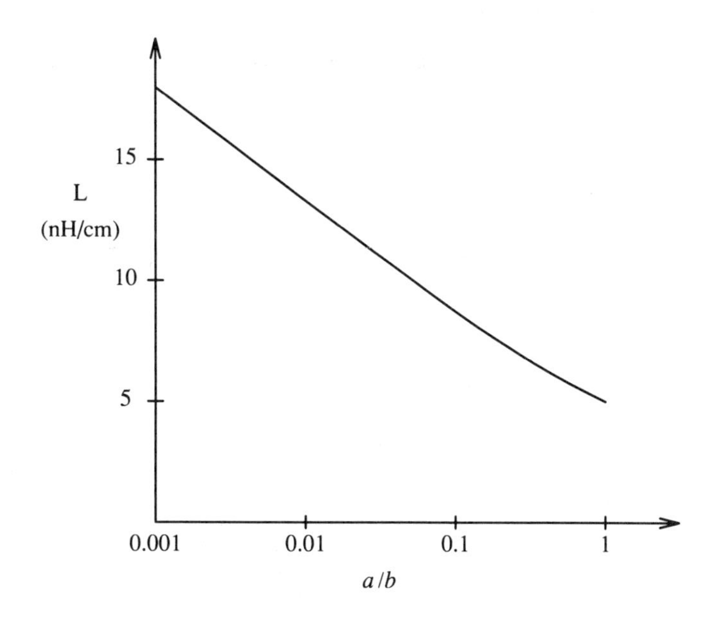

Figure 1.27 Inductance of metal wire versus a/b.

many chemical reagents. Therefore, the most widely used high-conductivity gate and interconnect materials are silicides of refractory metals such as Mo, Ta (tantalum), W, and Ti (titanium). They are formed on the top of the doped polysilicon layer by co-sputtering refractory metal and silicon. Titanium silicide ($TiSi_2$) interconnects offer the lowest sheet resistance, approximately 1 Ω/square, followed by tantalum silicide ($TaSi_2$) at about 3 Ω/square, and molybdenum silicide at 5 Ω/square. Tantalum and molybdenum silicide processes are directly compatible with standard silicon gate technology. Titanium silicide is reactive to hydrofluoric acid, and development is behind Mo and Ta silicides.

Low-resistivity gate material is essential in PLA, ROM, RAM, and datapath designs. The feature size allowed for silicides are usually much smaller than that of second-level metal [72] [73].

Two-level metal CMOS technology emerged recently. The two metal layers are insulated by low-temperature silicon dioxide, and sophisticated planarization techniques, including via-hole filling, are employed [74]. Planarization is especially necessary for multilevel metal interconnects in gate-array devices [75]. The feature size allowed for the second-level metal is often larger than the feature size allowed for the first-level metal and polysilicon. The best way to use the multilevel metal technology for an efficient VLSI layout has yet to be worked out. Reduced thickness of the first-level metal prevents use of the first-level metal for a very high power V_{SS} or V_{DD} routing. The increased capacitance of the polysilicon and the first-level metal must be tolerated.

1.16 PN JUNCTION PARASITICS

Capacitance of source and drain diffused islands of a FET is the major contributer to capacitive load of a gate. The capacitance of a reverse-biased PN junction, whose doping profile is shown in Fig. 1.28, is derived in this section. The formula has wide application in this book.

In an abrupt PN junction shown in Fig. 1.28 widths $\overline{\alpha J}$ and $\overline{J\delta}$ are completely depleted of electrons and holes, respectively. The field increases linearly with distance from zero at point α to the maximum at point J, and then linearly decreases to zero at point δ. Electrical neutrality requires that

$$N_D W_D = N_A W_A$$

and the potential sustained by the pair of the space-charge layers V_J is

$$V_J = \frac{q}{2\varepsilon_S}(N_D W_D{}^2 + N_A W_A{}^2)$$

From the two equations,

$$W_A = \left\{ \frac{2\varepsilon_S V_J}{q(N_A + N_D)} \left[\frac{N_D}{N_A} \right] \right\}^{1/2}$$

$$W_D = \left\{ \frac{2\varepsilon_S V_J}{q(N_A + N_D)} \left[\frac{N_A}{N_D} \right] \right\}^{1/2}$$

and the positive charge stored in the capacitor is

$$Q_P = qN_D W_D = \left\{ \frac{2q\,\varepsilon_S V_J}{N_A + N_D} N_A N_D \right\}^{1/2}$$

Even if no potential is applied to the PN junction from the external source, the potential V_J is not zero. Holes of the P-type material diffuse into the N-type material, and the electrons in the N-type material into the P-type material. The potential difference established in this way originates from the difference in the affinity of the doped materials to electrons and holes, and is called a diffusion potential. The diffusion potential is related to the band gap of the semiconductor, and for silicon it is about 0.7 V. The external bias voltage is in series with this potential. Therefore,

$$V_J = V_{dif} + V$$

where V is the voltage applied of the external source. Then the PN junction capacitance is given by

$$C_J = \frac{\partial Q_P}{\partial V} = \left\{ \frac{q\,\varepsilon_S N_A N_D}{2(N_A + N_D)(V + V_{dif})} \right\}^{1/2} \text{ F/cm}^2$$

$$= \frac{C_J(0)}{\sqrt{1 + (V/V_{dif})}} \tag{1.34}$$

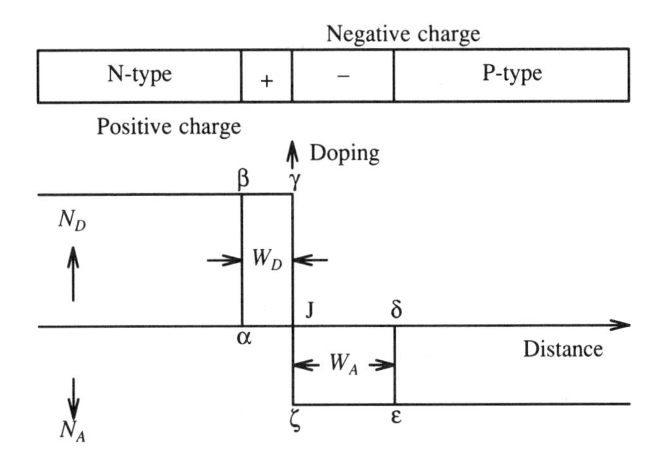

Figure 1.28 Doping profile of a PN junction.

where

$$C_J(0) = \left[\frac{q\varepsilon_S N}{2V_{dif}}\right]^{1/2} = 3.4 \times 10^{-8}\left[\frac{N}{10^{16}\text{cm}^{-3}}\right]^{1/2} \text{ F/cm}^2$$

and where

$$N = \frac{N_A N_D}{N_A + N_D}$$

Assuming that $N = 2 \times 10^{16}\,\text{cm}^{-3}$, $C_J(0) = 4.8 \times 10^{-8}\,\text{F/cm}^2$. The flat-bottom capacitance [see C_A of Eq. (1.32)] follows the 1/2-power law of voltage dependence of Eq. (1.34) accurately. The sidewall capacitance [see C_p of Eq. (1.32)] depends less on voltage, reflecting the graded junction at the surface of the semiconductor [69].

A comparison with the gate oxide capacitance is informative. Assuming that the gate oxide thickness is 250 angstroms,

$$C_G = \frac{\varepsilon_{OX}}{T_{OX}} = 13.8 \times 10^{-8}\,\text{F/cm}^2 \tag{1.35}$$

Since the drain area of an ordinary FET is typically two to three times larger than the gate area, the drain-diffused island capacitance from Eq. (1.34) and the gate capacitance from Eq. (1.35) are comparable. Therefore, a large diffused area is very undesirable in a high-performance CMOS VLSI chip. SOS (SOI) technology has the advantage of a small source/drain capacitance. An interesting technique of isolating source/drain area from silicon substrate by oxide has been reported [76]. Oxide

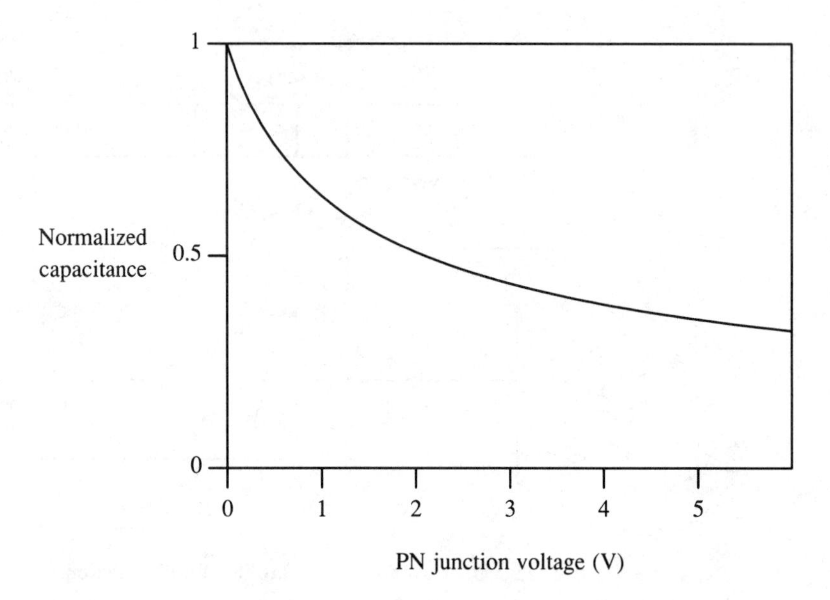

Figure 1.29 PN junction capacitance versus junction voltage.

sidewall isolation to reduce capacitance contributed by periphery of small islands is a new technology trend [77].

Drain island capacitance decreases with reverse bias, as shown by Eq. (1.34). Figure 1.29 shows normalized capacitance versus voltage. By applying about 1 V of back bias, the drain capacitance decreases by as much as 50%. A technique to reduce the effective drain capacitance is substrate biasing. Application of reverse bias, and the technique of generating voltages outside the range of V_{DD} and V_{SS}, are now well established, especially in NMOS VLSI. If the reverse bias is controlled by feedback, the threshold of the FETs can be controlled simultaneously. A problem of this technique is that ground and substrate connection must be separated.

1.17 PARAMETERS CHARACTERIZING THE SUBSTRATE

One of the most neglected problems regarding CMOS VLSI devices relates to the electrical phenomena in the substrate. In addition to working as the mechanical support, as the insulating barrier between components, and as the path for heat conduction, the substrate is a medium for AC and DC current flow. As CMOS VLSI technology advances, the problems of power bus noise, of accurately estimating parasitic capacitances, and of latch-up demand knowledge of electrical phenomena in the substrate. Studies of the substrate electrical phenomena have just begun [78]. Some early works are intended to study leakage mechanisms of DRAMs [79] [80] [81] or latch-up [82].

TABLE 1.2 SUBSTRATE PARAMETERS

Definition	Symbol	Value
N-tub-epi capacitance	C_{NE}	$0.3 - 0.5 \times 10^{-8}$ (F/cm^2)
Metal-sub. capacitance	C_{MS}	0.3×10^{-8} (F/cm^2)
Poly-sub. capacitance	C_{PS}	0.968×10^{-8} (F/cm^2)
Diffusion (P+)	C_{P+}	2.47×10^{-8} (F/cm^2)
Diffusion (N+)	C_{N+}	3.19×10^{-8} (F/cm^2)
Gate capacitancc	C_G	1.38×10^{-7} (F/cm^2)
N-tub sheet resistance	R_N	1.67×10^3 (Ω/square)
N-tub vert. resistance	r_N	1.5×10^{-4} (Ω cm^2)
P-tub sheet resistance	R_P	6.51×10^3 (Ω/square)
P-tub vert. resistance	r_P	$\approx 1.6 \times 10^{-2}$ (Ω cm^2)
P+ sub. sheet resistance	R_{P+}	$0.06 - 0.2$ (Ω/square)
P+ sub. vert. resistance	r_{P+}	$1.5 - 5 \times 10^{-4}$ (Ω cm^2)

Let us consider a twin-tub CMOS device, fabricated on a P+ substrate. The substrate is connected to V_{SS} via P+ diffused contacts in the P-tubs, and the back metal contact. The currents are injected into the substrate or into the N-tub by capacitive coupling. Current injected into the substrate can be estimated as follows. Wiring parasitics of metal are typically 2.5 pF/cm. The switching time of a hard-driven node in a 1-μm CMOS VLSI circuit is typically 2.5 ns or less. Then the current is given by

$$2.5 \times 10^{-12} \frac{5\text{V}}{2.5 \times 10^{-9}\text{s}} = 5 \text{ mA/cm}$$

Suppose that the wire is on an isolated N-tub. The current flows into the N-tub from V_{DD} and accumulates positive charge on the silicon surface directly under the wiring. This positive charge attracts negative charge on the undersurface of the wiring, which is supplied by current of the NFET that pulls down the wire. If a 2.5-mm-long wire runs along the 2.5-mm-long edge of 2.5 mm × 1 mm rectangular N-tub, and if the N-tub is contacted to V_{DD} on the other 2.5-mm-long edge, and if the N-tub sheet resistance is $1\,k\Omega$/square, a voltage of 500 mV is developed across the N-tub for 2.5 ns. This is a large potential. Whether or not the large potential *really* develops requires examination based on semiconductor device physics.

In order to understand the substrate electrical phenomena, parameter values characterizing the substrate must be determined. The parameter values summarized in Table 1.2 are from a twin-tub CMOS technology of about 1 to 2 μm feature size. The numbers are an example, and they depend on the particular technology considered.

Discussions in the following two sections are on specialized subjects. These sections can be skipped if the reader is not interested in the details.

1.18 CONTACTS TO THE SUBSTRATE

It is a common practice in modern VLSI layout to use only standardized contacts everywhere on a chip. A contact hole is the smallest feature, and small contacts restrict current and result in significant contact resistance. We study here the contact resistance originating from the restricted current flow in the semiconductor. Components of contact resistance originating from interface imperfections are not considered.

(A) N-Tub Contact Resistance

Current flow within an isolated N-tub is approximately two-dimensional. Until the current spreads out to the area of one-half the typical contact-to-contact distance the current is restricted by the contact. Then a simple field analysis gives

$$\text{N–tub contact resistance} = \frac{R_N}{2\pi}\log\left[\frac{b}{2a}\right] \tag{1.36}$$

where b is the distance between the two nearby contacts and a is an equivalent contact radius defined by $\pi a^2 = a_0^2$, where a_0 is the edge of a square contact hole. When $a_0 = 1.5 \times 10^{-4}$cm, $a = 0.846 \times 10^{-4}$cm. If $b = 40 \times 10^{-4}$cm,

$$\text{N–tub contact resistance} = \frac{1.67 \times 10^3}{2 \times 3.1416}\log\frac{40 \times 10^{-4}}{2 \times 0.846 \times 10^{-4}} = 0.84 \times 10^3\Omega$$

(B) P-Tub Contact Resistance

This resistance is defined by the voltage developed when a unit current generator is connected between the contact and the grounded P+ substrate. Because the epitaxial layer under the diffused P-tub has high resistivity, the current spreads out first from the contact horizontally, along the conductive P-tub surface, and then the current heads down to the substrate, as shown in Fig. 1.30. If the equivalent contact size is defined in the same way as in the N-tub contact,

$$\text{P–tub contact resistance} = \left\{\pi\left[\frac{1}{R_P} + \frac{2a}{\sqrt{R_P l \rho_e}}\right] + \frac{\pi a^2}{l \rho_e}\right\}^{-1} \tag{1.37}$$

where l is the thickness of the high resistivity P-type epitaxial layer and ρ_e is the resistivity.

This equation is derived as follows. With reference to Fig. 1.30, the potential within the P-tub $\phi(r)$ and current $I(r)$ are functions of distance r from the center of the contact. The problem has an axial symmetry around the dotted center line of the contact. These quantities satisfy the following equations:

$$\frac{d\phi}{dr} = -\frac{R_P}{2\pi r}I$$

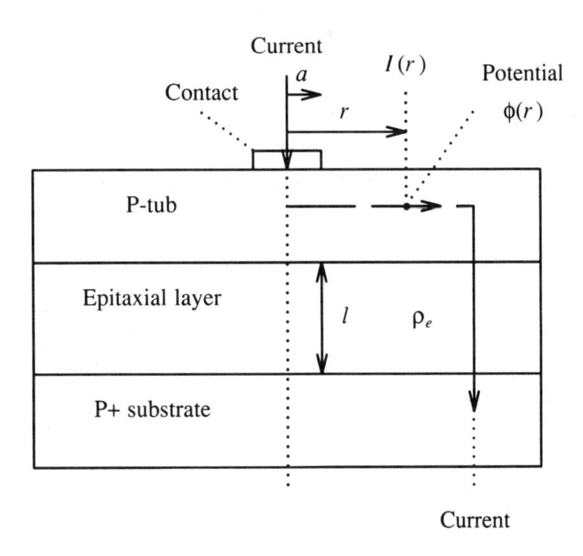

Figure 1.30 Current flow from P-tub contact.

and

$$\frac{dI}{dr} = -\frac{2\pi r}{l\rho_e}\phi$$

The P+ substrate is assumed to be held at ground potential. By eliminating I, potential ϕ satisfies

$$\frac{d^2\phi}{dr^2} + \frac{1}{r}\frac{d\phi}{dr} - k^2\phi = 0$$

where $k^2 = R_P/(l\rho_e)$. There is a solution of the form

$$\phi = e^{-kr} r^n$$

when $r \to \infty$. By substitution $n = -1/2$. Then

$$\phi = \phi_0 \sqrt{\frac{a}{r}} e^{-k(r-a)} \tag{1.38}$$

$$I = -\frac{2\pi r}{R_P}\frac{d\phi}{dr}$$

From this solution the resistance of the contact is derived as

$$\left\{\frac{I(a)}{\phi(a)} + \frac{\pi a^2}{l\rho_e}\right\}^{-1} = \left\{\pi\left[\frac{1}{R_P} + \frac{2a}{\sqrt{R_P l\rho_e}}\right] + \frac{\pi a^2}{l\rho_e}\right\}^{-1}$$

If $l = 8 \times 10^{-4}$, $\rho_e = 16.5\,\Omega\,\text{cm}$, $R_P = 6.51 \times 10^3$/square, and $a = 0.716 \times 10^{-4}$, the P-tub contact resistance is $1.77 \times 10^3\,\Omega$. The calculation depends on the simplifying assumptions and its accuracy is not very high. If there are many contacts on P-tub surface, and current flows predominantly within the P-tub surface, Eq. (1.36) with R_N replaced by R_P gives the better estimate of contact resistance.

1.19 ELECTRICAL PHENOMENA IN THE SUBSTRATE

Since V_{SS} and V_{DD} buses that are connected to the P- and N-tubs, respectively, are held at constant potential, currents injected into the tubs are by capacitive coupling from the internal logic nodes. As for the gate capacitance C_G, the effect is more complex. Current is injected from the gate directly into the substrate if the channel is not formed. When the channel is fully formed, the current is first injected to the channel, and then to the source, to the drain, or to the substrate. Voltage change of the wires and diffusion islands inject currents into the P- and N-tubs that are determined using the capacitances C_{MS}, C_{PS}, C_{P+}, and C_{N+}. Using the typical switching time for a 1.25-to 1.75-μm CMOS (2.5 ns) and the capacitance values of the wiring and the diffusion islands, the current density into the substrate is estimated to be

$$J = (0.3 \text{ to } 3) \times 10^{-8}\ \text{F/cm}^2\ \frac{5\text{V}}{2.5 \times 10^{-9}\text{s}} = (6 \text{ to } 60)\ \text{A/cm}^2$$

Using the tabulated parameter values and the current, we study the substrate phenomena in the following.

(A) Uniform Current Injection into a Large Area

The currents injected from a large parasitic capacitor create a vertical potential in the substrate. An estimate of the potential is as follows:

P-tub: $r_P J = 7.75 \times 10^{-3} \times (6 \text{ to } 60) = 46.5 \text{ to } 465\ \text{mV}$

N-tub: $r_N J = 1.50 \times 10^{-4} \times (6 \text{ to } 60) = 0.9 \text{ to } 9\ \text{mV}$

P+sub: $r_{P+} J = (1.5 \text{ to } 5) \times 10^{-4} \times (6 \text{ to } 60) = 0.9 \text{ to } 30\ \text{mV}$

From this estimate, the P+ substrate is very close to constant potential if the backside metallization is well grounded. Parameter r_P is significantly higher than r_N and r_{P+} due to the high-resistivity P-type epitaxial layer (this estimate depends critically on the details of technology). A large circuit built on a P-tub suffers from the voltage across the epitaxial layer: When a PLA precharges, the P-tub surface potential may be driven up by several hundred millivolts from ground. A large circuit on an N-tub is coupled primarily to the P+ substrate capacitively through N-tub-epi capacitance. This problem will be discussed later in this section.

Let C stand for any of the parasitic capacitances, C_{MS}, C_{PS}, C_{P+}, or C_{N+}. In a capacitor on a P-tub, time constant is $(r_P + r_{P+})C = (2.4 \text{ to } 26.3) \times 10^{-11}\text{s}$. The longest time constant occurs when the capacitor is an N+ diffusion island on P-tub (due to large capacitance C_{N+}). The time constant is, however, at least an order of magnitude less than the typical switching time. Therefore, the surface potential of the P-tub follows the voltage change of the upper plate of the capacitor.

The parasitic capacitance of wires is conventionally determined from the structure of the wires and of the surface of the P- or the N-tubs. For capacitances on a P-tub the values determined in this way are adequate, based on the time constant estimate of the preceding paragraph. As for the capacitances on an N-tub the values are suspected to be an overestimate, since N-tub-epi capacitance exists in series. The effect should be quite significant, amounting to a factor of 2 overestimate of the parasitic capacitance. This conclusion, however, is subject to how an N-tub works in the direction perpendicular to the surface, and a more detailed discussion is given in Section (B).

(B) Transverse Spreading of Electrical Disturbances in the Substrate

When a current is injected into a P-tub the potential at the point that is distance r away from the point of injection is given by

$$\phi = \phi_0 \left[\frac{a}{r} \right]^{1/2} e^{-k(r-a)}$$

where $k = \sqrt{R_P / l \rho_e}$ [see Eq. (1.38)]. If $R_P = 6.51 \times 10^3\ \Omega$/square, $l = 8 \times 10^{-4}$ cm, $\rho_e = 16.5\ \Omega$ cm, and $1/k = 14.2\ \mu$m. This is the characteristic distance that an electrical disturbance of one point influences the nearby point.

The equivalent circuit model of an N-tub on P-substrate is a two-dimensional RC network consisting of a sheet resistance R_N and N-tub-epi capacitance C_{NE}. The equation satisfied by the N-tub potential ϕ is

$$\frac{\partial \phi}{\partial t} = D \nabla^2 \phi \tag{1.39}$$

where $D = 1/C_{NE} R_N$ and ∇^2 is a two-dimensional Laplacian. D is given by the parameter values listed in Table 1.2 as

$$D = \frac{1}{(0.3 \text{ to } 0.5) \times 10^{-8} \times 1.67 \times 10^3} = (1.2 \text{ to } 2) \times 10^5\ \text{cm}^2/\text{s}$$

Figure 1.31(a) shows a metal wire on an N-tub. Let the wire be driven from 0 to V_{DD} volts within a negligibly short time at time $t = 0$, with the surface of the N-tub originally at V_{DD}. The N-tub surface directly underneath the wire is driven up to

$$V_{DD} + V_{DD} \frac{C_{MS}}{C_{MS} + C_{NE}}$$

since $r_N C_{NE} \approx 10^{-12}$ s, which is very small. The potential spike $V_{DD} C_{MS}/(C_{MS} + C_{NE})$ decays with time as shown in Fig. 1.31(b). The height of the potential spike decreases with time as $(4D(t + t_0))^{-1/2}$, and the width of the spike increases as $(4D(t + t_0))^{1/2}$, where t_0 depends on the initial width of the spike. Since the initial width of the spike should be the width of the wire W, $W = 2 \times \sqrt{4Dt_0}$, or $t_0 = W^2/16D$.

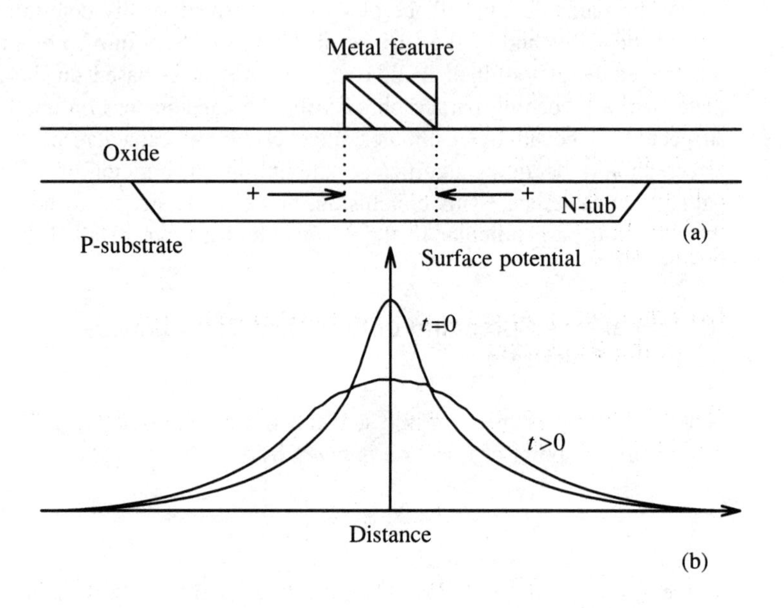

Figure 1.31 Mechanism of relaxation of the charge induced on N-tub.

The width and height of the potential spike at time t are given, respectively, by

$$\text{width} = \sqrt{16Dt + W^2}$$

$$\text{height} = V_{DD}\frac{C_{MS}}{C_{MS}+C_{NE}}\frac{1}{\left(1+\dfrac{16Dt}{W^2}\right)^{1/2}} \tag{1.40}$$

Within a typical gate switching time of 2.5 ns, the spike spreads out to the distance

$$\sqrt{4Dt_{sw}} = \sqrt{4 \times 1.6 \times 10^5 \times 2.5 \times 10^{-9}} = 4 \times 10^{-2} = 400\ \mu\text{m}$$

The problem is relevant to the accuracy of the computed capacitance of a feature on an N-tub. The limit of error in capacitance of a square feature having size $W \times W$ cm^2, by the effect of the series connected N-tub-epi capacitance, is estimated from

$$CW^2\left[1 - \frac{C}{C_{NE}}\frac{W^2}{16Dt_{SW}}\right]$$

where C is any of C_{MS}, C_{PS}, *or* C_{P+}. CW^2 is the capacitance computed in a conventional way, and the term $(C/C_{NE})(W^2/16Dt_{SW})$ is the first-order correction of the effect of the N-tub-epi series capacitance, derived from the ratio of the area involved in the electrical phenomena. The effective capacitance is less than CW^2, but the correction term is small. Since $(C/C_{NE}) \leq 3$, the correction term is at the most $3(W/8 \times 10^{-2})^2$, which is negligible unless the structure is several hundred micrometers in size. This conclusion is fortunate for the present practice of wiring capacitance computation, but the way the conclusion is reached is not at all obvious.

Electrical disturbance in an N-tub decays as fast as that in a P-tub. In a P-tub, however, the potential spike decays by moving holes in the vertical direction. Within an N-tub a potential spike decays just as fast, by moving electrons in the horizontal direction. This horizontal motion of electrons cannot be fast enough, however, if the structure is large. Present VLSI chips are designed such that the objects like FETs and wires are very small. If a designer neglects this subtle point and builds a huge structure on a chip, weird phenomena are likely to occur.

Equation (1.39) does not include the effects of series resistance that is in series with N-tub-epi capacitance. For precision analysis the effects must be included.

(C) The Way That Substrate Contacts Work

1. Contacts to the P-type substrate. When the back metal is connected solidly to the ground, P-tub contacts on the surface of the chip play a relatively minor role. Since each contact has about $2\,k\Omega$ of contact resistance, even 10^4 contacts still contribute $0.2\,\Omega$ of resistance. The large area contact to the back metal has orders of magnitude less contact resistance. In order to reduce the ground bus noise of a VLSI chip (Section 9.5), use of the back metal contact is the promising way. Back metal is, however, not grounded during wafer probing. Then, since conductivity of the P+ substrate is very high, each P-tub contact carries about the same current, wherever the contact may be.

2. Contact to the N-tub. N-tub contacts close to the point of current injection work in the same way as P-tub contacts. The distance from the current source to the contact is then less than

$$\sqrt{4Dt_{sw}} = \sqrt{4 \times 1.6 \times 10^5 \times 2.5 \times 10^{-9}} = 400\ \mu\text{m}$$

This is the requirement that the current that flows through the N-tub-epi capacitance is not a significant fraction of the current injected into the N-tub. If distance to the nearest N-tub contact is more than that, a substantial fraction of the injected current flows to the P+ substrate. Furthermore, the effects of time-averaging the injected currents and the effects of compensation of the positive and the negative injected currents from different sources of injection become significant. The current that flows through an N-tub contact is the averaged current of the sources within distance L, and is average over time interval L^2/D, where L is the size of the area covered by the contact. If $L = 500\ \mu$m, $L^2/D = 15.6$ ns. The time is longer than a typical gate switching time and is different for different contacts.

The substrate of a CMOS VLSI chip has been considered trouble-free. Indeed, there is no very serious problems at present, but there are some areas that call for attention: If the surface of the tubs are driven very hard, the voltage swing may influence nearby devices. When the substrate is biased to a breakdown condition, however, many practical problems, such as overloading of the substrate bias generator, FET threshold voltage variations, and excess minority carriers in the substrate, occur. The substrate current in this condition was studied by Hsu [83].

1.20 REFERENCES

01. S. M. Sze "Physics of semiconductor devices" John Wiley & Sons, New York, 1981.

02. J. R. Brews "Physics of the MOS transistor" Applied solid state science, supplement 2A, Academic Press, New York, 1981.

03. E. H. Nicollian and J. R. Brews "MOS (Metal-Oxide-Semiconductor) physics and technology" John Wiley & Sons, New York, 1982.

04. A. S. Grove "Physics and technology of semiconductor devices" John Wiley & Sons, New York, 1967.

05. J. T. Wallmark and H. Johnson "Field-effect transistors, physics, technology and applications" Prentice-Hall, Inc., Englewood Cliffs, New Jersey, 1966.

06. P. Richman "MOS field-effect transistors and integrated circuits" John Wiley & Sons, New York, 1973.

07. W. Fichtner "Physics and simulation of small MOS devices" IEDM82 Digest, pp. 638-641, December 1982 and W. Fichtner "Very short channel MOSFET's-physics and technology" Extended abstracts of the 16th International Conference on Solid State Devices and Materials, Kobe, Japan, 1984, pp. 77-81.

08. L. W. Nagel "SPICE 2: A computer program to simulate semiconductor circuits" Memo ERI-M520, University of California, Berkeley, Calif., May 1975.

09. P. Richman "Complementary MOS field-effect transistors on high-resistivity silicon substrates" Solid-State Electronics, vol.12, pp. 377-383, May, 1969 and M. H. White and J. R. Cricchi "Complementary MOS transistors" Solid-State Electronics, vol.9, pp. 991-1008, 1966.

10. K. Yu, R. J. C. Chwang, M. T. Bohr, P. A. Warkentin, S. Stern and C. N. Berglund "HMOS-CMOS—a low power high performance technology" IEEE J. Solid-State Circuits, vol.SC-16, pp. 454-459, October 1981.

11. O. Minato, T. Masuhara, T. Sasaki, H. Nakamura, Y. Sakai, T. Yasui and K. Uchibori "2K x 8 bit Hi-CMOS static RAMs" IEEE J. Solid-State Circuits, vol. SC-15, pp. 656-660, August 1980.

12. I. Fukushima, K. Kuwahara, K. Itoigawa, M. Nagata, K. Hoya, N. Horie and S. Ichimura "A BiMOS FET processor for VCR audio" ISSCC83 Digest, pp. 242-243, February 1983.

13. R. Hoshikawa, H. Kikuchi, S. Baba, S. Sato, K. Kawato, N. Inui and O. Wada "A 10, 000-gate CMOS LSI processor" ISSCC80 Digest, pp. 106-107, February 1980.

14. R. R. Troutman "VLSI device phenomena in dynamic memory and their application to technology development and device design" IBM J. Res. and develop., vol. 24, pp. 299-309, May 1980.

15. References to SOI technology are "Comparison of thin film transistor and SOI technologies" edited by H. W. Lam and M. J. Thompson, North Holland, New York 1984.

16. H. Hatano, Y. Uchida, M. Isobe, K. Maeguchi and H. Tango "Floating substrate effects in SOS VLSIs" IEDM81 Digest, pp. 359-362, December 1981.
17. A. K. Rapp "SOS: New design tools and techniques" ISSCC73 Digest, pp. 170-171, February 1973.
18. J. F. Allison, J. R. Burns and F. P. Heiman "Silicon on sapphire complementary MOS memory systems" ISSCC67 Digest, pp. 76-77, February 1967.
19. W. Fischer, E. P. Jacobs, I. Eisele and G. Dorba "Electron mobility in Si-MOSFETs with an additional implanted channel" Solid-State Electronics, vol. 22, pp. 225-228, March 1979.
20. K. Nishiuchi, H. Shibayama, T. Nakamura, T. Hisatsugu, H. Ishikawa and Y. Fukukawa "A Gb MOS logic circuit with buried channel MOSFETs" ISSCC80 Digest, pp. 60-61, February 1980.
21. L. C. Parillo "Process and device considerations for micron and submicron CMOS technology" IEDM85 Digest, pp. 398-402, December 1985.
22. R. W. Ahrons and R. D. Gardner "The interaction of technology and performance in complementary-symmetry MOS integrated circuits" ISSCC69 Digest, pp. 154-155, February 1969.
23. L. C. Parrillo, R. S. Payne, R. E. Davis, G. W. Reutlinger and R. L. Field "Twin-tub CMOS— A technology for VLSI circuits" IEDM80 Digest, pp. 752-755, December 1980.
24. J. Agraz-Guerena, R. A. Ashton, W. J. Bertram, R. C. Melin, R. C. Sun and J. T. Clemens "Twin-tub III—a third generation CMOS technology" IEDM84 Digest, pp. 63-66, December 1984.
25. H. J. Geipel, Jr., and A. G. Fortino "Process modeling and design procedure for IGFET thresholds" IEEE J. Solid-State Circuits, vol. SC-14, pp. 430-434, April 1979.
26. P. J. Coppen, K. G. Aubuchon, L. O. Bauer and N. E. Moyer "A complementary MOS 1. 2 volt watch circuit using ion implantation" Solid-State Electronics, vol. 15, pp. 165-175, 1972 and E. C. Douglas and A. G. F. Dingwall "Ion implantation for threshold control in COSMOS circuits" IEEE Trans. on Electron Devices, vol. ED-21, pp. 324-331, June 1974.
27. R. W. Ahrons and P. D. Gardner "Interaction of technology and performance in complementary symmetry MOS integrated circuits" IEEE J. Solid-State Circuits, vol. SC-5, pp. 24-29, February 1970.
28. J. A. Geurst "Calculation of high-frequency characteristics of thin-film transistors" Solid-State Electronics, vol. 8, pp. 88-90, January 1965.
29. M. Shoji "Analysis of high-frequency thermal noise of enhancement mode MOS field-effect transistors" IEEE Trans. on Electron Devices, vol. ED-13, pp. 520-524, June 1966.
30. H. Borkan and P. K. Weimer "An analysis of the characteristics of insulated-gate thin-film transistors" RCA Review, vol. 24, pp. 153-165, June 1963.
31. Y. El-Mansy "MOS device and technology constraints in VLSI" IEEE J. Solid-State Circuits, vol. SC-17, pp. 197-203, April 1982.
32. M. Aoki, K. Yano, T. Masuhara, S. Ikeda and S. Meguro "Optimum crystallographic orientation of submicron CMOS devices" IEDM85 Digest, pp. 577-580, December 1985.
33. G. Cheroff, D. L. Crithlow, R. H. Dennard and L. M. Terman "IGFET circuit performance n-channel versus p-channel" IEEE J. Solid-State Circuits, vol. SC-4, pp. 267-271, October 1969.
34. H. C. Pao and C. T. Sah "Effects of diffusion current on characteristics of metal-oxide semiconductor transistors" Solid-State Electronics, vol. 9, pp. 927-937, October 1966.
35. J. A. Geurst "Theory of insulated-gate field-effect transistors near and beyond pinch-off" Solid-State Electronics, vol. 9, pp. 129-142, February 1966.

36. J. R. Brews, W. Fichtner, E. H. Nicollian and S. M. Sze "Generalized guide for MOSFET miniaturization" IEEE Electron Device Letters, vol. EDL-1, pp. 2-4, 1980.
37. W. Fichtner, E. N. Fuls, R. L. Johnson, R. K. Watts and W. W. Weik "Optimized MOSFETs with subquartermicron channel lengths" IEDM83 Digest, pp. 384-387, December 1983.
38. Y. P. Tsividis "Design considerations in single-channel MOS analog integrated circuits-A tutorial" IEEE J. Solid-State Circuits, vol. SC-13, pp. 383-391, June 1978.
39. P. Grignoux and R. L. Geiger "Modeling of MOS transistors with nonrectangular-gate geometries" IEEE Trans. on Electron Devices, vol. ED-29, pp. 1261-1269, August 1982.
40. T. Y. Huang "Effects of channel shapes on MOSFET hot-electron resistance" Electronics Letters, vol. 21, pp. 211-212, February 1985.
41. T. Sakurai and T. Iizuka "Gate electrode RC delay effects in VLSIs" IEEE J. Solid-State Circuits, vol. SC-20, pp. 290-294, February 1985.
42. A. G. F. Dingwall, R. E. Stricker and J. O. Sinniger "A high speed bulk CMOS C^2L microprocessor" IEEE J. Solid-State Circuits, vol. SC-12, pp. 457-462, October 1977.
43. R. E. Luscher and J. S. De Zaldivar "A high density CMOS process" ISSCC85 Digest, pp. 260-261, February 1985.
44. W. Fichtner, R. K. Watts, D. B. Fraser, R. L. Johnston and S. M. Sze "0. 15 μm channel-length MOSFETs fabricated using E-beam lithography" IEDM82 Digest, pp. 722-725, December 1982.
45. M. Kanuma and K. Hashimoto "High performance silicided source/drain CMOSFET without parasitic effects" 1984 Symposium on VLSI Technology Digest, pp. 34-35, August 1984.
46. G. Merckel "A simple model of the threshold voltage of short and narrow channel MOSFETs" Solid-State Electronics, vol. 23, pp. 1207-1213, December 1980.
47. P. P. Wang "Device characteristics of short-channel and narrow width MOSFETs" IEEE Trans. on Electron Devices, vol. ED-25, pp. 779-786, July 1978.
48. B. Hoeneisen and C. A. Mead "Current-voltage characteristics of small size MOS transistors" IEEE Trans. on Electron Devices, vol. ED-19, pp. 382-383, March 1972.
49. G. F. Newmark "Theory of the influence of hot electron effects on insulated gate field effect transistors" Soild-State Electronics, vol. 10, pp. 169-175, March 1967.
50. B. T. Murphy "Unified field-effect transistor theory including velocity saturation" IEEE J. Solid-State Circuits, vol. SC-15, pp. 325-328, June 1980.
51. J. A. Cooper, Jr., and D. F. Nelson "Measurement of the high-field drift velocity of electrons in inversion layers on silicon" IEEE Electron Device Letters, vol. EDL-2, pp. 171-173, July 1981.
52. R. W. Coen and R. S. Muller "Velocity of surface carriers in inversion layers on silicon" Solid-State Electronics, vol. 23, pp. 35-40, January 1980.
53. T. Sato, Y. Takeishi, H. Tango, H. Ohnuma and Y. Okamoto "Drift velocity saturation of holes in Si inversion layers" J. Phys. Soc. Japan, vol. 31, pp. 1846, 1971.
54. F. F. Fang and A. B. Fowler "Hot electron effects and saturation velocities in silicon inversion layers" J. Appl. Phys., vol. 41, pp. 1825-1831, March 1970.
55. L. M. Dang, H. Iwai, Y. Nishi and S. Taguchi "P-channel versus N-channel in MOS-ICs of submicron channel lengths" Japanese J. Appl. Phys., vol. 19, pp. 107-112 1980.
56. W. Fichtner, R. M. Levin and G. W. Taylor "Experimental results on submicron-size p-channel MOSFETs" IEEE Electron Device Letters, vol. EDL-3, pp. 34-37, February 1982.
57. R. R. Troutman "Subthreshold design constraints for insulated gate field-effect transistors"

ISSCC73 Digest, pp. 108-109, February 1973.
58. J. J. Sparks "Junction transistors" Pergamon Press, Oxford, 1966.
59. R. R. Troutman and S. N. Chakravarti "Subthreshold characteristics of insulated-gate field-effect transistors" IEEE Trans. on Circuit Theory, vol. CT-20, pp. 659-665, November 1973.
60. J. R. Pfiester, J. D. Shott and J. D. Meindl "Performance limits of NMOS and CMOS" ISSCC84 Digest, pp. 158-159, February 1984.
61. J. T. Nelson "Hot carriers and trapping" Wafer level reliability assessment workshop Digest, pp. 33-57, Stanford Sierra Lodge, Lake Tahoe, October 24-27, 1982.
62. J. Y. Chi and R. P. Holmstrom "Constant voltage scaling of FETs for high frequency and high power applications" Solid-State Electronics, vol. 26, pp. 667-670, July 1983.
63. P. K. Chatterjee, W. R. Hunter, T. C. Holloway and Y. T. Lin "The impact of scaling laws as the choice of n-channel or p-channel for MOS VLSI" IEEE Electron Device Letters, vol. EDL-1, pp. 220-223, October 1980.
64. R. H. Dennard, F. H. Gaensslen, H. N. Yu, V. L. Rideout, S. Bassous and A. R. Leblanc "Design of ion-implanted MOSFETs with very small physical dimension" IEEE J. Solid-State Circuits, vol. SC-9, pp. 256-268, October 1974.
65. K. C. Saraswat and F. Mohammadi "Effect of scaling of interconnections on the time delay of VLSI circuits" IEEE J. Solid-State Circuits, vol. SC-17, pp. 275-280, April 1982.
66. H. B. Bakoglu and J. D. Meindl "Optimal interconnect circuits for VLSI" ISSCC84 Digest, pp. 164-165, February 1984.
67. M. Fukuma and R. H. Uebbing "Wiring capacitance simulation in two and three dimensions" 1984 Symposium on VLSI Technology Digest, pp. 24-25, August 1984.
68. P. Subramanian "Modeling MOS VLSI circuits for transient analysis" IEEE J. Solid-State Circuits, vol. SC-21, pp. 276-285, April 1986.
69. H. Iwai, K. Taniguchi, M. Konaka, S. Maeda and Y. Nishi "Two dimensional nature of diffused line capacitance in coplanar MOS structure" IEDM80 Digest, pp. 728-731, December 1980.
70. H. Hasegawa, M. Furukawa and H. Yanai "Properties of microstrip line on $Si{-}SiO_2$ system IEEE Trans. on Microwave Theory and Techniques, vol. MTT-19, pp. 869- 881, November 1971.
71. J. A. Stratton "Electromagnetic Theory" McGraw-Hill Book Company, New York, 1941.
72. S. Murarka, D. B. Fraser, A. K. Sinha and H. J. Levinstein "Refractory silicides of titanium and tantalum for low-resistivity gates and interconnects" IEEE J. Solid-State Circuits, vol. SC-15, pp. 474-482, August 1980.
73. T. Mochizuki, T. Tsujimaru, M. Kashiwagi and Y. Nishi "Film properties of $MoSi_2$ and their application to self-aligned $MoSi_2$ gate MOSFET" IEEE J. Solid-State Circuits, vol. SC-15, pp. 496-500, August 1980.
74. H. Okabayashi "Multilevel metallization technology for VLSIs" 1984 Symposium on VLSI Technology, pp. 20-23.
75. T. Kobayashi, H. Tago, T. Moriya and S. Yamamoto "A 6K-gate CMOS gate array" ISSCC82 Digest, pp. 174-175, February 1982.
76. J. Sakurai "A new buried-oxide isolation for high-speed high-density MOS integrated circuits" IEEE J. Solid-State Circuits, vol. SC-13, pp. 468-471, August 1978.
77. C. W. Teng, G. Pollack and W. R. Hunter "Optimization of sidewall masked isolation process" IEEE J. Solid-State Circuits, vol. SC-20, pp. 44-50, February 1985.
78. T. A. Johnson, R. W. Knepper, V. Marcello and W. Wang "Chip substrate resistance

modeling technique for integrated circuit design" IEEE Trans. on Computer-Aided Design, vol. CAD-3, pp. 126-134, April 1984.

79. O. Kudo, M. Tsurumi, H. Yamanaka and T. Wada "Influence of substrate current on hold-time characteristics of dynamic MOS ICs" IEEE J. Solid-State Circuits, vol. SC-13, pp. 235-239, April 1978.
80. B. Eitan, D. Frohman-Bentchkowsky and J. Shappir "Holding time degradation in dynamic MOS RAM by injection-induced electron currents" IEEE Trans. on Electron Devices, vol. ED-28, pp. 1515-1519, December 1981.
81. L. S. White, G. R. M. Rao, P. Linder and M. Zivits "Improvement in MOS VLSI device characteristics built on epitaxial silicon" Silicon processing symposium (San Jose, Calif.), edited by D. C. Gupta, pp. 190-203, 1983.
82. Y. Niitsu, G. Sasaki, H. Nihira and K. Kanzaki "Resistance modulation effect in n-well CMOS" IEEE Trans. on Electron Devices, vol. ED-32, pp. 2227-2231, November 1985.
83. F-C Hsu and K. Y. Chiu "Hot-electron substrate current generation during switching transients" Digest of 1984 Symposium on VLSI Technology, pp. 86-87, 1984.

2

CMOS Static Gates

2.1 INTRODUCTION

In a present-day CMOS VLSI chip, the most important functional components are CMOS static gates. In gate-array devices, CMOS static gates are used almost exclusively. In microprocessors and supporting circuits, most of the control interface logic is implemented using CMOS static gates. CMOS dynamic gates (Chapter 5) are generally used only where the logic delay is critical, in such applications as data manipulation and state-vector generation.

Preeminence of CMOS static gates originates from the merits of simplicity and reliability. Therefore, a family of CMOS static gates attracts our attention first. In this chapter we review various CMOS static gates and we understand how they execute logic functions. Details of circuit operation and the problems of switching delay will be discussed in the following chapters.

We use, however, several features of CMOS static gate circuits that make the technology outstanding: A CMOS static gate uses enhancement-mode NFETs for pulldown, and enhancement mode PFETs for pullup. A CMOS static gate can be made more symmetrical for pullup and pulldown operations than any other gates. A CMOS static gate carries no DC current when the gate is in a steady state. Theoretically, it is feasible to construct similar logic gates using bipolar PNP and NPN transistors [01], but the scheme has technical problems and is therefore not widely used. In this respect CMOS static gates are truly unique. A consequence of the basic characteristic that a CMOS static gate carries no quiescent current is that the CMOS logic levels are equal to the two power supply voltages. The difference in the voltage levels representing logic 0 and 1 is the maximum attainable.

Another consequence is that a CMOS static gate is ratioless. Independent of the relative size of PFETs and NFETs the gate operates correctly. An improperly scaled gate may be slow and may be noise sensitive, but the gate still executes correct logic operations. Together with pullup-pulldown symmetry and the absence of standby current, being ratioless is a great advantage of CMOS static gate circuits.

In this chapter we study the basic circuit structures of CMOS static gates, always calling attention to the important point that a gate in steady state carries no DC current. Behind this useful property of a CMOS static gate are the dual and symmetrical schemes of implementing logic operations, a normal and an inverted logic. We first study CMOS logic inverters and a few other basic circuits, and then we introduce the concept of duality.

2.2 INVERTERS, TRANSMISSION GATES, AND TRISTATABLE INVERTERS

Several different FET symbols are used in CMOS circuit publications. In this book the symbols in Fig. 2.1(a) will be used. A PFET has an outgoing arrow from channel, and NFET an arrow directed to channel. The direction of the arrow indicates the forward conduction of the channel-to-substrate PN junction (in PFET, from a P-type channel to an N-type substrate). The FET symbols are used more frequently than any other symbols at the time of writing (about 40% of recent publications use the symbols). Furthermore, the symbols are the only ones that represent an FET as a four-terminal device. For circuit-level figures four-terminal representation is required.

A CMOS inverter is a combination of a PFET and an NFET, as shown in the circuit diagram and logic diagram of Fig. 2.1(a). When input I is high, NFET MN1 conducts but PFET MP1 is turned off. Output node O is connected to the ground via the channel of NFET MN1, and the voltage level is ground (logic level 0 or low). When input I is low, PFET MP1 conducts and NFET MN1 is turned off. The output is then at V_{DD} (logic level 1 or high). Output O and input I are always the logical complement of each other. The logic symbol of an inverter, a triangle with a bubble, is common to TTL, or ECL logic.

A tristatable inverter is shown in Fig. 2.1(b). Clocks CK and $\overline{CK}$ in the circuit diagram correspond to the logic diagram, as indicated. When clock CK is high (V_{DD} potential) and clock $\overline{CK}$ is low (ground potential), the circuit is equivalent to an inverter (transmitting state). When CK is low and $\overline{CK}$ is high, however, FET MP2 and MN2 are both turned off. The output node is disconnected, and therefore the node is in a high-impedance state, normally called a tristate. A tristatable inverter is used to drive a communication channel such as a databus, to build a multiplexer, or as the means of temporary storage of data.

The logically equivalent but circuitwise-different gate shown in Fig. 2.1(c) is not desirable from the circuit operation's viewpoint [02]. In this circuit, if input I switches when the gate is in a tristate (MP1 and MN1 turned off), a capacitive coupling of the input signal to the output through gate-drain capacitances of FETs MP2

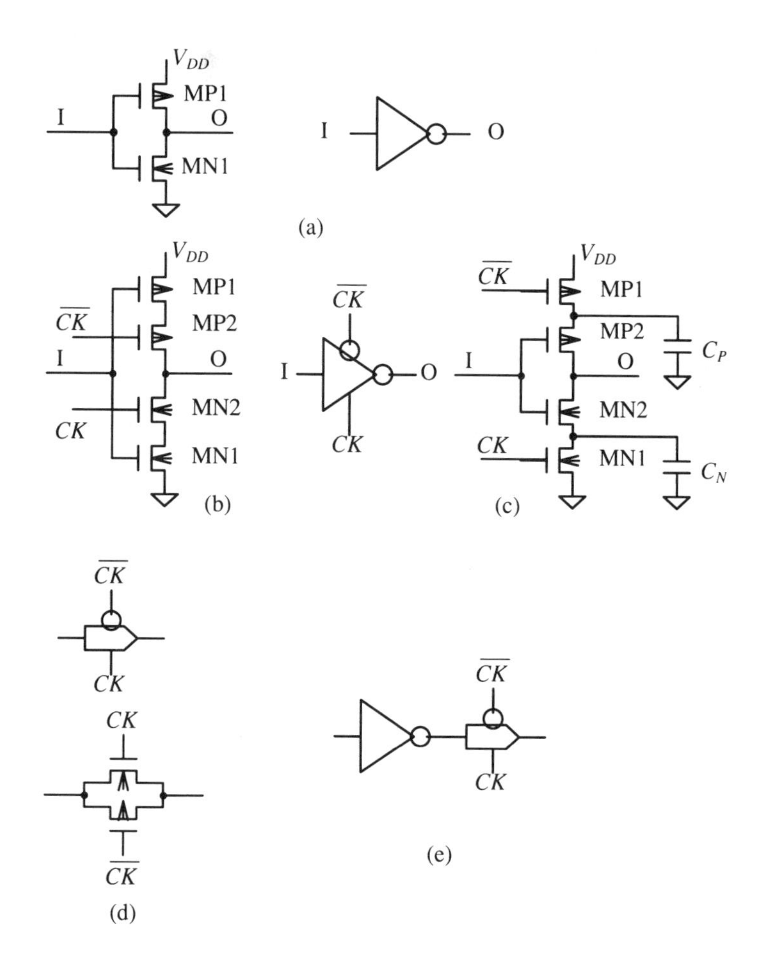

Figure 2.1 CMOS inverter, tristatable inverter, and transmission gate.

and MN2 occurs, or the circuit transmits digital noise. Furthermore, charge stored on node capacitances C_N and C_P may be gated to the output and "shared" with charge on the output capacitance. Suzuki, Odagawa, and Abe studied this problem in detail [03]. A tristatable inverter can also be used to store data temporarily in the capacitance of the output node. For this application the circuit must be designed carefully to avoid charge-sharing problems and to secure noise immunity.

A transmission gate is a passive analog switch as shown in Fig. 2.1(d) [04] [05]. A PFET and an NFET are connected in parallel, and are driven by a pair of complementary clocks, CK and $\overline{CK}$. When CK is high and $\overline{CK}$ is low, the transmission gate passes correct CMOS logic levels to the output. A single pass transistor (e.g., an

NFET) is unable to drive to a voltage higher than V_{DD} minus the threshold voltage of the NFET, and therefore correct CMOS logic levels cannot be transmitted (such a "transmission gate" is called a pass transistor, and it can be used if an extra care is paid in the circuit design). When CK is low and $\overline{CK}$ is high, the two terminals of a transmission gate are disconnected.

A combination of an inverter and a transmission gate is equivalent to a tristatable inverter, as shown in Fig. 2.1(e). The circuit in Fig. 2.1(b) can be obtained from the one in Fig. 2.1(e) by deleting the (unneeded) connection between the PFETs and the NFETs.

To operate a transmission gate or a tristatable inverter, a pair of positive-going and negative-going clock pulses are required. This requirement originates from the fundamental symmetry of a CMOS circuit with respect to PFETs and NFETs. The complexity of clocking is, however, a cost well paid to secure reliability in logic operations and conveniences in logic and circuit design.

The circuit diagrams of Figs. 2.1(a) to (e) suggest that there are symmetries in CMOS static gates. A gate consists of an equal number of NFETs and PFETs. The circuit that connects the output node to V_{DD} and the circuit that connects the output node to V_{SS} (or the ground) are related by a simple topological symmetry. This symmetry will be discussed in the next section.

2.3 NORMAL AND INVERTED LOGIC

The logic operation of ANDing three logic variables A, B, and C is equivalent to ORing the complements of the three variables $\overline{A}$, $\overline{B}$, and $\overline{C}$, and then inverting the result, as shown in Fig. 2.2.

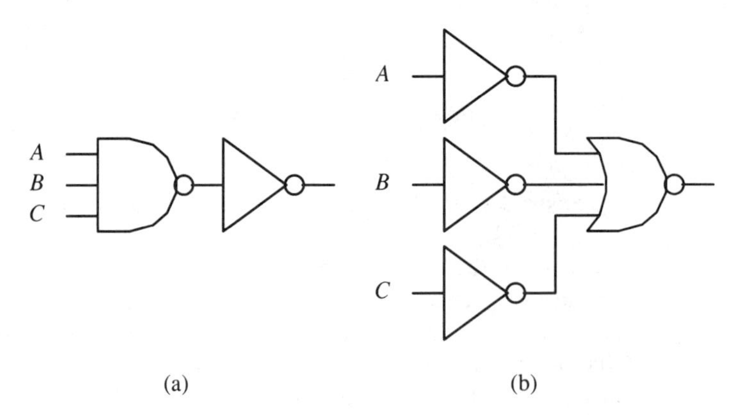

Figure 2.2 Normal and inverted logic operations.

In Boolean algebra, the equivalence is described by De Morgan's rule [06],

$$ABC = \overline{\left[\bar{A} + \bar{B} + \bar{C}\right]} \tag{2.1}$$

When logic is executed by the circuit of Fig. 2.2(a), the operation is called normal logic, and when it is executed by the circuit of Fig. 2.2(b), inverted logic. The definition is somewhat arbitrary and ambiguous. The definition is based on the immediate availability of logic variables A, B, and C, rather than $\bar{A}$, $\bar{B}$, and $\bar{C}$. In Fig. 2.2(b) the complementary variables must be generated first for the purpose of executing the logic, and this is the reason why the operation is called inverted logic. When a signal and its complement are both available, the definition is ambiguous.

Within a single CMOS static gate, however, the concepts of normal and inverted logic have the following meanings. A CMOS static gate is a circuit consisting of a group of NFETs and a group of PFETs; the NFETs function is to connect the output node to ground, and that of the PFETs to connect the node to V_{DD}. A signal input to a CMOS static gate goes to one or more PFETs and to one or more NFETs of the respective groups. A CMOS static gate works such that if the group of NFETs connect the output node to the ground, the group of PFETs disconnect the output node from V_{DD}, and vice versa. This happens for any combination of logic levels to the inputs.

Construction of such a circuit is guided by De Morgan's rule. We consider two NFET circuits: the first connects the output node to the ground if and only if the three variables A, B, and C are all high, or $ABC = H$, and the second connects the output node to the ground if and only if any one of the three variables A, B, and C is high, or $A + B + C = H$. Suppose that the NFETs of the second circuit are all replaced by PFETs and the ground by V_{DD}. Then the PFET circuit connects the output node to V_{DD} only when $\bar{A} + \bar{B} + \bar{C} = H$, because PFETs conduct when the gate voltage is low. De Morgan's rule, Eq. (2.1), shows that if $\bar{A} + \bar{B} + \bar{C} = H$, $ABC = L$, and vice versa. Therefore, only V_{DD}, or ground, but not both, is connected to the output node. The pullup and the pulldown circuits of a CMOS static gate have a symmetry based on the normal and inverted logic. The pullup circuit of a NOR(X) gate is structurally the same circuit as the pulldown circuit of a NAND(X) gate: The difference is that the NFETs and the PFETs are exchanged and the power supply connection is flipped.

2.4 NAND AND NOR GATES

The most widely used CMOS static gates are NAND and NOR gates with small number of inputs. Figure 2.3 shows a three-input NAND gate (often abbreviated as NAND3) [04] and two-input NOR gate (NOR2). Practically, NAND gates with more than six inputs, and NOR gates with more than four inputs are used very infrequently, because they have relatively long delays. In the circuit of NAND3 gate of Fig. 2.3(a), node O is pulled down only if all the inputs A, B, and C are high. Then all three PFETs are off, and therefore no DC current flows from V_{DD} to the ground, as discussed in the preceding section. The circuit diagrams show this fact clearly: If the circuit connected between the output node O and ground conducts, the circuit connected between the output node and the power bus (V_{DD}) does not, and if the circuit between O and V_{DD} conducts, the circuit between O and ground does not. Since all the circuits driven by output node O are capacitive, no DC current flows from V_{DD} to the ground when the gate is in a steady state.

X inputs of NAND(X) or NOR(X) gates are all logically equivalent. The three inputs A, B, and C of NAND3 gates are, however, circuitwise not equivalent inputs. In order to indicate the difference, the inputs are labeled in alphabetical order starting from the ground side of NAND gates and from the V_{DD} side of NOR gates. The difference shows up in the switching delay of the gate (Section 4.4): The switching delay from input A is the longest. For some applications this difference is inconvenient. NAND and NOR gates can be made absolutely symmetrical by connecting the FETs as shown in the example of symmetrized NAND2 gate of Fig. 8.25(b). Symmetrized NAND2 gates have applications in clock decoding circuits.

Whether NAND or NOR gates are preferred is determined by the specific details of the technology. NOR and OR gates are the choice in bipolar ECL ICs. In CMOS static logic, however, NAND gates are preferred to NOR gates, since NAND gates use more conductive NFETs in series, and therefore the sizes of the NFETs and PFETs in a single gate are better balanced. If a standard size ratio of PFET to NFET (about 2.0) is used in a multi-input NOR gate, the pullup delay is much longer than the pulldown delay, and therefore the gate is rather inconvenient for timing optimization. It is a customary practice of designers to convert many NOR gates into NAND gates.

In a CMOS static logic gate, any internal node that takes voltages other than 0 or V_{DD} in a steady state of the gate does not represent a Boolean function of input logic variables. We assume that FETs have no leakage current. Here we note that not only the output node, but also some internal nodes of a CMOS static gate represent Boolean functions of input variables (see Section 2.7). As discussed in detail in Section 3.4, the following peculiarity exists in a MOSFET circuit. If input signal A to the NOR2 gate of Fig. 2.3(b) is high, and input signal B is low, node O is at ground potential but node X voltage is uncertain. Suppose that node X is charged to V_{DD} by some external means, and is then released. The node is discharged only down to voltage V_{THP}, where V_{THP} is the threshold voltage of the PFET including the back-bias effect. The node can never be discharged all the way down to ground, since the PFET is cut off. If node X is charged by external means to voltage V satisfying $0 < V < V_{THP}$, the node voltage stays unchanged. These statements are valid unless the PFET is leaky. Since node X can be brought to a voltage level other than 0 or V_{DD}, the node does not represent a Boolean function of the input variables. A node like X retains the information about the recent history of switching of the gate.

A CMOS static gate consists of two circuits, circuit P connected between V_{DD} and the output node, and circuit N connected between the output node and ground, as shown in Fig. 2.4. Circuit P consists of PFETs only, and circuit N of NFETs only. If a PFET existed in a pulldown current path (within circuit N), the voltage of the output node, at least for a certain set of input variables, cannot be less than the threshold

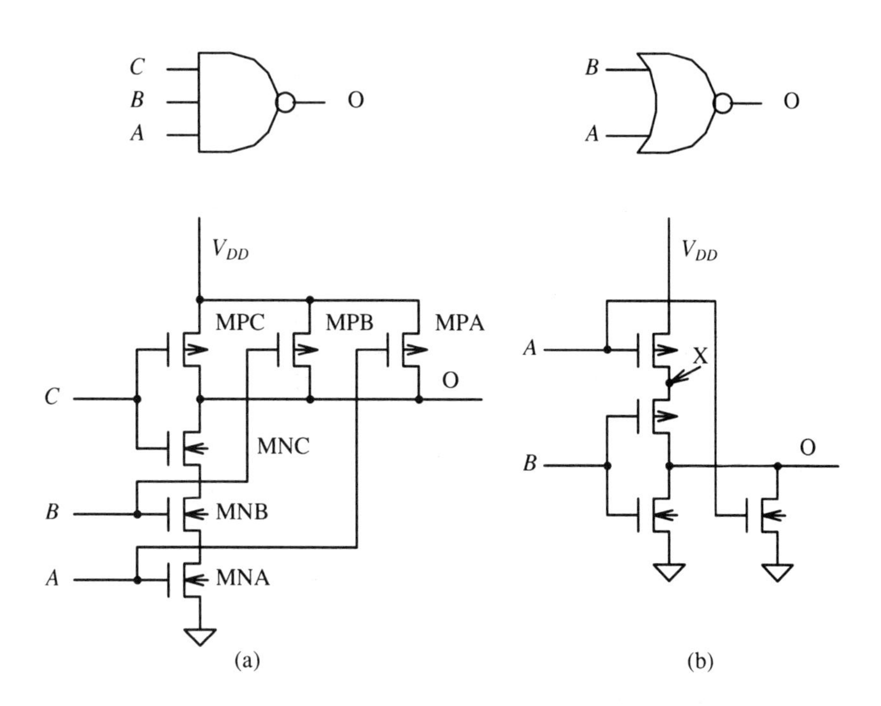

Figure 2.3 CMOS static NAND3 and NOR2 gates.

voltage of the PFET, V_{THP}. This contradicts the requirement that the output node represents a Boolean function of the input logic variables.

The FETs in circuit P and in circuit N are connected, reflecting the symmetry originating from the normal and the inverted logic. This symmetry is revisited here, by considering NFET MNA and PFET MPA of Fig. 2.3(a) as a pair. For a stable input logic signal level, if MNA is on, MPA is off, and vice versa. They are never on or off simultaneously. Therefore if they are considered together, the pair is an analog of a traffic intersection, shown in Fig. 2.5. In the model each of the intersecting roads is a one-way street, and left or right turns at the intersections are forbidden. Then if any one of the three intersections is "stop" for the south-to-north traffic, the intersection is "go" for the east-to-west traffic. In this traffic model, once the output node is disconnected from ground, the node is connected to V_{DD}. In the other case when all the intersections are "go" for the south-to-north traffic, all the east-to-west traffic is "stop": Once the output node is connected to ground, the node is disconnected from V_{DD}. The circuit of NAND3 gate shown in Fig. 2.3(a), which consists of the three series-connected NFETs and the three parallel-connected PFETs, is derived directly by replacing the traffic intersections of Fig. 2.5 by NFETs and PFETs.

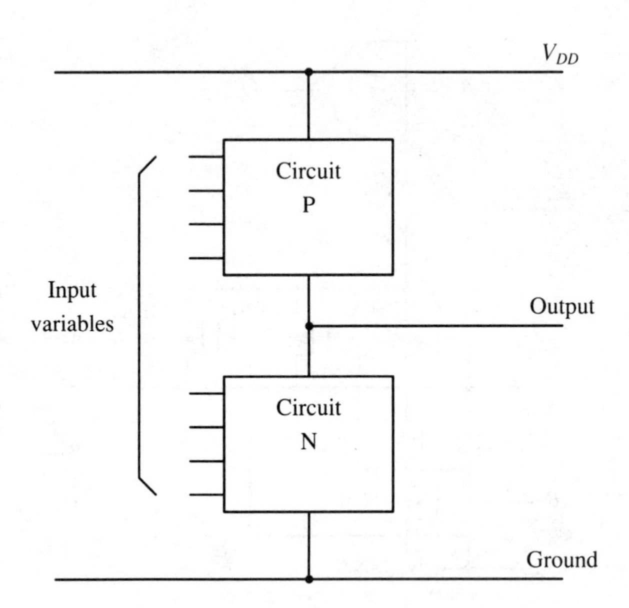

Figure 2.4 Structure of the circuit of CMOS gate.

This traffic model is the most convenient method of understanding the operations of a complex CMOS static gate, and is also a practical way to draw FET-level circuit diagrams. It is often confusing to draw FET-level circuits. In practice it is easy to construct the NFET part of the gate, because the circuit is the normal logic. The PFET circuit is confusing. Once the NFET circuit is known, the PFET circuit can be drawn easily using the traffic model.

There is an alternative way of drawing the PFET circuit. Two series-connected NFETs are substituted by two parallel-connected PFETs, and two parallel-connected NFETs are substituted by two series-connected PFETs.

In CMOS static NAND, NOR, AOI, and OAI gates the technique is straightforward. In a more complex gate shown later in Fig. 2.10(a) and (b), however, the substitution process becomes confusing, since how to substitute NFETs MNE and MNH of Fig. 2.10(b) is not clearly prescribed. The traffic model, which is an adaptation of the method of drawing dual circuits [07] [08], is convenient practically for any CMOS gate. Furthermore, the method shows clearly that CMOS static gates carry no standby current.

2.5 COMPOSITE GATES

It is possible to integrate more than one logic operation into a single CMOS gate [05]. Especially important are composite CMOS gates that produce an inverted sum of

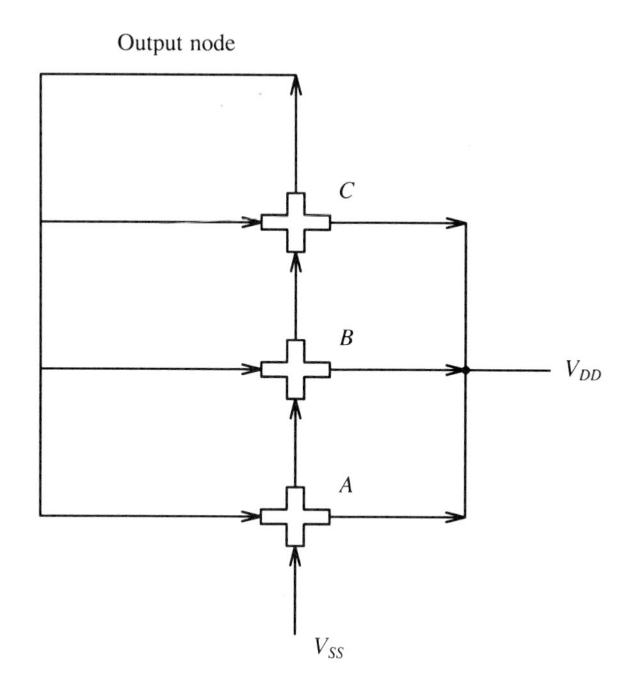

Figure 2.5 Traffic model of CMOS NAND3 gate.

products of the form $\overline{(AB + CD)}$, or an inverted product of sums like $\overline{(A + B)(C + D)}$. The gates are called AOI (And Or Invert) gates and OAI (Or And Invert) gates, respectively. They are shown in Figs. 2.6(a) and 2.6(b), respectively.

When the NFET circuit of a composite gate is given, the corresponding PFET circuit can be drawn using the traffic model shown in Fig. 2.7(a) and (b). In an AOI22 gate [Fig. 2.7(a)], two (east-west) "roads" α and α' run parallel in the enclosed area. There is no roadblock between the roads α and α'. Just as with real auto traffic, the traffic of road α is able to proceed on road α', and vice versa: The two roads α and α' are considered connected together by a crossroad β. In an OAI22 gate [Fig. 2.7(b)], the traffic of road α cannot merge with the traffic of road α' since there is a roadblock, γ. Because of the roadblock the PFET circuit of an OAI22 becomes different from that of an AOI22, as shown in Fig. 2.6(a) and (b).

In AOI gates, any number of ANDed terms can be NORed, and the ANDed terms may have any number of inputs. The same is true for OAI gates. Therefore, AOI and OAI gates may become quite complex. It is not a good design practice, however, to use very complex composite gates. In practice, AOI or OAI gates of more than 10 inputs are used only in exceptional cases. Composite CMOS gates can be decomposed into cascaded simple CMOS gates as shown in Fig. 2.8. The equivalent

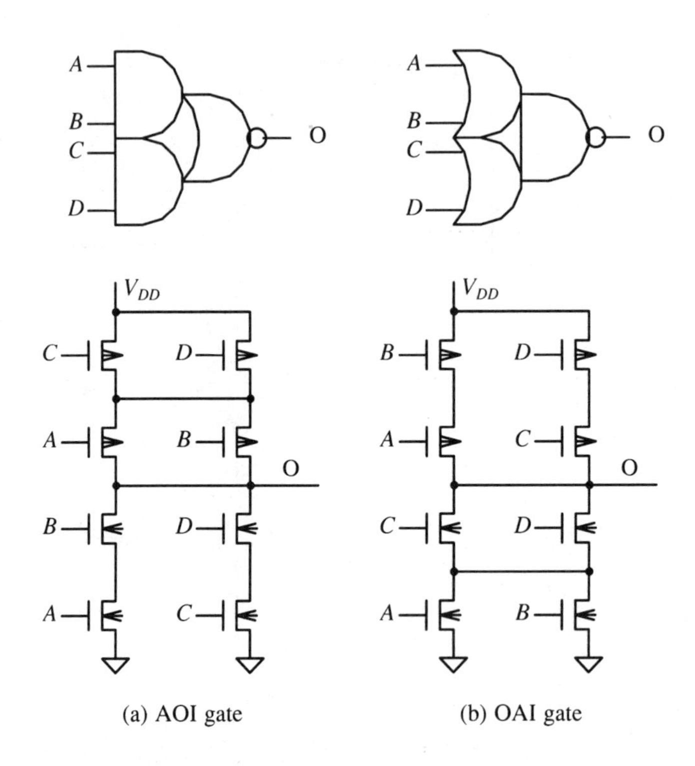

(a) AOI gate (b) OAI gate

Figure 2.6 CMOS AOI and OAI gates.

decomposed gate chain is faster, and it has more load-driving capability than does a very complex gate.

A multiplexer function can be implemented, in principle, using an AOI22...2 gate and an inverter. A gate of this type, with more than three pairs of data and control inputs, is very slow because of the pullup chain consisting of PFETs. The multiplexer function is customarily implemented using tristatable inverters or transmission gates.

Composite CMOS gates are not limited to AOI or OAI gates. If more than one input of a composite AOI gate receives the same logic signal as shown in Fig. 2.9(a), the gate can be simplified by factoring, as shown in Fig. 2.9(b). The factored gate has shorter delay than the original gate, and therefore it has practical applications. The logic diagram of Fig. 2.9(b) is a composite AND-OR-AND-OR-AND-INVERT gate.

Although the gate was introduced here by factoring the complex AOI gates of Fig. 2.9(a), such a composite gate is certainly a fully acceptable member of a static CMOS gate family. The logic symbol used in Fig. 2.9(b) is now frequently used [09].

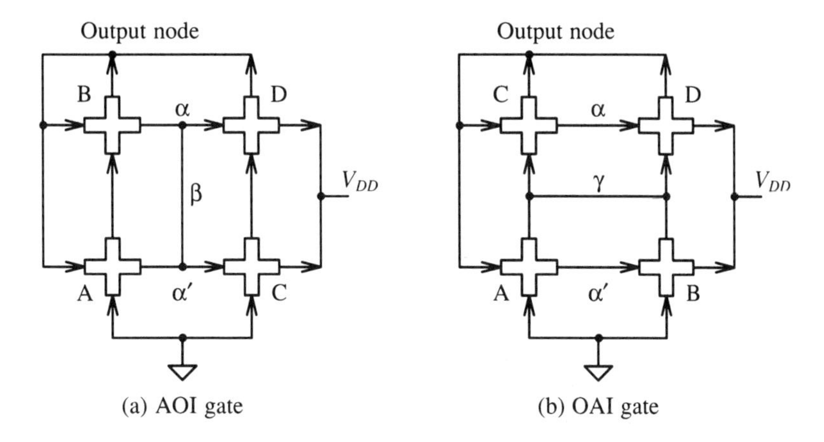

Figure 2.7 Traffic models of AOI and OAI gates.

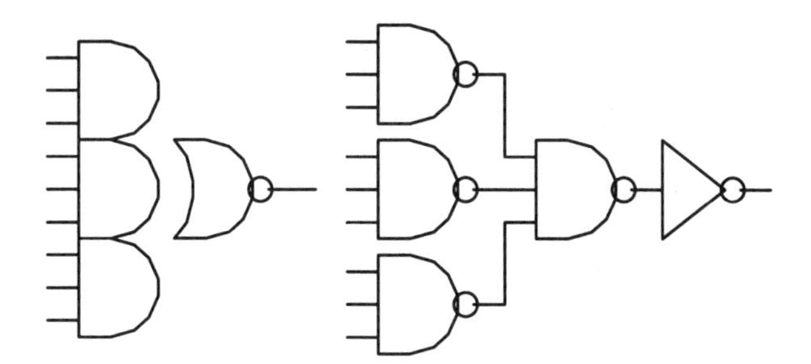

Figure 2.8 Decomposition of complex gate into cascaded simple gates.

2.6 ANALYSIS AND SYNTHESIS OF CMOS STATIC GATES

Composite gates of the preceding section are not yet the most general CMOS static gates. When any multi-NFET circuit that has two terminals (one for the output node, the other for the ground) is given, the circuit can be the pulldown circuit of a CMOS static gate, and a corresponding PFET circuit can be found. The PFET circuit may not be unique, however; there can be more than one PFET circuit that can be the partner of the NFET circuit.

When an arbitrary NFET circuit is given, we wish to know (1) what logic function the completed CMOS static gate executes, and (2) what is a corresponding PFET circuit. This problem is relevant to a reverse engineering of an already existing CMOS chip. In this section the two problems are discussed, using an example shown in Fig. 2.10(a).

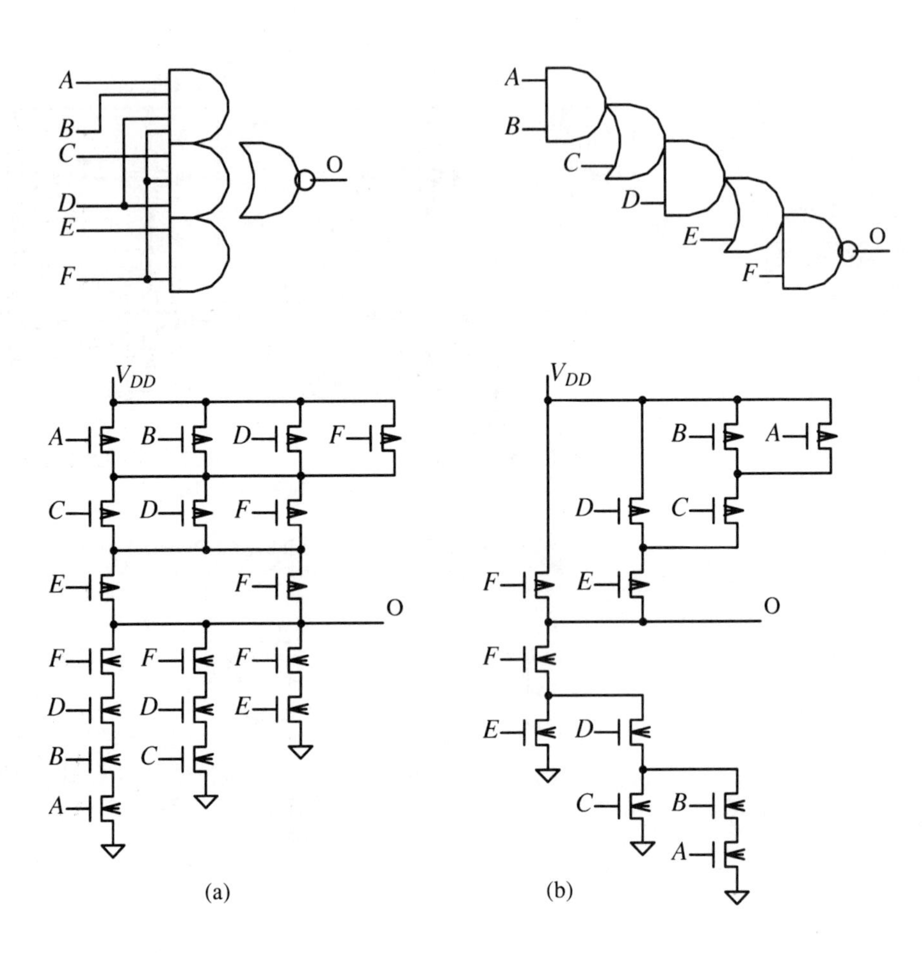

Figure 2.9 Complex CMOS gate derived by collating AOI gate.

(A) Analysis of Functionality

The functionality of the circuit of Fig. 2.10(a) is not obvious. The circuit can be simplified by substitution of FET MNA and MND by MNAD, and FET MNC and MNI by MNCI, as shown in Fig. 2.10(b). A fictitious gate ORs signals A and D producing $A + D$, and signal $A + D$ drives the gate of NFET MNAD. The simplified circuit is still too complex to determine the functionality by inspection. The best way is then to examine under what conditions the output node is connected to the ground by the NFET circuit. There are two NFETs that are connected directly to the ground, MNCI and MNF. If input F is high, the output node is connected to the ground only when any of the following Boolean variables are high:

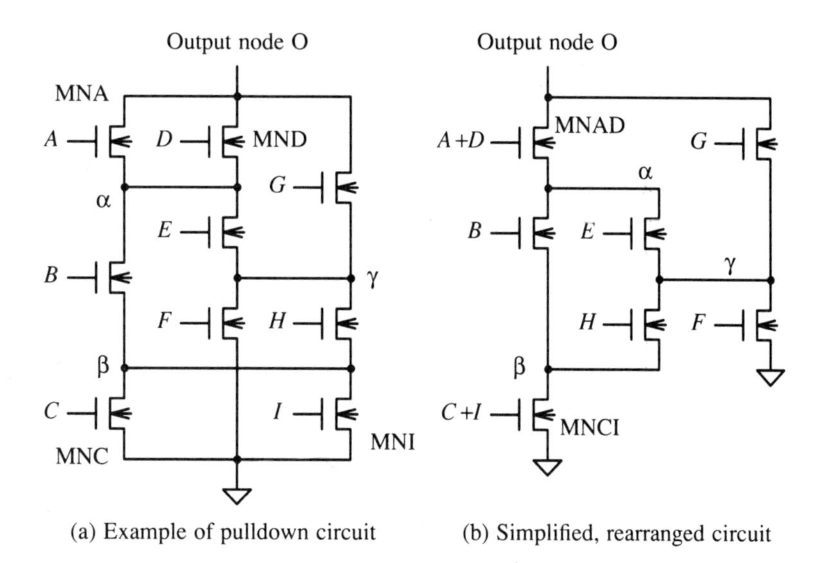

(a) Example of pulldown circuit (b) Simplified, rearranged circuit

Figure 2.10 Analysis of functionality of CMOS gate.

$$G\,,\ H\cdot B\cdot(A+D)\ \text{and}\ E\cdot(A+D) \tag{2.2}$$

This conclusion is reached by exhaustively searching for the paths that connect node γ to output node O. Similarly, if input CI is high, the output node is grounded only if any of the Boolean variables

$$B\cdot(A+D)\,,\ \ B\cdot E\cdot G\,,\ \ H\cdot G,\ \text{and}\ \ H\cdot E\cdot(A+D) \tag{2.3}$$

is high. Then the functionality of the completed CMOS static gate that has the circuit of Fig. 2.10(a) as the pulldown circuit is

$$O=\overline{F\cdot[G+(H\cdot B+E)\cdot(A+D)]+(C+I)\cdot[B\cdot E\cdot G+H\cdot G+(H\cdot E+B)\cdot(A+D)]} \tag{2.4}$$

This expression can then be simplified using Boolean algebra.

(B) Synthesis of a CMOS Static Gate

A traffic model of the simplified circuit of Fig. 2.10(b) is shown in Fig. 2.11(a). From the model the PFET circuit can be drawn, as shown in Fig. 2.11(b). Reflecting the symmetry of the NFET circuit, the PFET circuit has a symmetry also.

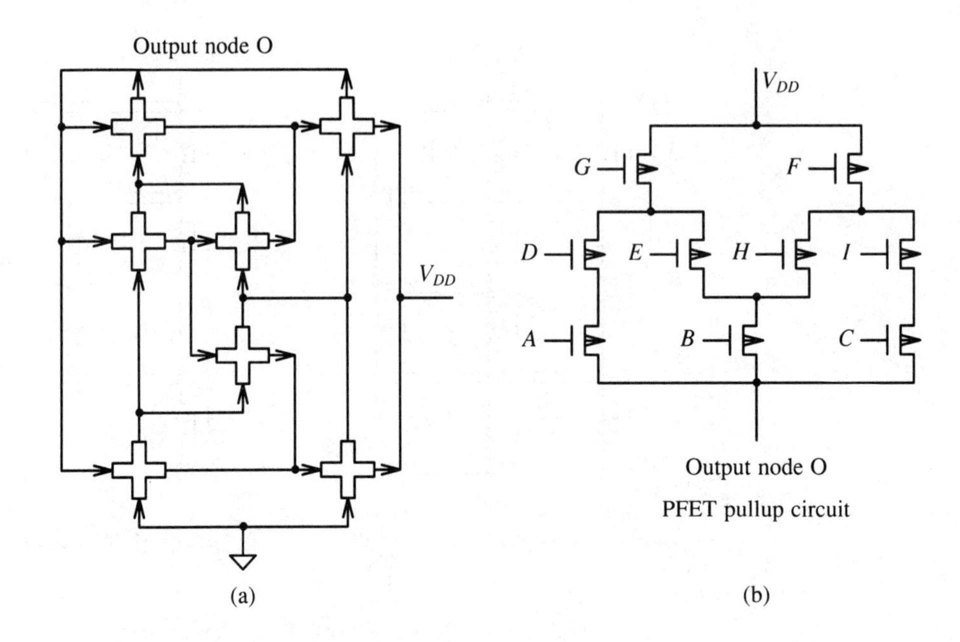

Figure 2.11 Synthesis of CMOS static gate using traffic model.

2.7 SYNTHESIS OF CMOS STATIC GATE FUNCTIONS AND SHARING OF FETS

A static multi-input NOR gate has a long pullup delay, especially if loading is heavy. To improve the delay time the series-connected PFETs can be replaced by a single PFET, and some additional logic is provided to drive the gate of the PFET. The FET-level circuit of Fig. 2.12, which involves three NFETs and a pullup PFET, is not a CMOS gate, but the entire circuit works equivalently to a static NOR3 gate and is faster than a conventional NOR3 gate for large load capacitance.

A generalization of this concept is the cascode voltage switch, shown in Fig. 2.13 [10]. In a cascode voltage switch, a logic variable and its complement are both assumed to exist and are made available after any logic operation is carried out. In Fig. 2.13 the circuit within box I pulls node Q down only when $A(B+C)=H$. Since $\overline{A(B+C)}=\overline{A}+\overline{B}\,\overline{C}$ and this is the circuit within box II, the circuit does not pull node $\overline{Q}$ down when $A(B+C)=H$. Therefore, node $\overline{Q}$ can be connected to the gate of PFET MP1 and node Q to the gate of PFET MP2. The circuit in box II is the complement of the circuit in box I. The circuit requires two more FETs than a conventional gate and requires both polarities of each input. For special applications like XOR logic, the scheme has an advantage. The circuit of Fig. 2.13 is the simplest

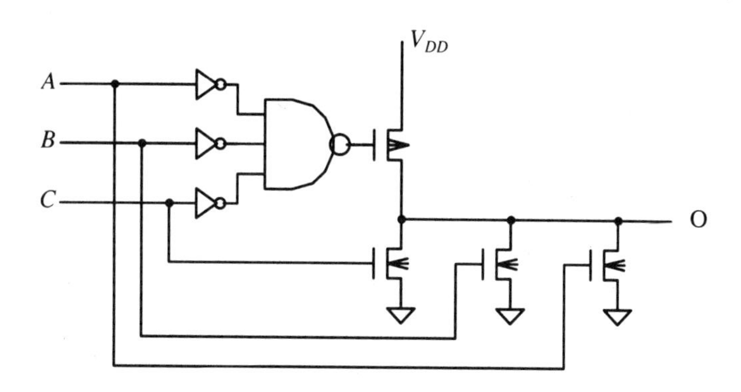

Figure 2.12 Synthetic CMOS gate.

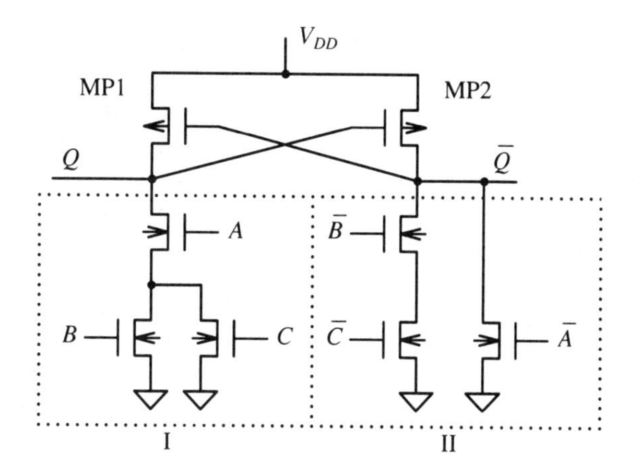

Figure 2.13 Cascode voltage switch.

when boxes I and II contain only one NFET each. This circuit is used to restore healthy bipolarity CMOS signals from deteriorated bi-polarity signals [11].

Differential split-level CMOS logic (DSLL) uses logic circuits similar to a cascode voltage switch, but the operational principles are different. Figure 2.14(a) shows a DSLL buffer, and Fig. 2.14(b) an example of a DSLL gate. The reference voltage, V_{REF}, of DSLL buffer is set at $(V_{DD}/2)+V_{THN}$, where V_{THN} is the threshold voltage of the NFET, including the back-bias effect. When node $\overline{AI}$ is pulled down by conventional CMOS gate, node AI is pulled up to $(V_{DD}/2)$. The same node AI is floated at about voltage $V_{DD}/2$ if the pair of signals AI and $\overline{AI}$ originate from another

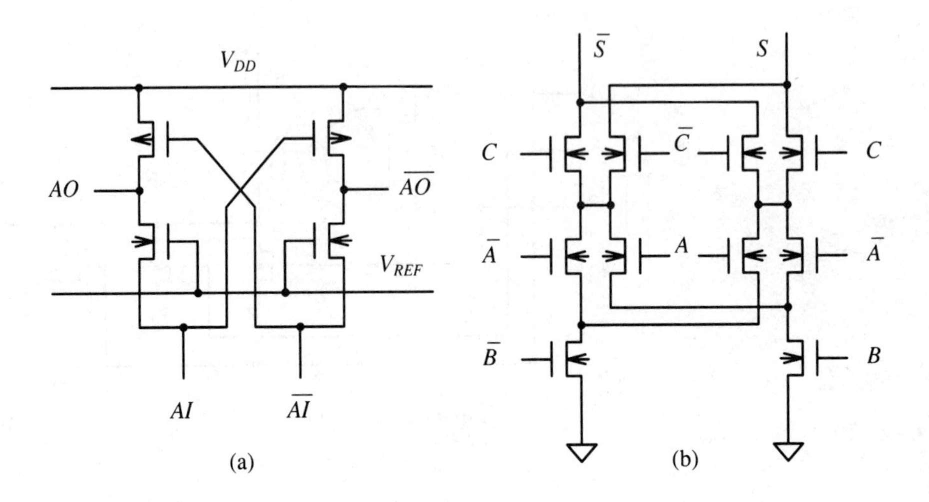

Figure 2.14 DSLL buffer and logic gate.

DSLL gate, and if $\overline{AI}$ output has a conducting path to ground. In either case node AO pulls up and node $\overline{AO}$ pulls down. The circuit of Fig. 2.14(a) works as a double-rail buffer. The circuit of Fig. 2.14(b) is a full adder of three logic variables A, B, and C (a parity circuit of the three variables). The outputs S and $\overline{S}$ are always buffered by the circuit of Fig. 2.14(a) [12].

Several CMOS gates can share FETs to reduce FET count. Figure 2.15(a) shows a combined tristatable inverter that works as a single-pole double-throw data switch. The circuit can be implemented by two independent tristatable inverters, but the two FETs that are driven by input data I can be shared. Such sharing can be done in general if some subsidiary functions of the gate are needed as separate outputs. The composite gate of Fig. 2.15(b) implements the AOI22 function, NAND2 function, and inverter function simultaneously. In this circuit, node O_3 is an internal node of AOI22 gate, but is the output node of the NAND2 gate. This node is able to take only V_{DD} or the ground potential level for any combination of the input variables, and therefore the node represents a valid combination of the input variables. The same is true for node O_1. In general, any gate that is not an inverter can share at least some FETs to construct an extra inverter. Such sharing is often useful to reduce the number of FETs.

Sharing and saving FETs can go to a very considerable degree to implement complex functions. The counter circuit by Oguey and Vittoz shown in Fig. 2.16(a) divides the input frequency by 2, as shown by the timing diagram of Fig. 2.16(b). In this example the number of inputs is only one, the clock. If the number of inputs is small there is flexibility to construct ingenious circuits to reduce the number of FETs. Saving FETs is crucial in CMOS devices used in small equipment like a wrist watch:

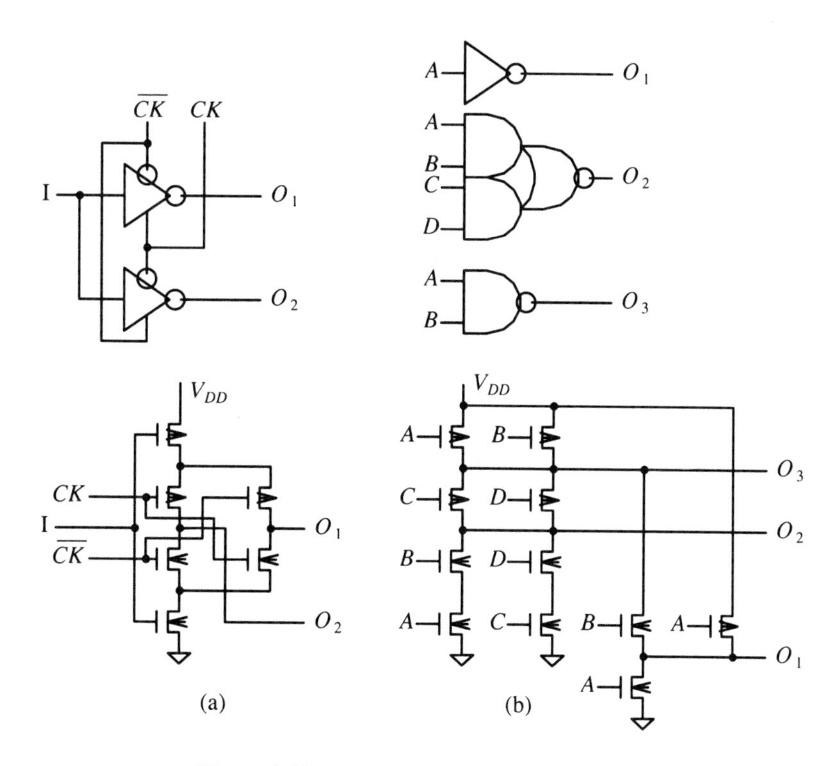

Figure 2.15 Sharing FETs among logic gates.

For such applications the circuit size as well as the power consumption must be minimized [13]. For a general-purpose design the circuit is rather inconvenient: Recent trend is toward more conventional use of gates. One problem with circuits of this type is design verification. The problem, however, can be resolved by new-generation logic simulators.

2.8 CMOS STATIC LOGIC OPTIMIZATION PROBLEMS

CMOS allows different methods of logic implementation. There are certain guidelines, however, to design high-performance (short-propagation-delay) circuits. In order to build a high-performance system, logic circuits are first designed following the guidelines, and FET sizes are then optimized following the techniques discussed later in Sections 8.6 to 8.10. Following is a summary of logic optimization methods.

(A) Proper Use of Gates

If more than five Boolean variables A_0–A_4 must be ORed, it is better to invert them first, and then NAND them. In a Boolean expression it is

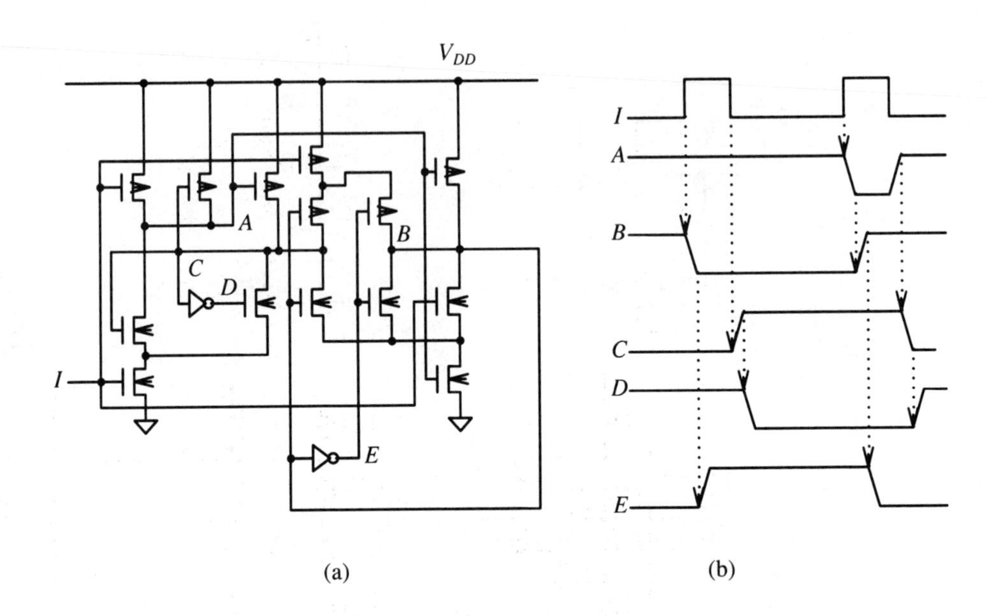

Figure 2.16 Compact binary counter circuit and waveforms.

$$A_0+A_1+A_2+A_3+A_4=\overline{\overline{A_0}\,\overline{A_1}\,\overline{A_2}\,\overline{A_3}\,\overline{A_4}} \tag{2.5}$$

and the logic of the right-hand side is implemented. Another implementation is to split the signals into the smaller groups, NOR them, and then NAND the results as

$$A_0+A_1+A_2+A_3+A_4=\overline{\overline{A_0+A_1}\cdot\overline{A_2+A_3}\cdot\overline{A_4}} \tag{2.6}$$

CMOS NOR gates having more than five inputs should not be used in high-performance circuits.

(B) Complex Gates versus Simple Gates

Since And-Or-Invert or Or-And-Invert gates are available in CMOS logic, designers sometimes abuse them without giving much consideration to delay. Practically, a complex gate having more than 10 inputs should be decomposed into several simpler gates. An example was shown before, in Fig. 2.8. Instead of a straightforward implementation of

$$\overline{\prod A_i + \prod B_i + \prod C_i} \tag{2.7}$$

the implementation

$$\overline{\overline{\overline{\prod A_i}\,\overline{\prod B_i}\,\overline{\prod C_i}}} \tag{2.8}$$

will have a shorter delay, and flexibility for adjusting FET sizes in case of heavy capacitive loading.

(C) Gate Configurations That Create Large Unbalance in Pullup and Pulldown Delay Times

When a NOR gate and NAND gate are cascaded and the PFET/NFET size ratio is in the normal range (1.5-2.5), the circuit responds rapidly to the input upgoing transition, but very slowly to the input downgoing transition. A logic circuit of this structure should either be avoided or an unusual FET scaling technique should be used. The circuit generates

1. $\Pi \overline{A_i}$ $\quad A_i$ = input to the first NOR gate
2. $\Sigma A_i + \Sigma \overline{B_i}$ $\quad B_i$ = input to the second NAND gate
3. $\Pi \overline{A_i} \Pi B_i \Pi \overline{C_i}$ $\quad C_i$ = input to the third NOR gate
4. $\Sigma A_i + \Sigma \overline{B_i} + \Sigma C_i + \Sigma \overline{D_i}$ $\quad D_i$ = input to the fourth NAND gate

and they are either straight sums or products of all the input variables, with proper inversion for each. This problem can be resolved by a circuit that sums a large number of variables, like

$$S = \sum_{i=1}^{N} A_i \tag{2.9}$$

A circuit that sums a large number of variables can be modified to a circuit that multiplies a large number of variables simply by inverting the inputs and the output:

$$P = \overline{\sum_{i=1}^{N} \overline{A_i}} = \prod_{i=1}^{N} A_i \tag{2.10}$$

A circuit that sums a large number of variables in a very short time is a domino CMOS OR gate, which is discussed in Section 5.6.

2.9 INPUT/OUTPUT BUFFERS

A conventional inverter is used to interface a CMOS level signal from the outside to the inside of a chip. If the outside signal is a TTL level signal, the inverter should be scaled to match the TTL threshold voltage. The details are discussed in Section 8.7.

As the speed of scaled-down CMOS devices approaches the speed of ECL devices, an ECL-to-CMOS interface is required. ECL devices use power supply voltages $V_{CC} = 0\,\text{V}$ (ground), and $V_{EE} = -5\,\text{V}$. To interface ECL signals to a CMOS chip, the circuit shown in Fig. 2.17(a) is used.

Communication between chips having different power supply voltages is inconvenient. CMOS chips may be powered using a 0 V and a -5 V power supply. Then an ECL-to-CMOS logic level converter shown in Fig. 2.17(b) [14] can be used. The

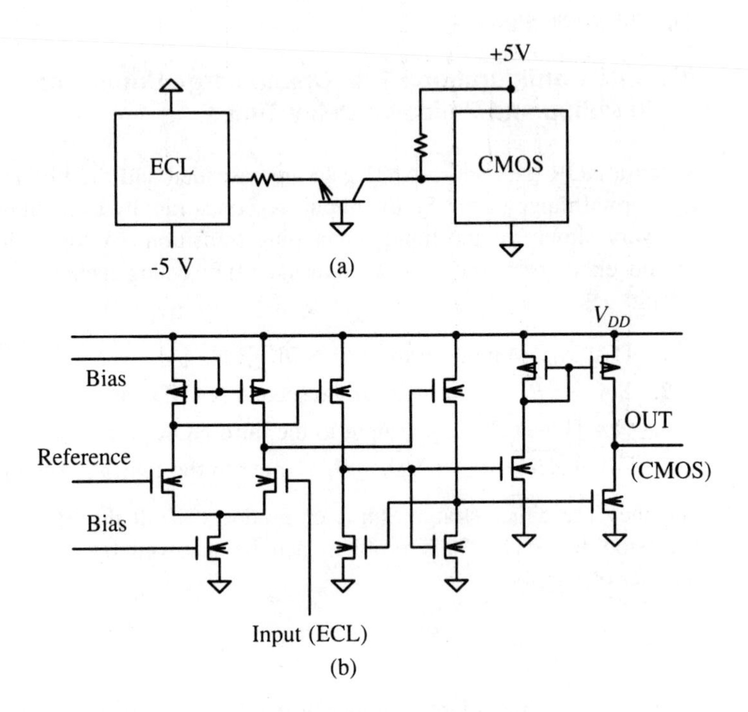

Figure 2.17 ECL-to-CMOS logic-level converter.

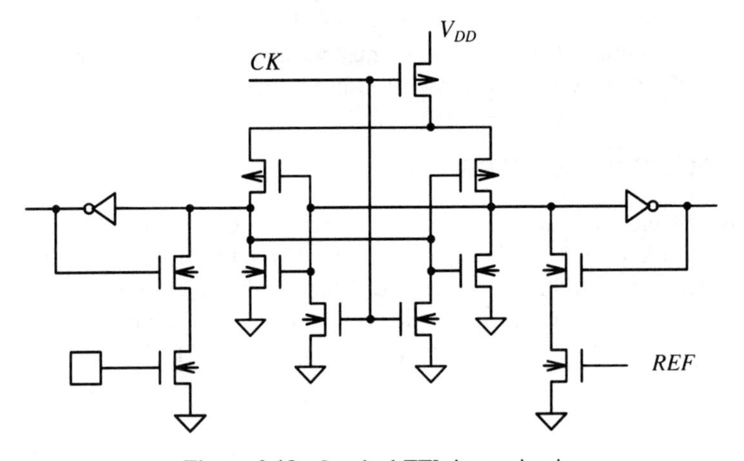

Figure 2.18 Latched TTL input circuit.

circuit is a CMOS differential amplifier driving an NFET source-follower level shifter and a current mirror CMOS output stage. The fast logic level converter consumes high power, but high power must be tolerated for high speed.

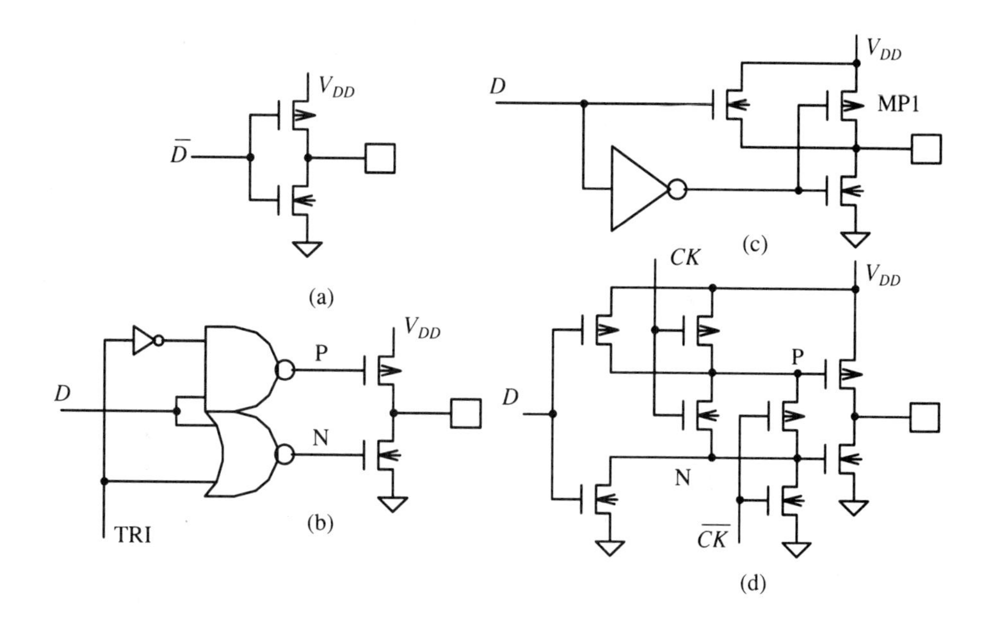

Figure 2.19 Various output buffers.

If the signal from the outside is to be latched by the input circuit of a chip, a level-sensitive latch is used. The circuit shown in Fig. 2.18 is the TTL input interface by Chan [15]. The reference voltage is set at the TTL threshold level (1.4 V above the ground). When the strobe is high the two current paths are established, whose conductivities depend on the voltage levels of the input and of the reference. When *CK* makes a high-to-low transition the more conductive current path pulls the corresponding output node of the cross-coupled latch down more strongly than the other output node, and once the voltage of the node begins to fall the other current path is shut off by a regenerative action. A CMOS level logic signal is generated and is latched.

The simplest output driver is the large CMOS inverter (wide FETs) shown in Fig. 2.19(a). Depending on the required output logic level, the FET size ratio is adjusted to equalize the delay. TTL level drivers have a PFET and an NFET that have approximately the same size.

Bidirectional I/O interface circuits require a tristatable driver. Figure 2.19(b) is a CMOS tristatable driver. When the tristate signal, TRI, is high, node N is low and node P is high, thereby turning both the NFET and the PFET off. When TRI is low,

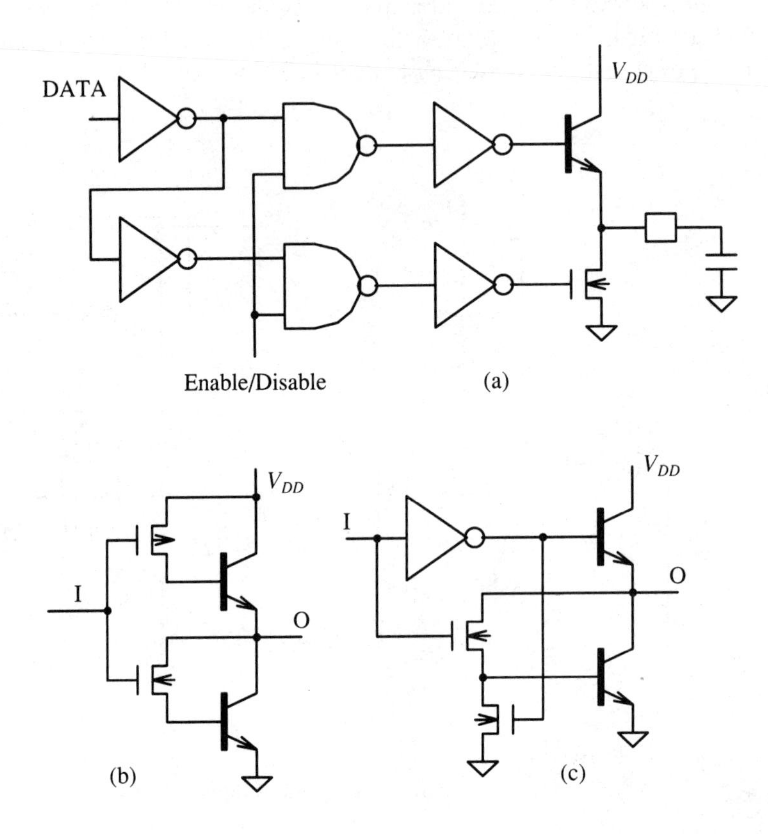

Figure 2.20 BiMOS buffers and output drivers.

however, the circuit operates as a noninverting buffer. A version of this circuit is shown in Fig. 2.19(d). When CK is high and $\overline{CK}$ is low, the circuit is equivalent to a cascaded two-stage inverter. When CK is low and $\overline{CK}$ is high, however, two nodes P and N are disconnected, and set at high and low logic levels, respectively, resulting in a tristate of the driver. In a TTL level output driver, an NFET is often used to pull up the output node. This is feasible since the TTL high-logic-level voltage is 2.4 V, sufficiently below $(V_{DD}-V_{TH})$ (approximately 3.5 V above ground), and therefore an NFET has adequate drive capability as a source follower. The driver has the advantage of reduced latch-up susceptibility. The circuit is shown in Fig. 2.19(c). A small PFET MP1 may be added to drive the output all the way to V_{DD}.

A recent trend is to use NPN bipolar junction transistors (BJT) to drive heavy capacitive loads. Such technology is called BiMOS. The output driver circuit of Fig. 2.20(a) uses an NPN BJT for pullup [16]. The NPN BJT is available free from an N-substrate CMOS technology. Figures 2.20(b) and (c) show the buffer circuits used to drive a heavily loaded internal node of a chip. The circuits require isolated (independent) BJTs. The buffers are slower for small capacitive loads, but for large capacitive loads they are significantly faster than a CMOS driver [17]. The buffers are not able to drive all the way to V_{DD} or V_{SS} levels, but for some applications this is tolerable.

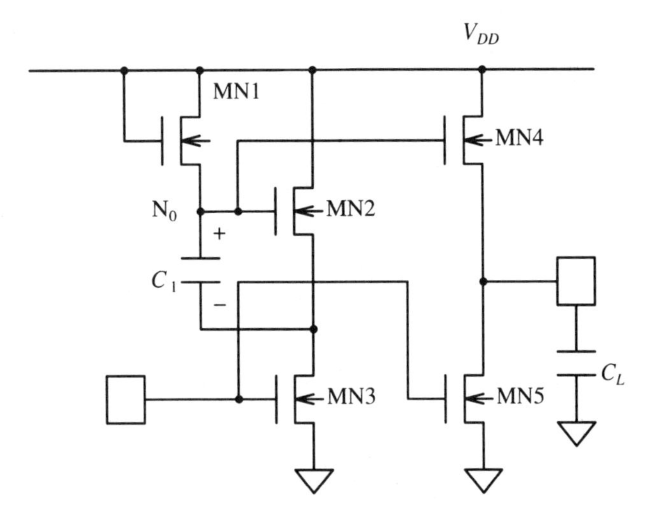

Figure 2.21 NMOS bootstrap driver.

A bootstrap driver in NMOS technology can be used to increase the drive capability of the pullup NFET. In the circuit shown in Fig. 2.21, capacitor C_1 is charged to the polarity indicated, when the input is high and MN3 is conducting. When MN3 is turned off by the input low-going transition, node N_0 is pulled up and MN1 shuts off. Then node N_0 is pushed up above V_{DD}, thereby driving the output transistor MN4 very hard. The back-bias effect of NFET MN4 can be compensated by this technique [18].

2.10 D-LATCHES, SET/RESET LATCHES, AND THE OTHER STORAGE ELEMENTS

To store information temporarily, a static latch is most frequently used. Figure 2.22(a) shows a standard CMOS D (for Delay) latch circuit. The circuit consists of two tristatable inverters TI1 and TI2, and an inverter, INV. When clock CK is high and clock $\overline{CK}$ is low, TI1 transmits the input data to output Q and TI2 is inactivated. When CK becomes low and $\overline{CK}$ becomes high, TI1 shuts off, and TI2 drives $\overline{Q}$, thereby applying a self-sustaining positive feedback to hold the data previously brought into the latch by TI1.

A CMOS D-latch requires a pair of synchronized positive-going and negative-going clock signals. Convention of the polarity of CK and $\overline{CK}$ is such that $CK = H$

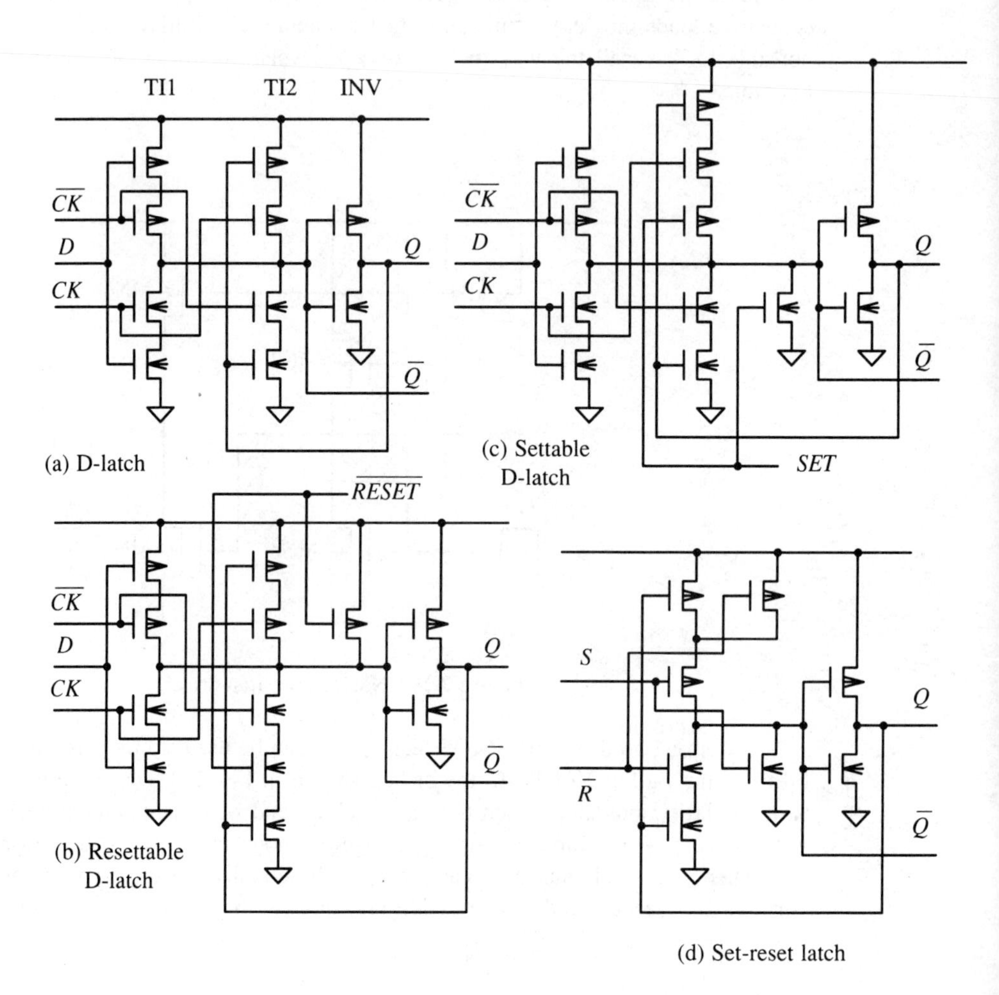

Figure 2.22 CMOS D-latches and set-reset latch.

and $\overline{CK} = L$ mean transmission of TI1. A reader needs to memorize this convention.

The data stored in a resettable [Fig. 2.22(b)] or settable [Fig. 2.22(c)] D-latch can be reset or set asynchronously, or independent of the state of the clocks. A latch is set if Q is high and $\overline{Q}$ is low. This capability is useful in providing initial values of the latch data. The settable D-latch is often undesirable, since the $\overline{Q}$ output that is driven by the series-connected three PFETs has very weak pullup capability.

Another temporary data storage device is an unclocked set-reset latch. In the circuit of Fig. 2.22(d), input S should normally stay low, and $\overline{R}$ normally stay high. The latch is set when S is momentarily pulled up to V_{DD}. When $\overline{R}$ temporarily drops to zero the latch is reset.

J-K flip flops are infrequently used in CMOS VLSI chips. This practice gives the impression that general-purpose storage devices belong to the bygone SSI days.

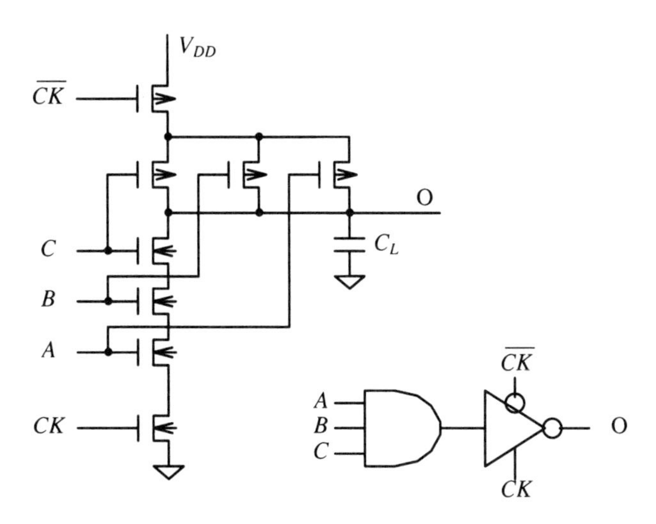

Figure 2.23 Combination of logic gate and tristatable inverter.

It is possible to integrate one or more CMOS static logic gates and a tristatable inverter, that is, a (dynamic) latch. The gate of Fig. 2.23 executes Boolean operation $\overline{ABC}$ when CK is high and $\overline{CK}$ is low, and the gate stores the result in output capacitor C_L when CK is low and $\overline{CK}$ is high. The circuit is equivalent to cascading an AND gate and a tristatable inverter, as shown in the figure [19]. This circuit must be designed with care: If one input switches when the latch is not transparent, charge sharing may influence the output node voltage [02]. The mechanism and the solution to this problem have been discussed in Section 2.2.

Yet another class of CMOS storage element is called lambda cell devices, shown in Fig. 2.24(a) [20]. In Fig. 2.24(a), if MP1 is turned on, node A is pulled up, and MN2 turns on. Then node B is pulled down, and PFET MP2 turns on. Node A is sustained high, thereby maintaining the conducting state of the cell. If MN1 turns on, MN2 turns off, and the cell returns to the nonconducting state. This circuit is convenient to drive light-emitting diode. For reliable operation of the memory cell, resistors r_N and r_P that compensate for the leakage currents of MP2 and MN2, respectively, are required. The high resistors can be fabricated using narrow tub (P-tub or N-tub) features. Another device that has memory is the CMOS lambda diode shown in Fig. 2.24(b). The terminal impedance at node C has negative resistance at the voltage where the inverter switches [21].

The circuit shown in Fig. 2.24(c) works as follows. Capacitors at the gates of NFET and PFET are charged to voltage V_N and V_P, respectively. Then voltage V_M is developed between terminals A and G. Decay-time constants of gate voltages V_N and V_P are adjusted to be the same. If PFET and NFET are regarded as a voltage divider of power supply voltage $E_N + E_P$, division ratio does not change with time, although the resistances of the FETs change. Then voltage V_M stays unchanged for a long time. The circuit is a memory of analog voltage [22].

Some may dismiss the circuits of Fig. 2.24 as intellectual curiosities. This view is not justifiable. No circuit is useless.

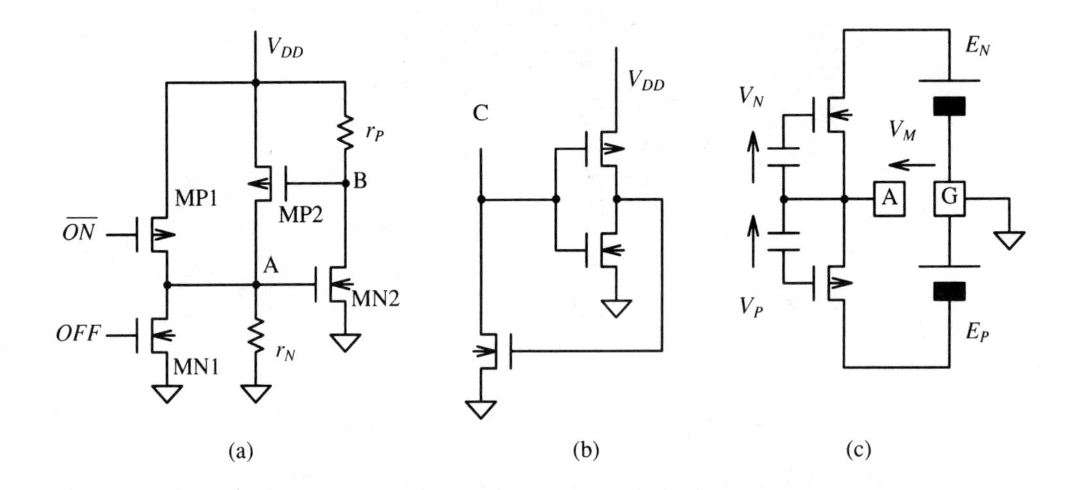

Figure 2.24 Simple memory cell, lambda diode, and analog memory cell.

2.11 SCHMIDT TRIGGERS

A Schmidt trigger is a gate that has a very abrupt input-output characteristic. The steep transition is made possible by positive feedback used in the circuit which also causes hysteresis.

The Schmidt trigger circuit shown in Fig. 2.25 works as follows. Suppose that V_I was low. When V_I makes a gradual low-to-high transition, the voltage of node N_0 decreases and at a certain voltage PFET MP1 turns on. Since the current path to ground through PFET MP1 involves only one PFET, the PFET pulls node N_P down very hard. The voltage of node N_P drops suddenly, and so does the node N_0 voltage. This results in a further decrease in node N_P voltage, and the effect is regenerative. Node N_0 voltage drops quickly down to zero. This occurs at the input voltage V_{IU}.

When V_I was originally at a high logic level, and V_I makes a high-to-low transition, the switching occurs when input voltage is V_{ID}. We then have $V_{IU} - V_{ID} = V_H > 0$, and V_H is the hysteresis voltage. That the circuit has hysteresis can be understood as follows. Suppose that the PFET/NFET size ratio is chosen so that the pullup and the

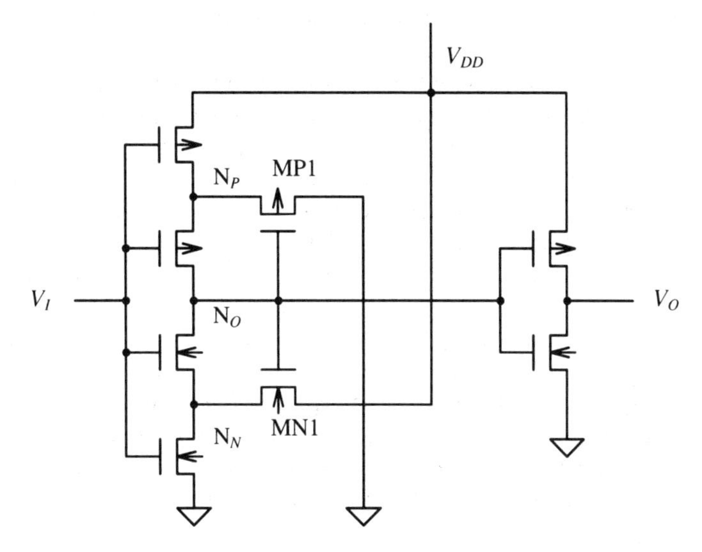

Figure 2.25 CMOS Schmidt trigger circuit.

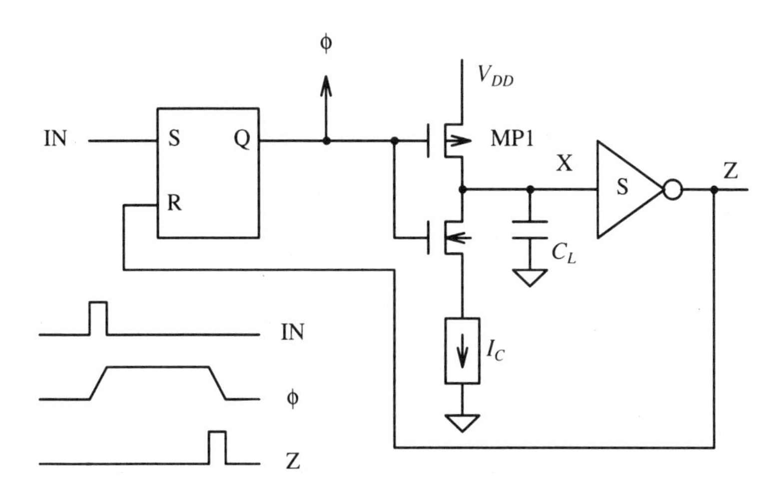

Figure 2.26 Delayed clock generator.

pulldown capabilities of the FETs are the same. When the input voltage V_I is zero, the gate voltage of NFET MN1 is V_{DD}, and PFET MP1 is off. Therefore, as the input voltage goes up, node N_N gets current from V_{DD}, and the current is more than the current node N_P loses to the ground. Therefore, when V_I reaches $V_{DD}/2$, node N_0 has not yet reached $V_{DD}/2$. Therefore the switching threshold is higher than $V_{DD}/2$. In the same way the switching threshold for the downgoing transition is less than $V_{DD}/2$. Thus there is hysteresis.

There are several versions of Schmidt trigger circuits. A simple Schmidt trigger circuit reported by Nagaraj and Satyam uses a depletion-mode PFET load device [23].

A Schmidt trigger is used to reshape signals whose transition is too gradual. A signal that makes very gradual transition (if 10% to 90% transition time is 100 times or more than a typical gate delay) is undesirable. Gates carry overlap current (Section 3.17) and gates become noise sensitive (Section 9.6).

The Schmidt trigger circuit allows design of a simple timer [24]. The circuit shown in Fig. 2.26 has a set-reset latch, current generator I_C, integrating capacitor C_L, and Schmidt trigger. If IN is low, the circuit is in the quiescent state. Node ϕ is low, and capacitor C_L is charged to V_{DD}. When an upgoing pulse comes in to terminal IN, the latch is set, and node ϕ goes up. PFET MP1 turns off, and C_L begins to discharge. Then after time $t_p = f \cdot C_L V_{DD}/I$ (where f is a numerical factor of the order of 1/2), the Schmidt trigger switches and the latch is reset. Since node X voltage decreases very slowly, a Schmidt trigger is required to reshape the waveform. Node Z generates a delayed pulse.

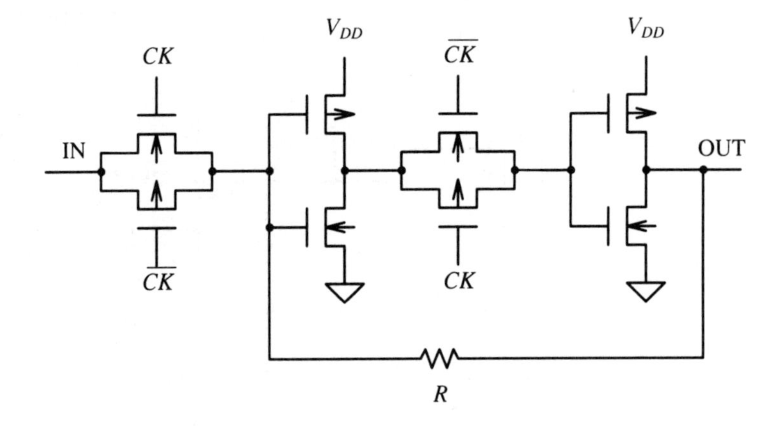

Figure 2.27 Simple CMOS shift register.

2.12 SHIFT REGISTERS

A shift register can be constructed by cascading conventional CMOS D-latches and by clocking them with a pair of two-phase, nonoverlapping clocks. A simpler circuit can be used, however. The circuit designed by Ipri and Sarace is shown in Fig. 2.27. The circuit has a feedback resistor R that ranges from 10^6 to $10^8\ \Omega$. The high resistors can be fabricated using undoped polysilicon features. The resistor compensates for the leakage current from the input transmission gate when the transmission gate is off. The resistor is so large that the transmission gate has no difficulty in driving the gate of the first inverter. With the feedback resistor, the shift register is able to retain the state indefinitely when CK is low and $\overline{CK}$ is high. If the delays through the inverters and the transmission gates are significantly long compared to clock skew and the transition time, the same pair of CMOS clocks can be wired in the opposite polarity to successive shift register stages [25].

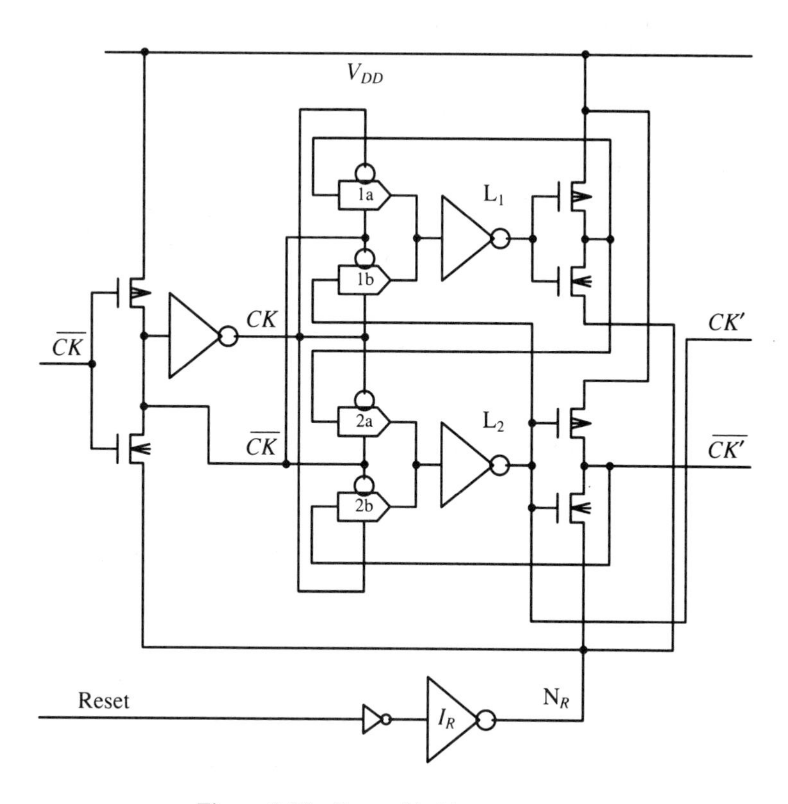

Figure 2.28 Resettable binary counter.

2.13 BINARY COUNTERS

A binary counter is two cascaded CMOS D-latches with feedback. The Q output of the first latch drives the input of the second latch, and the $\overline{Q}$ output of the second latch drives the input of the first latch. The clocks of the two latches are cross-connected so that if the first latch is transparent, the second latch is not, and vice versa. The data in a latch that were clocked in by a clock edge stay unchanged until the next clock edge of the same transition polarity arrives, and at that time the complementary value is clocked in. Therefore, the circuit divides the clock by 2. The divided clock pulse has an exactly 50% duty cycle, whatever the duty cycle of the input clock pulse may be.

In a sequential circuit such as a counter, resetting the circuit is a practical problem. The circuit shown in Fig. 2.28 is used in the RCA CD4020 binary counter. The circuit consists of two latches, L_1 and L_2, built from CMOS transmission gates and inverters [04]. When the CK input is high and the $\overline{CK}$ input is low, transmission gate 1b and transmission gate 2b are transmitting. Latch L_1 is transmitting and latch L_2 retains the previous data.

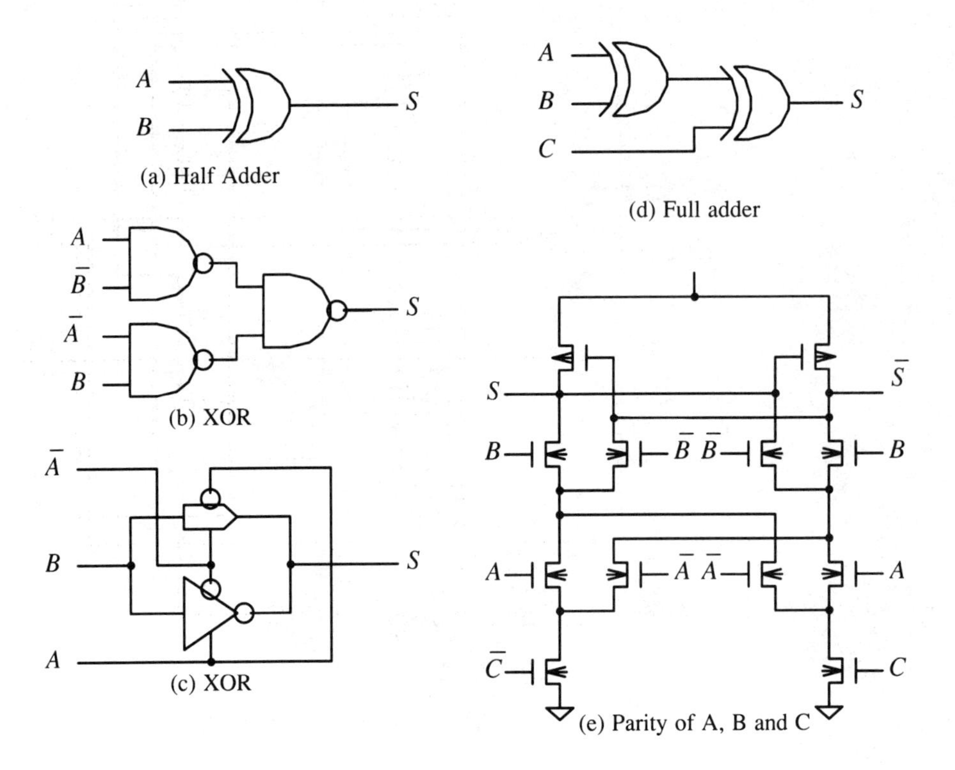

Figure 2.29 Exclusive ORs and full adders.

When a positive-going reset pulse arrives, transmission gate 1a and transmission gate 2a transmit, and Q outputs of the both latches are set to logic level high, since a large inverter, I_R, drives node N_R up to V_{DD}. When the reset pulse is turned off latch L_1 is transmitting and latch L_2 now retains the high logic level. In this way the latches were cleared. After the reset a pair of half-frequency clocks having a definite time relationship appear at the outputs CK' and $\overline{CK'}$ as the circuit is clocked. This pair of clocks can be used to drive the cascaded next stage, in order to generate a clock of 1/4 frequency.

In a CMOS circuit the initial condition of the circuit can be set or reset by turning the power supply voltage off by a power driver, as shown in Fig. 2.28. Driver I_R must be large (typically equal or larger than the sum of the sizes of the gates whose power is supplied through the driver) to guarantee fast set or reset time.

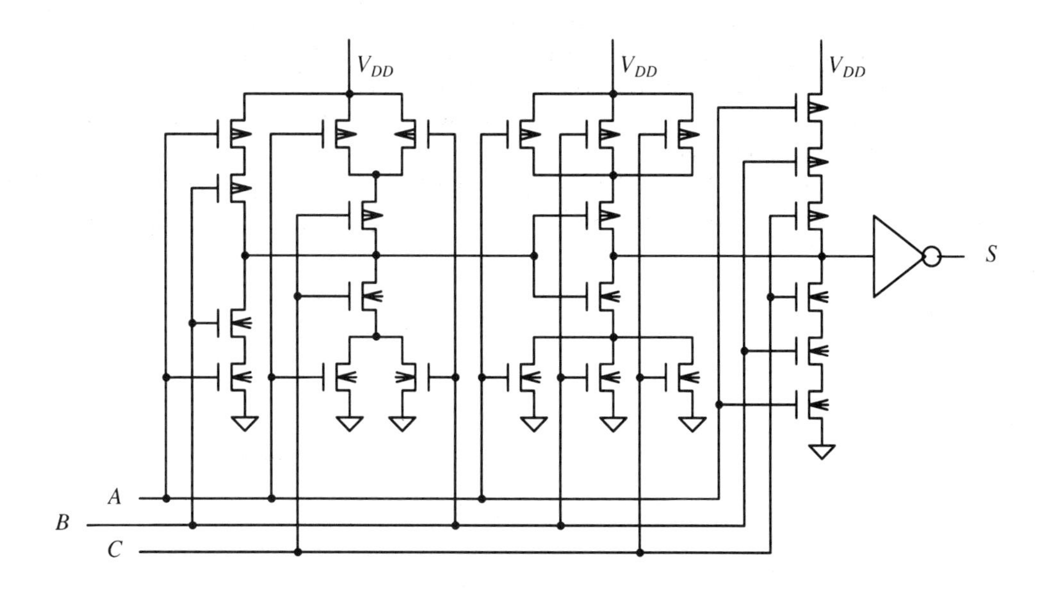

Figure 2.30 Compact full adder circuit.

2.14 ADDERS

A logic circuit that adds a pair of 1-bit numbers A and B and generates the sum S is a half adder. A half adder is an exclusive-OR (XOR) gate, shown in Fig. 2.29(a). The XOR function can be implemented in several different ways. Two examples are shown in Fig. 2.29(b) and (c). The implementation shown in Fig. 2.29(c) [26] is based on the idea that an XOR function of variables A and B is

$$XOR(A, B) = B \quad \text{if} \quad A = \text{Low} \tag{2.11}$$

$$XOR(A, B) = \overline{B} \quad \text{if} \quad A = \text{High} \tag{2.12}$$

This circuit should be used with care since the delay of signal that propagates through cascaded transmission gates is proportional to the square of the number of gates, rather than proportional to the number of gates (Section 6.8).

A circuit that adds a pair of 1-bit numbers A and B and the carry C, and that generates the sum S is a full adder. A full adder is two cascaded exclusive-OR gates shown in Fig. 2.29(d), whose Boolean expression is

$$S = XOR(XOR(A, B), C) \tag{2.13}$$

A full adder can be implemented using the XOR gates of Fig. 2.29(b) or (c). A DSLL implementation of a full adder is shown in Fig. 2.29(d). Yet another implementation of a full adder is shown in Fig. 2.30 and used in the RCA CD4008A 4-bit adder. This implementation requires only 26 FETs as compared to 32 by the implementation of Fig. 2.29(b) including inverters. Since the sum output is available from an inverter, interfacing the adder to the other functional blocks of the chip is easier than the circuit using the XOR of Fig. 2.29(c). The circuit cannot be decomposed into individual CMOS gates, but it is easy to understand that the circuit carries no steady-state current, by an exhaustive check.

An N-bit adder is built from N full adders and a circuit that generates carries for each bit. The $(i+1)$st carry bit $C_i + 1$ is generated recursively from the carry of the ith bit C_i as

$$C_i + 1 = (A_i + B_i)C_i + A_i B_i \tag{2.14}$$

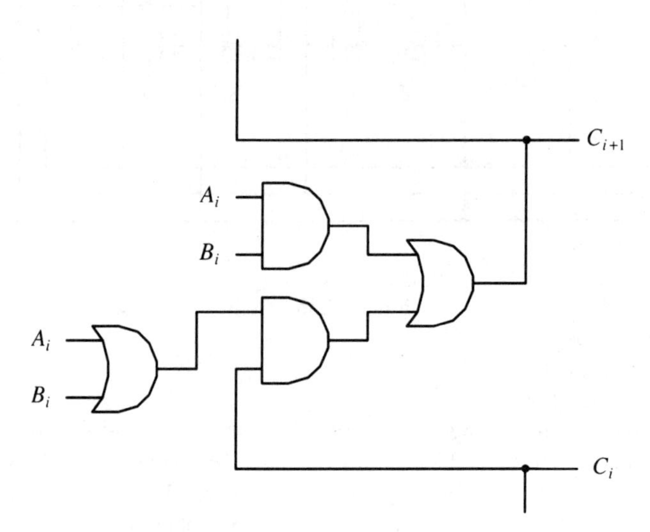

Figure 2.31 Ripple carry generator.

The logic circuit is shown in Fig. 2.31. N such circuits are stacked. The carry entered from the LSB (least significant bit) ripples to the MSB (most significant bit) and the carries at each bit position are generated sequentially. Adders of 4- and 8-bit numbers use ripple carry generation. For 16- and 32-bit adders, ripple carry generation takes too much time. There are several schemes to speed up carry generation, as discussed in Sections 7.14 and 7.15.

2.15 ANALOG AND FUNCTIONAL CIRCUIT BLOCKS

The analog circuits discussed in this section are not conventional gates for digital signal processing, but are expected to play an important role in future CMOS VLSI technology.

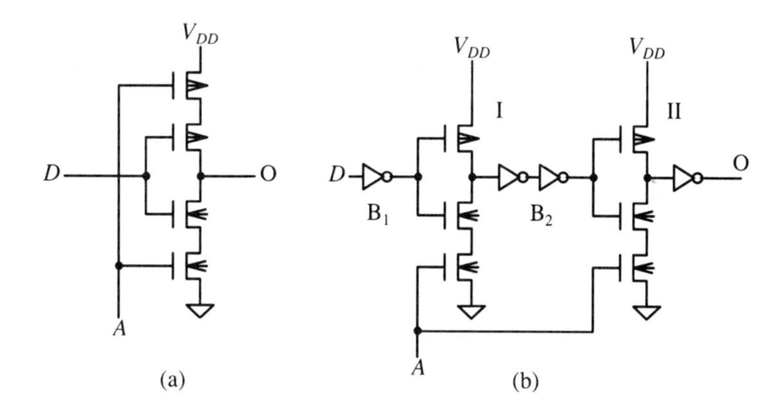

Figure 2.32 Pulse-reshaping circuits.

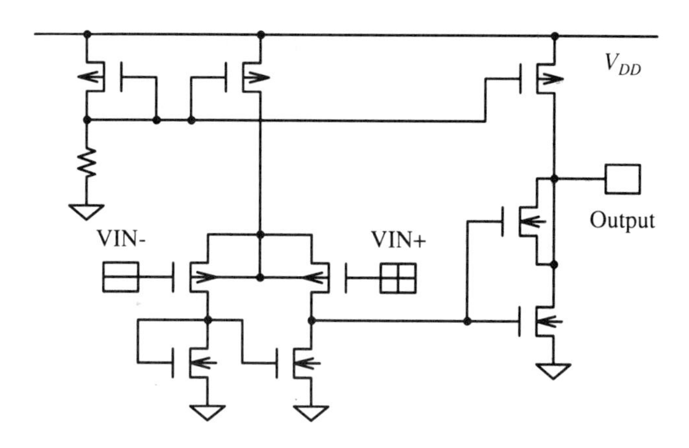

Figure 2.33 Operational amplifier.

Except for application to analog-digital combined circuits, analog technique offers powerful solution in reducing sensitivity of digital integrated circuits to process, temperature and voltage variations, and also to set the operating point of critical digital circuits. Digital circuit techniques are practically powerless in solving this type of problem. Introduction of analog circuit techniques into CMOS VLSI is an important new trend for the future. A comprehensive discussion of the subject of MOS analog circuit technology was given by Hodges [27]. CMOS analog circuits have advantages of small power, and have a nearly perfect analog switch (transmission gate). CMOS analog circuits have, however, the disadvantages of larger offset voltage for differential amplifiers, and a limited voltage swing caused by the back-bias effect compared to bipolar circuit [28]. The two problems can be resolved by using BiMOS technology. Analog circuits discussed in this section were selected for usefulness in digital CMOS design, and no claim is made of a balanced review of analog CMOS technology.

(A) Pulse-Width Adjustment Circuit

The circuit shown in Fig. 2.32(a) has a digital input D and an analog control input A. If the control voltage A is at $V_{DD}/2$ and the FET sizes are chosen appropriately, the digital signal from input D is inverted, with equal pullup and pulldown delays. If voltage A is lower than $V_{DD}/2$, the circuit takes more time to pull down than to pull up. Therefore, a wide upgoing pulse at input D emerges as a narrow downgoing pulse at output O. If voltage A is higher than $V_{DD}/2$, a wide downgoing pulse at input D emerges as a narrow upgoing pulse at output O. The circuit allows voltage control of the width of a pulse [29]. The circuit is useful in reshaping clock waveforms when used in a feedback circuit including a pulse-width detector and an operational amplifier (Section 7.10).

(B) Pulse-Delay Circuit

The circuit of Fig. 2.32(b) delays the pulse at input D by the amount controlled by the analog voltage A [29]. The input upgoing edge of the pulse is delayed by gate II, and the downgoing edge by gate I, by exactly the same amount. The buffers B_1 and B_2 (pullup-pulldown symmetrical) make the pulse rise and fall times at the inputs of control gates I and II the same.

(C) Operational Amplifier

The circuit of Fig. 2.33 works as a general-purpose operational amplifier [30] [31]. To design precision analog circuits like operational amplifiers, relaxed design rules are used for better matching of the components and for reducing the effects of process variations: Generally, gate lengths of 7 to 8 μm are used in analog devices, instead of 1 to 2 μm for digital devices. Since the offset voltage of an MOS operational amplifier is larger than that of bipolar operational amplifiers, dynamic offset compensation techniques are often used. If analog circuits are included as a part of digital circuits, careful layout and shielding the sensitive high-gain circuits (e.g., operational amplifiers) from noise is required [32].

(D) Reference Voltage Source

The circuit of Fig. 2.34 generates the temperature-independent band-gap voltage of silicon (=1.1 V) at the output [33]. The negative temperature coefficient of the base-emitter voltage drop of the bipolar transistor is compensated by the positive temperature coefficient of a second voltage, which is the difference of voltages of two sets of three series-connected MOS diode chains, whose current densities are different. Currents I_1 and I_2 satisfy $I_1 \gg I_2$, and the diode-connected FETs are in the subthreshold region. The voltage difference $V_1 - V_2$ has a positive temperature coefficient. The currents I_1 and I_2 are generated by the ratioed current mirror circuit. This circuit nulls the first-order temperature coefficient of the reference voltage. A very sophisticated circuit that compensates the second-order temperature coefficient has been developed [34].

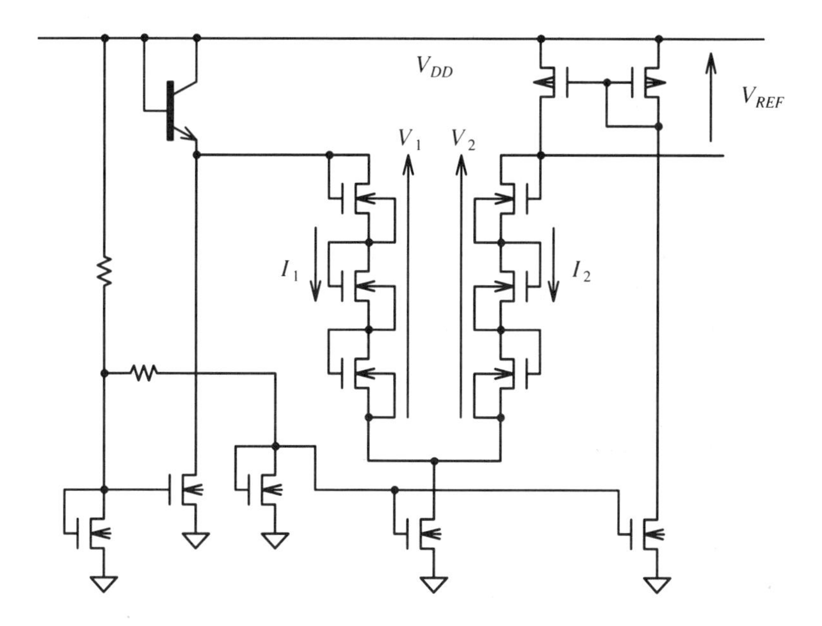

Figure 2.34 Band-gap reference voltage generator.

The circuit shown in Fig. 2.35 [35] generates the band-gap voltage V_{REF}, using NFET MN1, whose gate is made of P-type heavily doped polysilicon, and MN2, whose gate is made of N-type heavily doped polysilicon. The difference of the threshold voltages of the two FETs equals the band-gap voltage of silicon. The other FETs bias the two FETs to equal currents, so that the band-gap voltage is generated at the source of MN2.

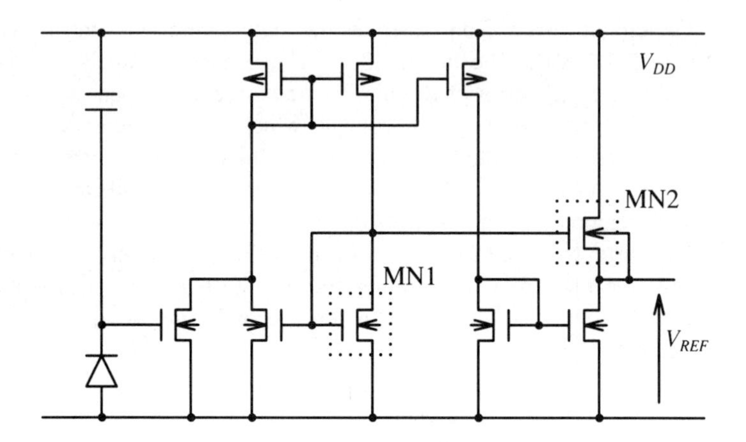

Figure 2.35 Band-gap reference voltage generator using gate work-function difference.

(E) Substrate Bias Generation

Sometimes it is advantageous to apply a negative bias to the P-type substrate, thereby reducing the parasitic capacitance of the diffused islands of NFETs and further preventing accidental forward biasing of the PN junctions [36]. A clocked circuit that generates DC voltage outside the range covered by the power supply is called a charge pump.

The circuit shown in Fig. 2.36(a) generates a negative voltage at the terminal SUB. All NFETs are on P-tubs that are connected to SUB. Q and $\overline{Q}$ are a pair of complementary clocks as shown in Fig. 2.36(b). When Q is high, node A is clamped to $+V_{TH}$ by a MOSFET in diode connection, MN1, whose source is grounded. When Q is low, node A is lower than V_{TH} by V_{DD}. At the same time node B is clamped to the voltage higher than the voltage of node A by V_{TH}, by a MOSFET in diode connection, MN2. This loading pulls the node A voltage above $-(V_{DD}-V_{TH})$. When clock Q becomes high again, the voltage of node B goes down by V_{DD}. At the same time, node C is clamped to the voltage of node B plus V_{TH}. When the voltage of node C, B, or A is lower than the substrate voltage, the MOS device MNC, MNB, or MNA, respectively, conducts, thereby pulling the substrate voltage down. With this mechanism a DC voltage lower than the ground potential is generated. This circuit is identical to the voltage doubler rectification circuit used in transformer-less vacuum tube radio [37].

The substrate voltage generator can be controlled by a feedback mechanism, to maintain the voltage within a narrow range. A technique to maintain the threshold voltage of a short-channel FET through negative feedback control exercised through

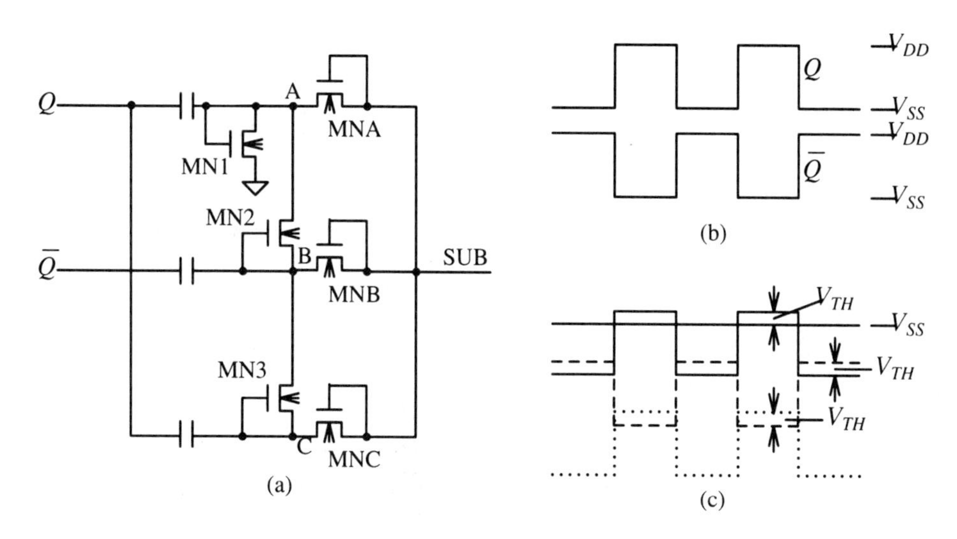

Figure 2.36 Charge pump circuit for substrate biasing.

the substrate potential has been reported [38] [39]. When substrate bias is applied, the ground and the substrate connections must be separated.

(F) A/D Conversion

CMOS VLSI chips sometimes include simple analog-to-digital conversion circuits. A simple example of an A/D converter is shown in Fig. 2.37. The 8X the minimum size FET of the reference port is driven by 20 μA of current. The 2X and 6X PFETs generate 5 μA and 15 μA of current, respectively. The input circuit, 10X NFET MN1, is driven by the unknown current source, and equal currents are generated by MN2 and MN3. If the input current is less than 5 μA, both OUT1 and OUT2 terminals are high. If the input current is in the range $5\,\mu A < I < 15\,\mu A$, OUT1 is low but OUT2 is high. If the current is more than 15 μA, both OUT1 and OUT2 terminals are low. This circuit works as a simple A/D converter for a small number of bits [40].

(G) Switched Capacitor Circuit

CMOS transmission gates are ideal analog switches. Using analog switches, the switched capacitor circuits shown in Fig. 2.38(a) can be made [41]. The switch alternately contacts poles A and B, at the rate of f times per second. Every time the switch contacts pole A, capacitor C is charged to voltage V_A. When the switch flips to pole B, capacitor C loses charge $C(V_A - V_B)$ to settle at voltage V_B. The net effect is a current from pole A to pole B given by $C(V_A - V_B)f$. The circuit is equivalent to a resistor connected between poles A and B of value $1/Cf$ [Fig. 2.38(b)], for frequencies much less than the switching frequency f. The resistor has applications in

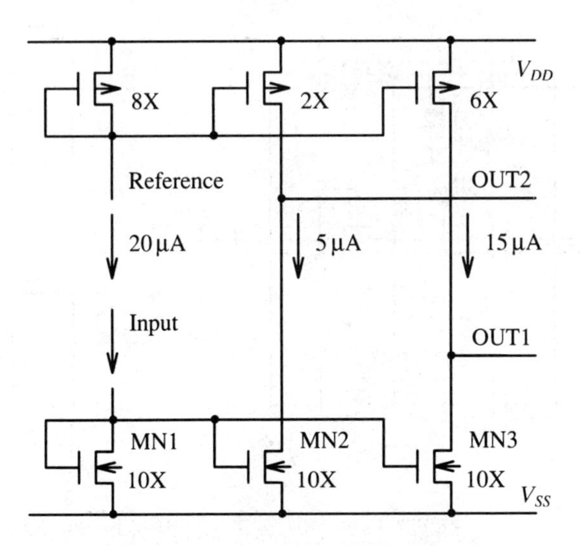

Figure 2.37 Simple A/D conversion circuit.

filtering. The switched capacitor circuit shown in Fig. 2.38(c) is equivalent to an inductance as shown in Fig. 2.38(d), for frequencies much lower than the switching frequency f. The inductance value is $L = 1/(4Cf^2)$.

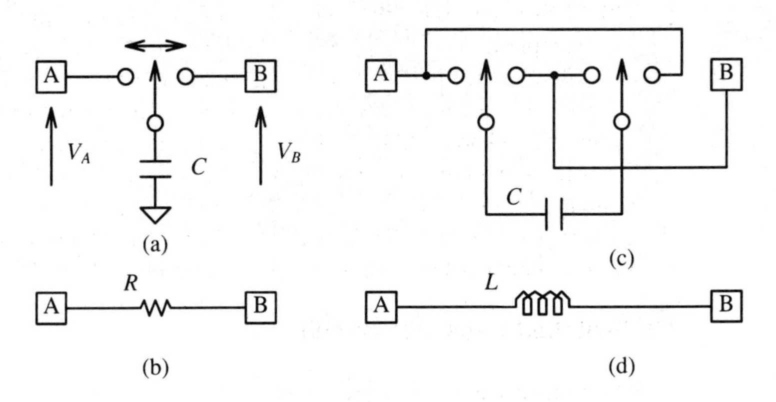

Figure 2.38 Switched-capacitor circuits.

Switched capacitors can perform many analog functions, such as adding and subtracting voltages. Such analog functions are expected to be an integral part of many future VLSI chips [42]. Many analog circuits developed in NMOS technology are also applicable to CMOS [43].

2.16 RC OSCILLATORS

The circuit shown in Fig. 2.39(a) is convenient to generate a clock signal for a CMOS circuit, whose frequency stability specifications are not stringent. Suppose that inverter 1 has a switching threshold voltage at $V_{DD}/2$. The inverter pulls node X down at time $t = 0$. If the delays of the inverters are negligibly small, node Y pulls up to V_{DD}, and therefore node Z is pulled up from $V_{DD}/2$ to $(3/2)V_{DD}$, all at time $t = 0$. The voltage of node Z decays exponentially with time constant RC.

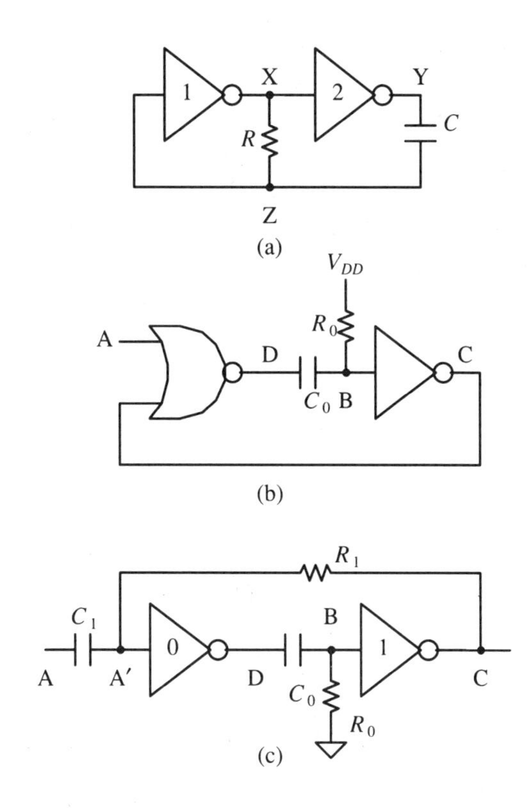

Figure 2.39 Relaxation oscillator and monostable trigger.

If the output resistance of inverter 1 is negligibly small, the voltage of node Z drops to $V_{DD}/2$ after time $(\log 3)RC$. Then inverter 1 pulls up, inverter 2 pulls down, and node Z voltage drops from $V_{DD}/2$ to $-V_{DD}/2$, all at time $t = (\log 3)RC$. The voltage of node Z is then pulled up by resistor R, and at $t = 2(\log 3)RC$ inverter 1 pulls down. This process is repeated, with period $2(\log 3)RC$. Since node voltages exceed the range between 0 (ground) and V_{DD}, oscillator circuits are built from discrete components.

When R and C are reduced the period of oscillation decreases. The calculation of period is valid when the period is much longer than the gate delay, and R is much larger than the output resistance of the inverter. If the effects of gate delay time and the output resistance are included in the analysis, the period of oscillation is longer than the period given by the calculation. The frequency of oscillation can be varied using an external variable resistor or voltage-controlled FET used as a variable resistor.

2.17 MONOSTABLE TRIGGER CIRCUIT

A monostable trigger is a circuit that generates a pulse of predetermined width every time the quiescent circuit is triggered by a narrow pulse. The circuit of Fig. 2.39(b) works as follows. In the quiescent state node B is high, node C is low, and since the trigger input A is low, D is high. When A is driven by a narrow positive pulse, node D and node B are pulled down to ground, and node C is pulled up. Assuming that the gate delays are negligible, node C is high before the trigger pulse at input A is turned off, as shown in Fig. 2.40 (curve C).

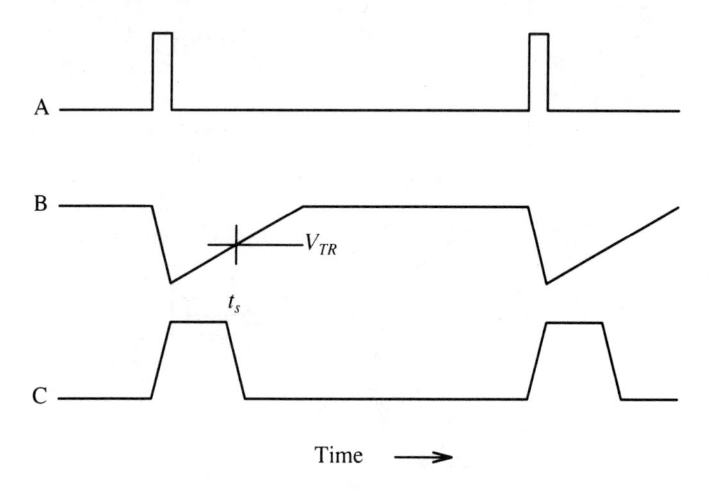

Figure 2.40 Waveforms of monostable trigger.

The voltage of node B increases as resistor R_0 pulls the node up with time constant $R_0 C_0$. At time t_s the voltage reaches the switching threshold of the inverter, and node C pulls down. The circuit returns to the quiescent state. The width of the upgoing pulse at node C is determined by the time constant $R_0 C_0$ and the threshold voltage of the inverter, V_{TR}.

The threshold voltage depends on processing conditions. The circuit shown in Fig. 2.39(c) partially compensates for the effects of threshold variation. In the quiescent state node B is low, node A′ is high, and node D is low. When a negative-going pulse arrives at the input node A, node A′ is driven down to ground potential and nodes D and B go up. Node C goes down, and jointly with the trigger pulse node A′ is kept at the logic low level. Node B is pulled down to the ground potential by resistor R_0, and takes time

$$T_0 = (R_0 C_0)\log\frac{V_{DD}}{V_{TR1}} \tag{2.15}$$

for inverter 1 to switch. Node C is then pulled up. Node A′ is pulled up by resistor R_1 and inverter 0 switches after time

$$T_1 = (R_1 C_1)\log\frac{V_{DD}}{V_{DD} - V_{TR0}} \tag{2.16}$$

The pulse width T_W is the sum of the two terms,

$$T_W = T_0 + T_1 \tag{2.17}$$

If resistances and capacitances can be matched closely,

$$R_0 C_0 = R_1 C_1 = RC$$

then

$$T_W = (RC)\log\frac{V_{DD}^2}{(V_{DD} - V_{TR0})V_{TR1}} \tag{2.18}$$

Therefore, if V_{TR1} and V_{TR0} are approximately $(V_{DD}/2)$, and if they track each other $(V_{TR1} = V_{TR0})$, T_W can be made insensitive to the small threshold variation.

Oscillator and monostable trigger circuits require external resistance or capacitance components. FETs can be either discrete components or so-called transistor building blocks. In case of building block FETs, gate input must be protected from the damage caused by electrostatic surges. Conventional protection circuitry is a voltage clamp circuit to V_{SS} and V_{DD} using a resistor and diodes. This type of protection circuit does not work since the input protection circuit must not clamp the voltage within the range $-V_{DD}/2 < V < (3/2)V_{DD}$. A protection circuit that satisfies this requirement can be designed as described in Section 9.9.

2.18 CONTROLLED OSCILLATORS

To generate an accurate clock, the crystal oscillator shown in Fig. 2.41 is used [44]. Resistor R is used to bias the inverting amplifier at the highest gain operating point (Sections 3.2 and 4.7). Resistor R is two to three orders of magnitude larger than the internal resistance of the inverting amplifier. The clock signal from the oscillator is amplified by four to six stages of scaled-up inverters to generate the CMOS clock. CMOS crystal oscillators are used only at relatively low frequencies. To generate crystal-controlled clocks at frequencies higher than 10 to 20 MHz, bipolar crystal oscillators are used exclusively. A crystal oscillator requires a high-gain small-signal amplifier. Bipolar transistors have higher transconductance than MOSFETs and therefore they are better for this application.

Suppose that a crystal-controlled clock at frequency f clocks a CMOS VLSI system. Sometimes clocks at different frequencies are required. The technique of generating a clock at frequency $f/2$ using countdown circuits was discussed in Section 2.13. The technique of generating a clock at frequency $2 \cdot f$ is more involved than that. There are two different techniques.

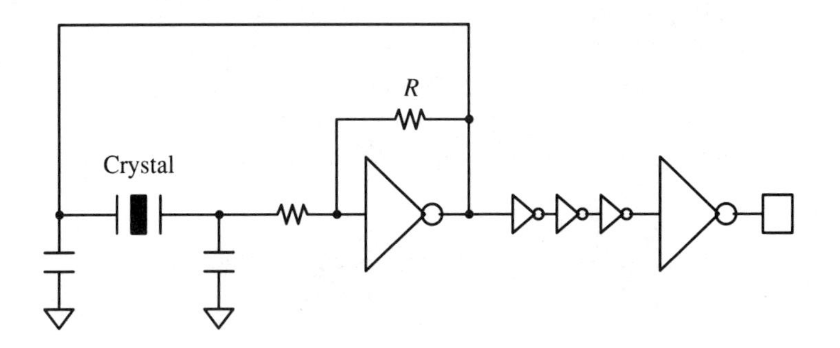

Figure 2.41 Crystal-controlled clock generator.

(A) Digital Frequency Multiplication

The clock at fundamental frequency f is first reshaped to a square pulse having exactly 50% duty cycle. The square pulse is delayed by time T_P, and the original and delayed signals are exclusive ORed. We obtain a positive-going pulse of width T_P at every transition of the fundamental clock.

(B) Analog Frequency Multiplication

Suppose that resistance R in the RC oscillator circuit of Fig. 2.39(a) is replaced by a MOSFET. The oscillation frequency can then be controlled by the gate voltage. By adjusting this gate voltage the oscillator frequency is tuned slightly lower than $2 \cdot f$. Using a small capacitance the signal at frequency f is coupled into any active node of the RC oscillator. Then the oscillator is synchronized: The oscillation frequency is locked exactly in $2 \cdot f$.

The other way is to count down the frequency of the voltage-controlled oscillator, and compare the phase of the counted-down clock with that of the clock at frequency f. If there is any difference, an error voltage is generated. The error voltage is fed back to the gate of the MOSFET that controls the frequency. The RC oscillator is locked at frequency $2 \cdot f$. This technique is called a phase-locked loop.

Dynamic logic circuits in advanced microprocessors require precision clock generators. Until recently, however, clock circuits of microprocessors had been quite unsophisticated. The technical gap will be filled by the introduction of advanced analog circuit design techniques.

2.19 VARIATIONS OF CMOS LOGIC CIRCUITS

It is possible to build NMOS-like logic circuits by substituting grounded gate PFETs for depletion-mode load devices in NMOS gates. Such gates have the advantage of reduced input capacitance resulting in higher switching speed at the cost of standby power. The gates have an application in address decoding of memory. Such a technique has been considered as nonstandard, but for future performance-sensitive CMOS VLSI, to improve critical circuit performance at the expense of power is a justifiable design alternative [45].

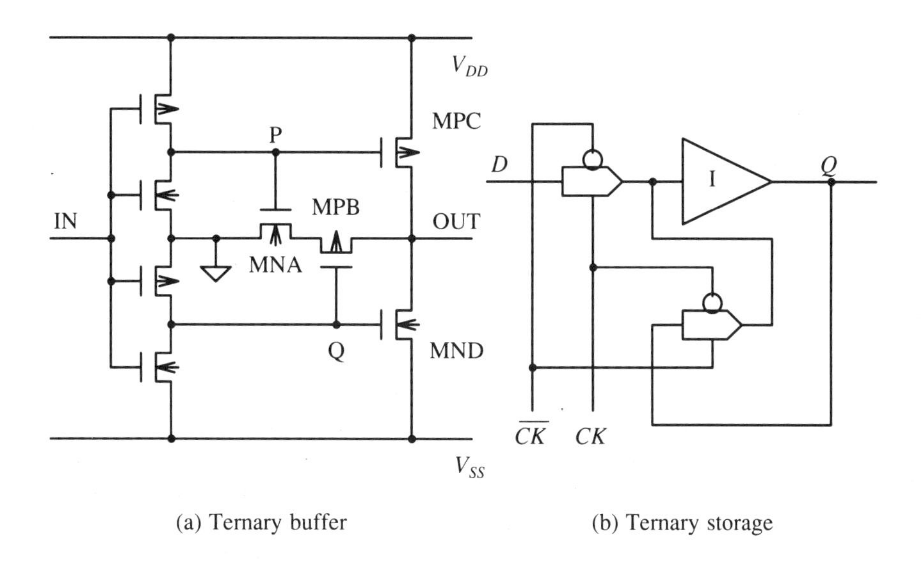

(a) Ternary buffer (b) Ternary storage

Figure 2.42 Ternary buffer and latch.

Such a circuit is already used in static column decoding of memory devices. The load device can be a JFET made between the P-tubs of an N-substrate CMOS process [40]. The PFET load can be switched to reduce power [47]. A static buffer is inserted in the logic chain to increase stability. CMOS static PLAs are designed using this technique. CMOS PLAs are two cascaded arrays of NOR gates. A single PFET load device is used. The assembly of the NOR gates is laid out in a regular rectangular shape. The first-stage NOR array is called a decoder. For logic flexibility both input signal and its complement are made available to the NOR gates in the decoder. The second-stage NOR array is called a ROM. A PLA is often used to generate state vectors of a microprocessor [48].

An exotic CMOS logic that uses a clocked power supply has been proposed [49] [50].

A ternary (three-value) logic circuit can be designed using CMOS [51]. The buffer circuit shown in Fig. 2.42(a) takes a logic variable whose values are 0, 1, and 2, corresponding to V_{SS} (<0), 0 (ground) and V_{DD} (>0), respectively. When the input variable takes one of these values, the output is the same value, available from the lower internal impedance source [52]. The states of the nodes are listed in Table 2.1.

TABLE 2.1 STATES OF TERNARY LOGIC BUFFER

IN	P	Q	MNA	MPB	MPC	MND	OUT
2	1	0	OFF	ON	ON	OFF	2
1	2	0	ON	ON	OFF	OFF	1
0	2	1	ON	OFF	OFF	ON	0

When this buffer is combined with two transmission gates, a storage device for the ternary variable can be designed, as shown in Fig. 2.42(b). A functionally complete set of ternary operations convenient for integrated circuit implementation is reported [53]. An interesting ternary Schmidt trigger [54] and a ternary storage cell [55] circuit have also been reported.

2.20 CIRCUIT SELECTION

In this chapter we reviewed many CMOS digital circuits and a few analog circuits useful in digital design. There are many other circuits that are not included. For one Boolean operation there are many circuits that produce the same result if all the requirements for correct circuit operation are satisfied, but if not, the circuits produce wrong results. Tolerance of circuit to nonideal operating conditions is the merit of the circuit. The chip designer selects the best circuit for the purpose and assembles the individual circuits into an *integrated* circuit. Whether or not the selection is correct determines the quality of the design. We give some guidelines for selection before ending this chapter.

Circuit selection is not just a technical issue. The experience, beliefs, and character of the designer, and the objective of the particular IC development project are factors in circuit selection. The author does not insist that the following typical guidelines for a conservative designer are totally inflexible.

1. Circuits and subsystems (assembly of gates such as an adder) that have been used before and that have no record of failure are always the best choice. The circuits and subsystems should be cataloged, stored in the library, and made available for future design. The library would be different for different design organizations, and the organization that has the best library is the winner in the IC design business. VLSI chips should be assembled using as many previously tested circuits and subsystems whose characteristics are well known. It is important to reduce the number of circuits that require invention to the absolute minimum; design efforts should be concentrated on critical circuits.

2. High-performance circuits often have reliability problems caused by process, temperature, and power supply voltage variations. In VLSI chip design the designer cannot afford to compromise reliability. Circuits should be screened first by these criteria.

3. When standard circuits are not adequate, they can still be improved, for example by adding a few additional components, by setting the operating point by applying analog feedback, and the like. Analog circuit design techniques are especially useful in reducing sensitivity of circuit to process, temperature, and voltage variations. Circuits obtained by improving established circuits are often more reliable and have higher performance ratings than those of newly invented circuits. An example is shown in Section 8.12 (a NAND2 gate in a clock generator can be symmetrized to reduce clock skew).

4. Circuits to execute elementary logic functions (e.g., NAND, XOR, ...) and standard analog functions (e.g., operational amplifier) are already well known. Reputable and reliable designs exist. Designers should select the best one for their purpose and should not waste time reinventing such elementary functions.

5. If an attractive new idea for a circuit emerges, the idea should be tested by actually making the circuits on silicon wafer (often called test pattern experiments). If wafer fabrication is not possible, a breadboard model circuit can be built using CMOS FET building blocks (e.g., the RCA CD4007 chip). Circuit simulation is not sufficient to detect all the problems, especially when feedback is involved.

6. Except for circuits included in a regular structure (e.g., memory cells and datapath circuits), the area occupied by a circuit is never very critical, because FET packing density of VLSI chip is not high. If additional FETs and increased-size of FETs improve performance and reliability, these are attractive options.

2.21 REFERENCES

01. R. A. Stehlin "The realization of 1.0-volt multivibrators using complementary integrated circuits" IEEE J. Solid-State Circuits, vol. SC-4, pp. 284-288, October 1969.
02. Y. Suzuki, M. Hirasawa and K. Odagawa "Clocked CMOS calculator circuitry" ISSCC73 Digest, pp. 58-59, February 1973.
03. Y. Suzuki, K. Odagawa and T. Abe "Clocked CMOS calculator circuitry" IEEE J. Solid-State Circuits, vol. SC-8, pp. 462-469, December 1973.
04. A. K. Rapp, L. P. Wennick, H. Borkan and K. R. Keller "Complementary-MOS integrated binary counter" ISSCC67 Digest, pp. 52-53, February 1967.
05. A. A. Alaspa and A. G. F. Dingwall "COS/MOS parallel processor array" IEEE J. Solid-State Circuits, vol. SC-5, pp. 221-227, October 1970.

06. E. J. McCluskey "Logic design principles" Prentice-Hall Inc., Englewood Cliffs, New Jersey 1986.
07. V. Belevitch "Classical network theory" Holden-Day, San Francisco, 1968.
08. S. H. Caldwell "Switching circuits and logical design" John Wiley & Sons, New York, 1965.
09. C. K. Erdelyi "Random logic design utilizing single-ended cascode voltage switch circuit in NMOS" IEEE J. Solid-State Circuits, vol. SC-20, pp. 591-593, April 1985.
10. L. G. Heller, W. R. Griffin, J. W. Davis and N. G. Thoma "Cascode voltage switch logic: a differential CMOS logic family" ISSCC84 Digest, pp. 16-17, February 1984 and C. K. Erdelyi "Random logic design utilizing single-ended cascode voltage switch circuits in NMOS" IEEE J. Solid-State Circuits, vol. SC-20, pp. 591-594, April 1985.
11. A. S. Shubat, J. A. Pretorius and C. A. Salama "Differential pass transistor logic in CMOS technology" Electronics letters, vol. 22, pp. 294-295, March 1986.
12. L. C. M. G. Pfennings, W. G. J. Mol, J. J. J. Bastiaens and J. M. F. van Dijk "Differential split-level CMOS logic for sub-nanosecond speeds" ISSCC85 Digest, pp. 212-213, February 1985.
13. H. Oguey and E. Vittoz "Resistance-CMOS circuits" IEEE J. Solid-State Circuits, vol. SC-12, pp. 283-285, June 1977 and E. Vittoz, B. Gerber and F. Leuenberger "Silicon-gate CMOS frequency divider for the electronic wrist watch" IEEE J. Solid-State Circuits, vol. SC-7, pp. 100-104, April 1972.
14. E. L. Hudson and S. L. Smith "An ECL compatible 4K CMOS RAM" ISSCC82 Digest, pp. 248-249, February 1982.
15. Y. F. Chan "A 4K CMOS erasable PROM" IEEE J. Solid-State Circuits, vol. SC-13, pp. 677-680, October 1978.
16. T. Masuhara, O. Minato, T. Sasaki, Y. Sakai and M. Kubo "A high-speed, low-power Hi-CMOS 4K static RAM" ISSCC78 Digest, pp. 110-111, February 1978.
17. S. C. Lee, W. Schucker and P. T. Hickman "Bi-CMOS circuits for high performance VLSI" 1984 Symposium on VLSI Technology Digest, pp. 46-47.
18. J. Koomen and J. van den Akker "A MOST inverter with improved switching speed" IEEE J. Solid-State Circuits, vol. SC-7, pp. 231-237, June 1972.
19. K. Manabe, N. Someya, M. Ueno, H. Neishi, M. Imai, S. Okamoto and K. Suzuki "A C^2MOS 16-bit parallel microprocessor" ISSCC76 Digest, pp. 14-15, February 1976.
20. A. C. Ipri "Lambda diodes utilizing an enhancement-depletion CMOS/SOS process" IEEE Trans. on Electron Devices, vol. ED-24, pp. 751-756, June 1977.
21. C. Y. Wu, K. N. Lai and C. Y. Wu "Generalized theory and realization of a voltage-controlled negative resistance MOS device (Lambda MOSFET)" Solid-State Electronics, vol. 23, pp. 1-7, January 1980.
22. A. Noda "Leak-compensated analog memory with pair MOSFETs" IEEE J. Solid-State Circuits, vol. SC-5, pp. 75-77, April 1970.
23. K. Nagaraj and M. Satyam "Novel CMOS Schmidt trigger" Electronics Letters, vol. 17, pp. 693-694, September 1981.
24. R. Woudsma and J. M. Noteboom "The modular design of clock generator circuits in a CMOS building block system" IEEE J. Solid-State Circuits, vol. SC-20, pp. 770-774, June 1985.
25. A. C. Ipri and J. C. Sarace "CMOS/SOS semi-static shift registers" IEEE J. Solid-State Circuits, vol. SC-11, pp. 337-338, April 1976.
26. K. LeClair, R. Bell, D. Breid, P. Torgerson, D. Fier and B. Jensen "A 32-bit CMOS

microprocessor using a semicustom cell library" CICC84 Digest, pp. 10-13, 1984.
27. D. A. Hodges, P. R. Gray and R. W. Broderson "Potential of MOS technologies for analog integrated circuits" IEEE J. Solid-State Circuits, vol. SC-13, pp. 285-294, June 1978.
28. L. Wofford "Silicon gates spur linear CMOS to bipolar speeds, hold offset stable" Electronics, vol. 55, pp. 137-140, March 1982.
29. R. H. Krambeck and M. Shoji "Skew-free clock circuit for integrated circuit chip" US patent No. 4, 479, 216, October 23 1984.
30. P. R. Gray and R. G. Meyer "MOS operational amplifier design—a tutorial overview" IEEE J. Solid-State Circuits, vol. SC-17, pp. 969-982, December 1982.
31. A. B. Grebene "The operational amplifier in custom LSI design" CICC84 Digest, pp. 540-545, 1984.
32. T. Takamizawa, Y. Hashimoto, M. Arakawa, K. Katco, Y. Tani, T. Satoh, N. Kitagawa and J. D. Bryant "A monolithic voice recorder" ISSCC82 Digest, pp. 264-265, February 1982.
33. Y. P. Tsividis and R. W. Ulmer "A CMOS reference voltage source" ISSCC78 Digest, pp. 48-49, February 1978 and Y. P. Tsividis and R. W. Ulmer "A CMOS voltage reference" IEEE J. Solid-State Circuits, vol. SC-13, pp. 774-778, December 1978.
34. B-S Song and P. R. Gray "A precision curvature-compensated CMOS band-gap reference" ISSCC83 Digest, pp. 240-241, February 1983.
35. H. J. Oguey and B. Gerber "MOS voltage reference based on polysilicon gate work function difference" IEEE J. Solid-State Circuits, vol. SC-15, pp. 264-269, June 1980.
36. J. Pathak, H. Kurowski, R. Pugh, R. Shrivastava and F. Jenne "A 19ns 250mW programmable logic device" ISSCC86 Digest, pp. 246-247, February 1986.
37. H. J. Reich "Principles of electron tubes" McGraw-Hill Book Company, New York, 1941.
38. M. Kubo, R. Hori, O. Minato and K. Sato "A threshold voltage controlling circuit for short channel MOS integrated circuit" ISSCC76 Digest, pp. 54-55, February 1976.
39. E. M. Blaser, W. M. Chu and G. Sonoda "Substrate and load gate voltage compensation" ISSCC76 Digest, pp. 56-57, February 1976.
40. D. A. Freitas and K. W. Current "CMOS current comparator circuit" Electronics Letters, vol. 19, pp. 695-697, August 1983.
41. D. Herbst, B. Hoefflinger, K. Schumacher and R. Schweer "MOS switched-capacitor filters" ISSCC79 Digest, pp. 74-75, February 1979.
42. R. Castello and P. R. Gray "Performance limitations in switched-capacitor filters" IEEE Trans. on Circuits and Systems, vol. CAS-32, pp. 865-875, September 1985.
43. Y. P. Tsividis "Design considerations in single-channel MOS analog integrated circuits—A tutorial" IEEE J. Solid-State Circuits, vol. SC-13, pp. 383-391, June 1978.
44. R. G. Meyer and D. C. F. Soo "MOS crystal oscillator design" IEEE J. Solid-State Circuits, vol. SC-15, pp. 222-228, April 1980.
45. H. Sakamoto and L. Forbes "Grounded load complementary FET circuits: Sceptre analysis" IEEE J. Solid-State Circuits, vol. SC-8, pp. 282-284, August 1973.
46. Y. Sakai, T. Masuhara, O. Minato and N. Hashimoto "MOS buried load logic" ISSCC80 Digest, pp. 56-57, February 1980.
47. S. Hanamura, M. Aoki, T. Masuhara, O. Minato, Y. Sakai and T. Hayashida "Low temperature CMOS 8x8b multipliers with sub 10ns speeds" ISSCC85 Digest, pp. 210-211, February 1985.
48. P. May and F. C. Schiereck "High-speed static programmable logic array in LOCMOS" IEEE J. Solid-State Circuits, vol. SC-11, pp. 365-369, June 1976.
49. B. J. Hosticka, U. Kleine, H. Vogt and G. Zimmer "DICMOS-novel MOS logic"

Electronics Letters, vol. 18, pp. 930-932, October 1982.
50. E. Harari "Novel dynamic merged load technology" IEEE J. Solid-State Circuits, vol. SC-20, pp. 537-541, April 1985.
51. J. L. Huertas, J. I. Acha and J. M. Carmona "Design and implementation of tristatables using CMOS integrated circuits" IEE J. Electronic Circuits and Systems, vol. 1, pp. 88-94, April 1977 and H. T. Mouftah "Design and implementation of tristatables using CMOS integrated circuits" IEE J. Electronic Circuits and Systems, vol. 2, pp. 61-62, March 1978.
52. A. Heung and H. T. Mouftah "An all-CMOS ternary identity cell for VLSI implementation" Electronics Letters, vol. 20, pp. 221-222, March 1984.
53. V. H. Tokmen "A functionally-complete ternary system" Electronics Letters, vol. 14, pp. 69-70, February 1978.
54. K. Ramkumar and K. Nagaraj "A ternary Schmidt trigger" IEEE Trans. on Circuits and Systems, vol. CAS-32, pp. 732-735, July 1985.
55. K. Nagaraj and K. Ramkumar "Static RAM cell for ternary logic" Proc. IEEE, vol. 72, pp. 227-228, February 1984.

3

Switching of CMOS Static Gates

3.1 INTRODUCTION

In this chapter we analyze characteristics of CMOS static gates, especially delay characteristics, from the working mechanisms of the circuit. Although CMOS gate delay time is the most frequently used parameter, it is often not precisely defined and is often not examined completely enough to avoid ambiguity. As we will see, a number of conditions must be specified exactly before the delay time of a gate is defined unambiguously.

One such basic ambiguity originates from the dependence of delay time on the input signal rate of change. This problem is closely related to a unique property of a CMOS gate—that a CMOS gate is ratioless: The output node of a CMOS static gate settles at the correct voltage level reflecting the states of the input logic variables, regardless of the FET size. All CMOS gates need not switch at the same input threshold voltage, as in TTL gates.

In order to define the delay time of a gate, however, a logic threshold voltage, at which the gate is considered to switch, is required. The absence of a well-founded switching threshold voltage leads to the situation that the delay time of a CMOS gate must be determined using a *defined* logic threshold voltage that is different from the actual switching threshold voltage of an individual gate. It is impracticable to define the delay time of a gate individually using the switching threshold voltage of the gate and the switching threshold voltage of the fanout gates, especially if the fanout gates do not have the same switching threshold voltage. The delay time of a CMOS gate is the interval elapsed from the moment when the input signal crosses the defined threshold voltage to the moment when the output voltage crosses the same threshold voltage.

Logic threshold voltage of a CMOS static gate is defined by focusing attention on the basic symmetry of a CMOS gate for pullup and pulldown, even though the symmetry may not be exploited for each gate. The defined logic threshold voltage is set halfway between V_{DD} and the ground voltages. The gate delays are defined using the moments of times when the input/output node voltages go across the defined threshold voltage.

This definition is convenient for circuit design and optimization. Since delay times are assigned for all gates, the delay times can be added to compute the delay of a long logic chain. It is easy to identify which gate must be redesigned to reduce the total delay of a logic chain. This definition is acceptable when the slew rate of the gate input voltage is more than a certain lower limit determined from the technology and from the gate design. Working within this restriction the defined threshold voltage ($V_{DD}/2$) is convenient and is practically useful.

There are certain gates in a CMOS VLSI chip, however, for which the slew rate of input voltage is smaller than the lower limit. Then a number of problems emerge from the use of a defined logic threshold voltage that is not based on the switching threshold voltage of each gate. Understanding these problems demands knowledge of the static switching characteristics of a CMOS static gate.

3.2 STATIC CHARACTERISTICS OF CMOS GATES

When the input voltage of a CMOS gate changes very slowly the instantaneous output voltage of the gate is determined by the instantaneous input voltage. The gate is then switching in a static mode. The relationship between the input and the output voltages can be found using the simplified FET characteristics of Section 1.6 [01]. With reference to Fig. 3.1(a), the current of NFET MN1, I_N, is given by

$$I_N = B_N \left\{ V_G - V_{THN} - (1/2)V_D \right\} V_D \quad \text{for} \quad V_D < V_G - V_{THN} \tag{3.1a}$$

NFET is then in the triode region. Otherwise, the NFET is in saturation, and

$$I_N = \frac{B_N}{2}(V_G - V_{THN})^2 \quad \text{for} \quad V_D > V_G - V_{THN} \tag{3.1b}$$

For the PFET MN2,

$$I_P = B_P \left\{ V_{DD} - V_G - V_{THP} - (1/2)(V_{DD} - V_D) \right\} (V_{DD} - V_D) \quad \text{for} \quad V_D > V_G + V_{THP}$$

PFET is then in triode region. Otherwise, the PFET is in saturation, and

$$I_P = \frac{B_P}{2}(V_{DD} - V_G - V_{THP})^2 \quad \text{for} \quad V_D < V_G + V_{THP}$$

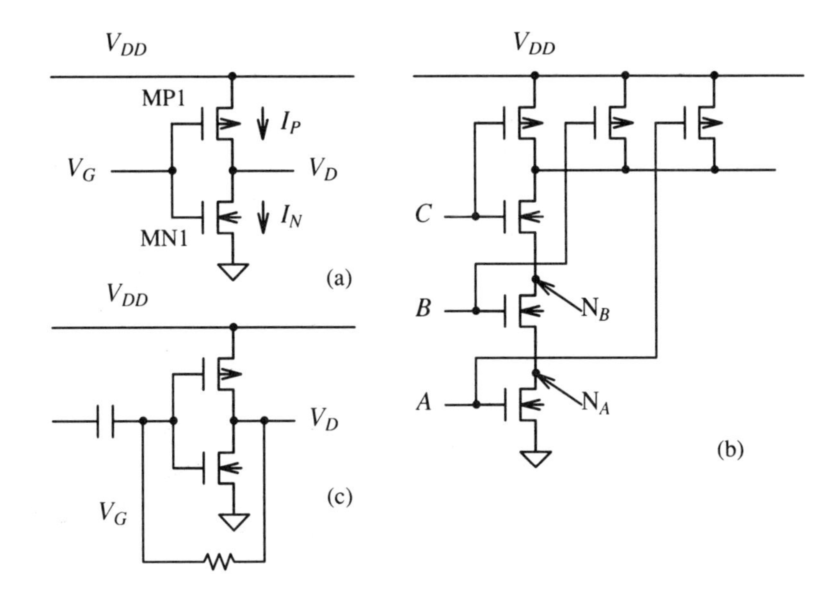

Figure 3.1 CMOS static inverter and NAND3 gate.

In this chapter we often use normalized voltages defined by

$$v_D = \frac{V_D}{V_{DD}}, \quad v_G = \frac{V_G}{V_{DD}}, \quad v_{THN} = \frac{V_{THN}}{V_{DD}}, \text{ and } v_{THP} = \frac{V_{THP}}{V_{DD}}$$

It is convenient to plot the channel current of the NFET in the $v_G - v_D$ coordinate system as shown in Fig. 3.2. This is a map of "equal height" contours. Above the dotted line the NFET is in saturation (current is independent of v_D). This two-dimensional plot was made as follows. Using the maximum drain current available for digital switching $I_{D\text{MAX}} = (B_N/2)(V_{DD} - V_{THN})^2$, the channel current I_D is normalized as

$$i_D = \frac{I_D}{I_{D\text{MAX}}}$$

The normalized channel current is given by

$$i_D = 0 \text{ if } v_G < v_{THN}$$

$$i_D = \frac{2\left\{v_G - v_{THN} - (1/2)v_D\right\}v_D}{(1 - v_{THN})^2} \quad \text{if } v_D < v_G - v_{THN}$$

and

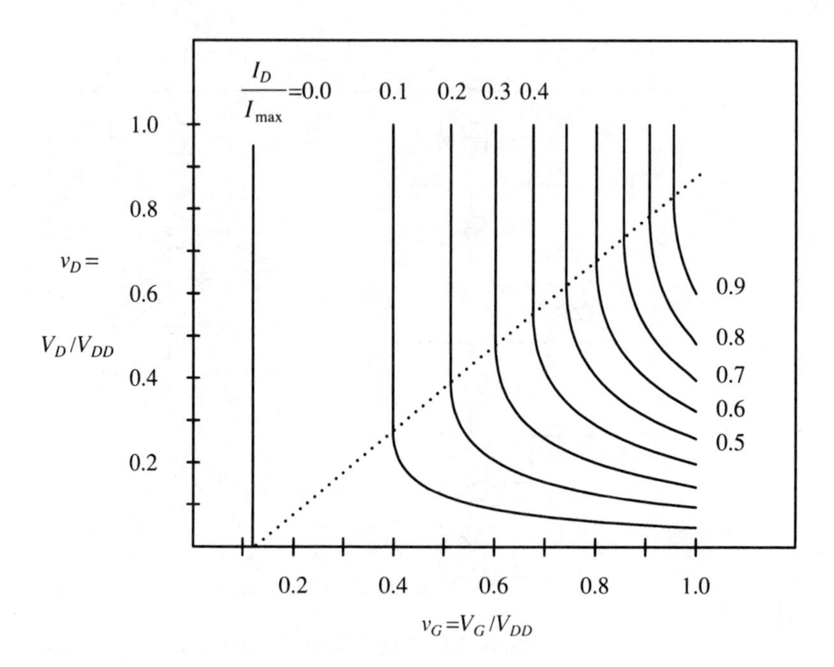

Figure 3.2 NFET current versus normalized gate and drain voltages.

$$i_D = \left[\frac{v_G - v_{THN}}{1 - v_{THN}} \right]^2 \quad \text{if} \quad v_D > v_G - v_{THN}$$

The normalized formula contains only one parameter, normalized threshold voltage v_{THN}, which was set at $v_{THN} = V_{THN} / V_{DD} = 0.12$ in Fig. 3.2. A similar plot for a PFET can be drawn. The PFET plot is obtained by rotating Fig. 3.2 by 180 degrees about its center. This plot is convenient for estimating the FET current during gate switching.

We now study static characteristics of the CMOS inverter. Since DC output current is zero, DC current of NFET, I_N equals DC current of PFET, I_P.

$$B_N \left\{ V_G - V_{THN} - (1/2) V_D \right\} V_D = \frac{B_P}{2} (V_{DD} - V_G - V_{THP})^2 \quad \text{for} \quad V_D < V_G - V_{THN}$$

and

$$B_P \left\{ V_{DD} - V_G - V_{THP} - (1/2)(V_{DD} - V_D) \right\} (V_{DD} - V_D) = \frac{B_N}{2} (V_G - V_{THN})^2$$

for $V_D > V_G + V_{THP}$. Solving these equations for V_D, we obtain

$$V_D = (V_G - V_{THN}) - \left\{ (V_G - V_{THN})^2 - \left[\frac{B_P}{B_N} \right] (V_{DD} - V_G - V_{THP})^2 \right\}^{1/2} \quad (3.2)$$

for $V_D < V_G - V_{THN}$ and

$$V_{DD} - V_D = (V_{DD} - V_G - V_{THP}) - \left\{ (V_{DD} - V_G - V_{THP})^2 - \left[\frac{B_N}{P_P} \right] (V_G - V_{THN})^2 \right\}^{1/2}$$

for $V_D > V_G + V_{THP}$

Since the two equations contain V_G within a square root sign, $\partial V_D / \partial V_G$ diverges when the content of the square root approaches zero. We have $\partial V_D / \partial V_G \rightarrow \infty$ when $V_G \rightarrow V_{GTR}$, where V_{GTR} is given by

$$V_{GTR} = \frac{\sqrt{B_P} V_{DD} + \sqrt{B_N} V_{THN} - \sqrt{B_P} V_{THP}}{\sqrt{B_N} + \sqrt{B_P}} \quad (3.3)$$

or by normalization,

$$v_{GTR} = \frac{V_{GTR}}{V_{DD}} = \frac{1 + \sqrt{\beta} v_{THN} - v_{THP}}{1 + \sqrt{\beta}}$$

where v_{THN} and v_{THP} are the normalized FET threshold voltages and where $\beta = B_N / B_P$ is the beta ratio of the FETs.

Equations (3.2) and (3.3) were derived by Burns [02]. Normalizing the voltages of Eq. (3.2) using $v_D = V_D / V_{DD}$ and $v_G = V_G / V_{DD}$, we have

$$v_D = (v_G - v_{THN}) - \left\{ (v_G - v_{THN})^2 - \left[\frac{1}{\beta} \right] (1 - v_G - v_{THP})^2 \right\}^{1/2} \quad \text{for} \quad v_G - v_{THN} > v_D \quad (3.4)$$

and

$$1 - v_D = (1 - v_G - v_{THP}) - \left\{ (1 - v_G - v_{THP})^2 - \beta (v_G - v_{THN})^2 \right\}^{1/2} \quad \text{for} \quad v_D > v_G + v_{THP}$$

The solid curves in Fig. 3.3 show v_D versus v_G for $\beta = 3$ (curve 1), 1 (curve 2), and 1/3 (curve 3), assuming the typical threshold voltages of $v_{THN} = 0.6/5 = 0.12$ and $v_{THP} = 1.0/5 = 0.2$.

In the triangular area above line P of Fig. 3.3, PFET of the inverter is in the triode region. Below line P the PFET is in saturation. Similarly, the NFET is in the triode region below line N and in saturation above it. In the region between lines N and P both PFET and NFET are in saturation. In the gradual and long-channel model, an FET in saturation works as a current source. Therefore, in the region between lines P and N, equilibrium can be reached only for a single gate voltage where the current of the NFET and that of the PFET are equal.

This voltage, the switching threshold voltage of the inverter V_{GTR}, is given by Eq. (3.3). The normalized gate threshold voltage $v_{GTR} = V_{GTR} / V_{DD}$ versus β is plotted in Fig. 3.4, for the typical case of $v_{THP} = 0.2$ and $v_{THN} = 0.12$ and for a special case,

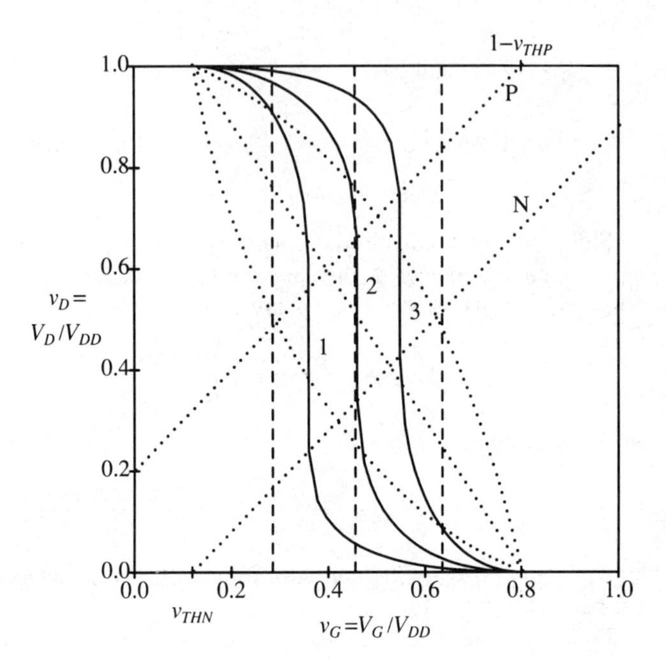

Figure 3.3 Static switching characteristics of CMOS inverter.

$v_{THP} = v_{THN} = 0$. In the latter case the curve is symmetrical with respect to $\beta = 1.0$ and $v_{GTR} = 0.5$. Unequal FET threshold voltages as well as unequal conductance parameters (B_P and B_N) cause asymmetrical switching characteristics. Burns pointed out that the switching threshold voltage is independent of β and is $V_{DD}/2$ if $v_{THN} = v_{THP} = 0.5$. This special case is interesting, but this case has not yet found application and could give very long propagation delays. Klein considered dependence of switching threshold voltage on process variation [03].

It is convenient to consider very simplified FET characteristics to make closed-form analysis possible. An example of such a characteristic is

$$I_N = 0 \quad \text{when} \quad V_G < V_{THN} \tag{3.5}$$

$$I_N = b_N(V_G - V_{THN}) \quad \text{when} \quad V_D > 0$$

$$\frac{dI_N}{dV_D} = \infty \quad \text{when} \quad V_D = 0$$

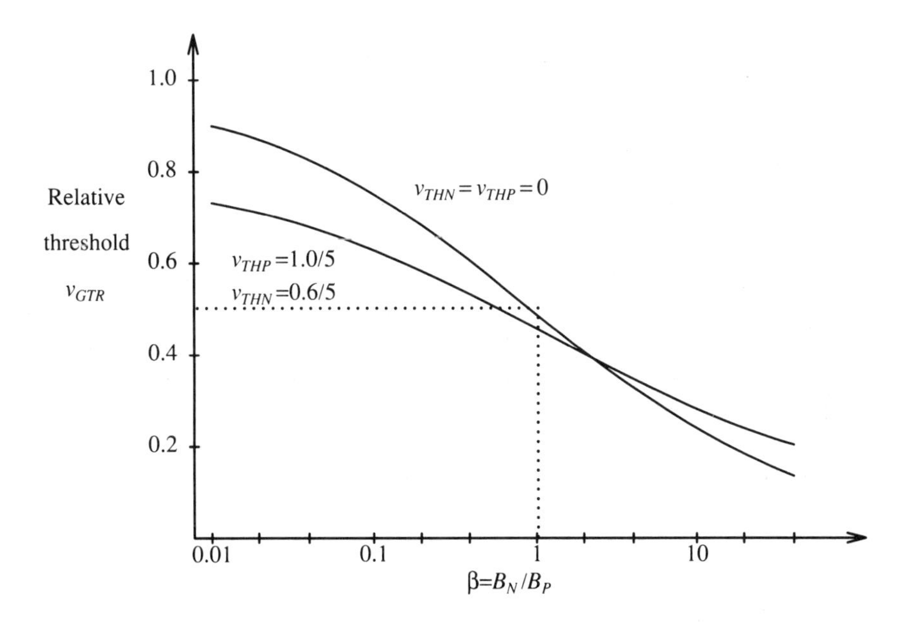

Figure 3.4 Switching threshold voltage of inverter versus beta ratio of FET.

and a similar characteristic for the PFET. In Eq. (3.5) we used parameter b_N, which has dimension Ω^{-1}, instead of B_N of Eq. (3.1), which has dimension $\Omega^{-1}V^{-1}$. The static characteristic of an inverter is

$$V_D = V_{DD} \text{ when } V_G < V_{GTR} \tag{3.6}$$

$$V_D = 0 \text{ when } V_G > V_{GTR}$$

where

$$V_{GTR} = \frac{b_P V_{DD} + b_N V_{THN} - b_P V_{THP}}{b_N + b_P}$$

or

$$v_{GTR} = \frac{1 + \beta v_{THN} - v_{THP}}{1+\beta}, \text{ where } \beta = \frac{b_N}{b_P}$$

Here we used β, the same symbol as Eq. (3.3), since β is the ratio of conductivity of NFET and PFET. The static characteristics are shown by the vertical dashed lines for three cases, $\beta = 3, 1$, and 1/3, in Fig. 3.3.

On the other extremum, when the NFET characteristic is like a linear voltage-controlled resistor,

$$I_N = 0 \text{ when } V_G < V_{THN} \tag{3.7}$$

$$I_N = B_N(V_G - V_{THN})V_D \quad \text{when} \quad V_G > V_{THN}$$

and if a similar characteristic is used for the PFET, the static characteristics of an inverter are given by

$$v_D = \frac{1 - v_{THP} - v_G}{1 - v_{THP} - v_G + \beta(v_G - v_{THN})} \tag{3.8}$$

The static characteristics are shown by the dotted curves for $\beta = 3, 1$, and $1/3$, in Fig. 3.3.

Characteristics of a real CMOS inverter are similar to the ones shown by the solid curves in Fig. 3.3, but the other simpler cases are useful for analyzing special problems. Equations (3.5) and (3.7) give linear dependence of FET conductivity on gate voltage, while Eq. (3.1) gives quadratic dependence. In a scaled-down CMOS FET where carrier velocity saturation is significant, the linear dependence is closer to reality. Absence of a simple but physically well-founded formula like Eq. (3.1) in the short-channel, high-field case makes analysis of scaled-down CMOS circuits difficult.

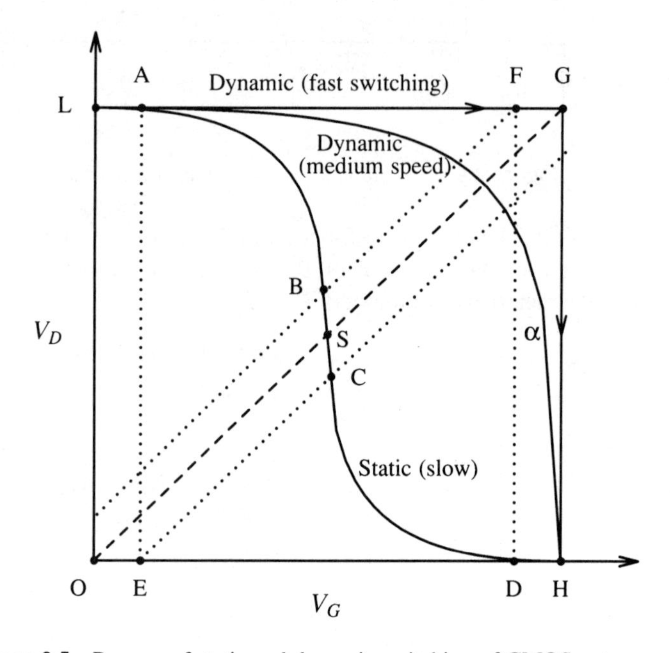

Figure 3.5 Process of static and dynamic switching of CMOS gate.

Simplified formulas such as Eqs. (3.5) and (3.7) are useful, by providing illuminating closed-form results.

Static characteristics of a three-input NAND gate, shown in Fig. 3.1(b), depend on the way the circuit is driven [02]. Let us define a reference inverter that has the same NFET size as the NAND3 gate, and also the same PFET size as the NAND3 gate. If only one input switches, and if the others stay at logic high (V_{DD}) voltage level, the switching threshold voltage of the NAND3 gate is close to, but lower than, the switching threshold voltage of an inverter whose beta ratio is reduced to one-third of the reference inverter. When all three inputs switch simultaneously, however, the switching threshold is close to that of an inverter whose beta ratio is reduced by a factor of $(1/3)^2$ of that of the reference inverter.

When the input signal of a gate switches very slowly from ground to V_{DD}, the state of the gate follows the static characteristic of Fig. 3.5, from point L to A, B, C, D, and finally to H. When the input signal switches very fast, however, the output node has no time to respond. The state of the gate follows points L, A, F, G, and H. The intermediate input switching speed curve α of Fig. 3.5 represents the states of the gate during switching. We call the curve a state-of-the-gate curve.

A CMOS static gate breaks into two pieces at the output node (Section 2.4). Between the output node and V_{DD} there are only PFETs. If the output node stays at V_{DD} while the input signal is in a rapid low-to-high transient, there is no current from the V_{DD} supply to ground. When the input signal settles at the high logic level the gate still draws no current, since all PFETs are off. Therefore, switching along line segments L, A, F, G, H dissipates no power from V_{DD}. All the dissipated power comes from energy stored in the capacitive load (capacitive load is assumed connected between the output node and ground). Any state of a gate represented by a point that falls within rectangle A, E, D, F carries extra current upon switching. As the state-of-the-gate curve deviates from the fast transit characteristic L, A, F, G, H the gate expends additional power. The state-of-the-gate curve is convenient to evaluate the power efficiency of the circuit.

3.3 SMALL-SIGNAL AND LARGE-SIGNAL SWITCHING DELAYS

The state-of-the-gate curve is useful to evaluate small-signal characteristics of CMOS circuits. An inverter DC-biased at the static operating point S of Fig. 3.5, which is the intersection of the static characteristic and line $\overline{OG}$, can be used as a small-signal amplifier. The circuit of Fig. 3.1(c) sets the inverter to operating point S.

The switching delay of CMOS inverters (large-signal) and small-signal amplifiers are correlated by the following observation [04]. Since the capacitive loads of the linear and the digital circuits are the same, the difference of the two circuits is the current available to drive the capacitive load. Let the current of the NFET when $V_G = V_D = V_{DD}$ be $I_{D\,\mathrm{MAX}}$. This is the maximum current available from the device. Digital switching delay time T_D is defined as follows: If input of the inverter switched at $t = 0$, the output of the inverter crosses $V_{DD}/2$ at $t = T_D$. The digital switching delay time is then given by

$$T_d(\text{digital}) = \frac{f\ C_L V_{DD}}{I_{D\,\mathrm{MAX}}} \tag{3.9}$$

where f is a factor needed to approximate the averaged current during switching of the FET, and is in the range 1 to 2.

The small-signal amplifier of Fig. 3.1(c) is biased at the operating point S where the amplifier has the maximum transconductance. The transconductance is the sum of the PFET and the NFET transconductance. Assuming that the PFET contributes as much as the NFET, and that the transconductance of the FET is proportional to the gate voltage as in a scaled-down CMOS FET, the small-signal transconductance of the amplifier is given by

$$g_m = g\frac{I_{D\,\mathrm{MAX}}}{V_{DD}}$$

where g is a factor on the order of unity. If the linear amplifier has a voltage gain G, an input signal of amplitude ΔV_G is amplified to $G\,\Delta V_G$. The current generator $g_m \Delta V_G$ deposits charge required to generate $G\,\Delta V_G$, $G\,\Delta V_G C_L$ within time

$$T_D\,(\mathrm{linear}) = G\frac{C_L V_{DD}}{g I_{D\,\mathrm{MAX}}} \tag{3.10}$$

Comparison of Eqs. (3.9) and (3.10) shows that the linear amplifier has essentially the same delay for unity gain, $G = 1$. If G is larger, however, the extra charge is expended to produce a larger signal amplitude at the output, and therefore the linear amplifier has a proportionally longer delay. This result is known as the gain-bandwidth limit of a linear amplifier [05]. This is an important point in CMOS digital circuit design. Since CMOS logic signal amplitude is much larger than that of TTL, ECL, or GaAs logic, logic level conversion from small-amplitude logic to CMOS logic takes a long time.

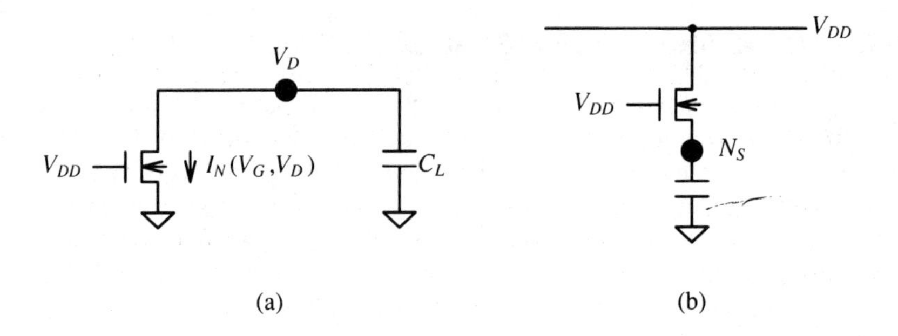

Figure 3.6 Basic process of switching.

3.4 BASIC SWITCHING-TIME ANALYSIS

There are two basic modes of circuit operation that are relevant to CMOS gates. They are the processes of charging and discharging a capacitor through an NFET, as shown schematically in Figs. 3.6(a) and (b). In this section the delay for both modes is found in closed form, using a simplified FET model [Eqs. (3.1a) and (3.1b)]. The complementary cases for PFETs have the same form.

(A) Discharging a Capacitor Through an NFET

In Fig. 3.6(a) we assume that the capacitor is originally charged to V_{DD} and that the gate voltage of the FET makes a step transition from zero to V_{DD} at time $t=0$. This problem was first analyzed by Burns [02]. The basic circuit equation satisfied by the voltage of the capacitor, V_D, is

$$C_L \frac{dV_D}{dt} = -I_N(V_{DD}, V_D) \tag{3.11}$$

The equation is solved subject to the initial condition

$$V_D(0) = V_{DD}$$

The charge deposited on C_L by the gate-to-drain capacitive coupling of the FET is neglected. This feedforward effect is discussed in Section 4.9. Current $I_N(V_{DD}, V_D)$ is given by the gradual, long-channel FET model [Eqs.(3.1a) and (3.1b)] as

$$I_N(V_{DD}, V_D) = B_N \left\{ V_{DD} - V_{THN} - (1/2)V_D \right\} V_D \quad \text{for } V_D < V_{DD} - V_{THN}$$

and

$$= (1/2)B_N(V_{DD} - V_{THN})^2 \quad \text{for } V_D > V_{DD} - V_{THN}$$

Until the output voltage drops from V_{DD} to $V_{DD} - V_{THN}$, $V_D(t)$ is determined from

$$V_D(t) = V_{DD} - \frac{B_N(V_{DD} - V_{THN})^2 t}{2C_L} \tag{3.12}$$

Equation (3.12) is valid when $0 \le t \le t_S$, where t_S is given by

$$t_S = \frac{2C_L V_{THN}}{B_N(V_{DD} - V_{THN})^2} = \left[\frac{2V_{THN}}{V_{DD} - V_{THN}} \right] t_0 \tag{3.12a}$$

where

$$t_0 = \frac{C_L}{B_N(V_{DD} - V_{THN})} \tag{3.12b}$$

For $t > t_S$, Eq. (3.11) is,

$$C_L \frac{dV_D}{dt} = -B_N \left\{ V_{DD} - V_{THN} - (1/2)V_D \right\} V_D$$

This equation is solved subject to the initial condition

$$V_D(t_S) = V_{DD} - V_{THN}$$

and the solution is

$$V_D(t) = (V_{DD} - V_{THN}) \frac{2\exp\left[\dfrac{t_S - t}{t_0}\right]}{1 + \exp\left[\dfrac{t_S - t}{t_0}\right]} \quad (3.13)$$

Equations (3.12) and (3.13) are plotted in Fig. 3.7 by the solid curve. When $t \to \infty$,

$$V_D(t) = 2(V_{DD} - V_{THN}) \exp\left[\frac{t_S - t}{t_0}\right]$$

The switching delay time is determined by solving $V_D(T_D) = (1/2)V_{DD}$ as

$$T_D = t_0 \left\{ 2\frac{v_{THN}}{1 - v_{THN}} + \log(3 - 4v_{THN}) \right\} \quad (3.14)$$

where $v_{THN} = V_{THN}/V_{DD}$.

(B) Charging a Capacitor Through FET

The problem of charging a capacitor by an NFET as shown in Fig. 3.6(b) is more involved, since the threshold voltage of an NFET depends on the source voltage relative to the substrate. To make the problem manageable, this back-bias effect is neglected in the following analysis. The result is therefore only semiquantitative. The gate voltage makes a low-to-high transition at $t = 0$, and the voltage of the capacitor is zero at $t = 0$. The equation satisfied by the source voltage V_S is

$$C_L \frac{dV_S}{dt} = I_N(V_{DD} - V_S, V_{DD} - V_S) = (1/2)B_N(V_{DD} - V_S - V_{THN})^2 \quad (3.15)$$

The solution of this equation that satisfies $V_S(0) = 0$ is

$$V_S(t) = (V_{DD} - V_{THN}) \frac{\dfrac{t}{2t_0}}{1 + \dfrac{t}{2t_0}} \quad (3.16)$$

where $t_0 = C_L / B_N(V_{DD} - V_{THN})$ was defined before. Equation (3.16) was derived by Crawford [06]. Equation (3.16) is plotted in Fig. 3.7 by the dotted curve. When the NFET has no leakage current the maximum voltage is $V_{DD} - V_{THN}$. If the back-bias effect is included, V_{THN} must be larger than the threshold voltage for zero source to substrate voltage. If $V_{THN}(V_S)$ is the threshold voltage of an NFET when the source voltage is V_S, the maximum voltage $V_{S\max}$ is determined from the implicit equation

$$V_{DD} - V_{THN}(V_{S\max}) = V_{S\max}$$

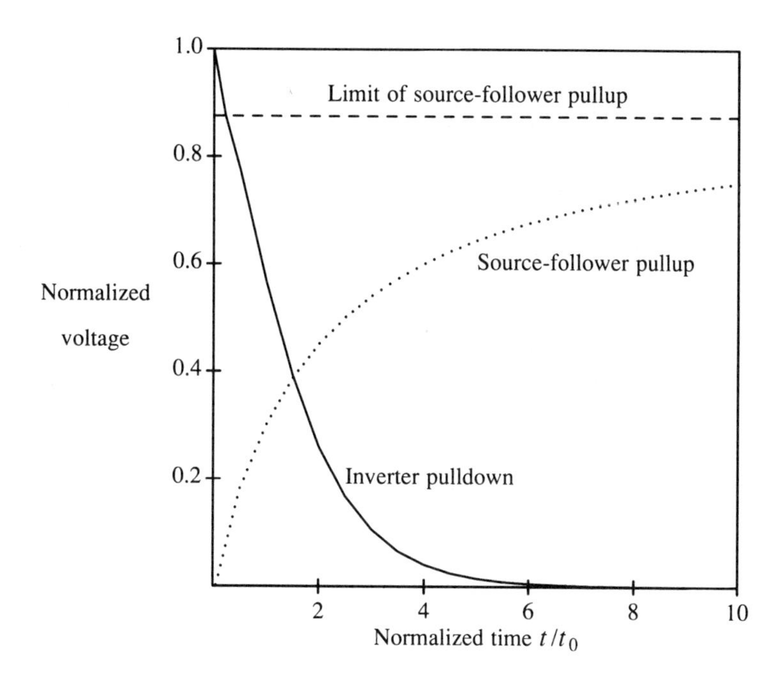

Figure 3.7 Charge-discharge waveform of capacitor through FET.

Note that charging and discharging a capacitor through a PFET are analogous to discharging and charging a capacitor through an NFET, respectively, and the same equations apply with substitution of the PFET parameters and V_D replaced with $V_{DD} - V_D$.

3.5 SWITCHING OF AN INVERTER

If the input voltage of the inverter does not switch instantly, the analysis becomes complex. Only an outline of the method of analysis is given here. The input voltage $V_G(t)$ is assumed to be

$$V_G(t)=0 \ \ (t<0), \ \ =\alpha t \ \ (0<t<t_{IS}=V_{DD}/\alpha), \text{ and } \ \ =V_{DD} \ \ (t>t_{IS}) \qquad (3.17)$$

The input voltage is shown by curve V_{IN} of Fig. 3.8. The five phases of switching; I, II, III, IV, and V follow in sequence. In phase I, the PFET is in the triode region, and the NFET is in saturation. In phase II, both FETs are in saturation. In phase III-V the NFET is in the triode region and the PFET is in saturation. Only the analysis in phase I is given to show the detail.

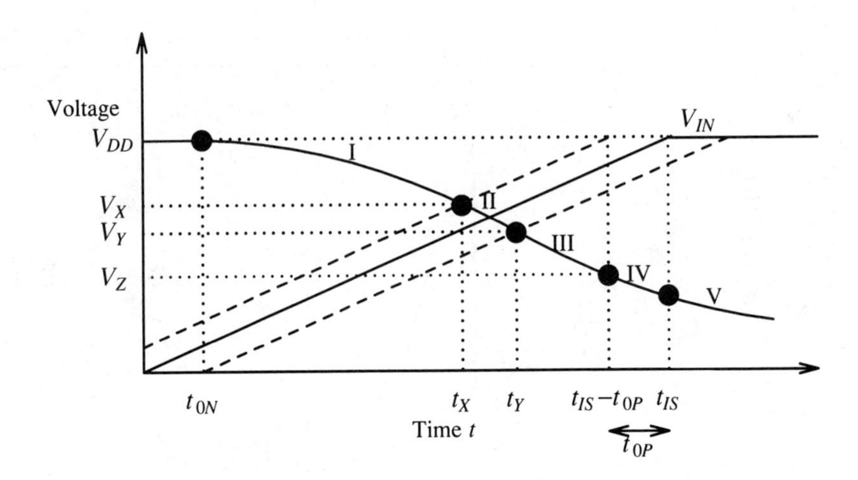

Figure 3.8 Analysis of gate delay, including input voltage waveform.

The circuit equation is

$$C_L \frac{dV_D}{dt} = -\frac{B_N \alpha^2}{2}(t - t_{ON})^2 + B_P \alpha \left\{ t_{IS} - t - t_{OP} - \frac{1}{2\alpha}(V_{DD} - V_D) \right\} (V_{DD} - V_D) \quad (3.18)$$

where $t_{ON} = V_{THN}/\alpha$ and $t_{OP} = V_{THP}/\alpha$. For convenience we substitute $\phi = V_{DD} - V_D$. The initial condition to Eq. (3.18) is

$$\phi = 0 \text{ when } t = t_{ON}$$

We use $\theta = t - t_{ON}$ as the independent variable,

$$C_L \frac{d\phi}{d\theta} = \frac{B_N \alpha^2}{2}\theta^2 - B_P \alpha (t_{IS} - t_{OP} - t_{ON} - \theta - \frac{\phi}{2\alpha})\phi \quad (3.19)$$

This equation is solved subject to the initial condition $\phi = 0$ when $\theta = 0$. We use the method of power-series expansion [07]. If we approximate ϕ by θ^ρ, the left-hand side of the equation is of the form $C_L \rho \theta^{\rho - 1}$, while the right-hand side consists of the terms of the form

$$\theta^2, \ \theta^\rho, \ \theta^{\rho+1}, \ \text{and } \theta^{2\rho}$$

Therefore, $(\rho - 1)$ can be $2, \rho, \rho + 1,$ *or* 2ρ, of which only 2 is consistent. We then have ϕ of the form

$$\phi = f_0\theta^3 + f_1\theta^4 + f_2\theta^5 + \cdots \quad (3.20)$$

By substituting ϕ into Eq. (3.19) and equating the coefficient of the terms having the equal powers of θ, we obtain the following results:

$$f_0 = \frac{B_N \alpha^2}{6C_L}$$

$$f_1 = -B_P \alpha (t_{IS} - t_{0P} - t_{0N}) \left(\frac{B_N \alpha^2}{24 C_L{}^2} \right)$$

and

$$f_2 = (B_P \alpha)^2 (t_{IS} - t_{0P} - t_{0N})^2 \left(\frac{B_N \alpha^2}{120 C_L{}^3} \right) + \frac{B_N B_P \alpha^3}{30 C_L{}^2}$$

and after some algebra,

$$V_D(t) = V_{DD} - \frac{V_{DD}}{6} \left(\frac{V_{DD}}{\alpha t_N} \right) \left[\left(\frac{\alpha \theta}{V_{DD}} \right)^3 - \frac{1}{4} \left(\frac{V_{DD}}{\alpha t_P} \right) \left(\frac{\alpha \theta}{V_{DD}} \right)^4 + \dots \right] \tag{3.21}$$

where $t_N = C_L / B_N V_{DD}$ and $t_P = C_L / B_P V_{DD}$, and assuming $V_{DD} \gg V_{THN}, V_{THP}$. The output node voltage decreases as $(t - t_{0N})^3$ (this result assumes that there is no gate-drain capacitive coupling). This dependence originates from the NFET current increasing with time as $(t - t_{0N})^2$. Switching period I ends when time t_X is reached, where t_X is determined by

$$V_X = V_D(t_X) = V_G(t_X) + V_{THP}$$

At this time the PFET and NFET are both in saturation, and phase II begins. Voltage V_X and time t_X are determined by solving the equations, and the values provide the initial condition to phase II. Analysis for phase II can be carried out in the same way. A power-series solution of a circuit equation is powerful in many problems of CMOS digital circuits, and therefore an example was shown here. The complexity of this particular result, however, prevents developing the gate delay theory by directly solving the circuit equations, except for a few special cases.

3.6 SWITCHING OF AN INVERTER IN THE CONDITION OF CARRIER VELOCITY SATURATION

In this section the switching delay problem is solved using a more realistic FET model than that used in Section 3.4. Carrier velocity saturation causes reduced saturation voltage and reduced saturation current of a short-channel FET (see Section 1.10) [08]. After the switching of the input voltage from ground to V_{DD} at time $t = 0$, the current of the NFET I_N is given by

$$I_N(V_{DD}, V_D) = I_{D\max} = \frac{B_N}{2} V_{D\max}{}^2 \quad \text{for } V_D > V_{D\max} \tag{3.22}$$

where

$$V_{D\max} = V_0 \left[\left(1 + 2 \frac{V_{DD} - V_{THN}}{V_0} \right)^{1/2} - 1 \right] \tag{3.23}$$

and

$$I_N(V_{DD}, V_D) = \frac{B_N}{1 + \left(\frac{V_D}{V_0}\right)} \left\{ V_{DD} - V_{THN} - (1/2)V_D \right\} V_D \quad \text{for } V_D < V_{D\max} \quad (3.24)$$

Inverter delay-time analysis based on Eqs. (3.22) to (3.24) was reported by Mayer [09]. They used approximation; here the problem is solved without approximation.

Since the switching delay is defined by the time for V_D to drop from V_{DD} to $V_{DD}/2$, whether or not $V_{D\max} > V_{DD}/2$ is important. If $V_{D\max} < V_{DD}/2$, the NFET is a time-independent current generator ($I_{D\max}$) through the initial phase of switching, and therefore the problem is simplified. Let $V_{D\max} = V_{DD}/2$ when V_0 is V_{0C}, where V_{0C} is determined from the equation

$$V_{D\max} = \frac{1}{2} V_{DD}$$

as

$$V_{0C} = \frac{1}{1 - 2v_{THN}} \left(\frac{V_{DD}}{4}\right) \quad (3.25)$$

1. When $V_0 < V_{0C}$ (strong velocity saturation), the drain voltage $V_D(t)$ is given by

$$V_D(t) = V_{DD} - \frac{I_{D\max}}{C_L} t$$

and the delay time is given by

$$T_D = \frac{1}{2} \frac{C_L}{I_{D\max}} V_{DD} = \left(\frac{C_L}{B_N V_{DD}}\right) \frac{1}{{v_0}^2} \frac{1}{\left[\left\{1 + 2\left(\frac{1 - v_{THN}}{v_0}\right)\right\}^{1/2} - 1\right]^2} \quad (3.26)$$

where $v_0 = V_0/V_{DD}$.

2. When $V_0 > V_{0C}$ (weak velocity saturation), the NFET is in saturation until time t_S is reached, where t_S satisfies $V_D(t_S) = V_{Dmax}$. Time t_S is given by

$$t_S = C_L \left(\frac{V_{DD} - V_{D\max}}{I_{D\max}}\right) = \left(\frac{C_L}{B_N V_{DD}}\right) \Phi_0(v_{D\max}) \quad (3.27)$$

where

$$\Phi_0(v_{D\max}) = \frac{2(1 - v_{D\max})}{{v_{D\max}}^2}$$

where $v_{D\max} = V_{D\max}/V_{DD}$. When $V_0 \to \infty$, $V_{D\max} \to V_{DD} - V_{THN}$, and $t_S \to (2C_L/B_N V_{DD})[v_{THN}/(1 - v_{THN})^2]$, which is the same result as in Section 3.4. After time t_S, V_D satisfies the equation

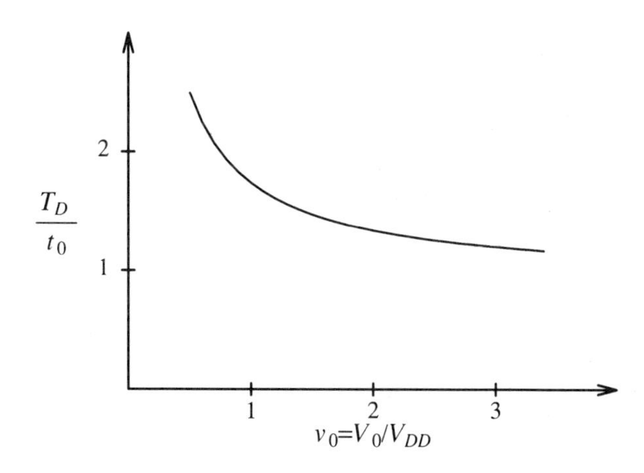

Figure 3.9 Effects of carrier velocity saturation on gate delay time.

$$C_L \frac{dV_D}{dt} = -I_N(V_{DD}, V_D) = -\frac{B_N}{1 + \left(\frac{V_D}{V_0}\right)} \left\{ V_{DD} - V_{THN} - (1/2)V_D \right\} V_D \quad (3.28)$$

This equation can be solved by separation of variables. The initial condition is $V_D = V_{D\max}$ at $t = t_S$ and the solution is

$$t - t_S = \left(\frac{C_L}{B_N V_{DD}} \right) \Phi_1(v_D, v_0, v_{D\max}, v_{THN}) \quad (3.29)$$

where

$$\Phi_1(v_D, v_0, v_{D\max}, v_{THN})$$

$$= \frac{1}{1 - v_{THN}} \log \left[\frac{\left\{ 2(1 - v_{THN}) - v_D \right\} v_{D\max}}{\left\{ 2(1 - v_{THN}) - v_{D\max} \right\} v_D} \right] + \left(\frac{2}{v_0} \right) \log \left[\frac{2(1 - v_{THN}) - v_D}{2(1 - v_{THN}) - v_{D\max}} \right]$$

The delay is obtained by setting $v_D = 1/2$ in Eq. (3.29). When $v_0 \to \infty$, Eq. (3.29) converges to Eq. (3.13). Figure 3.9 shows the normalized delay time versus v_0. Delay time is normalized to t_0, given by Eq. (3.12b). Parameter v_{THN} is 0.12. When $v_0 \to 0$ the delay is proportional to $1/v_0$.

3.7 GATE CONFIGURATIONS

Before we study the delay time of CMOS static gates in detail, several points must be clearly stated.

1. A CMOS static gate is not uniquely defined when a logic symbol, as shown in Fig. 3.10(a), is given. The gate has two structural variations, shown in Figs. 3.10(b) and 3.10(c). The two variations have different delays. A circuit-level diagram must be specified.

2. Logically equivalent inputs are different in terms of circuits. The two inputs A_1 and A_2 of Fig. 3.10(a) are logically equivalent, but the delays from the two inputs are different.

3. Pullup and pulldown delays of a gate are different.

4. When the AOI23 gate of Fig. 3.10(b) or (c) pulls up by switching one of the B inputs, the pullup delay depends on whether A_1 or A_2, or both of them, are low. The delay is shorter if both A_1 and A_2 are low. Therefore, the states of the other inputs must be specified, in a multi-input gate.

5. If more than one input switches simultaneously, the delay is different from the delay when only one input switches and the other inputs stay at the stable logic levels. In the case of simultaneous switching, the delay depends on the mutual timing of the two inputs.

6. The delay depends on the transition time and waveform of the input.

7. In Section 2.4 we noted that a CMOS static gate has a number of internal nodes that do not represent Boolean functions of the input variables. These nodes reflect the input variables of the past as well as the present. The definition of delay time is subject to the specification of the state of the gate before switching. This problem will be discussed in the next section.

3.8 PRE-SWITCHING STATES OF A GATE

Let us consider the state of internal node X of the CMOS NOR2 gate shown in Fig. 3.11(a). This node is able to take a very wide range of potential, depending on the past history of the gate.

Node X can be V_{DD} volts if input B is high and input A is low. Starting from this state, if input A makes a low-to-high transition, the capacitive coupling from the gate to the drain of PFET MPA pulls the voltage of node X above V_{DD}.

If PFET MPB is leaky but MPA is not, node X discharges to the voltage that depends on the time elapsed since the switching of input A: The minimum is the ground potential. Suppose that input B makes a high-to-low transition after the discharge. Node X is then pulled down at least to V_{THP} (the threshold voltage of PFET MPB, including back-bias effect), but the voltage can be lower.

As shown in this example, voltages of internal nodes prior to switching are very uncertain [10]. If the gate is complex, the uncertainty in the voltages of internal nodes is significant, and the uncertainty results directly in an uncertainty of the switching delay. It is obviously impossible to analyze all the possibilities. The conventional

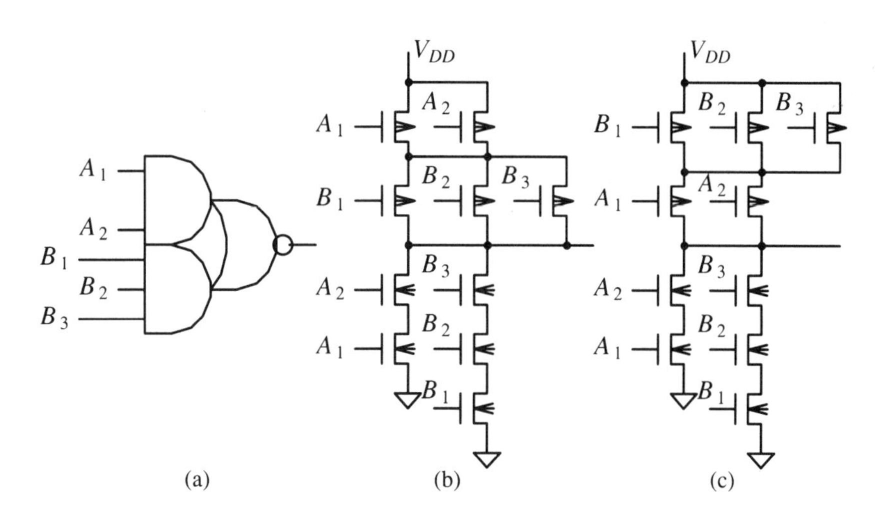

Figure 3.10 Two different structures of AOI23 gate.

approach to avoid this problem is to seek the worst case. Although worst-case analysis is sometimes acceptable, the worst-case results are often useless, especially when precision timing design is required. This important point is not universally recognized. Furthermore, worst-case analysis often disregards important details of gate switching mechanisms, and therefore it is not a productive way of fundamental research. To carry out a realistic delay analysis, the most realistic initial state for each problem must be selected from the set of possible initial states. The following line of thought leads to the most realistic choice.

Suppose that a CMOS static gate pulled up and the gate delay time was T_D. The gate subsequently pulls down, after time T_C. In a high-performance CMOS VLSI chip, $T_C > T_D$, but T_C may be only one or two orders of magnitude longer than T_D. This is equivalent to saying that a gate is busy. This condition is well satisfied in clock circuits that require precision delay analysis, but not in a datapath logic (that can be designed using worst-case delay analysis). Since the leakage current is controlled to be three or more orders of magnitude less than the channel current of an FET (Section 1.11), the effects of leakage current can be neglected in a busy gate. Then the initial condition can be determined without including the effects of leakage current.

Let us study how the state of a complex gate before switching depends on the way the input signals were set up. An example of a complex gate is shown in Fig. 3.11(b). Some of the representative initial conditions are listed in Table 3.1. In the Table "L" stands for ground, "H" for V_{DD}, "X" for $V_S = (V_{DD} - V_{THN})$, and "?" for an unknown voltage that depends on the past history of the gate. The postfixed (+, –) symbols and "0" are explained shortly. Let us first ignore the effects of capacitive coupling.

TABLE 3.1 STATES OF INTERNAL NODES

	Input State							States of Nodes						
Case	A	B	C0	C1	C2	D	E	N0	N1	N2	N3	N4	N5	O
1	L	L	L	L	L	H	0	?	?	L	H	H	H	H
2	L	L	L	L	L	0	H	?	?	X	H	H	H	H
3	L	L	L	L	L	0	0	?	?	?	H	H	H	H
4A	L	H	H	H	H	L	L	X	X	?	H	?	?	H
4B	L	L	H	H	H	L	L	X-	X	?	H	?	?	H
4C	L	L	L	L	L	L	L	X-	X-	?	H	H	H	H
4D	L	L	H	L	L	L	L	X-	X	?	H	H	H	H
4E	L	L	H	L	H	L	L	X-	X	?	H	H+	H+	H
4F	L	L	H	H	H	L	L	X-	X+	?	H	H+	H+	H
4G	0	H	H	H	H	L	L	X+	X+	?	H	H+	H+	H

Case 1. Input signals A to C_2 are low, and D is high. Table entry "0" for input E represents that input E is logic level low, and the input is about to make a low-to-high transition. The gate works as an inverter from input E to output O. The node N2 voltage is definite and is zero. Although the voltages of nodes N0 and N1 are uncertain, that does not matter for this particular switching delay. There are many irrelevant nodes of this kind when a complex gate switches from only one of the inputs.

Case 2. This is the same as case 1, except that inputs D and E are exchanged. The voltage of node N2 is $V_S = (V_{DD} - V_{THN})$, where V_{THN} is the threshold voltage, including the back-bias effect. Voltage V_S is the highest voltage an internal node N2 is able to take if leakage current does not exist.

Case 3. This is the initial condition before inputs D and E make a low-to-high transition simultaneously. The node N2 voltage is uncertain. If this initial condition is set up first by effecting a high-to-low transition of input D, and then that of input E, the voltage of node N2 is V_S. If the order is reversed, the voltage of node N2 is zero.

The voltages of nodes N0 and N1 can be V_S or zero for the same reason: The voltage of node N0 is zero if input B goes low before input A, and the voltage is V_S if input B and at least one of the inputs C0 to C2 goes low after input A.

Cases 1 to 3 neglect leakage currents and capacitive coupling.

Case 4. If the effect of capacitive coupling from the gate of the FET is included, the initial condition becomes more complex. Assume that inputs B, $C0$, $C1$, and $C2$ are high, inputs A, D, and E are low, and input B makes a high-to-low transition. Node N0 was originally at potential V_S. If input B makes a high-to-low transition, node N0 is partially discharged by the capacitive coupling. This reduced voltage is

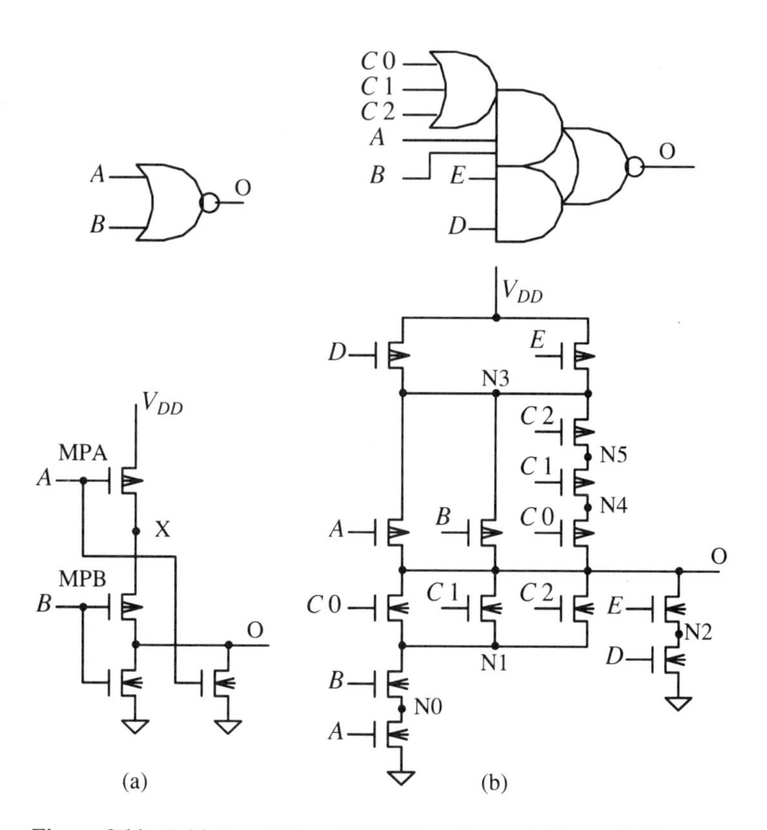

Figure 3.11 Initial condition of CMOS static gate before switching.

shown by "X-" in Table 3.1. If the transition occurs slowly, however, the lost charge is partially recovered by the channel current that flows before the FET turns off completely. If the gate of a FET that is completely isolated from the rest of the circuit by nonconducting FETs is driven up or down, the source and the drain node voltages are influenced significantly. Nodes N4 and N5 could suffer from this effect. The sequence 4A to 4G in the table provides an initial condition to upgoing transition at input A. In this example nodes N0 and N1 can be driven above V_S and nodes N4 and N5 above V_{DD}.

We observe many such cases, and consider what is the typical initial condition for the pulldown of a gate. We arrive at the following conclusion. All the nodes in the PFET circuit are charged to V_{DD}, and all the nodes in the NFET circuit that are not directly connected to the ground by a chain of conducting NFETs are charged to voltages in the range from V_S to V_{DD}. The delay of a gate in a VLSI chip cannot be defined more precisely than the uncertain initial conditions dictate. Therefore, this typical case provides practically the best idea about the delay.

3.9 THEORY OF DELAY TIME OF CMOS STATIC GATES

As was observed in the preceding two sections, gate-delay analysis involves many parameters, and the mathematical procedure is very complex. The accurate theory of Sections 3.4 to 3.6 gives unmanageably complicated results that can be handled only by a computer. Work using numerical analysis has been reported [11].

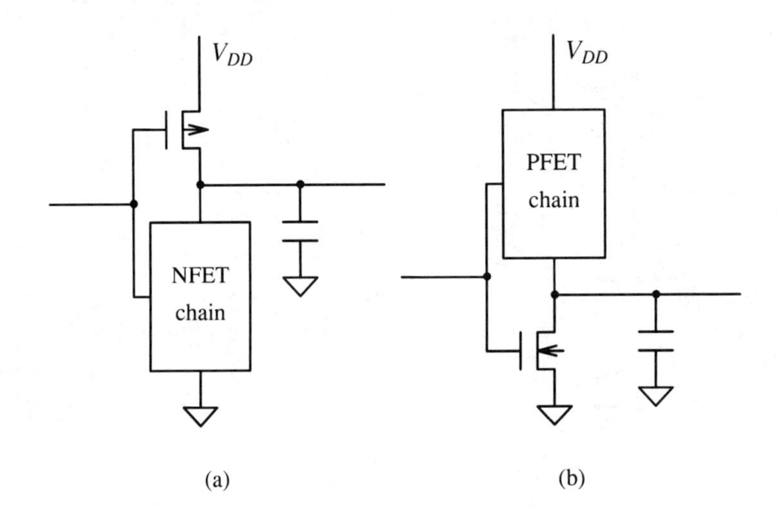

Figure 3.12 Two models of CMOS static gate.

Practically, the delays were calculated by a circuit simulator. What happens in a circuit, however, cannot be directly understood from the simulator's numerical results. Here we follow yet another circuit-theoretical approach that correlates the circuit structure of a gate (the ways the FETs are connected) directly to the delay. This theory is useful in developing the algorithm of a gate-level delay simulator, but the present objective is to understand the mechanisms of switching of a gate in detail.

Work by Wu et al. uses basically the same approach [12]. Their results are, however, very complex because they attempted to correlate gate delay directly with device parameters. This is a refinement we do not need now, since here we wish to study how gate delay is determined by gate circuit structure.

We consider the pulldown delay only, since the pullup delay is calculated in exactly the same manner. The two gate structures to be distinguished are (1) the case where the pulldown NFET chain is quite long but the pullup PFET chain is short, as shown in Fig. 3.12(a), and (2) the case where the pulldown NFET chain is quite short but the pullup PFET chain is long, as shown in Fig. 3.12(b). Two distinct theories are needed, and they are discussed in the following sections.

The basic idea for simplification is to approximate the nonlinear, time-dependent current-voltage characteristics of a FET by a simple, linear, time-independent relationship. This is equivalent to substituting a linear resistor for the device. Although this approach is an intuitive one, a detailed examination shows that there are a number of fundamental problems that must be resolved before we have a well-defined model. Above all, how to define the equivalent linear resistance value of a FET must be clearly stated, and an estimate of the accuracy must be obtained. The problem becomes too complex if the dependence of gate delay on the rate of change of the input voltage is included in the analysis. Therefore, we assume that the gate input voltage switches instantly. Delay dependence on the input slew rate is discussed in Section 3.16.

3.10 MECHANISM OF PULLDOWN OF AN NFET CHAIN

The mechanism of pulldown of a long NFET chain is as follows. In Fig. 3.13(a), the voltage of node C can be V_{DD}, $V_S = V_{DD} - V_{THN}$, or zero, depending on the initial condition. When the voltage is zero, NFET MND is in state 1 shown in the I-V curve of Fig. 3.13(b). The FET MND conducts current that moves the charge of node O to node C. Switching of the output node starts by redistributing the charge between these nodes of the gate.

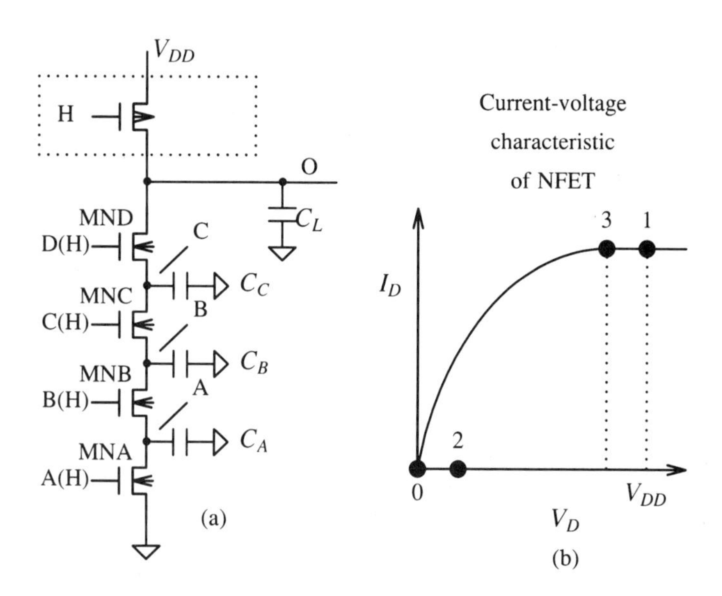

Figure 3.13 State of FET before switching.

When the voltages of nodes A, B, ..., C are V_{DD} or $V_{DD} - V_{THN} = V_S$, MND is in state 0 or in state 2 of Fig. 3.13(b), respectively. State 0 is a cutoff (nonconducting) state: Even if there is a small change in the source voltage, FET MND remains in the cutoff state. State 2 is a state of saturation, but the saturation current is zero. If the source voltage decreases, FET MND conducts immediately. In either state, FET MND does not conduct until node C is pulled down below voltage V_S. Until MND conducts, the voltage of node O is unchanged.

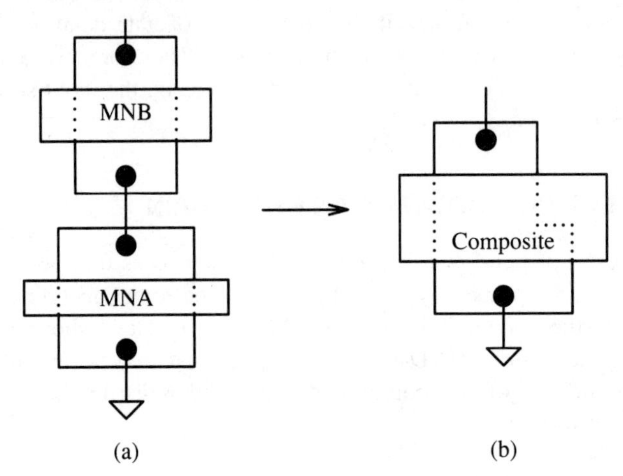

Figure 3.14 Merging FET in the process of pulldown switching.

The worst-case (slowest switching) initial condition is that all internal nodes (A to C) are charged to V_{DD}. If the NFET chain begins to discharge by pulling up the gate of NFET MNA, this is the initial condition. In Fig. 3.13(a), NFET MNA is in state 1, conducting current. Until node A is pulled down to V_S by NFET MNA, NFET MNB, which was in state 0, does not conduct. This process takes time T_{A1}, given by

$$T_{A1} = \frac{C_A (V_{DD} - V_S)}{I_D(A)} \tag{3.30}$$

where $I_D(A)$ is the saturation current of NFET MNA when the gate voltage is V_{DD}, and C_A is the capacitance assigned to node A. C_A consists of such components as the gate capacitance of MNA (including the channel-to-substrate capacitance), the drain island capacitance of MNA, and the source island capacitance of MNB. When the voltage of node A is V_S, NFET MNB starts conducting and node B begins to discharge. Node B is pulled down to V_S after time T_{B1}, where

$$T_{B1} = \frac{C_B (V_{DD} - V_S)}{I_D(A, B)} \tag{3.31}$$

where $I_D(A, B)$ is the saturation current of a FET that is a composite of the two NFETs MNA and MNB, as shown in Fig. 3.14(a) and (b).

Equation (3.31) neglects the effects of capacitance C_A that is internal to the composite FET. This approximation is good, since the current that flows in the upper half-channel of the composite FET (where the channel was originally pinched) mainly determines delay T_{B1}. After time T_{B1}, node C begins to discharge. In the same way node D begins to discharge after time

$$T_{C1} = \frac{C_C(V_{DD} - V_S)}{I_D(A, B, C)} \tag{3.31a}$$

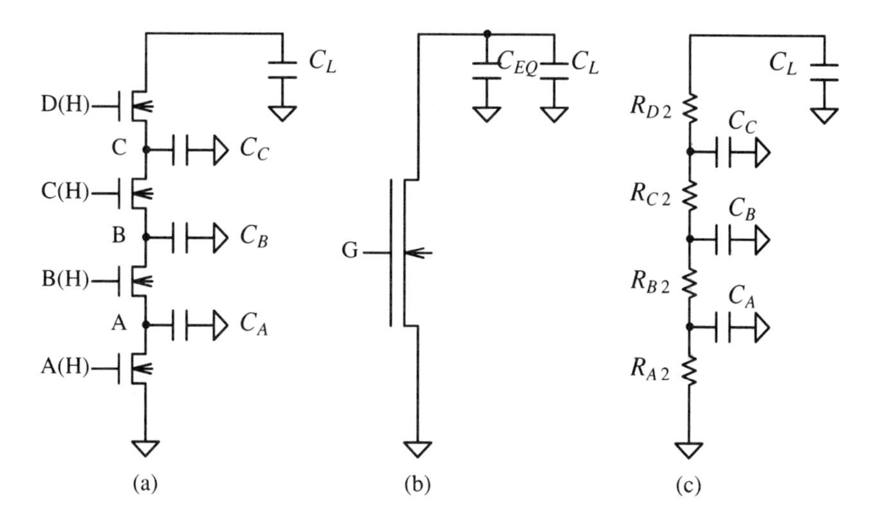

Figure 3.15 Switching process of NFET pulldown chain.

where $I_D(A, B, C)$ is the saturation current of the composite FET constructed by joining MNA, MNB, and MNC. Therefore, the output node O of an FET chain shown in Fig. 3.13(a) discharges to V_S when

$$T_D(1) = T_{A1} + T_{B1} + T_{C1} + \ldots + T_{X1} \tag{3.32}$$

where the suffix X stands for the topmost NFET of the FET chain.

Let the width of the FETs MNA, MNB, ... be $W_A, W_B, \cdots$ and the channel length $L_A, L_B, \cdots$, respectively. Then, from Eqs. (1.12) and (1.13),

$$\frac{1}{I_D(A, B, C, \ldots)} = \frac{2T_{OX}}{\varepsilon_{OX}\mu(V_{DD} - V_{THN})^2}\left[\frac{L_A}{W_A} + \frac{L_B}{W_B} + \cdots\right] \tag{3.32a}$$

This relationship is valid if the carriers in the channel have a field-independent mobility, or if the conduction is ohmic. We set

$$R_{A1} = \rho_1 \frac{L_A}{W_A}, \quad R_{B1} = \rho_1 \frac{L_B}{W_B}, \quad \cdots$$

where

$$\rho_1 = \frac{2T_{OX}(1 - v_S)}{\varepsilon_{OX}\mu V_{DD}(1 - v_{THN})^2}$$

and we note that $1 - v_S > v_{THN}$ due to the back-bias effect. We define

$$T_{A1} = R_{A1}C_A$$

$$T_{B1} = (R_{A1} + R_{B1})C_B$$

....

$$T_{X1} = (R_{A1} + R_{B1} + \ldots + R_{X1})C_X$$

where C_X is the load capacitor to the gate $C_L = C_X$. We wrote C_X here to avoid confusion. Then we have

$$T_D(1) = T_{A1} + T_{B1} + \ldots \tag{3.33}$$

$$= R_{A1}C_A$$

$$+(R_{A1} + R_{B1})C_B$$

$$+(R_{A1} + R_{B1} + R_{C1})C_C$$

....

$$+(R_{A1} + R_{B1} + \cdots + R_{X1})C_L$$

We now wrote C_L instead of C_X in Eq. (3.33).

Equation (3.33) has the same structure as Elmore's delay formula [13] [Eq. (6.20)]. This formula is a direct consequence of the linear circuit theory that gives the delay of a cascaded RC chain consisting of resistance $R_{A1}, R_{B1}, \cdots$ and $C_A, C_B, \cdots$. What Eq. (3.33) shows is the following. A set of linear, time-independent resistance values $R_{A1}, R_{B1}, \cdots$ can be defined, which are proportional to the number of squares of their respective FET channels. The resistances and the node capacitances $C_A, C_B, \cdots$ give the delay time until the output node begins to pull down, using conventional linear circuit theory.

The process of arriving at Eq. (3.33) includes merging the discrete FETs of Fig. 3.15(a) one by one from the grounded end until all the conducting NFETs are integrated into the single long-channel NFET shown in Fig. 3.15(b). As the output node begins to pull down, the FET chain is represented by a single, long-channel NFET and the π equivalent capacitance C_{EQ} placed at the output node, which takes the effects of the internal capacitances $C_A, C_B, \cdots$. Capacitance C_{EQ} is used here for convenience of analysis. To carry out the analysis it is only necessary to assume that an equivalent capacitance exists. Practically, it is difficult to define this equivalent capacitance precisely. The delay of the circuit of Fig. 3.15(b) can then be derived in a closed form, using the theory of Section 3.4. The result is given in Eq. (3.14). In the equation, C_L is the total load capacitance,

$$C_L \rightarrow C_L + C_{EQ}$$

and B_N of the long-channel FET is

$$B_N = \frac{\varepsilon_{OX}\mu}{T_{OX}} \left(\frac{L_A}{W_A} + \frac{L_B}{W_B} + \cdots \right)^{-1}$$

and therefore the delay is given by Eq. (3.14) of Section 3.4,

$$T_D(2) = (C_L + C_{EQ})(R_{A2} + R_{B2} + \cdots) \tag{3.34}$$

where

$$R_{A2} = \rho_2 \frac{L_A}{W_A}, \quad R_{B2} = \rho_2 \frac{L_B}{W_B}, \ldots$$

and where

$$\rho_2 = \frac{T_{OX}}{\varepsilon_{OX}\mu(V_{DD} - V_{THN})} \log(3 - 4v_{THN})$$

This equation can be derived easily from Eq. (3.14), by eliminating the first term in the braces.

Equation (3.34) is based on an approximation of the lumped capacitance C_{EQ} at the output node of the gate. The method of determining the equivalent capacitance is an integral part of the circuit model, and the prescription involves approximation. The accuracy of the model can be improved by returning the capacitance $C_A, C_B, \cdots$ to the original node of the resistors, as shown in Fig. 3.15(c). Then,

$$T_D(2) = T_{A2} + T_{B2} + T_{C2} + \ldots + T_{X2} \tag{3.35}$$

$$T_{A2} = R_{A2} C_A$$

$$T_{B2} = (R_{A2} + R_{B2}) C_B$$

$$T_{C2} = (R_{A2} + R_{B2} + R_{C2}) C_C$$

....

$$T_{X2} = (R_{A2} + R_{B2} + \cdots + R_{X2}) C_L$$

Then by adding Eqs. (3.33) and (3.35), we obtain

$$T_D = T_D(1) + T_D(2) \tag{3.36}$$

$$= R_A C_A$$

$$+ (R_A + R_B) C_B$$

$$+ (R_A + R_B + R_C) C_C$$

....

$$+ (R_A + R_B + \ldots + R_X) C_L$$

where

$$R_A = (\rho_1 + \rho_2)\frac{L_A}{W_A}, \quad R_B = (\rho_1 + \rho_2)\frac{L_B}{W_B} \quad \ldots$$

and in this way the equivalent resistance of the NFETs is defined. This analysis shows the nature of approximation involved in the linear resistor model very clearly. To determine the actual value of the equivalent resistor from this analysis, however, may be inconvenient.

3.11 BASIC DELAY DATA

In order to develop the RC chain model of CMOS gates still further, we need to examine extreme gate structures: NAND or NOR gates that have many inputs X. Practically, NAND(X) with $X > 6$ and NOR(X) with $X > 5$ are not used in CMOS VLSI design. If we study only the narrow range of X from 1 to 6, however, the theory inevitably becomes a curve fitting. Our primary objective in the following sections is to understand the switching mechanisms, not to construct a computer model [14].

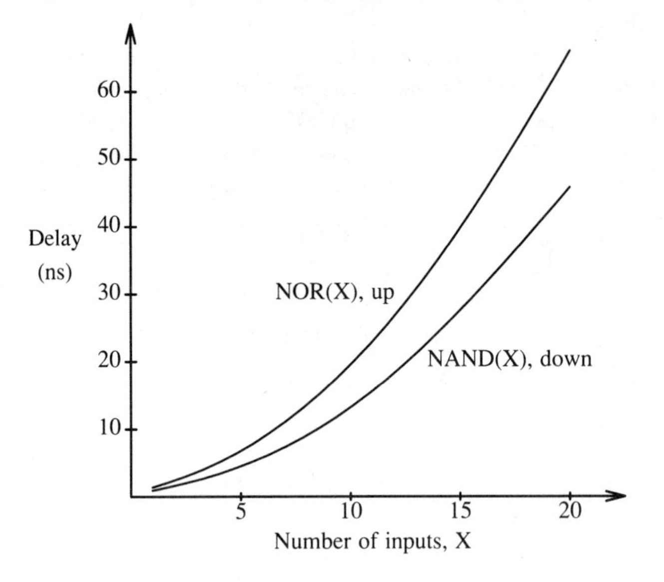

Figure 3.16(a) Pullup delay of NOR gates and pulldown delays of NAND gates versus X.

Figures 3.16(a) and (b) show the delay data of NAND(X) and NOR(X) gates. The delay data were obtained by ADVICE simulations using 1.75-μm CMOS, middle current process, at 105 degrees centigrade, 5.0 volts power supply voltage, and fast input switching speed (0.2 ns from 0 to 100% switching). The NFET width is 4 μm, and the PFET width is 7.25 μm. Three inverters, all of the same FET size, load the output node. NAND(X) with large X requires more time to pull the output node down than to pull it up and NOR(X) requires more time to pull up than to pull down. The delay data for this switching polarity are plotted in Fig. 3.16(a). The input signal drives the gate of the FET closest to the V_{SS} or V_{DD} bus, and all other inputs are set at the logic level that enables the gate. For the other switching polarity the gates switch faster. The data for this switching polarity are plotted in Fig. 3.16(b).

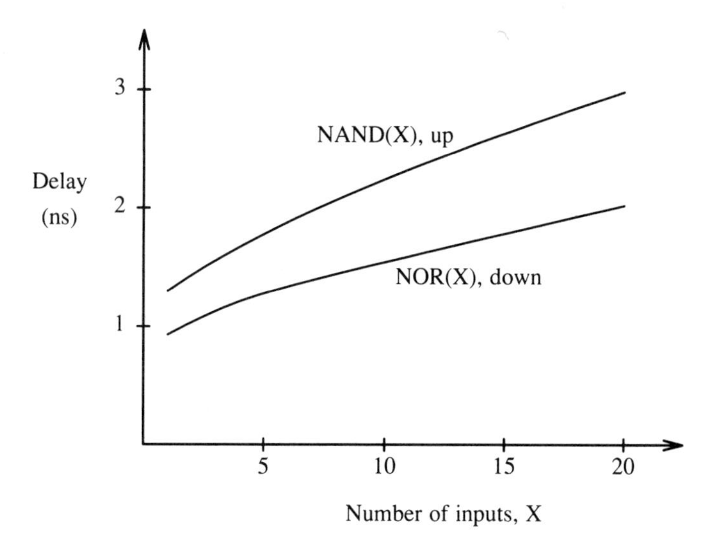

Figure 3.16(b) Pullup delay of NAND gates and pulldown delays of NOR gates versus X.

Two significant characteristics are obvious from Figs. 3.16(a) and 3.16(b). In Fig. 3.16(a), the delay increases quadratically with X, and in Fig. 3.16(b) the delay increases very slowly with X, and almost linearly. The slow increase in the delay in Fig. 3.16(b) is surprising: In NOR(X) of 20 inputs, how is the charge stored in the 20 series-connected PFETs all discharged by a single minimum-size NFET within only 2 ns? The quadratic dependence of Fig. 3.16(a) and the slow dependence of Fig. 3.16(b) will be explained from the theory based on the RC chain model of the gates.

TABLE 3.2 GATE/DRAIN CAPACITANCE (pF)

FET	Gate	Drain
PFET (7.25 μm)	0.0223	0.0185
NFET (4 μm)	0.0113	0.0126

The capacitances of the FETs are tabulated in Table 3.2. The capacitance values are the average values over the range of voltages. The DC characteristics of the FETs

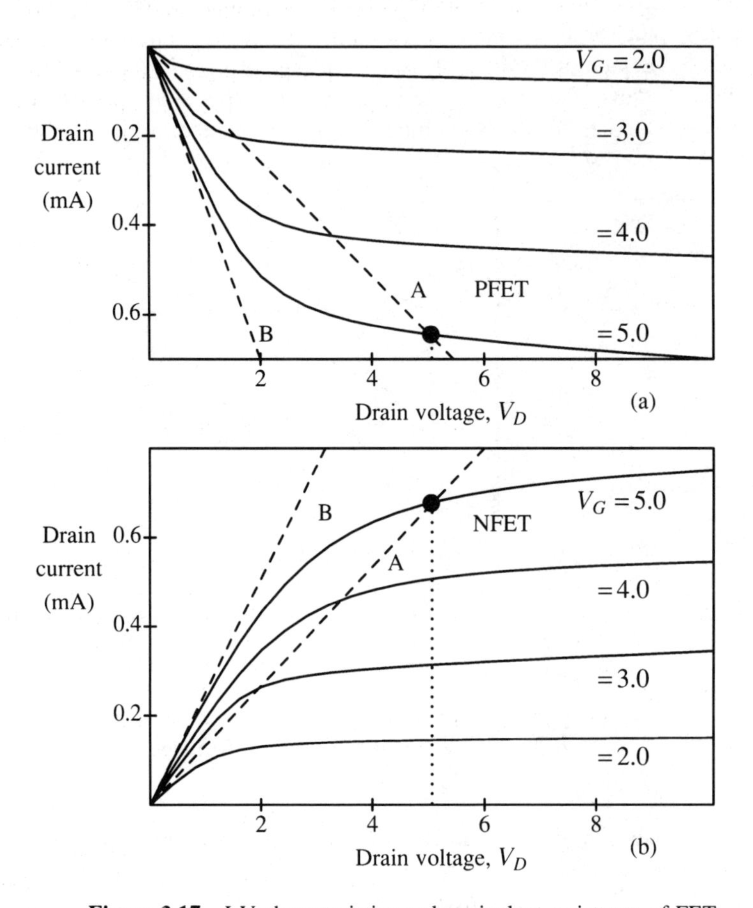

Figure 3.17 I-V characteristics and equivalent resistance of FET.

are shown in Figs. 3.17(a) and 3.17(b). The saturation current and the DC resistances are tabulated in Table 3.3. If a linear resistor is substituted for a FET, the resistance value should be of the order of 10 k Ω. To determine the resistance value from the I-V characteristics of FET is a complicated matter, and this problem is discussed in the next section.

3.12 PHYSICAL MODEL OF DISCHARGE THROUGH AN FET CHAIN

When a NAND(X) gate pulls down, charge on the output node is drained through the chain of series-connected NFETs. A physical model appropriate to predict the discharge process is shown in Fig. 3.18. The NFET chain of Fig. 3.18(a) can be approximated by the cascaded RC chain of Fig. 3.18(b), where R is the equivalent

TABLE 3.3 FET SATURATION CURRENT AND RESISTANCE

	PFET		NFET	
V_G	$I_D(V_D=5)$ (μA)	V_D/I_D (kΩ)	$I_D(V_D=5)$(μA)	V_D/I_D (kΩ)
5	648	7.71	698	7.16
4	431	11.6	508	9.84
3	232	21.6	316	15.8
2	75	66.6	139	35.9
1	0	∞	16	312

resistance of the NFET, and C is the sum of the parasitic capacitances of an NFET given by

$$C = 2C_D + C_G$$

where C_D is the capacitance of the drain diffused island (assumed equal to the capacitance of the source diffused island; drain and source islands of different FETs are accounted for separately), and C_G is the sum of the gate-to-channel capacitance and the channel-to-substrate capacitance. Substrate capacitance was neglected in the following calculation (Section 1.5). Strictly speaking, where and how to assign the parasitic capacitances is a complicated matter, because a FET is a nonlinear device. A detailed analysis based on device physics has recently been published [15]. We do not go into the details here, assuming that the different ways of assigning capacitance creates small errors. C_L is the sum of the capacitive load at the output node, consisting of fanout gate capacitance and the drain capacitances of the pullup PFETs (total number = X) that are all turned off. We assume that the slew rate of the input signal is very high (i.e., the input voltage makes a transition in zero time) and that the input signal is connected to the gate of MN1 of Fig. 3.18(a) (this is the slowest case input). Capacitive coupling from the gate of MN1 to node 1 is neglected, and all the internal nodes of the FET chain are initially charged to V_{DD}. All NFETs have the same size (4 μm), as do the PFETs (7.25 μm).

The time required to discharge node X to $V_{DD}/2$ in the equivalent circuit of Fig. 3.18(b) is given by

$$T_D = \tau_0 \left\{ \left[0.693 \frac{C_L}{C} + 0.323 \right] X + 0.370 X^2 \right\} \tag{3.37}$$

where

$$\tau_0 = CR$$

This formula is derived in Section 6.8.

To use Eq. (3.37), the FET equivalent resistance must be determined. We go back to the definition of equivalent resistance. The equivalent resistance is the DC resistance of the FET averaged over the interval of switching. In an inverter (where the FET chain contains only one FET), the resistance of a FET at the early phase of switching (R_0) is determined by line A of Fig. 3.17, and the resistance at the last phase of switching (R_∞) by line B. R_0 is more relevant than R_∞ to delay-time analysis. For a PFET the resistance ranges over

$$R_0(7.71\,\text{k}\Omega) \rightarrow R_\infty(2.92\,\text{k}\Omega) \tag{3.38a}$$

and for an NFET,

$$R_0(7.16\,\text{k}\Omega) \rightarrow R_\infty(4.20\,\text{k}\Omega) \tag{3.38b}$$

The equivalent resistance falls within the range from R_∞ to R_0. The gradual, long-channel FET model gives $R_0 = 2R_\infty$. In this model the average of R_0 and R_∞ is $(3/4)R_0$. Therefore, the best first estimate of the equivalent resistance is to use $(3/4)R_0$, irrespective of the ratio of R_0 and R_∞ in real short-channel FETs, which may not be 2. This choice reflects the fact that R_0 is more important than R_∞ in delay determination. It would be informative to study various mechanisms that influence selection of the equivalent resistance value, and we study this problem in this section.

1
2
Z
X-1
X
MN1
MN2
MNX
C_L
(a)

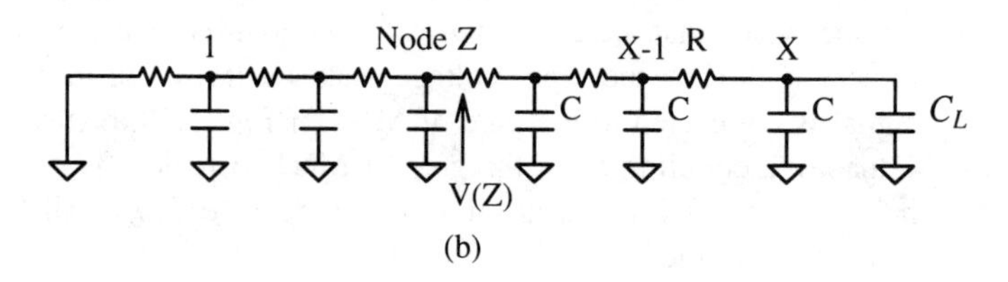

(b)

Figure 3.18 *RC* chain model of a CMOS gate (pulldown).

Using this definition, we have for PFETs $R = 0.75 \times 7.71 = 5.78$ kΩ and for NFETs $R = 0.75 \times 7.16 = 5.37$ kΩ. Using these parameters the gate delays are calculated as follows: For a PFET chain, per stage capacitance is given by

$$C = 2C_D + C_G = 2 \times 0.0185 + 0.0223 = 0.0593\text{pF}$$

and for an NFET chain, per stage capacitance is

$$C = 2C_D + C_G = 2 \times 0.0126 + 0.0113 = 0.0365\text{pF}$$

The time constant per stage of the chain is given by, respectively,

$$\text{For PFET chain: } \tau_0 = 5.78 \times 10^3 \times 0.0593 \times 10^{-12} = 0.343\text{ns}$$

$$\text{For NFET chain: } \tau_0 = 5.37 \times 10^3 \times 0.0365 \times 10^{-12} = 0.196\text{ns}$$

For a NAND gate with a load three times the minimum-size gate, the output node capacitance is given by

$$C_{D(PFET)} X + 3 \left\{ C_{G(NFET)} + C_{G(PFET)} \right\}$$

Since the output node has only one NFET drain diffused island, $C_{D(NFET)}$ must be subtracted from the sum,

TABLE 3.4(a) NAND GATES

X	Delay(theor.)	Delay(sim.)	D(sim.)/D(theor.)
3	2.45	2.2	0.89
4	3.72	3.2	0.86
5	5.48	4.3	0.78
10	18.2	13.3	0.73
15	37.2	27.3	0.73
20	67.6	45.8	0.68

TABLE 3.4(b) NOR GATES

X	Delay(theor.)	Delay(sim.)	D(sim.)/D(theor.)
3	3.09	3.2	1.03
4	4.80	4.8	1.00
5	7.03	6.7	0.95
10	23.3	20.2	0.87
15	46.3	40.0	0.87
20	78.9	66.0	0.84

$$C_L = C_{D(PFET)} X + 3 \left\{ C_{G(NFET)} + C_{G(PFET)} \right\} - C_{D(NFET)}$$

$$= (0.0185X + 0.0880)\text{pF}$$

In the same way, for a NAND gate with three minimum-size fanouts,

$$C_L = (0.0126X + 0.0821)\text{pF}$$

From the equations and the parameter values the gate delays are calculated as shown in Tables 3.4(a) and 3.4(b).

This result shows that NOR gate delays agree better than the NAND gate delays. The difference is, however, due to the crude definition of the equivalent resistance. For the integrity of the model, it is more important that the difference can be explained, thereby revealing any missing detail. The NFET equivalent resistance should have been about 10 to 15% less, to obtain comparable agreement. The reason for the smaller NFET equivalent resistance is that the threshold voltage of the NFET is significantly less than that of the PFET, and therefore the NFET is much more conductive than the PFET at the small gate voltages, as observed from Figs. 3.17(a) and 3.17(b). Because the equivalent resistance depends on the details of the I-V characteristic, the resistance should be determined by seeking the best consistency between the simulated and calculated gate delays.

For large X, the calculated delays are significantly more than the simulated delays. The difference can also be explained by the definition of equivalent resistance. When more FETs are connected in series, the voltage each FET sustains is less. The field in the channel is less, and the effects of carrier velocity saturation are less. This has the effect of increasing the conductivity of the FETs. Since the equivalent resistance used in the present analysis was determined using a single FET whose drain is biased to full V_{DD} (5 V), the equivalent resistance determined in this way is too high to represent a multimember FET chain, and this error is greater for NFETs than for PFETs.

Output voltage waveform of a gate that has a long NFET chain is shown in Fig. 3.19. Three phases, 1, 2, and 3, are identified in the waveform. In phase 1, internal nodes of the long NFET chain are discharged progressively from the grounded end to the output node, but the output node stays unchanged at V_{DD}. This phase lasts about time

$$T_D(1) \approx 0.4(NR)(NC) \tag{3.39}$$

This formula approximates the last term of Eq. (3.37).

In phase 2, the NFET chain discharges load capacitance C_L. Analysis of this phase is complicated. In the theory of this section, the RC chain model (loaded with capacitor C_L) and its interpolative delay formula, Eq. (3.37), were used to simplify analysis. When the load capacitor is very large ($C_L \gg NC$), the capacitance of the internal nodes may be neglected and we have

$$T_D(2) \approx 0.7(NR)C_L \tag{3.40}$$

In phase 3, the NFET chain is approximately a linear resistor. The value of the linear resistor, R_0, is less than that of the equivalent resistor of NFET, R, since all the NFETs sustain small source-drain voltages. The voltage waveform of the output node is given approximately by

$$V_{OUT} \approx \text{constant} \times \exp(-t/NR_0C_L) \tag{3.41}$$

when $C_L \gg NC$.

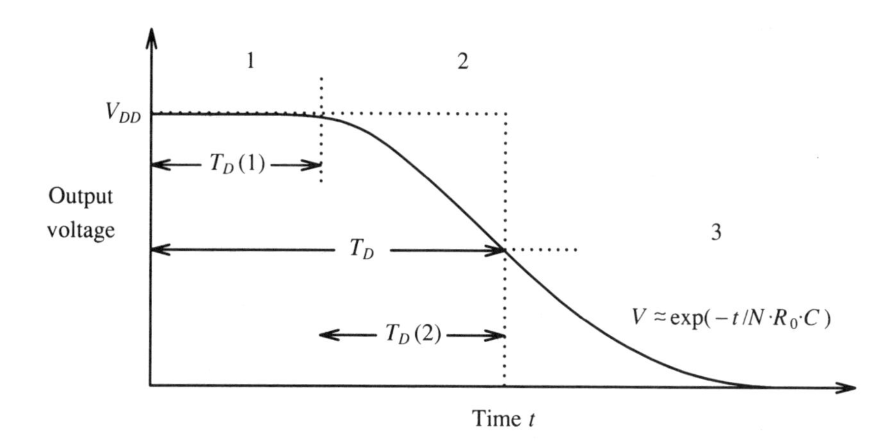

Figure 3.19 Output voltage of gate.

3.13 PHYSICAL MODEL OF DISCHARGE FROM THE END OF A PFET CHAIN

In this section we explain why the pulldown delay time of a multi-input NOR gate and pullup delay of a multi-input NAND gate are quite short. The NOR gate shown in Fig. 3.20(a) can be modeled using the RC chain circuit of Fig. 3.20(b). When NFET MN1 is turned on, the NFET can be approximated by resistor R_D. This resistor discharges load capacitor C_L, as well as capacitors $C_A, C_B, C_C, \ldots$. Resistors $R_A, R_B, \ldots$ are , however, very much larger than R_D, because they are channel resistances of PFETs that suffer from the back-bias effect: As capacitance C_A is discharged, the channel resistance of MPA, R_A, increases. Therefore, when the voltage of output node O is well below $(1/2)V_{DD}$, the capacitors $C_A, C_B, C_C, \ldots$ have not yet discharged significantly. The charge of capacitors $C_A, C_B, \ldots$ trickles out slowly through resistors $R_A, R_B, \ldots$, and it takes a long time to discharge the entire chain. But the time needed to pull node O from V_{DD} to $V_{DD}/2$ is quite short. We study this model using the following two closed-form theories.

For convenience of analysis, the discrete RC chain of Fig. 3.20(b) is replaced by a distributed RC chain that has C F/cm of capacitance and R (Ω/cm) of series resistance. Let T_D be the delay time of the gate. During time interval $0 \rightarrow T_D$, the electrical signal propagates distance l on the distributed RC chain, where l is determined by

$$T_D = RCl^2 \tag{3.42}$$

as derived in Section 6.8. When the RC chain is observed from the input terminal the chain looks like a capacitor whose value is

$$lC = \left[T_D \frac{C}{R} \right]^{1/2} \tag{3.43}$$

T_D is determined from the requirement of self-consistency,

$$R_D \left\{ C_L + \left[T_D \frac{C}{R} \right]^{1/2} \right\} = T_D \tag{3.44}$$

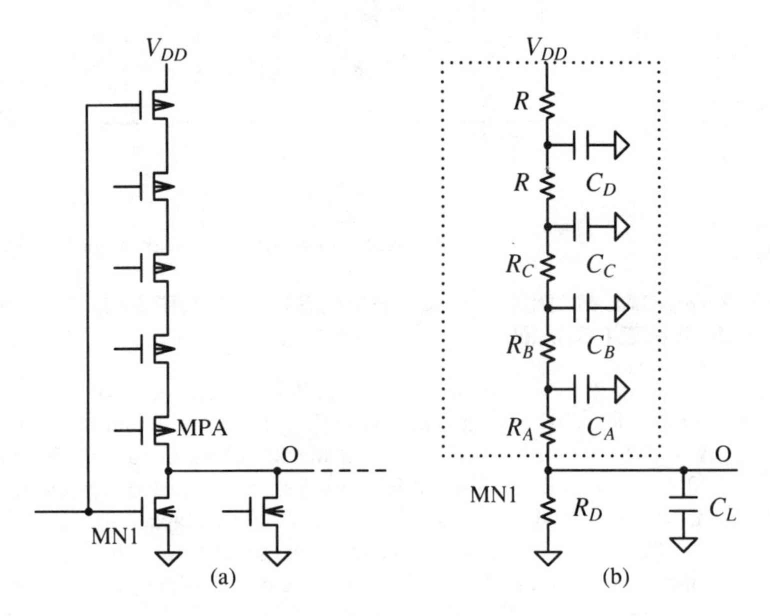

Figure 3.20 Model of CMOS gate that has long FET chain as load.

This equation is solved for T_D as

$$T_D = \frac{R_D{}^2}{2} \left\{ \frac{2C_L}{R_D} + \frac{C}{R} \pm \left\{ \left[2\frac{C_L}{R_D} + \frac{C}{R} \right]^2 - 4\frac{C_L{}^2}{R_D{}^2} \right\}^{1/2} \right\}$$

where the minus sign is not relevant because this solution has an unrealistic limit [$T_D \to 0$ when $(C/R) \to \infty$]. Then we have

$$\frac{T_D}{R_D C_L} = 1 + \frac{\lambda^2}{2} + \left[\lambda^2 + \frac{1}{4}\lambda^4 \right]^{1/2} \tag{3.45}$$

where

$$\lambda = \left[\frac{R_D}{C_L} \frac{C}{R} \right]^{1/2}$$

When $R \to \infty, \lambda \to 0$, and therefore $T_D \to R_D C_L$. In this limit the capacitance of the PFET chain is invisible to the pulldown NFET. When $R \to 0$, $\lambda \to \infty$, and

$$\frac{T_D}{R_D C_L} \to \lambda^2 \quad \left[= \frac{R_D C}{C_L R} \right]$$

or

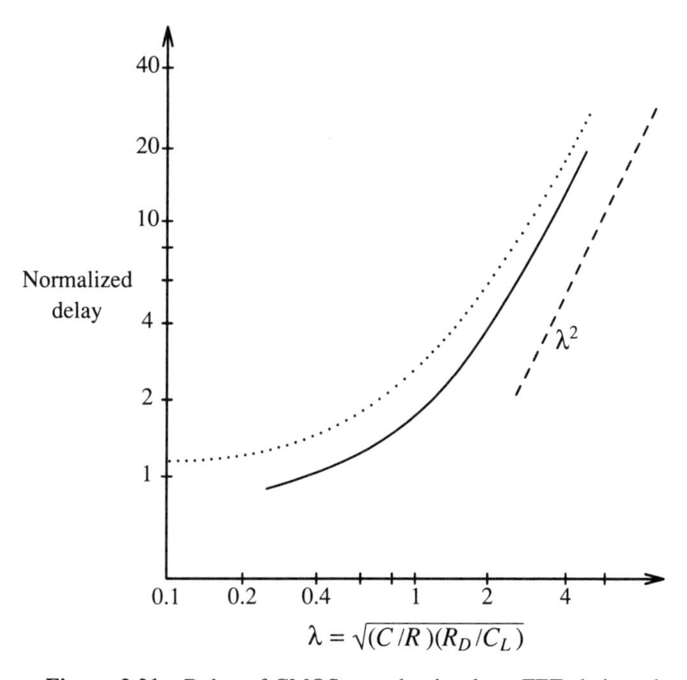

Figure 3.21 Delay of CMOS gate that has long FET chain as load.

$$T_D \to \frac{R_D}{R} R_D C = R_D{}^2 \left(\frac{C}{R} \right)$$

The limit is also reasonable. Equation (3.45) contains the ratio (C/R) only; the length of the RC chain, l, does not appear in the result. The switching time is determined by R_D, C_L, and (C/R) only. Large parasitic capacitances of the series-connected PFETs never determine the delay time. The dotted curve in Fig. 3.21 shows $T_D/R_D C_L$ versus λ.

This mode of gate operation is acceptable if the gate is required to respond quickly to only one polarity of transition (Section 8.11). This mode is not, however, an ideal mode of operation of CMOS static gates. CMOS static gates working in this mode have many undesirable characteristics, such as noise sensitivity and unpredictable delay (delay heavily dependent on the initial condition of the gate).

3.14 SWITCHING WAVEFORM

With reference to Fig. 3.20(b), the voltage of the output node $V(t)$ satisfies

$$\frac{1}{R_D}\int_0^t V(t)dt = \left\{ C_L + \left[\frac{Ct}{R} \right]^{1/2} \right\} \left\{ V_{DD} - V(t) \right\} \quad (3.46)$$

where the PFET chain was described equivalently by the time-dependent capacitor $\sqrt{Ct/R}$ of Eq. (3.43). This is equivalent to saying that the capacitors of the RC chain that exists within distance $l = \sqrt{t/(RC)}$ from the output node of the gate were assumed to have the same voltage as the output node. In Eq. (3.46), the variables are transformed using

$$V(t) = V_{DD}\,\phi(t) \quad \text{and} \quad t = R_D C_L \theta$$

and we have

$$\frac{d\phi}{d\theta} + \frac{1 + \frac{\lambda}{2}\frac{1}{\sqrt{\theta}}}{1 + \lambda\sqrt{\theta}}\phi = \frac{\frac{\lambda}{2}\frac{1}{\sqrt{\theta}}}{1 + \lambda\sqrt{\theta}} \quad (3.47)$$

Equation (3.47) is solved subject to

$$\phi = 1 \quad \text{when} \quad \theta = 0$$

Solving Eq. (3.47) is straightforward but tedious. The integral factor is found as

$$e^{\zeta(\theta,\lambda)}$$

where

$$\zeta(\theta,\lambda) = \frac{2}{\lambda}\left\{ \sqrt{\theta} + \left[\frac{\lambda}{2} - \frac{1}{\lambda} \right] \log(\sqrt{\theta} + \frac{1}{\lambda}) \right\}$$

and after some algebra, we obtain

$$\phi = \frac{e^{-\frac{2}{\lambda}\sqrt{\theta}}}{\left[\lambda\sqrt{\theta} + 1\right]^{1-\frac{2}{\lambda^2}}} \left\{ 1 + \left[\frac{\lambda^2}{2} \right]^{1-\frac{2}{\lambda^2}} \int_0^{\frac{2}{\lambda}\sqrt{\theta}} e^{\zeta} \left[\zeta + \frac{2}{\lambda^2} \right]^{-\frac{2}{\lambda^2}} d\zeta \right\} \quad (3.48)$$

Figure 3.22 shows plots of normalized output node voltage versus normalized time θ. The effect of trickle discharge of the PFET chain capacitances is quite significant.

When $\lambda \to 0$, Eq. (3.48) is not convenient for numerical computation. We then have

$$\phi \to e^{-\theta} \quad (3.49)$$

From Eqs. (3.48) and (3.49), the normalized delay time versus parameter λ can be determined as shown in Fig. 3.21 by the solid curve. The dotted curve shows Eq. (3.45), the result of the simplified theory. Except for a factor of about 1.4, which is the limit of accuracy originating from crude definition of R and R_D, the two curves are the same. Using

$$C_L = 3\left\{C_{G(NFET)} + C_{G(PFET)}\right\} + C_{DN} + XC_{DP} = 0.113 + 0.0185X \text{ pF}$$

$$R_D = 5.78\text{k}\Omega \quad \text{and} \quad R = 5.37\text{k}\Omega$$

$$C = 2C_{DN} + C_{GN} = 0.0365\text{pF}$$

for NAND(X) gates and

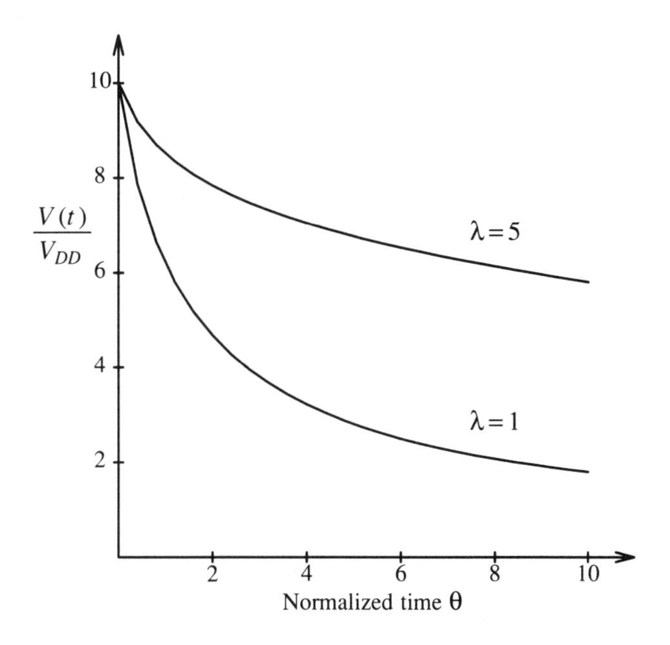

Figure 3.22 Discharge waveform of CMOS gate that has long FET chain as load.

$$C_L = 3\left\{C_{G(NFET)} + C_{G(PFET)}\right\} + C_{DP} + XC_{DN} = 0.119 + 0.0126X \text{ pF}$$

$$R_D = 5.37\text{k}\Omega \quad \text{and} \quad R = 5.78\text{k}\Omega$$

$$C = 2C_{DP} + C_{GP} = 0.0593\text{pF}$$

for NOR(X) gates, we obtain the results on Table 3.5, where

$$\lambda = \left[\frac{0.0551}{C_L} \right]^{1/2}$$

for NOR(X) and

$$\lambda = \left[\frac{0.0393}{C_L} \right]^{1/2}$$

for NAND(X). Using Fig. 3.21, the delays listed in Tables 3.5(a) and 3.5(b) were obtained.

TABLE 3.5(a) NOR GATES

X	Delay(theor.)	Delay(sim.)	D(sim.)/D(theor.)
3	1.36	1.12	0.823
4	1.46	1.21	0.828
5	1.55	1.29	0.832
10	1.93	1.54	0.797
15	2.32	1.79	0.771
20	2.68	2.03	0.757

TABLE 3.5(b) NAND GATES

X	Delay(theor.)	Delay(sim.)	D(sim.)/D(theor.)
3	1.46	1.57	1.07
4	1.59	1.68	1.05
5	1.71	1.79	1.05
10	2.30	2.24	0.973
15	2.89	2.62	0.905
20	3.39	2.98	0.878

As for NOR gates, the calculated delays are about 20% too large. This excess is again due to the crude definition of the equivalent resistance that leads to an overestimate of the NFET equivalent resistance, as discussed in Section 3.12 [Table 3.4(b)].

After this correction, the ratio of the simulated delay to calculated delay decreases with increasing X. The reason is that for a very short RC chain, Eq. (3.43) underestimates the capacitance contributed by the RC chain. The capacitance tends to zero when T_D tends to zero in Eq. (3.43). This is an obvious underestimate: The capacitance of the first PFET of the chain, MPA of Fig. 3.20(a), should be credited to the output node of the gate directly. Since the model is more accurate for larger X, the equivalent resistance used to compute Table 3.5 had been about 10 to 20% too large.

Tables 3.4 and 3.5 show that if corrections for the secondary effects are made to the equivalent resistances, the RC chain model is able to give a gate delay within about 10%.

The pulldown of an AOI gate has a closely related problem. Consider the AOI gate shown in Fig. 3.10. If only one of the NFET chain pulls down the output node, the charge stored in the nodes of the inactive NFET chain do not discharge until the output node is pulled below $V_{DD} - V_{THN}$. The pulldown waveform of such a complex gate has a bend at the output voltage where the extra capacitive load shows up. At that voltage the pulldown decelerates.

Practically, the equivalent resistance must be determined by seeking overall consistency in delays computed from the theory and determined from simulations for many different gates. The accuracy improves if different equivalent resistance values are used for pulldown RC chains and for RC chains that work as loads. The two types of RC chains work by significantly different mechanisms.

The two macromodels of NAND and NOR gates explain the switching characteristics of the gates very well. The models have clear physical meanings: The table lookup model does not have this advantage. The model of sequential discharge of FET chain discussed in Section 3.10 becomes important in estimating delays in reduced-voltage, submicron CMOS gates.

3.15 MODES OF SWITCHING AT HIGH AND LOW POWER SUPPLY VOLTAGES

The power supply voltage of early CMOS ICs was not standarized. Some used 15 V, while others used 12 V. Higher power supply voltages were advantageous to reduce the switching delay time of logic gates. The saturation current of a long-channel FET operating at low fields increases proportional to the square of the power supply voltage. The drain and source parasitic capacitances decrease with increasing junction voltage. The two effects overcome the increased voltage swing necessary to switch from one logic state to the other, and the gate delay decreases very significantly with increasing power supply voltage [16].

When the technology was scaled below 5 μm, the 5-V power supply voltage became a standard. As the FETs approached the short channel regime by further scaledown, the saturation current became linearly dependent on the power supply voltage. Then drastic delay reduction by increasing the power supply voltage does not happen any more (Section 4.2). Furthermore, the higher power supply voltage created many undesirable problems such as punch-through, breakdown, and oxide polarization. These undesirable side effects became so serious that in a submicron CMOS technology V_{DD} will be reduced below 5 V, typically to 3.5 V [17] (Section 9.18) and down to 2.5 V in superfine-line CMOS [18].

As the power supply voltage decreases, the threshold voltages of both PFET and NFET cannot be decreased proportionally and maintain satisfactory operation. A certain minimum FET threshold voltage is required to control subthreshold leakage current and to reduce the noise sensitivity of circuits. These requirements are, from the circuit-integrity viewpoint, essential. With scale-down, the ratios V_{THP}/V_{DD} and V_{THN}/V_{DD} inevitably must increase. Under this condition CMOS gate switching becomes more sequential.

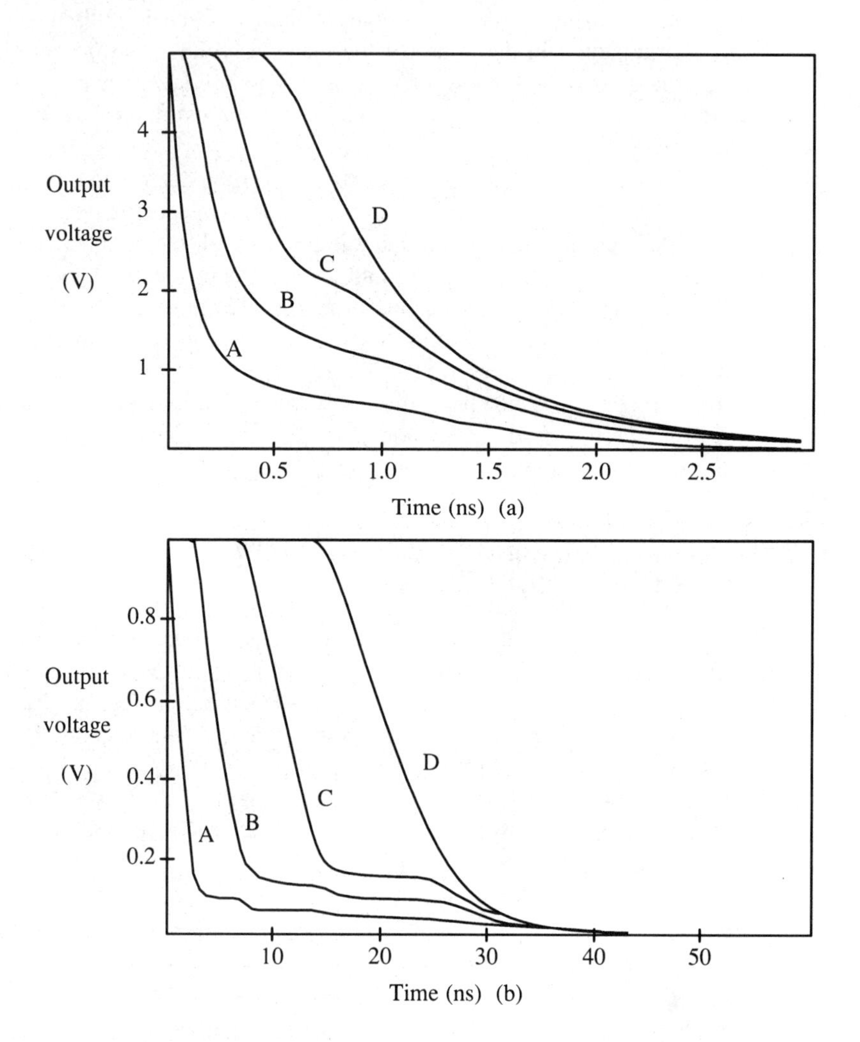

Figure 3.23 Sequential discharge of FET chain at low power supply voltage.

This phenomenon can be understood as follows. With reference to Fig. 3.18, as V_{DD} approaches V_{TH}, FET MN2 does not conduct until FET MN1 discharges node 1 practically down to ground potential. A back-bias effect on FET MN2 intensifies the effect. Discharge of node 2 does not begin until node 1 is almost at ground potential. In turn, discharge of node 3 begins only after node 2 is discharged almost to ground potential, and so on. Discharge of the FET chain occurs sequentially. At high power supply voltage, more than one FET discharges simultaneously and therefore the time to discharge the output node is significantly less.

Evidence of the sequential discharge of FET chain at low power supply voltages is given by the simulation results shown in Fig. 3.23, which were carried out using 1.75-μm CMOS with the medium-current process at 65 degrees centigrade. Four series-connected 4μm-wide NFETs discharge a 0.02 pF capacitor [circuit shown in Fig. 3.18(a) with $X = 4$]. A, B, and C are the internal nodes of the NFET chain labeled from the grounded end, and D is the output node. In Fig. 3.23(a) (high V_{DD} - 5 V) all the NFETs discharge the nodes simultaneously after about 0.6 ns. dV/dt of all the nodes are nonzero. In Fig. 3.23(b) (low V_{DD}, 1.0 V) two FETs never discharge simultaneously. The chain discharge is strictly sequential. This result shows that a CMOS gate operates in two distinct modes at high and at low power supply voltages.

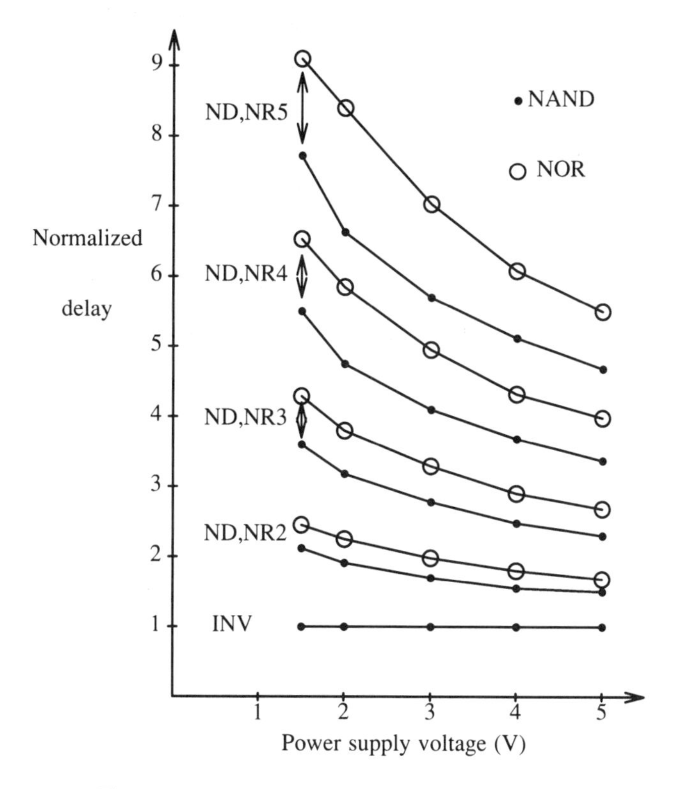

Figure 3.24 Gate delay versus power supply voltage.

Gate delays versus power supply voltage are shown in Fig. 3.24, with power supply voltage from 1.5 V to 5.0 V. The gate drives three times the minimum-size inverters. The delay times of the gates were normalized to the delay time of the inverter, since we wish to compare delays of different gates. Normalized delays of NAND gates (pulldown) and NOR gates (pullup) are plotted versus V_{DD}. For NAND gates (pulldown) and for NOR gates (pullup), the delay of the gate normalized to that of the inverter increases as the power supply voltage decreases. Practically, the effect is relevant in estimating the gate delay of VHSIC. If the delay of a VHSIC (submicron feature size with 2 to 3 V supply voltage) is estimated by comparing the time constant of FETs or the inverter delay to those of the larger-feature-size technology, the conclusion will be too optimistic.

3.16 DEPENDENCE OF GATE DELAY ON INPUT SIGNAL SLEW RATE

The gate-delay calculations to this point have assumed that the input signal switches within a much shorter time than that required for the output node of the gate to change state. When this assumption is not satisfied, new phenomena occur. In this section the problem is studied in detail, using an inverter as the example. Some recent gate-delay models include the effects of input slew rate [19] [20].

Currents I_P and current I_N of Fig. 3.1(a) depend on the input voltage V_G and output voltage V_D. We consider the input pullup (output pulldown) transient only. The other polarity transition can be studied, by exchanging the roles of the NFET and PFET. All the capacitances are lumped into a single load capacitance C_L.

The logic threshold voltage is $V_{DD}/2$: the gate-delay time T_D is defined as the interval from the time when the input voltage crosses the threshold to the time the output voltage crosses the threshold. With reference to Fig. 3.25(a) and 3.25(b), the input signal V_G is a function of time given by

$$V_G(t)=0 \qquad (t<0) \tag{3.50}$$

$$V_G(t)=\alpha t \qquad (0<t<t_{IS})$$

$$V_G(t)=V_{DD} \qquad (t>t_{IS})$$

where $\alpha=V_{DD}/t_{IS}$ is the input voltage slew rate. The output voltage reaches $V_{DD}/2$ when $t=t_S$. The delay time of the inverter is given by

$$T_D=t_S-\frac{1}{2}t_{IS} \tag{3.51}$$

If the input slew rate increases, T_D approaches t_∞. We define

$$t_\infty=\lim_{\alpha\to\infty} T_D$$

If the slew rate, α, is much less than V_{DD}/t_∞ three different cases are expected. Let the current of the NFET be a function of the gate and the drain voltages, $I_N(V_G,V_D)$, and the current of the PFET be $I_P(V_{DD}-V_G,V_{DD}-V_D)$.

1. If $I_N(V_{DD}/2,V_{DD}/2)<I_P(V_{DD}/2,V_{DD}/2)$, the output node voltage is still higher than $V_{DD}/2$ when the input voltage is $V_{DD}/2$, and therefore switching has not yet

occurred. For switching to occur, we must wait at least until the time when $I_N(V'_G, V_{DD}/2) = I_P(V_{DD} - V'_G, V_{DD}/2)$ is satisfied, where $V'_G > V_{DD}/2$. Since the input slew rate is α the delay of the gate is at least

$$T_D = \frac{V'_G - (V_{DD}/2)}{\alpha}$$

and T_D approaches ∞ as $\alpha \to 0$: In the limit of slow switching the delay time becomes infinite.

2. $I_P(V_{DD}/2, V_{DD}/2) = I_N(V_{DD}/2, V_{DD}/2)$, the output node voltage is close to $V_{DD}/2$, and the gate is ready to switch. The delay time is not zero, however. To determine the exact delay time the circuit equation must be solved.

3. If $I_P(V_{DD}/2, V_{DD}/2) < I_N(V_{DD}/2, V_{DD}/2)$, the output node voltage is less than $V_{DD}/2$ and the inverter has switched already. The delay time in this case may be negative. When $\alpha \to 0$ the delay time diverges to negative infinity.

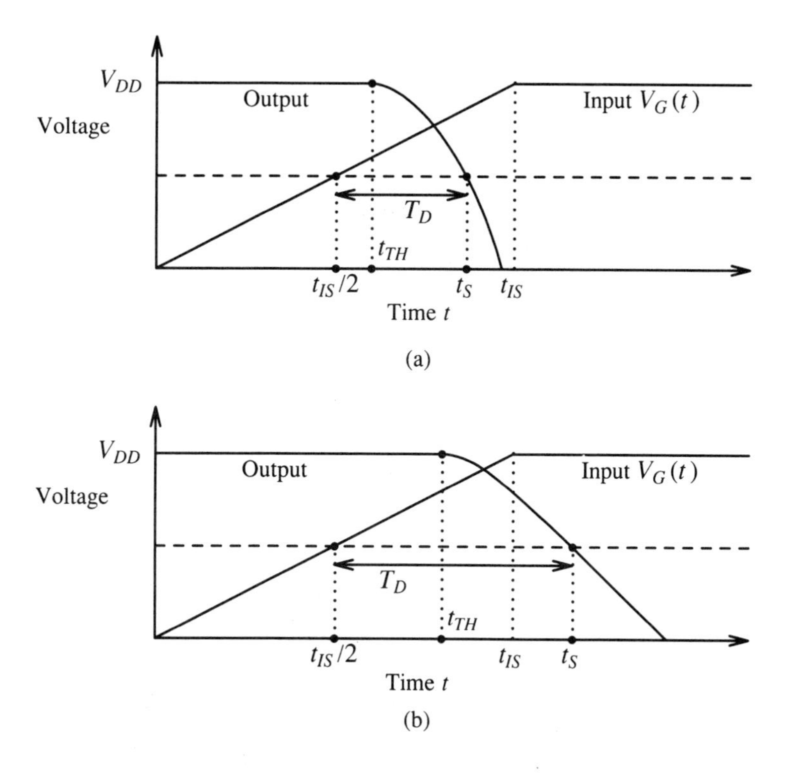

Figure 3.25 Analysis of gate delay versus input slew rate.

An infinitely long delay time in case 1 and a negative delay time in case 3 originate from the definition of delay time. To analyze this problem we need to study a circuit model of the inverter that allows inclusion of another parameter of the switching process, the input slew rate. The model should be simple so that closed-form results are available.

A FET has a current-voltage characteristic given by the gradual channel approximation, Eq. (3.10) or (3.20). The characteristics are, unfortunately, too complex to be usable in a closed-form analysis including the input voltage waveform. We use here a very simple FET model, where drain current of an NFET I_N is given by

$$I_N = b_N V_G \quad \text{if } V_D > 0$$

and when $V_D = 0$, current less than I_N can flow without developing voltage. The case of $V_D < 0$ never occurs. The PFET has a similar characteristic, given by

$$I_P = b_P (V_{DD} - V_G) \quad V_D < V_{DD}$$

and if $V_D = V_{DD}$, current less than I_P can flow without developing voltage.

The FET threshold voltages are not included in the analysis. By setting

$$I_P = I_N$$

we obtain

$$V_{GTR} = \frac{b_P}{b_P + b_N} V_{DD}$$

and therefore the static characteristic of the inverter consists of the two straight line segments,

$$V_D = V_{DD} \quad (0 < V_G < V_{GTR})$$

$$V_D = 0 \quad (V_{GTR} < V_G < V_{DD})$$

If this simplified FET model is used, the resultant equations can be solved in closed form and the dependence of the gate delay on input slew rate and on the inverter switching threshold voltage can be derived.

Using the input voltage waveform of Eq. (3.50), we define t_{TH} by $V_{GTR} = \alpha t_{TH}$. At time t_{TH} the input voltage is at the inverter's threshold,

$$t_{TH} = \frac{b_P}{b_N + b_P} \left(\frac{V_{DD}}{\alpha} \right) = \frac{1}{\beta + 1} \left(\frac{V_{DD}}{\alpha} \right) \tag{3.52}$$

where $\beta = (b_N / b_P)$. Until time t_{TH} is reached, V_D remains at V_{DD}. When $t > t_{TH}$ the current continuity law requires that

$$C_L \frac{dV_D}{dt} = -(b_N + b_P)\alpha(t - t_{TH})$$

By solving this equation with reference to Fig. 3.25(a), we obtain

$$V_D = V_{DD} - \frac{(b_N + b_P)\alpha}{2C_L}(t - t_{TH})^2$$

The time when the logic threshold voltage $V_{DD}/2$ is reached, t_S, is given by

$$t_S = t_{TH} + \left\{\frac{V_{DD}C_L}{\alpha(b_N + b_P)}\right\}^{1/2} \tag{3.53}$$

The delay time T_D is given by subtracting the input switching time, $t_{IS}/2 = V_{DD}/2\alpha$, as

$$T_D = t_S - \frac{V_{DD}}{2\alpha} = \frac{b_P}{b_N + b_P}\left(\frac{V_{DD}}{\alpha}\right) + \left\{\frac{C_L V_{DD}}{\alpha(b_N + b_P)}\right\}^{1/2} - \frac{V_{DD}}{2\alpha} \tag{3.54}$$

This result is valid when

$$t_S < \frac{V_{DD}}{\alpha} \quad \text{or} \quad \alpha < \alpha_0 = \frac{b_N}{b_N + b_P}\frac{V_{DD}}{\frac{C_L}{b_N}}$$

is satisfied.

When this condition is not satisfied, switching of the inverter occurs after time $t_{IS} = V_{DD}/\alpha$. With reference to Fig. 3.25(b), voltage V_D when $t = t_{IS} = V_{DD}/\alpha$ is given as

$$V_D\left(\frac{V_{DD}}{\alpha}\right) = V_{DD} - \frac{(b_N + b_P)\alpha}{2C_L}\left(\frac{V_{DD}}{\alpha} - t_{TH}\right)^2$$

and when $t > V_{DD}/\alpha$ the switching mechanism is that capacitance C_L charged to $V_D\left(\frac{V_{DD}}{\alpha}\right)$ discharges by the NFET current generator $b_N V_{DD}$, as

$$V_D(t) = V_D\left(\frac{V_{DD}}{\alpha}\right) - \left(\frac{b_N V_{DD}}{C_L}\right)\left(t - \frac{V_{DD}}{\alpha}\right)$$

The requirement that $V_D(t_S) = V_{DD}/2$ determines the switching time t_S as

$$t_S = \left(\frac{V_{DD}}{\alpha}\right)\left\{1 - \frac{b_N}{2(b_N + b_P)}\right\} + \frac{C_L}{2b_N}$$

and the delay time T_D is given by

$$T_D = t_S - \frac{V_{DD}}{2\alpha} = \left(\frac{V_{DD}}{2\alpha}\right)\left(\frac{b_P}{b_N + b_P}\right) + \frac{C_L}{2b_N} \tag{3.55}$$

When $\alpha \to \infty$, $T_D \to t_\infty = C_L/2b_N$. If T_D of Eqs. (3.54) and (3.55) are normalized by t_∞, we obtain

$$\frac{T_D}{t_\infty} = 1 + \left(\frac{1}{\beta}\right)\left(\frac{\alpha_0}{\alpha}\right) \quad (\alpha > \alpha_0) \tag{3.56}$$

$$\frac{T_D}{t_\infty} = 2\sqrt{\frac{\alpha_0}{\alpha}} + \left(\frac{1-\beta}{\beta}\right)\left(\frac{\alpha_0}{\alpha}\right) \qquad (\alpha < \alpha_0)$$

The slew rate, α, has the dimensions of [voltage]/[time]. It is convenient to use the unit of time t_∞ (the delay time of the inverter when the input switches instantly) everywhere. Then we define unit slew rate α_0 by

$$\alpha_0 = \frac{b_N}{2(b_N + b_P)}\left(\frac{V_{DD}}{t_\infty}\right) = \frac{\beta}{2(1+\beta)}\alpha_1$$

where $V_{DD}/t_\infty = \alpha_1$ and therefore,

$$\frac{\alpha_0}{\alpha} = \frac{\beta}{2(1+\beta)}\left(\frac{\alpha_1}{\alpha}\right) = \frac{\beta}{2(1+\beta)}\left(\frac{1}{S_I}\right) \tag{3.57}$$

where S_I is the input slew rate normalized to α_1. Using these relations, the normalized input slew rate S_I versus (T_D/t_∞) is as plotted in Fig. 3.26.

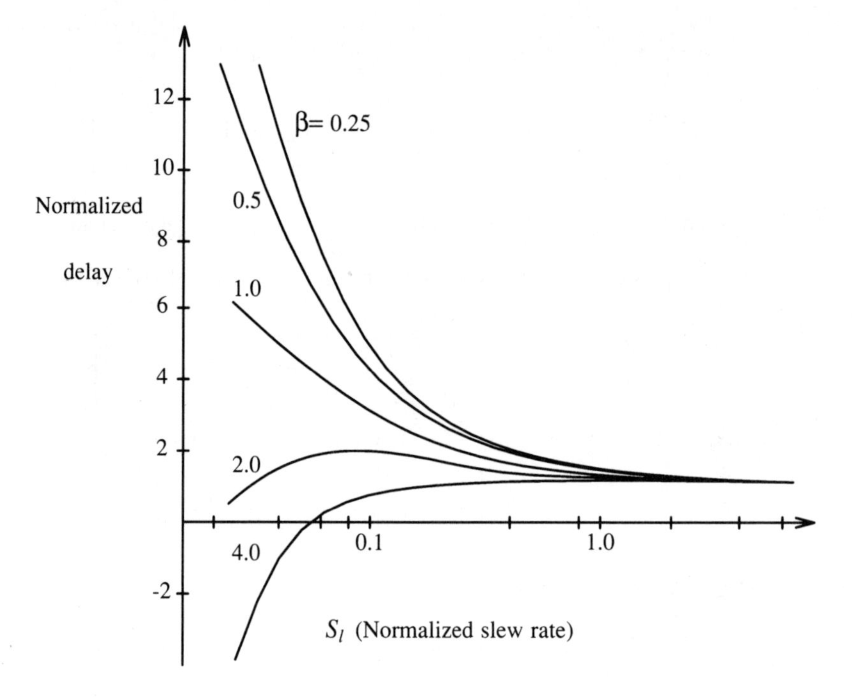

Figure 3.26 Gate delay versus input slew rate.

When $\beta < 1$, T_D/t_∞ diverges to positive infinity as $S_I \to 0$. When $\beta > 1$, T_D/t_∞ attains a maximum at

$$S_I = \frac{(1-\beta)^2}{2\beta(1+\beta)}$$

and then diverges to negative infinity as $S_I \to 0$. The delay maximum occurs because the conductance of the inverter decreases with decreasing input slew rate, and this effect shows up before the effects of the switching threshold show up. When $\beta = 1$, T_D / t_∞ diverges to positive infinity but by a reduced degree of divergence (1/2 power rather than the first power of $1/S_I$).

The dependence of the gate delay on the input slew rate, Eq. (3.56), was derived assuming a very simplified FET model, but the general structure of the formula is independent of the details of the FET characteristics. The gate delay always diverges when $S_I \to 0$ if $\beta \neq 1$. When $\beta \neq 1$ the divergence is as S_I^{-1}. With a different FET model the divergence when $\beta \neq 1$ would remain the same. The divergence as S_I^{-1} is due to the difference between the switching threshold of the gate and the CMOS logic threshold set at $V_{DD}/2$.

Divergence as $S_I^{-1/2}$ when $\beta = 1$ is the result of the assumed, perfectly saturating characteristic of the FETs. At the moment when the input voltage reaches $V_{DD}/2$, the output node is still at V_{DD}, the current of the PFET and the NFET are equal, and therefore there is no excess current available to discharge the load capacitance. The difference in the currents increases linearly with time. The voltage of the output node decreases proportional to the square of the time. The time required to drop the voltage by $V_{DD}/2$ is therefore proportional to the inverse of the square root of the slew rate. In a more realistic FET model, divergence of this type may not exist.

3.17 OVERLAPPING CURRENT

The overlapping current is the component of the gate current that flows directly from V_{DD} to V_{SS} [21]. This current does not contribute to gate switching, but contributes to power dissipation, and is therefore undesirable. When a CMOS gate switches very slowly the overlapping current becomes significant. In Fig. 3.27, the input voltage V_G is a uniform ramp

$$V_G = \alpha t$$

where α is the slew rate of the input voltage. Three parameters, t_{IS}, t_N, and t_P, are defined as follows:

$$V_{DD} = \alpha t_{IS} \quad V_{THN} = \alpha t_N \quad V_{DD} - V_{THP} = \alpha t_P$$

An inverter is considered as an example. When V_G is low, the NFET of the inverter is in saturation, and the overlap current I_O equals the NFET current, I_N, which is given by

$$I_O = I_N = \frac{B_N}{2}(V_G - V_{THN})^2 \tag{3.58}$$

When V_G is high, the PFET of the inverter is in saturation, and the overlap current equals the PFET current, I_P, which is given by

$$I_O = I_P = \frac{B_P}{2}(V_{DD} - V_G - V_{THP})^2 \tag{3.59}$$

Under no condition are both the NFET and the PFET in the triode region simultaneously. When the inverter switches, both FETs are in saturation. Then by requiring that $I_P = I_N$ the switching voltage V_{GTR} was determined previously as [Eq. (3.3)],

$$V_{GTR} = \frac{\sqrt{B_P}V_{DD} + \sqrt{B_N}V_{THN} - \sqrt{B_P}V_{THP}}{\sqrt{B_N} + \sqrt{B_P}}$$

The maximum overlap current $I_{O\max}$ occurs when the input voltage equals V_{GTR}, at time $t = t_{TH}$. Time t_{TH} is given by

$$t_{TH} = \frac{V_{GTR}}{\alpha}$$

and the maximum overlap current $I_{O\max}$ is given by

$$I_{O\max} = \frac{B_N B_P}{2(\sqrt{B_N} + \sqrt{B_P})}(V_{DD} - V_{THP} - V_{THN})^2$$

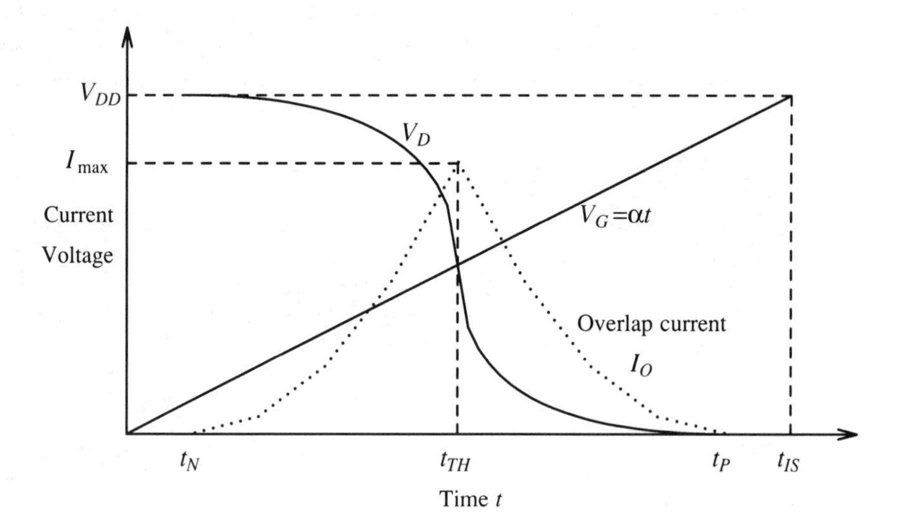

Figure 3.27 Overlap current of CMOS static gate.

The overlap current I_O is shown by the dotted curve in Fig. 3.27. The charge Q_N that flows in the interval from $t = t_N$ to $t = t_{TH}$ is given by

$$Q_N = \int_{t_N}^{t_{TH}} I_O(t)dt = \frac{B_N \alpha^2}{6}(t_{TH} - t_N)^3$$

and the charge Q_P that flows in the interval from $t = t_{TH}$ to $t = t_P$ is

$$Q_P = \frac{B_P \alpha^2}{6} (t_P - t_{TH})^3$$

and therefore the total charge that flows during the switching, Q_O, is

$$Q_O = Q_N + Q_P = \frac{1}{6\alpha} \frac{B_N B_P}{(\sqrt{B_N} + \sqrt{B_P})^2} (V_{DD} - V_{THP} - V_{THN})^3$$

The total charge is inversely proportional to slew rate α. The charge stored in the load capacitor and that is drained by switching the gate very fast, Q_C, is given by

$$Q_C = C_L V_{DD}$$

and if Q_O is more than two times Q_C, the overlapping current is considered quite significant. This condition occurs when

$$\frac{1}{12} \frac{1}{2 + \sqrt{B_P / B_N} + \sqrt{B_N / B_P}} \left[1 - \frac{V_{THP}}{V_{DD}} - \frac{V_{THN}}{V_{DD}} \right] (V_{DD} - V_{THP} - V_{THN})$$
$$\times \frac{\sqrt{B_N B_P} (V_{DD} - V_{THP} - V_{THN})}{C_L} > \alpha$$

or, for $B_P = B_N = B$ and negligible threshold voltages,

$$\frac{1}{48} \frac{V_{DD}}{t_0} > \alpha \qquad (3.60)$$

where $t_0 = C_L / BV_{DD}$ is the time constant of switching of the inverter. The large numerical factor in the denominator, 48, originates from the fact that a FET is a non-linear device and the current becomes quite small if both the drain and the gate voltages are halved. Small overlap current is an advantage of CMOS gates over bipolar gates. To design a low-power complementary circuit using PNP and NPN BJTs is difficult because the overlap current is large.

Two gates with different delays can be connected in parallel without creating too much overlap current if the delay difference is less than the switching time of the voltage (defined at 10 to 90% basis).

3.18 REFERENCES

01. H. Borkan and P. K. Weimer "An analysis of the characteristics of insulated-gate thin-film transistors" RCA Review, vol. 24,pp. 153-165,June 1963.
02. J. R. Burns "Switching response of complementary-symmetry MOS transistor logic circuits" RCA Review, vol. 25,pp. 627-661,December 1964.
03. T. Klein "Technology and performance of integrated complementary MOS circuits" IEEE J. Solid-State Circuits, vol. SC-4,pp. 122-130,June 1969.
04. J. K. Hawkins "Circuit design of digital computers" John Wiley & Sons, New York, 1968.
05. F. K. Manasse "Semiconductor electronics design" Prentice-Hall, Inc., Englewood Cliffs,

N. J., 1977.
06. R. H. Crawford "MOSFET in circuit design" McGraw-Hill Book Company, New York, 1967.
07. E. D. Rainville and P. E. Bedient "Elementary differential equations" MacMillan Company, New York, 1969.
08. B. T. Murphy "Unified, field-effect transistor theory including velocity saturation" IEEE J. Solid-State Circuits, vol. SC-15,pp. 325-328.
09. D. C. Mayer and W. E. Perkins "Analysis of the switching speed of a submicrometer-gate CMOS/SOS inverter" IEEE Trans. on Electron Devices, vol. ED-28,pp. 886-888,July 1981.
10. P. Subramanian "Modeling MOS VLSI circuits for transient analysis" IEEE J. Solid-State Circuits, vol. SC-21,pp. 276-285,April 1986.
11. N. B. Rabbat, W. D. Ryan and S. Q. A. M. A. Hossain "A computer modeling approach for LSI digital structures" IEEE Trans. on Electron Devices, vol. ED-22, pp. 523-531, August 1985.
12. C. Y. Wu, J. S. Hwang, C. Chang and C. C. Chang "An efficient timing model for CMOS combinatorial logic gates" IEEE Trans. on Computer-Aided Design, vol. CAD-4, pp. 636-650, October 1985.
13. W. C. Elmore "The transient response of damped linear networks with particular regard to wideband amplifiers" Journal of Applied Physics, vol. 19,pp. 55-63,January 1948.
14. M. H. White "Modelling of VLSI minicells" CICC82 Digest, pp. 180-1831982.
15. H. Iwai, M. R. Pinto, C. S. Rafferty, J. E. Oristrass and R. W. Dutton "Velocity saturation effect on short-channel MOS transistor capacitance" IEEE Electron Device Letters, vol. EDL-6,pp. 120-122,March 1985.
16. K. D. Wagner and E. J. McCluskey "Effect of supply voltage on circuit propagation delay and test applications" ICCAD85 Digest, pp. 42-44,1985.
17. S. Chou "Implementing the VLSI transition to 3 volts" ISSCC86 Digest, pp. 224-225, February 1986 (panel discussion).
18. W. Fichtner, E. N. Fuls, R. L. Johnston, R. K. Watts and W. W. Weick "Optimized MOSFETs with subquartermicron channel lengths" IEDM83 Digest, pp. 384-387, December 1983.
19. D. Auvergne, G. Cambon, D. Deschacht, M. Robert, G. Sagnes and V. Tempier "Delay-time evaluation in ED MOS logic LSI" IEEE J. Solid-State Circuits, vol. SC-21,pp. 337-343, April 1986.
20. J. L. Wyatt, Jr. "The sensitivity of inverter delay to details of the input waveform: a variational approach" MIT VLSI Memo No. 82-126, November 1982.
21. D. A. Hodges and H. G. Jackson "Analysis and design of digital integrated circuits" McGraw-Hill, New York, 1983 and D. Frohman-Bentchkowsky and L. Vadasz "Computer-aided design and characterization of digital MOS integrated circuits" IEEE J. Solid-State Circuits, vol. SC-4,pp. 57-64,April 1969.

4

CMOS Static Gates in Integrated Circuits

4.1 INTRODUCTION

In this chapter we summarize the gate-delay data obtained from circuit simulation, discuss the power-delay product of gates, and study the properties of CMOS static gates that originate from feedback mechanisms. The Miller effect occurs from negative feedback through the inevitable gate-drain parasitic capacitance. A ring oscillator, a CMOS D-latch, a negative resistance, and a negative capacitance use positive feedback as their essential working mechanisms.

4.2 GATE-DELAY DATA

Delays of CMOS static gates depend on a number of electrical, thermal, design-dependent and process-dependent parameters. From a practical point of view, some of the parameters, especially process-dependent parameters, must be assumed to represent the real chip environment. Delays at typical conditions are of special interest to designers.

This section is a compilation of data obtained by circuit simulations using ADVICE circuit simulator. Gate-delay data of a 1.75-μm CMOS, at 4.75 V at 105 degrees centigrade and at typical process (or medium-current process) are given. This is considered typical for VLSI design, since process variation is beyond a designer's control, the power supply voltage tends to be lower than 5 V by DC loading, and the temperature is higher than (standard) 65 degrees centigrade (often used for specification). Loading of a gate is specified by the number of minimum-size inverters connected to the output. Delay times versus loading are plotted. Gate types examined are INV, NAND2 to 5, NOR2 to 5, and AOI22 to AOI22222. The gates have logically equivalent but circuitwise different inputs (e.g., two inputs of a NAND2 gate are logically equivalent, but the input to the NFET connected directly to the output node pulls the output node down faster than the other input). In this case the worst-case input condition was chosen, the NFET closest from ground for NAND pulldown and the PFET closest to V_{DD} for NOR pullup. Pullup and pulldown delays are given separately. Figure 4.1 shows the pulldown delays and Fig. 4.2 the pullup delays. FET sizes used are as follows: NFETs are the minimum-size NFETs (4 μm wide) and PFETs are the second smallest (7.25 μm). All the source and drain parasitic capacitances were included separately (FETs are all separated and connected by short wires, whose capacitance is negligible).

The gate delays were defined as the interval between the time when the input signal reached $V_{DD}/2$ to the time when the output signal reached $V_{DD}/2$. A gate delay defined in this way is insensitive to the rate of input voltage change as long as the rate is not very slow, as was discussed in Section 3.16. We use a very short input transition time (0.2 nsec 0 to 100%). Then the total delay of a logic chain can be approximated by the sum of the delays of individual gates. The delays are summed by taking into account the pullup-pulldown transition polarity. One should note, however, that the summed delays of the gates are always significantly smaller than the delays of the real circuit, since wiring parasitic capacitances are not included. In practical design the following guidelines are convenient. In a 1.75-μm CMOS process, approximately 150-μm-long metal or polysilicon wire is equivalent to one minimum-size inverter load. The effects of polysilicon series resistance should be included for long polysilicon wires whose resistance exceeds 10% of the channel resistance of FET measured at the maximum gate voltage (V_{DD}) and minimum drain voltage (0). Due allowance for parasitics is essential in estimating the delay of a real circuit.

Figures 4.1(a) and (b) show that the delay time of a gate, T_D, is approximately a linear function of loading f, given by

$$T_D = \tau_0 + \tau_1 f \tag{4.1}$$

where f is the number of the minimum-size inverters loading the gate, and τ_0 and τ_1 are constant parameters characteristic of the gate. Parameter τ_0 is due to the capacitance of gate itself, and τ_1 is due to the load capacitance of the gate.

From Figs. 4.1(a) and 4.1(b) the following may be concluded. Curves for NAND2, AOI22, ..., and AOI22222 gates in Fig. 4.1(a) have identical slope, because their pulldown circuits consist of two series-connected NFETs. Curves for INV, NAND2, ..., and NAND5 gates in Fig. 4.1(a) have successively steeper slopes because the number of FETs in the pulldown circuits increases in that order. Section 3.12 explains the dependence.

The pulldown delay times of NOR gates depend weakly on the number of inputs. Similarly, pullup delay times of NAND gates in Fig. 4.1(b) depend weakly on the number of inputs. Section 3.13 explains this. The pullup delay time of AOI22222 is very long, since five series-connected PFETs pull up large capacitances of the gate's internal nodes. Large internal node capacitances also cause long delay at no load.

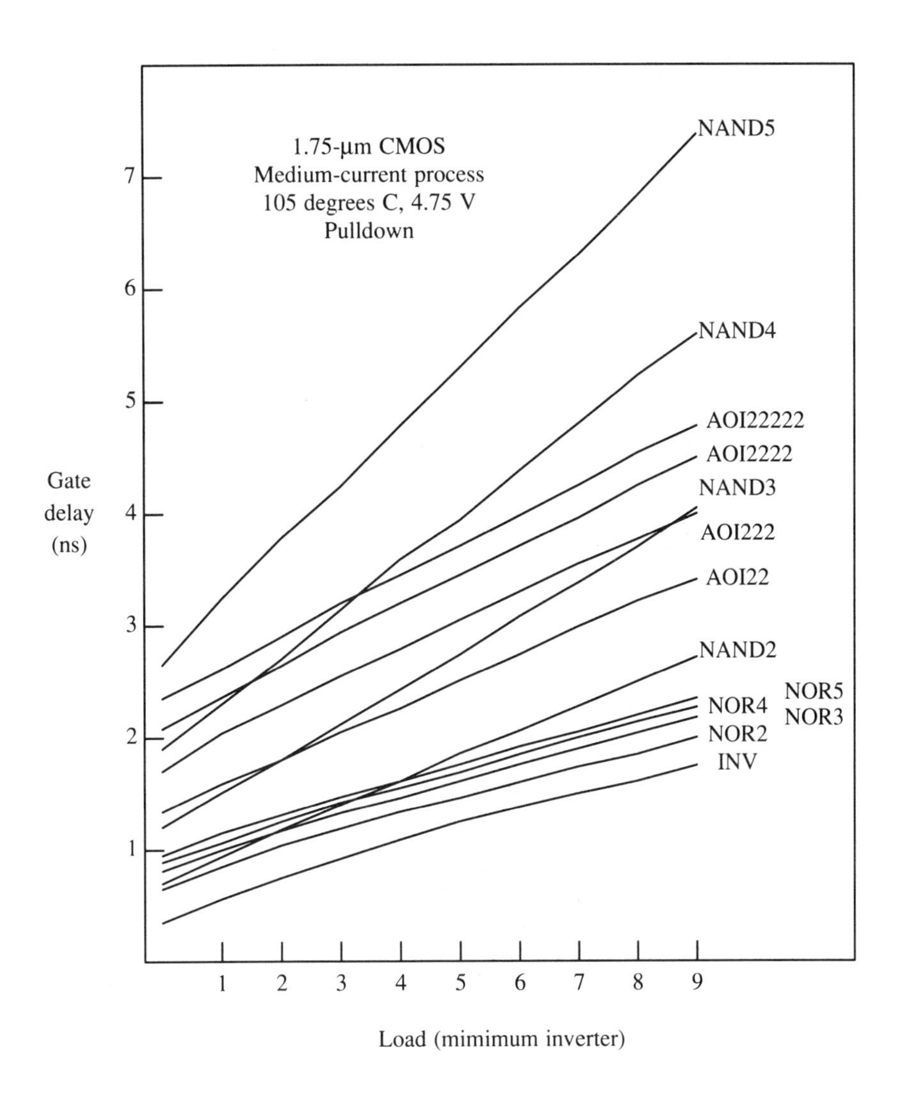

Figure 4.1(a) Gate delay versus loading (pulldown).

Dependence of the gate delay on the feature size of CMOS technology is shown in Fig. 4.2(a). Gate delays decrease approximately proportional to feature size. As we discussed in Sections 1.12 and 1.13, the gate and drain capacitances of FETs decrease approximately proportionally to the 1.5th power of the feature size. The saturation current of FETs reaches the maximum at about 1.75-μm feature size, and then decreases slowly as the feature size decreases. Therefore gate delays decrease approximately proportionally to feature size when it is less than about 1.75-μm.

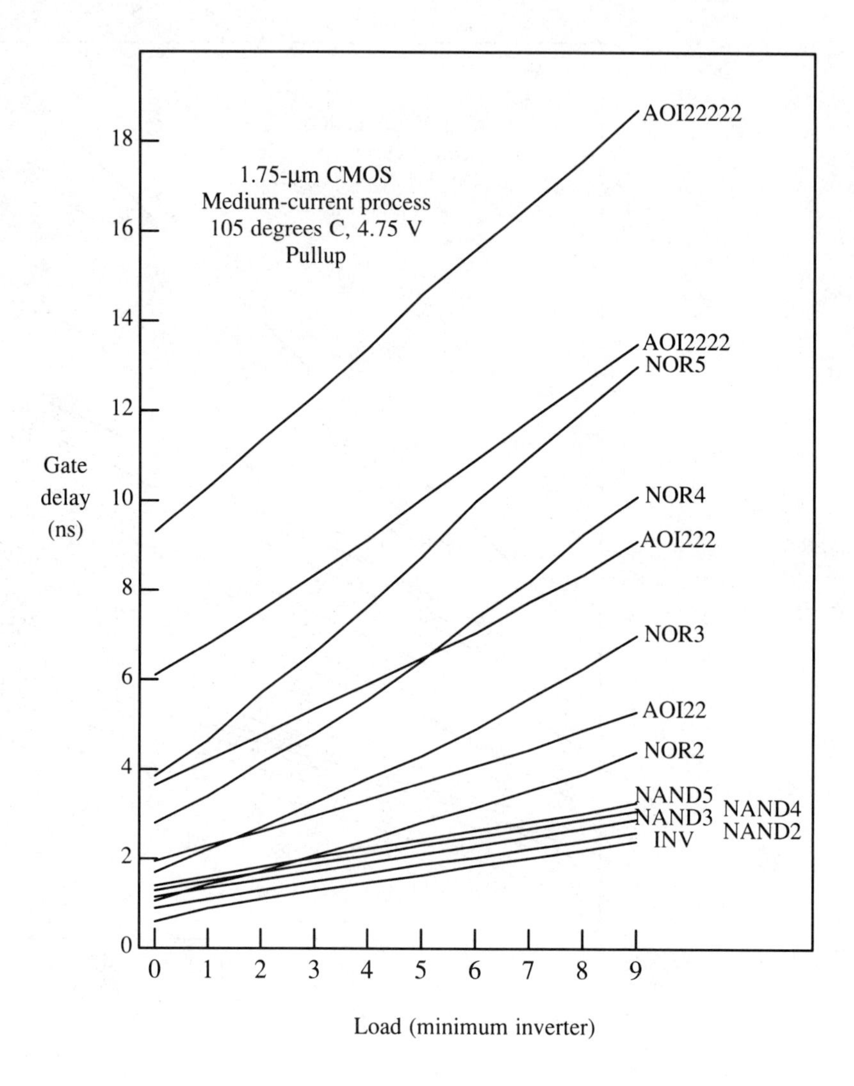

Figure 4.1(b) Gate delay versus loading (pullup).

Figure 4.2(a) shows CMOS technology that uses $V_{DD} = 5$ V. For feature size less than 1 μm, V_{DD} is reduced, and gate-delay dependence on the feature size will be different from that of the 5 V CMOS technology shown in Fig. 4.2(a).

Dependence of the gate-delay time on the power supply voltage has been discussed in Section 3.15. The relative delay times of gates normalized to that of an inverter were studied there. The delay time of inverter (which was used to normalize the other gate delay times) varies with power supply voltage as shown in Fig. 4.2(b).

A rapid increase in the inverter delay for power supply voltages of lower than 3 V is due to rapid increase in the (V_{TH}/V_{DD}) ratio. For V_{DD} higher than 5 V, the delay decreases slowly with increasing V_{DD}. This is because saturation current of FET (effective gate length of about 1µm) depends linearly on gate voltage (drift velocity saturation effect, Section 1.10).

The dependence of gate delay on temperature is shown in Fig. 4.2(c) which shows the normalized delay of the inverter (average of pullup and pulldown delays) versus temperature. The temperature dependence of the delay originates entirely from the temperature dependence of FET current which is approximately proportional to the inverse of the absolute temperature.

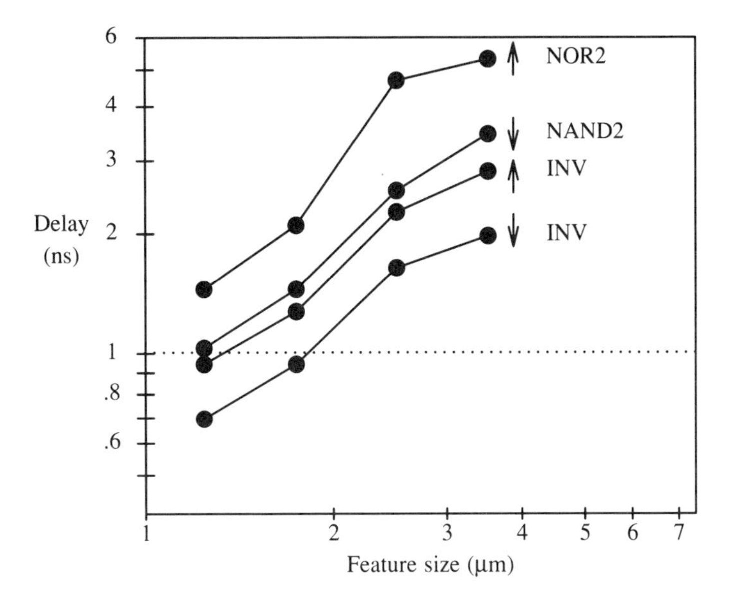

Figure 4.2(a) Gate delay versus feature size of technology.

4.3 SYMMETRIC GATES

When a certain PFET/NFET size ratio is chosen, the pullup and the pulldown delays of a gate can be made equal. The PFET/NFET size ratio for symmetry is tabulated in Table 4.1.

The size ratio was determined by ADVICE simulation, assuming a 1.75-µm CMOS at the medium current process at 65 degrees centigrade and 5.0 V of supply voltage. Delays from the slowest inputs (input closest to the power or the ground rails) were symmetrized, assuming that the other logically equivalent inputs are set at the logic level that enables the gate. Three times the minimum-size symmetric inverter (4-µm wide NFET and 8.9-µm PFET make a minimum-size symmetrical

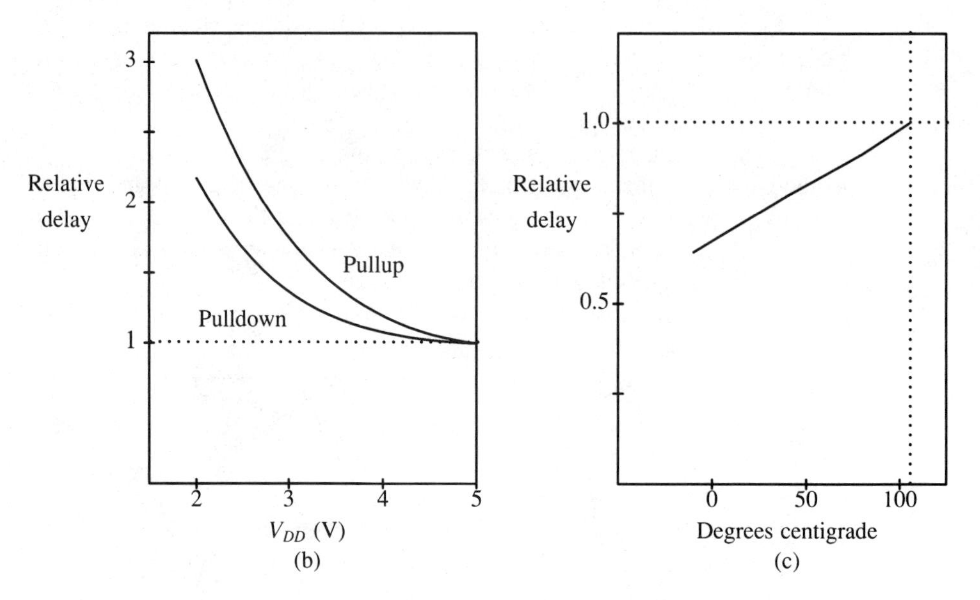

Figure 4.2(b) and (c) Gate delay versus power supply voltage and temperature.

TABLE 4.1 SYMMETRIC GATES

Gate	PFET/NFET	Delay at symmetry
INV	2.22	0.968 (ns)
NAND2	1.48	0.852
NAND3	1.15	1.12
NAND4	0.93	1.47
NAND5	0.750	1.88
NOR2	3.35	1.13
NOR3	4.40	1.64
NOR4	5.50	2.27
NOR5	6.70	3.00

inverter) was used as the load. The NFET size of INV and all the NOR gates was set at the minimum-size (4.0 μm wide). In an X-input NAND gate, the NFET size was set at X times the minimum-size. The PFET size for symmetry was determined by simulations, subject to these conditions. An inverter is symmetrical for pullup and pulldown delays when the PFET/NFET size ratio is 2.22. A NAND(X) gate is symmetrical at a ratio higher than $2.22/X$, and a NOR(X) gate at a FET size ratio less than $2.22X$. Note that usually we are not interested in symmetric gates to obtain the minimum delay (Section 8.9), but symmetrized gates have an important application in clock generation circuits.

4.4 SWITCHING DELAYS FROM DIFFERENT INPUTS

A CMOS NAND3 gate has three logically equivalent inputs. Since the three pulldown NFETs are connected in series, the pulldown delays from the three inputs are all different. The same difference in pullup delays exists in a NOR3 gate. The difference in delays is not always small. In a complex NAND gate (pullup) or NOR gate (pulldown), the difference can amount to 30% or more. The delay from the input nearest the output node is the shortest, and the delay from the input nearest ground or V_{DD} is the longest. Table 4.2 shows the simulated delays of NAND and NOR gates that have up to five inputs.

TABLE 4.2 DELAY FROM EQUIVALENT INPUTS (ns)

—	A	B	C	D	E
NAND2	1.072	1.024	—	—	—
NAND3	1.740	1.640	1.492	—	—
NAND4	2.568	2.468	2.256	2.016	—
NAND5	3.564	3.456	3.248	2.948	2.556
NOR2	1.568	1.416	—	—	—
NOR3	2.628	2.480	2.072	—	—
NOR4	3.872	3.704	3.304	2.668	—
NOR5	5.444	5.260	4.840	4.216	3.360

Identification of the different inputs is as follows: Input A is nearest the power or the ground bus, and the inputs are arranged in the alphabetical order from that end. The simulations were carried out assuming 1.75-μm CMOS at the medium current process, at 65 degrees centigrade and at V_{DD} = 5.0 V. The NFETs are 4 μm wide, and the PFETs are 7.25 μm wide. The difference in delays from the different inputs amounts to 16% for a NAND3 gate to 61% in a very complex NOR5 gate. From this result it is clear that a critical path signal should be introduced to a complex gate from the input closest to the output node.

The difference in delay time from the logically equivalent but circuitwise different inputs originates from two basically different mechanisms that are operative in the process of switching. For a NAND gate, when the FET closest to the ground switches, the charge stored in the output node must be drained to the ground through all the NFETs of the FET chain. When the FET closest to the output node switches, the charge stored in the output node is first shared among the other internal nodes of the gate, and the charge sharing first reduces the output node voltage. This charge redistribution process occurs relatively fast, and therefore the delay time from the input nearest the output node is the shortest. A similar effect in dynamic gates is reported [01]. Since SOS FETs have smaller source-drain capacitances and since the back-bias effect is absent, the difference in the switching time from the different inputs is less [02].

The delays in the preceding sections are for the input signal driving the slowest input of the gate, with the rest of the inputs set to the logic level that enables the gate. This delay is most often the realistic delay of a gate in a CMOS VLSI chip, but it is not the absolute worst-case delay. When more than one signal switches simultaneously, the delay of the gate becomes longer. A gate in a critical path often combines signals that arrive at almost the same time. Then the delay obtained by the conventional method of simulating a single change at a time is too optimistic. Circuit-level simulation is unable to include this effect completely. Gate-level simulation, including simultaneous switching, can be more accurate than circuit-level simulation (provided that a good gate delay model is available).

4.5 CMOS TRANSMISSION GATES

A CMOS transmission gate is a parallel connection of a PFET and an NFET, as shown in Fig. 2.1(d). When the gate of the NFET is at V_{DD} and the gate of the PFET is at ground, the transmission gate is on, and the signal at one terminal is transmitted to the other terminal. When the gate of the NFET is at ground and that of the PFET is at V_{DD}, the transmission gate is off, or the terminals are disconnected. When a CMOS transmission gate is on, the gate is practically equivalent to a linear resistor. When the transmission gate is off, the two terminals are coupled only by a small parasitic capacitance.

An NFET and a PFET connected in parallel make the on-resistance almost independent of the input and the output voltages and provide the capability to drive the output of the transmission gate all the way up to V_{DD} and all the way down to the ground. If only an NFET or a PFET is used instead, the voltage of the far end of the gate does not reach V_{DD} or ground, respectively [03]. This problem is tolerable in some digital circuits, but not in analog circuits.

Characteristics of a transmission gate when the gate is on can be analyzed using the characteristics of FETs from Chapter 1 [Eqs. (1.10a) and (1.10b)]. When input voltage V_1 is higher than output voltage V_0, current I, which flows from node 1 to node 0, is given by

$$I = I_N + I_P \tag{4.2}$$

where

$$I_N = 0 \text{ if } V_{DD} - V_0 - V_{THN}(V_0) \le 0 \tag{4.3}$$

$$I_N = B_N \left\{ V_{DD} - V_0 - V_{THN}(V_0) - (1/2)(V_1 - V_0) \right\} (V_1 - V_0)$$

$$\text{if} \quad V_{DD} - V_1 - V_{THN}(V_0) \geq 0$$

$$I_N = \frac{B_N}{2}\left\{V_{DD} - V_0 - V_{THN}(V_0)\right\}^2 \quad \text{otherwise}$$

$$I_P = 0 \quad \text{if} \quad V_1 - V_{THP}(V_{DD} - V_1) \leq 0 \tag{4.4}$$

$$I_P = B_P\left\{V_1 - V_{THP}(V_{DD} - V_1) - (1/2)(V_1 - V_0)\right\}(V_1 - V_0)$$

$$\text{if} \quad V_0 - V_{THP}(V_{DD} - V_1) \geq 0$$

$$I_P = \frac{B_P}{2}\left\{V_1 - V_{THP}(V_{DD} - V_1)\right\}^2 \quad \text{otherwise}$$

In Eqs. (4.2)-(4.4) the back-bias effect is included by allowing the threshold voltage to depend on the FET source voltage like $V_{THN}(V_0)$. This notation means the function of argument V_0, not the product. In Fig. 4.3 the states of a conducting transmission gate are represented within the square OXPY on the $V_0 - V_1$ coordinate system. In Fig. 4.3 curve α is a plot of

$$V_1 = V_{DD} - V_{THN}(V_0)$$

Above curve α the NFET is in saturation, since we assumed that $V_1 > V_0$. Within the area PQRS the NFET is nonconducting. In a similar manner curve β is a plot of

$$V_0 = V_{THP}(V_{DD} - V_1)$$

and on the lower left side of the curve the PFET is in saturation. Within the square OABC the PFET is nonconducting.

Since a transmission gate is symmetrical with respect to input and to output, Fig. 4.3 is symmetrical with respect to diagonal $\overline{OP}$. Within the region surrounded by curves α and β both the PFET and the NFET are in the triode region. In this region the DC conductance of the transmission gate is defined by the ratio of the current and the voltage as

$$G = \frac{I}{V_1 - V_0} \tag{4.5}$$

$$= B_N V_{DD} + \frac{1}{2}(V_0 + V_1)(B_P - B_N) - B_N V_{THN}(V_0) - B_P V_{THP}(V_{DD} - V_1)$$

$$\text{for} \quad V_1 > V_0$$

and

$$= B_N V_{DD} + \frac{1}{2}(V_0 + V_1)(B_P - B_N) - B_N V_{THN}(V_1) - B_P V_{THP}(V_{DD} - V_0)$$

$$\text{for} \quad V_0 > V_1$$

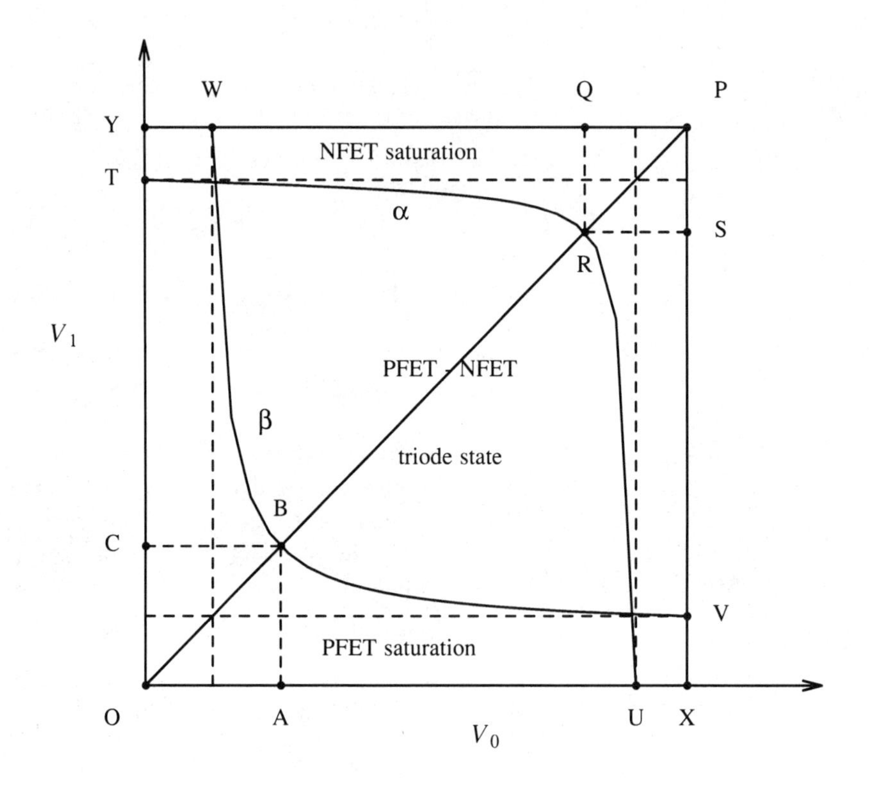

Figure 4.3 State of CMOS transmission gate on $V_0 - V_1$ plane.

In a symmetric transmission gate satisfying $B_P = B_N = B$,

$$G = B\left\{V_{DD} - V_{THN}(V_0) - V_{THP}(V_{DD} - V_1)\right\} \text{ when } V_1 > V_0 \tag{4.6}$$

and

$$= B\left\{V_{DD} - V_{THN}(V_1) - V_{THP}(V_{DD} - V_0)\right\} \text{ when } V_0 > V_1$$

Equation (4.6) shows that the DC resistance depends only weakly on voltages V_0 and V_1: The dependence originates only from the back-bias effect. The DC resistance of a transmission gate made using a 7.25-μm-wide NFET and a 17-μm-wide PFET of a 1.75-μm CMOS, which approximately satisfies the symmetry condition, falls within the range 1.2 to 1.75 kΩ. In a wide voltage range centered at $V_0 = V_1 = V_{DD}/2$ the resistance stays within the range 1.5 to 1.7 kΩ. The simulation assumes a 1.75-μm CMOS with the medium-current process, with 5.0 V power supply voltage and with 65 degrees centigrade. The on-resistance of a symmetrical transmission gate may be considered practically a linear resistance, for digital circuits.

If an NFET and PFET are designed to have the same size, capacitance coupling from the gate to the source and drain cancels. However, such a design is not symmetrical.

4.6 DEFINITION OF POWER-DELAY PRODUCT

The power-delay product is a fundamental parameter determining the quality of a CMOS process and gate design. Physically, the parameter is the average energy required for a gate to switch from low to high and high to low. The parameter has this simple definition related to switching, but this definition is not precise enough to be useful for circuit designers. In this section a precise definition is given by specifying the method of measurement, and using this definition, the power-delay products of present-day CMOS gates are determined.

Consider the inverter of Fig. 4.4(a), where we assume that the input voltage is originally high. MN1 is then on. MP1 is off, and therefore the voltage of capacitor C_L is zero, or no electrostatic energy is stored in the capacitor. C_L is a load capacitance of the inverter, including the drain island capacitances and the gate capacitances. The power is supplied, because of convenience in theoretical analysis, from a large capacitor C_D that is charged to V_{DD}. Some readers may wonder why a capacitor, rather than a voltage source, is used. If a voltage source is used, we must first solve for the current and voltage waveforms, and then we must integrate ohmic loss over time. We must assume appropriate current-voltage characteristics of FETs to do so (there is no universally acceptable closed-form formula). These details are, however, not relevant at all to the present problem. Using a capacitor as a power source, and regarding the switching as the process of establishing voltage equilibrium between capacitors, we are able to determine the energy loss associated with switching without going into the cumbersome details. When the gate voltage of the inverter makes a high-to-low transition, NFET MN1 turns off, and PFET MP1 turns on. In the course of switching, some overlap current flows. If the overlap current is neglected, however, the net effect of input switching is that some charge is transferred from capacitor C_D to capacitor C_L so that a voltage equilibrium between the capacitors is reached. The amount of charge transferred Q is given by

$$Q = C_L V_F \tag{4.7}$$

where the equilibrium voltage V_F is determined from conservation of charge as

$$C_D V_{DD} = (C_D + C_L) V_F$$

The energy loss of the process, ΔE, is given by

$$\Delta E = \frac{1}{2} C_D V_{DD}^2 - \frac{1}{2}(C_D + C_L) V_F^2 = \frac{1}{2} C_L V_{DD}^2 \frac{C_D}{C_L + C_D} \tag{4.8}$$

$$\rightarrow \frac{1}{2} C_L V_{DD}^2 \text{ (when } C_D \rightarrow \infty)$$

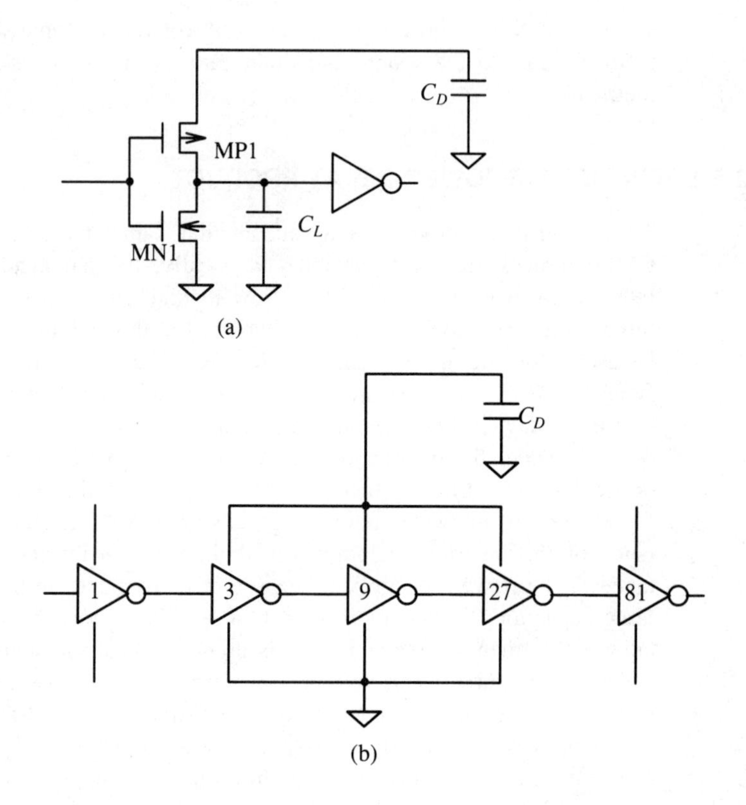

Figure 4.4 Method of measurement of power-delay product.

In the limit of very large C_D the lost energy equals the energy stored in the output capacitor. It is important to note that the energy is actually lost, not just transferred from C_D to C_L: The energy is lost as heat generated when the current flowed through the resistance of the channel of PFET MP1. Energy is also lost as electromagnetic radiation from the circuit.

When the input voltage makes a low-to-high transition, and if the overlap current is neglected, only the energy of the charge stored in capacitor C_L is lost as heat in the channel of NFET MN1. The energy is $(1/2)C_L V_{DD}^2$. This is the same amount of energy lost when the input voltage made a high-to-low transition previously. After the second transition, capacitor C_D lost energy given by

$$\frac{1}{2}C_D V_{DD}^2 - \frac{1}{2}C_D V_F^2 = \frac{C_D C_L}{2} \frac{2C_D + C_L}{(C_D + C_L)^2} V_{DD}^2 \rightarrow C_L V_{DD}^2 \qquad (4.9)$$

$$(\text{when } C_D \rightarrow \infty)$$

and this is the sum of the energies lost by the two transitions. In a real gate, overlap current flows and the loading gates have Miller effect (Sections 3.17 and 4.8). Further, the drain parasitic capacitances depend on the drain voltage. Therefore, the circuit of Fig. 4.4(a) must be simulated by a circuit simulator.

This straightforward procedure, however, still is not precise enough. There is an inevitable gate-to-drain capacitive coupling from the input voltage source that creates ambiguity. To determine the power-delay product accurately, we must use a reasonable number of fanouts (three), and we must also use a typical input rise time to include typical overlap current. Figure 4.4(b) shows a test circuit that satisfies these requirements.

In the circuit, the gates scale up by a ratio of 3 per stage. This is considered typical loading. Although this is only a loosely agreed-upon scale-up ratio, many practicing designers compare their notes assuming the ratio. The ratio is the closest whole number to the optimum scale-up ratio (2.718) discussed in Section 8.8. The first gate of the chain is driven by a pulse voltage source, and the typical loading of 3 from the second stage determines the proper rise-fall time of the signal. Inverters having size 3, 9, and 27 are supplied with voltage V_{DD} from a large capacitor C_D whose voltage is examined after the chain is driven by the pulse. The last, size 81 gate, provides proper loading to the last measured stage. Size 1 and size 81 gates are supplied by an independent DC power supply.

If the small energy loss of capacitor C_D is written as $\Delta E = C_D V_D \Delta V_D$, the power-delay product of a single gate is given by

$$\frac{\Delta E}{39 \times 2} = \frac{C_D V_D \Delta V_D}{78} \tag{4.10}$$

Although inverters are shown in Fig. 4.4(b), any gate may be substituted for the inverter.

The power-delay product of INV, NAND3, and NOR2 gates were determined by simulation using the circuit, with $V_{DD} = 5\,V$, temperature 65 degrees centigrade and the medium-current process, for recent generations of CMOS technologies. The results are plotted in Fig. 4.5.

In the 1.75-μm CMOS medium-current process for an inverter, C_L is given by

$$C_L = C_{DN} + C_{DP} + 3\left\{C_{G(NFET)} + C_{G(PFET)}\right\} = 0.137\,\text{pF}$$

where

$$C_{DN} = 0.0127\,\text{pF}, \; C_{GN} = 0.0113\,\text{pF}$$

$$C_{DP} = 0.0185\,\text{pF}, \; C_{GP} = 0.0223\,\text{pF}$$

Then an estimate of the power-delay product of an inverter is given by

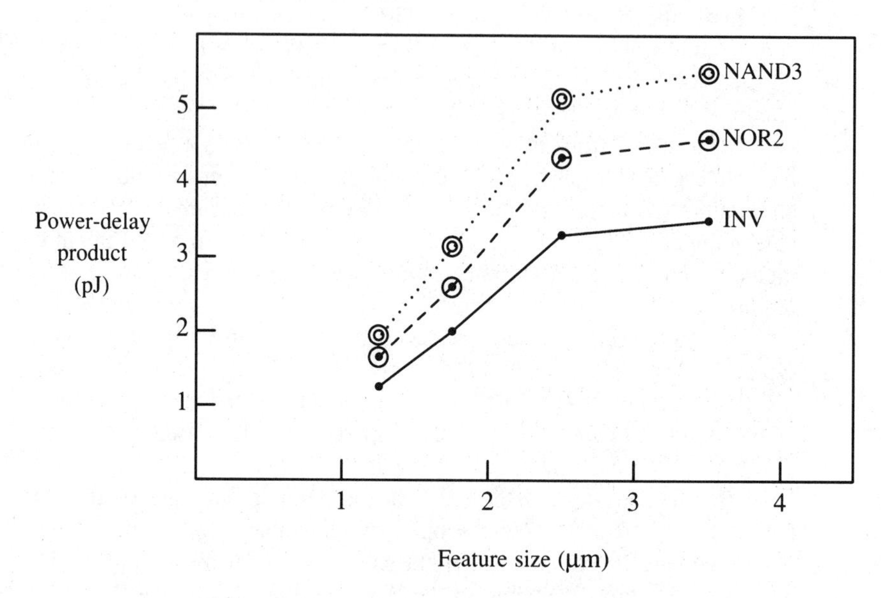

Figure 4.5 Power-delay product of CMOS gate.

$$(1/2)C_L V_{DD}{}^2 = (1/2) \times 0.137 \times 5^2 = 1.71 \text{ pJ}$$

The simulated power-delay product of 2.06 pJ is close but is larger than the estimated value because of the overlap current.

The power-delay product is not only useful in characterizing the CMOS process; it also has an application in estimating the power consumption of a CMOS VLSI chip. The power-delay product of a CMOS gate is especially clearly defined since the standby power is negligible. Mueller et al. analyzed the general case where the standby power is not zero [04]. It is important that the definition of power-delay product is clearly given before comparing different technologies with CMOS.

4.7 RING OSCILLATORS

A ring oscillator frequency measurement is often used to determine gate delays of CMOS circuits. A ring oscillator is N stages of cascaded inverting gates (N is an odd number) as shown in Fig. 4.6(a). We need to understand how oscillation occurs, and we need to correlate the ring oscillator frequency with gate delay.

Let us consider the case $N = 1$ in Fig. 4.6(a). The node settles at DC voltage V_E. If the input and output of an inverter are connected, the node is at voltage V_E. If the number of stages N is 3, 5, ..., all the nodes may be at voltage V_E, but the state is unstable. The N-stage ring oscillator shown in Fig. 4.6(a) can be modeled, in the vicinity of the DC bias point, by the linearized equivalent circuit of Fig. 4.6(b). We use the small deviation of the ith node voltage from V_E defined by $v_i = V_i - V_E$.

Parameters g_m, r_d, and C are the transconductance, the load resistance, and the load capacitance, respectively, of each small-signal amplifier. In a large-amplitude oscillation, the three linearized parameters r_d, g_m, and C are functions of the amplitude of oscillation, A. For convenience, the amplitude dependence is not explicitly written. The circuit equations are

$$C\frac{dv_1}{dt} = -\frac{v_1}{r_d} - g_m v_0 \tag{4.11}$$

$$C\frac{dv_2}{dt} = -\frac{v_2}{r_d} - g_m v_1$$

$$\cdots\cdots\cdots\cdots$$

$$C\frac{dv_{N-1}}{dt} = -\frac{v_{N-1}}{r_d} - g_m v_{N-2}$$

$$C\frac{dv_0}{dt} = -\frac{v_0}{r_d} - g_m v_{N-1}$$

where N is the number of cascaded stages (an odd integer). Substituting $j\omega$ for the differential operator, rearranging the equations, and then multiplying all the equations together, we get

$$\left(j\omega C + \frac{1}{r_d}\right)^N v_0 v_1 \ldots v_{N-1} = -g_m{}^N v_0 v_1 \ldots v_{N-1} \tag{4.12}$$

For an oscillation to start,

$$\left(j\omega\frac{C}{g_m} + \frac{1}{r_d g_m}\right)^N = -1 \tag{4.13}$$

must be satisfied. Therefore,

$$j\omega\frac{C}{g_m} + \frac{1}{r_d g_m} = \cos\left(k\frac{\pi}{N}\right) + j\sin\left(k\frac{\pi}{N}\right) \tag{4.14}$$

where $k = 0, 1, 2, \ldots, N-1$. When N is large and k is small, Eq. (4.14) gives the angular frequency as a function of k,

$$\omega \approx \frac{g_m}{C}\frac{\pi}{N}k$$

Equation (4.12) may be rewritten as

$$\left|\frac{v_1}{v_0}\right|\left|\frac{v_2}{v_1}\right| \ldots \left|\frac{v_0}{v_{N-1}}\right| = \frac{\alpha}{\left(1 + \omega^2 C^2 r_d{}^2\right)^{N/2}} \tag{4.15}$$

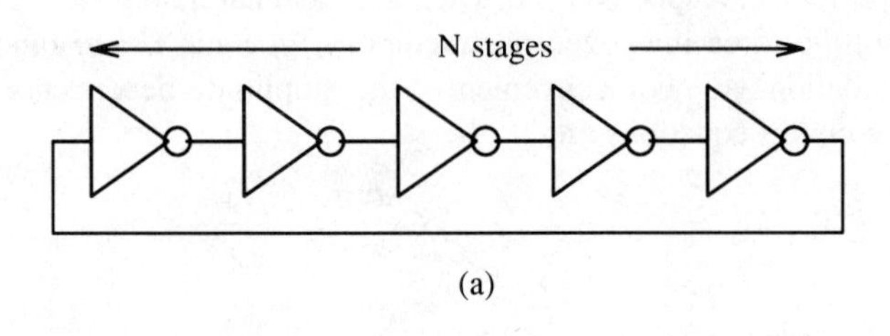

(a)

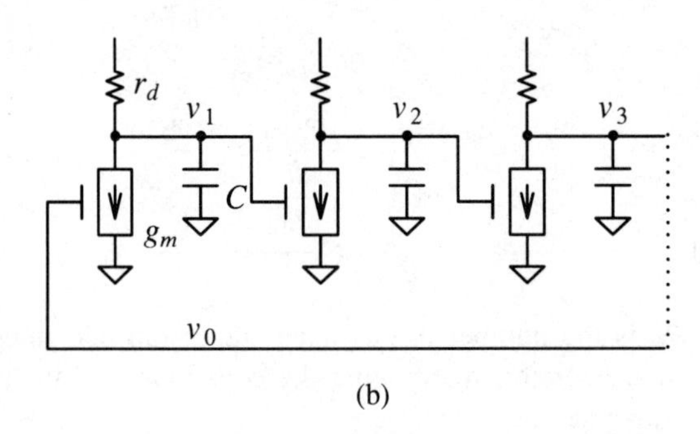

(b)

Figure 4.6 Analysis of a ring oscillator.

where the left-hand side is the open-loop gain, $\alpha = G^N$, and $G = g_m r_d$ is the DC gain of a single stage. Because

$$\omega C r_d \approx k \frac{\pi}{N} g_m r_d$$

the right-hand side of Eq. (4.15) is estimated to be

$$\frac{\alpha}{\left[1 + \omega^2 C^2 r_d{}^2\right]^{N/2}} \approx \alpha \exp\left[-\frac{\pi^2}{2} \frac{G^2}{N} k^2\right] \quad (N \to \infty) \tag{4.16}$$

The open-loop gain of Eq. (4.15) is a decreasing function of k. Therefore, an oscillation occurs almost always at $k = 1$. The frequency is given by

$$\omega = \frac{g_m}{C} \sin\left[\frac{\pi}{N}\right] \approx \frac{\pi g_m}{CN} \tag{4.17}$$

and the amplitude of oscillation is determined from

$$\frac{1}{G(A)} = \frac{1}{r_d(A) g_m(A)} = \cos\frac{\pi}{N} \tag{4.18}$$

Parameters g_m *and* r_d are both a decreasing function of the amplitude of oscillation, A. Since $G(A)$ is the voltage gain of a single stage, the gain saturation at large amplitude determines the final amplitude of oscillation. When the number of stages N is large, the minimum and the maximum voltages are V_{SS} and V_{DD}, respectively.

The one-stage circuit ($N=1$) is absolutely stable as observed from Eq. (4.18). The phase shift available by one-stage is not enough to create oscillation. The one-stage circuit is often used to generate a DC bias point for a linear CMOS amplifier. The circuit with $N=3$ shows either an oscillation or a damped oscillation. If $N=5$ or more, a stable oscillation is observed. In practice many more stages (the author once used 13 stages) are cascaded. As the number of stages N increases, the waveform changes from a nearly sinusoidal oscillation to a nearly square-wave oscillation. In the limit of square-wave oscillation, each inverter has the delay for driving a single load. A ring oscillator that has more than one load is designed by adding "dummy" load to each stage. One of the dummy load (inverter) can be used to observe oscillation. The period of oscillation is proportional to the number of cascaded stages. If the period of $k=1$ oscillation is divided by N, the pair delay of a gate (sum of pullup and pulldown delays) is determined.

Oscillations for $k>1$ are observed under certain conditions [05].

4.8 MILLER EFFECT IN CMOS GATES

Capacitive load to a CMOS gate is, in most cases, equivalent to a capacitor connected between the output node of the gate and ground (or V_{DD}). In the special cases we are going to study now, the other end of the capacitor is not connected to ground or V_{DD} (or any other constant-voltage source) but to the time-dependent voltage source $V_0(t)$ as shown in Fig. 4.7(a). Then the load impedance the CMOS driver sees becomes different from capacitance C_m. Current $I_1(t)$ and voltage $V_1(t)$ of Fig. 4.7(a) satisfy the equation

$$I_1(t)=C_m\frac{d}{dt}\left\{V_1(t)-V_0(t)\right\}$$

Voltage $V_0(t)$ consists of the part independent of $V_1(t)$, $V_0^*(t)$, and the part determined from $V_1(t)$. This latter part may be complex. Therefore, we consider a simple and manageable case: The dependent part is proportional to the value of V_1 at time τ before, $V_1(t-\tau)$, where τ is a small delay time.

$$V_0(t)=-AV_1(t-\tau)+V_0^*(t)$$

where $-A$ is a constant whose significance will be clear shortly. We have (for small τ)

$$V_0(t)=-A\left\{V_1(t)-\frac{dV_1(t)}{dt}\tau\right\}+V_0^*(t)$$

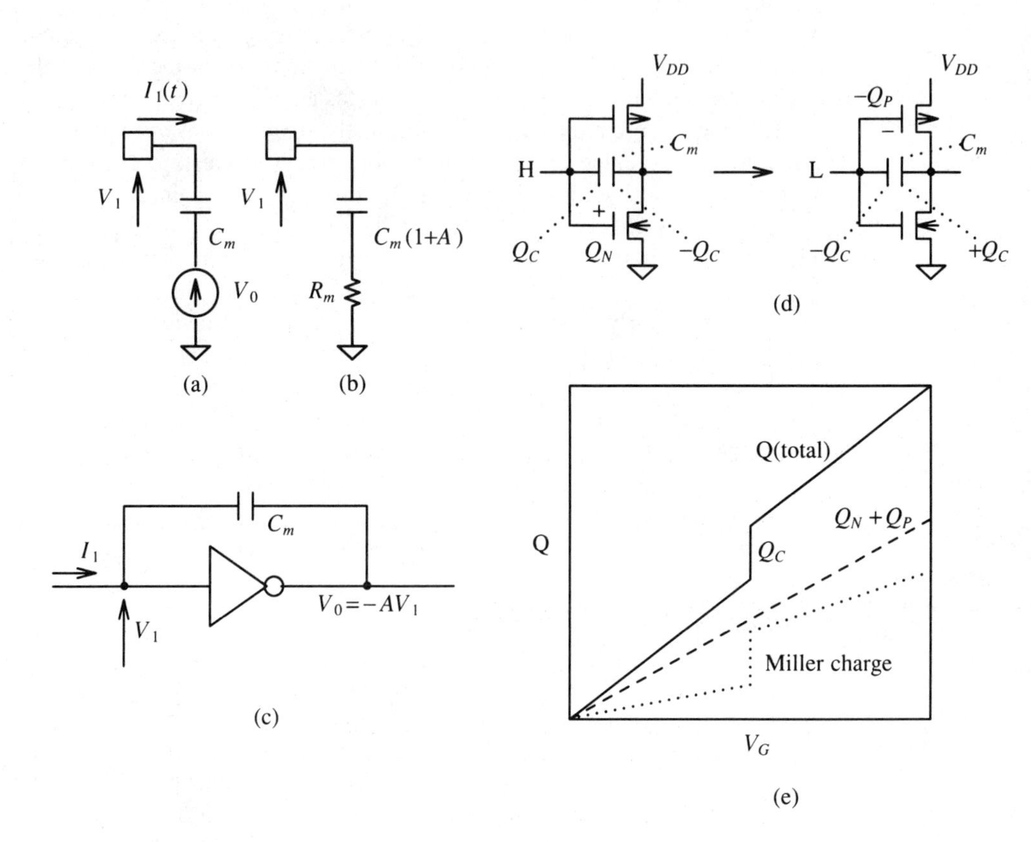

Figure 4.7 Mechanism of Miller effect in CMOS gate.

Then

$$I_1(t) = C_m(1+A)\frac{d}{dt}\left\{V_1(t) - R_m C_m(1+A)\frac{dV_1(t)}{dt}\right\} - C_m\frac{dV_0^*(t)}{dt}$$

where $R_m = A\tau/C_m(1+A)^2$. Since τ is small, R_m is small. This equation can be interpreted as follows. The last term is the current supplied from the independent part of voltage source $V_0^*(t)$. The first term is equivalent to the circuit shown in Fig. 4.7(b). The capacitance C_m is changed to $C_m(1+A)$ and a series resistance R_m that can be either positive or negative depending on τ, appeared in series with the capacitor C_m.

As we see, the effects of the dependent voltage source at the far end of capacitor is complex, and many interesting effects occur. We begin with a simple example of an inverting amplifier, and we examine the problem in great detail.

Suppose that the inverting amplifier of Fig. 4.7(c) has voltage gain A, infinite input impedance and zero output impedance, and zero delay, and that the input and output of the inverter are connected by a capacitor C_m. Then the input current I_1 at angular frequency ω is given by

$$I_1 = j\omega C_m(1+A)V_1$$

and therefore the input admittance of the amplifier, Y_1, is given by [06],

$$Y_1 = \frac{I_1}{V_1} = j\omega C_m(1+A) \tag{4.19}$$

Equation (4.19) shows that because of the voltage gain of the inverter, capacitance C_m was increased $(1+A)$ times to determine the input admittance of the inverter. Any voltage amplifier has this effect, which is called the Miller effect.

In this section we examine the mechanism of Miller effect in a CMOS gate in detail. The gate-drain coupling capacitance, called the Miller capacitance, originates from the gate-drain overlap capacitance of an FET, from gate-to-channel capacitance that is further coupled to drain, and also from the parasitic wiring capacitance. Classification of gate capacitance to individual components is discussed in Section 1.4. The effect of the rest of the gate capacitance is just to add extra capacitive admittance, and therefore the effect is trivial. The gate capacitance is not included in the following analysis for simplicity.

When a CMOS inverter switches from the high to the low input state, the states of charge of the NFET gate, PFET gate, and the Miller capacitance C_m change as shown in Fig. 4.7(d). Before complete the switching the NFET gate charge Q_N must be drained, negative charge $-Q_P$ must be supplied to the PFET gate, charge Q_C, must be drained from the Miller capacitor, and further, charge $-Q_C$ must be supplied. The total charge that must be moved by the driver is

$$Q_N - (-Q_P) + Q_C - (-Q_C) = Q_N + Q_P + 2Q_C \tag{4.20}$$

As the input voltage changes, the gate charge varies continuously. As for the charge stored in the Miller capacitance, half the change (Q_C) occurs abruptly at the time when the output of the inverter switches, as shown by the dotted curve of Fig. 4.7(e). The change is quite crisp since a CMOS inverter has the highest small-signal gain at the switching point, and the output voltage changes with very small input voltage change.

The Miller effect occurs in any voltage amplifier. The effect is, essentially, a small-signal, linear amplifier problem, and therefore the analysis of the effect does not depend very much on the model of the active device. Therefore, we use the simplest MOSFET model, assuming the gradual, long-channel approximation. To understand the Miller effect in digital switching, however, the effects of many small-signal Miller effects must be combined to find the entire picture. It is convenient to use the static characteristics of the inverter from Section 3.2. Referring to Fig. 4.8(a), the current of NFET MN1 is given by

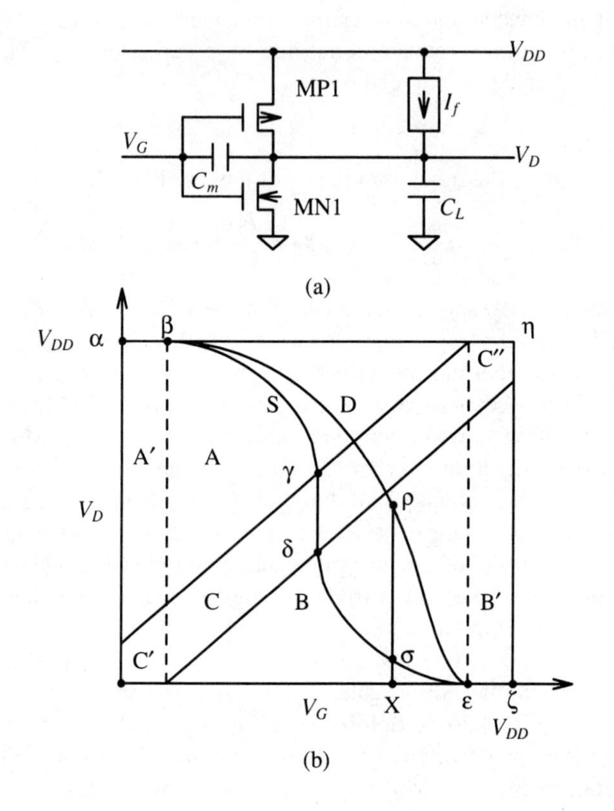

Figure 4.8 Analysis of Miller effect using clamped inverter model.

$$I_N = B_N \left\{ V_G - V_{THN} - (1/2)V_D \right\} V_D \quad V_D < V_G - V_{THN} \tag{4.21}$$

$$I_N = \frac{B_N}{2}(V_G - V_{THN})^2 \quad V_D > V_G - V_{THN}$$

and the current of PFET MP1 is

$$I_P = B_P \left\{ V_{DD} - V_G - V_{THP} - (1/2)(V_{DD} - V_D) \right\} (V_{DD} - V_D) \tag{4.22}$$

$$V_D > V_G + V_{THP}$$

$$I_P = \frac{B_P}{2}(V_{DD} - V_G - V_{THP})^2 \quad V_D < V_G + V_{THP}$$

The static characteristic of the inverter is determined by setting

$$I_N = I_P$$

The state of the inverter is represented by the square region of Fig. 4.8(b), which consists of the three regions A, B, and C. In region A, PFET MP1 is in the triode region and NFET MN1 is in saturation. In region C, both PFET MP1 and NFET MN1 are in saturation, and in region B, NFET MN1 is in the triode region and PFET MP1 is in saturation. We have in region B (see Section 3.1),

$$V_D = V_G - V_{THN} - \left\{ (V_G - V_{THN})^2 - \frac{B_P}{B_N}(V_{DD} - V_G - V_{THP})^2 \right\}^{1/2}$$

In region A a corresponding relationship exists. In region C the static characteristic is a vertical-line segment given by

$$V_{GTR} = \frac{\left[\dfrac{B_P}{B_N}\right]^{1/2} (V_{DD} - V_{THP}) + V_{THN}}{\left[\dfrac{B_P}{B_N}\right]^{1/2} + 1}$$

When the rate of change of the input voltage is much smaller than

$$\frac{V_{DD}}{\tau_0}(V/s) \text{ where } \tau_0 = \left[\frac{C_L + C_m}{B_N V_{DD}}\right] \tag{4.23}$$

and where C_m and C_L are the Miller and the load capacitances, respectively, the inverter switches following the static characteristic. With reference to Fig. 4.8(b), when input voltage V_G changes from 0 to V_{DD}, the output voltage V_D changes following the solid curve S, from point α to β, γ, δ, ε, and ζ. When the input voltage changes rapidly, however, the inverter switches following curve D. The instantaneous output voltage $V_D = \overline{\rho X}$ is unable to catch up to the steady-state voltage $\overline{\sigma X}$, except before (point α) and after (point ζ). During switching the inverter is not in the steady state.

In rapid switching, the inverter goes through states like ρ in Fig. 4.8(b), in which NFET MN1 and PFET MP1 of Fig. 4.8(a) carry different currents. Therefore the state, if left alone, returns to steady state σ. It is possible, however, to maintain the state ρ by injecting the difference of the current of PFET and NFET (I_f) by the fictitious current generator shown in Fig. 4.8(a). The current generator has an infinite internal impedance, and therefore the small-signal operation of the inverter at state ρ is not disturbed. The Miller effect of the inverter during switching can be studied from the small-signal characteristics of this "clamped" inverter, using the equivalent circuit of Fig. 4.8(a).

The transconductance and the drain conductance of the FETs are defined by

$$G_{mP} = -\frac{\partial I_P}{\partial V_G} \quad G_{mN} = \frac{\partial I_N}{\partial V_G} \tag{4.24}$$

$$G_{DP} = -\frac{\partial I_P}{\partial V_D} \quad G_{DN} = \frac{\partial I_D}{\partial V_D} \tag{4.25}$$

Using Eqs. (4.21) and (4.22), the results shown in Tables 4.3 and 4.4 are derived.

TABLE 4.3 TRANSCONDUCTANCE

Region	G_{mP}	G_{mN}
A	$B_P(V_{DD} - V_D)$	$B_N(V_G - V_{THN})$
A′	$B_P(V_{DD} - V_D)$	0
B	$B_P(V_{DD} - V_G - V_{THP})$	$B_N V_D$
B′	0	$B_N V_D$
C	$B_P(V_{DD} - V_G - V_{THP})$	$B_N(V_G - V_{THN})$
C′	$B_P(V_{DD} - V_G - V_{THP})$	0
C″	0	$B_N(V_G - V_{THN})$

TABLE 4.4 DRAIN CONDUCTANCE

Region	G_{DP}	G_{DN}
A	$B_P(V_D - V_G - V_{THP})$	0
A′	$B_P(V_D - V_G - V_{THP})$	0
B	0	$B_N(V_G - V_{THN} - V_D)$
B′	0	$B_N(V_G - V_{THN} - V_D)$
C	0	0
C′	0	0
C″	0	0

Using Tables 4.3 and 4.4 and the equivalent circuit of Fig. 4.8(a), the small-signal circuit equation of the clamped inverter is written as

$$C_m \frac{d}{dt}(\Delta V_G - \Delta V_D) - (G_{DP} + G_{DN})\Delta V_D - (G_{mP} + G_{mN})\Delta V_G - C_L \frac{d}{dt}\Delta V_D = 0 \tag{4.26}$$

Assuming sinusoidal variation the operator (d/dt) can be replaced by $j\omega$ and we obtain

$$\Delta V_D = \left[\frac{j\omega C_m - G_m}{j\omega C_T + G_D} \right] \Delta V_G \tag{4.27}$$

where

$$G_m = G_{mP} + G_{mN}, \quad G_D = G_{DP} + G_{DN}$$
$$C_T = C_L + C_m$$

From Eq. (4.27) the input current ΔI_G is found as

$$\Delta I_G = C_m \frac{d}{dt}(\Delta V_G - \Delta V_D) \tag{4.28}$$
$$= j\omega C_m \frac{j\omega C_L + G_D + G_m}{j\omega C_T + G_D} \Delta V_G$$

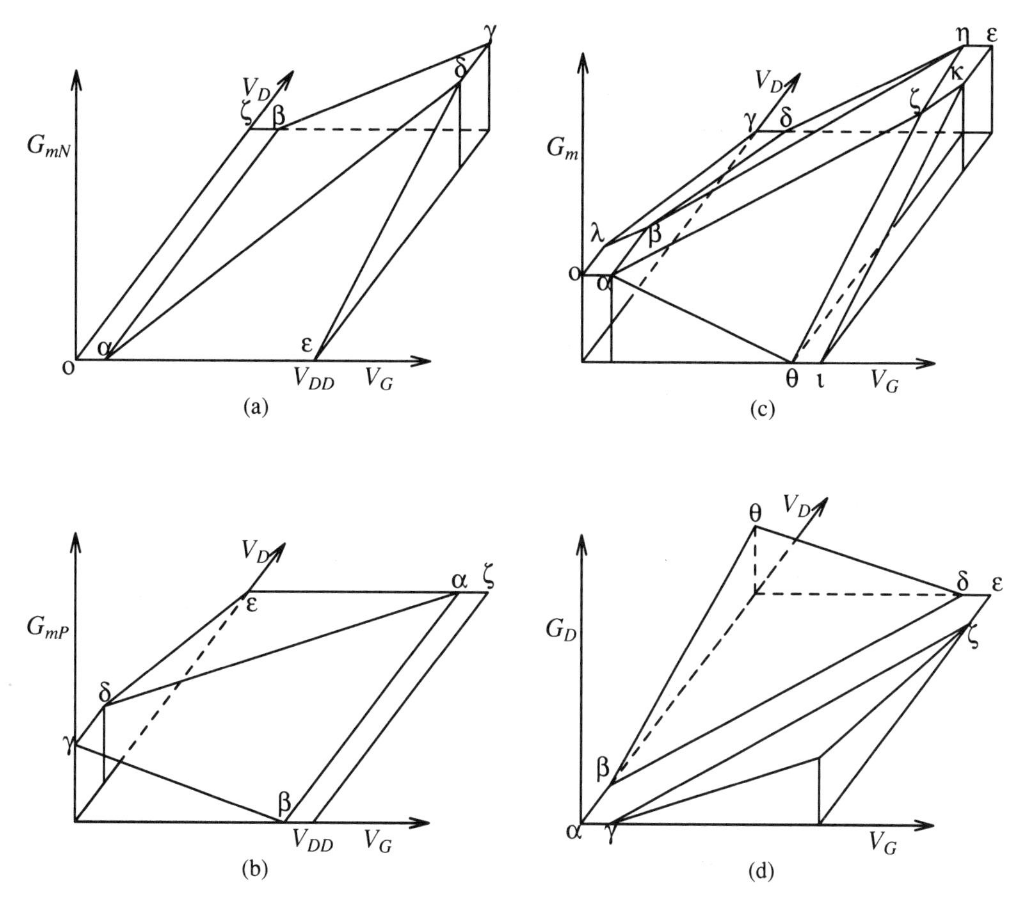

Figure 4.9 Transconductance and drain conductance of CMOS inverter.

Remember that this equation gives the input current due to the Miller capacitance only. The current of the gate capacitance is not included. If $\omega C_L \ll G_D + G_m$ *and* $\omega C_T \ll G_D$,

$$\Delta I_G = j\omega C_m \left(1 + \frac{G_m}{G_D}\right) \Delta V_G \tag{4.29}$$

Since G_m/G_D is the small-signal gain of the inverter, Eqs. (4.29) and (4.19) are identical. This is the case where the conductance of the amplifier dominates over the capacitive load at the operating frequency.

Equation (4.29) contains two conductance parameters G_D and G_m. Figure 4.9(a) and (b) show plots of G_{mN} and G_{mP} versus voltages V_G and V_D on the $V_G - V_D$ coordinate plane [the same square region as shown in Fig. 4.8(b)]. The plot can be made directly using the formula in Tables 4.3 and 4.4. The two-dimensional plot of G_{mN} consists of a rectangle $(0\alpha\beta\zeta)$, a triangle $(\alpha\delta\varepsilon)$, and a trapezoid $(\alpha\beta\gamma\delta)$. Since

$$G_m = G_{mN} + G_{mP} \tag{4.30}$$

The plot of Fig. 4.9(c), G_m, is a superposition of the two plots of Fig. 4.9(a) and (b). The plot is a ridge-shaped surface, high along the diagonal of the square area, consisting of four trapezoids, $(0\alpha\beta\lambda), (\beta\delta\gamma\lambda), (\theta\iota\kappa\zeta)$ and $(\varepsilon\kappa\zeta\eta)$; two triangles, $(\alpha\theta\zeta)$ and $(\beta\delta\eta)$; and a rhombic, $(\alpha\beta\eta\zeta)$. Similarly, Fig. 4.9(d) shows the two-dimensional plot of G_D.

From Figs. 4.9(c) and 4.9(d), G_m/G_D is large in the area centered along the diagonal of the square area of the plot. When $V_G - V_{THN} < V_D < V_G + V_{THP}$, $G_D = 0$, and therefore $G_m/G_D = \infty$. Then the capacitance defined by Eq. (4.28) becomes very large but not infinity. When $G_D = 0$, Eq. (4.28) becomes

$$\Delta I_G = \frac{C_m}{C_T}(j\omega C_L + G_m)\Delta V_G \tag{4.31}$$

$$\rightarrow \frac{C_m}{C_T} G_m \Delta V_G \quad (\text{when } \omega \rightarrow 0) \tag{4.32}$$

Equation (4.32) shows that when $V_G - V_{THN} < V_D < V_G - V_{THP}$, the input admittance of the inverter becomes resistive due to the Miller effect.

This effect occurs because the FETs work as ideal current generators if $G_D = 0$. The current generator is in series with Miller capacitance C_m, and therefore the capacitance becomes invisible. The effect is especially significant when $C_m \approx C_T$. Equation (4.32) shows that Miller "capacitance" is an incorrect term. The input impedance of a Miller effect dominated inverter in this case is a resistance, R_m. The resistance R_m is estimated as

$$R_m = \frac{C_m + C_L}{C_m} \frac{1}{G_m} \geq \frac{1}{2\left(B_N \dfrac{V_{DD}}{2}\right)} = \frac{1}{B_N V_{DD}} \tag{4.33}$$

assuming that $G_{mN} = G_{mP}$ and $C_m = C_L$. The Miller effect resistance R_m is the inverse of the transconductance of the amplifier. When the input impedance becomes resistive, the input of the inverter is effectively tied to the bias voltage at that moment by resistance R_m, and therefore the input voltage is momentarily clamped.

The switching of a Miller effect dominated inverter is illustrated in Fig. 4.10. In Fig. 4.10(a) an inverter with Miller effect is driven by a step voltage source that has

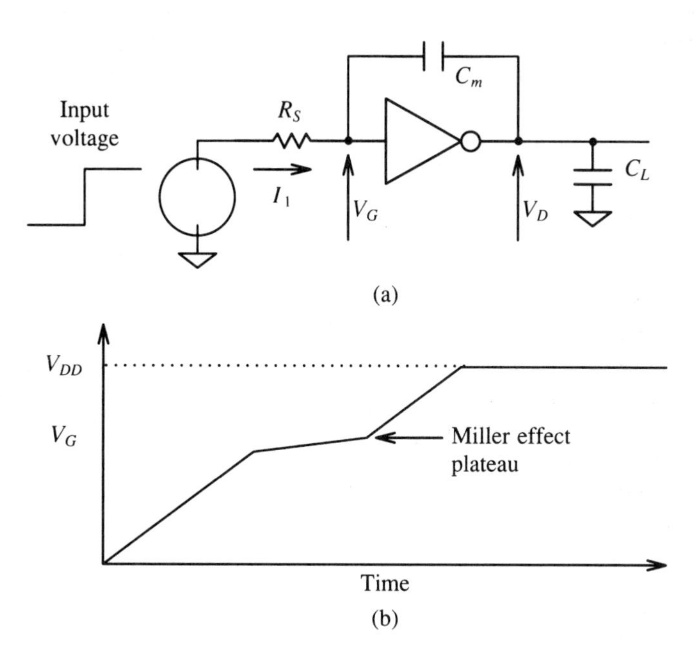

Figure 4.10(a) and (b) Switching waveform distortion by Miller effect.

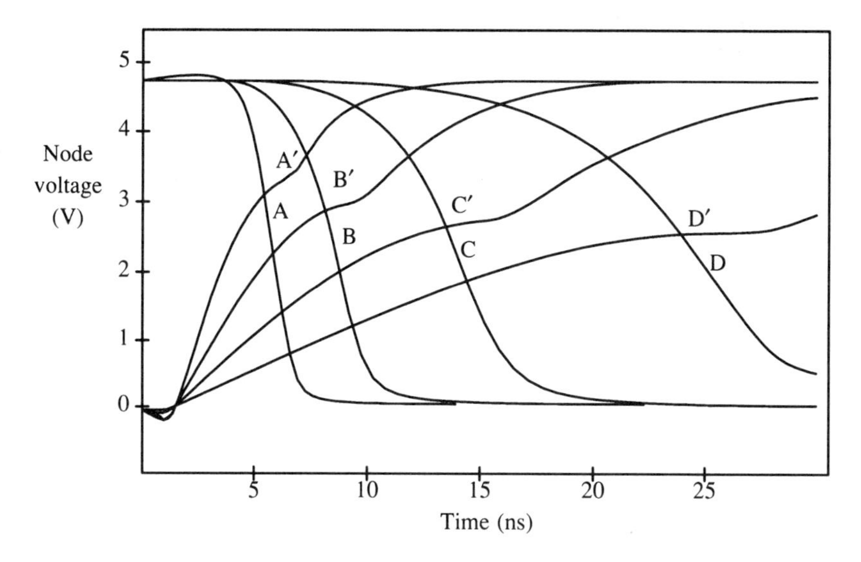

Figure 4.10(c) Miller effect plateau.

internal resistance R_S. When R_S is small and the rise time of the input voltage is very short, switching of the gate follows the inverse L-shaped track of Fig. 4.8(b), ($\alpha\beta\eta\zeta$). The Miller capacitance C_m acts parallel to load capacitance C_L. The capacitance from the input, C_m, dumps charge $C_m V_{DD}$ to the output node, and therefore the inverter begins to switch from output voltage $(1+(C_m/C_T))V_{DD}$. This feedforward is always associated with Miller effect.

When the input voltage rise time is comparable to the switching time of the inverter, the input/output voltages follow curve D of Fig. 4.8(b), and if the rise time is very slow, the voltages follow curve S of Fig. 4.8(b). When the point on the curve approaches area C, the input capacitance increases due to the Miller effect, and the rate of increase of input voltage V_G declines. When the point goes across area C, the input impedance of the inverter contributed by the Miller effect becomes resistive, and the rate of increase of input voltage V_G declines even further. As the point moves out from area C of Fig. 4.8(b), input voltage V_G increases faster again.

The waveform of input voltage V_G is shown in Fig. 4.10(b). A plateau of the input voltage is observable when the representative point of the inverter goes across area C of Fig. 4.8(b). The time required for the point to move across area C is estimated to be

$$f\frac{C_m V_{DD}}{I_1} \approx f\frac{C_m V_{DD}}{\frac{1}{R_S}\frac{V_{DD}}{2}} = 2fR_S C_m$$

where I_1 is the current the driver supplies, and f is a numerical factor of the order of unity. If switching threshold voltage of the inverter is V_{GTR}, $I_1 \approx (V_{DD} - V_{GTR})/R_S$ from Fig. 4.10(a). The waveform distortion by the Miller effect is more significant at slow input switching. Simulation results (using 2.5-μm CMOS, low current process) are shown in Fig. 4.10(c). Curves A, B, C, and D show V_D and A′, B′, C′, and D′ show V_G of Fig. 4.10(a). A′–A has load/driver size ratio 4, B′–B has 8, C′–C has 16, and D′–D has 32. With increasing size ratio the input transition time increases. As the input transition time increases, the voltage at which the plateau occurs decreases, and the plateau voltage approaches the static switching threshold voltage. The theory explains all Miller effect phenomena, and therefore the theory is a very useful guide in circuit design.

The waveform distortion shown in Fig. 4.10(b) creates uncertainty in the time when the input voltage to a gate crosses the switching threshold voltage of the gate, thereby causing inaccuracy in delay time predicted by simulation of the gate. Let us consider an inverter driving two inverters, one of which is very much larger than the other. The signal delay through the critical path through the smaller inverter depends on the voltage of the Miller effect plateau caused by the larger inverter. To avoid significant waveform distortion the input node should be driven hard: Fanout should be limited to less than about 6.

It should be noted that ΔI_G of Eqs. (4.28), (4.29), and (4.31) does not include the gate capacitances of the FETs. The total gate current, ΔI_G(total), is given by

$$\Delta I_G(\text{total}) = \Delta I_G + j\,\omega C_{FET}\,\Delta V_G \tag{4.34}$$

$$= j\,\omega \left\{ C_{FET} + C_m(1 + A) \right\} \Delta V_G$$

where $A = G_m/G_D$ and where C_{FET} is the sum of the gate capacitances of the PFET and NFET. A CMOS inverter in digital circuit has unity gain (large signal gain = 1). To determine the effective capacitive load contributed by the Miller effect, we separate the FET capacitance and the Miller capacitance of an inverter, by specifying the following measurement method. From Eq. (4.20) and Fig. 4.7(d) we have

$$C_{GATE}V_{DD} = Q_N + Q_P + Q_C \quad \text{and} \quad C_m V_{DD} = Q_C$$

where $C_{GATE} = C_{FET} + C_m$. We measure charges and from the measured charges we determine the capacitances. This procedure, however, is not straightforward because gate capacitance also depends on gate and drain voltages. As it is shown in Sections 1.4 and 6.3, gate capacitance decreases slightly as FET operating point moves from triode region to saturation region, reflecting decrease in carrier density at the drain's end of the channel.

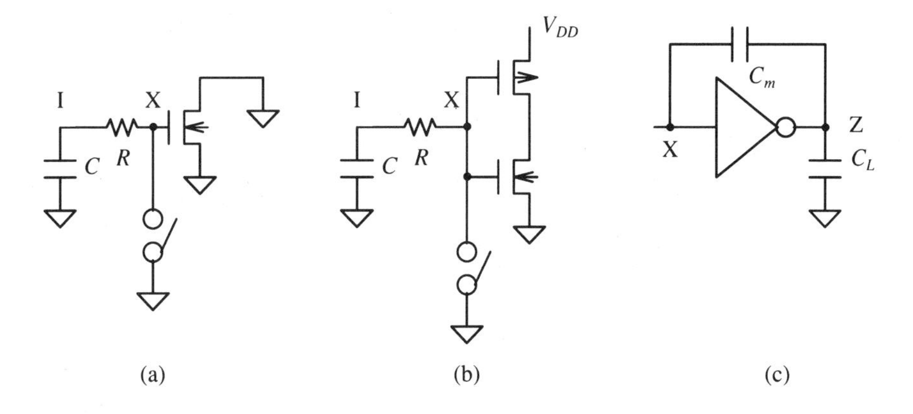

Figure 4.11 Method of measurement of Miller capacitance.

In the present-day CMOS technology (1.75μm feature size), however, gate capacitance includes gate to source, and gate to drain overlap capacitance (Section 1.4) that amounts to as much as 40% of the total capacitance. The overlap capacitance is voltage-independent. Then small voltage dependence of gate to channel capacitance becomes insignificant. This is not true in submicron CMOS that uses advanced processing technology to produce very small gate to source/drain overlap. Subject to the assumption that gate capacitance is approximately voltage-independent, C_{GATE} and C_m can be defined as follows. The circuits of Fig. 4.11(a) and (b) are used to determine the capacitance. In Fig. 4.11(a) and (b), capacitance C is charged to V_1, and node X is at ground potential until time $t = 0$ is reached. At that moment the switch opens and node X is released. After some time, nodes I and X approach the same voltage. If the voltage is V_0 we have

$$CV_1 = (C_X + C)V_0 \quad \text{or} \quad C_X = \left[\frac{V_1}{V_0} - 1\right] C \tag{4.35}$$

where C_X is the capacitance to be measured (either the gate capacitance of the FETs or the input capacitance of the inverter). The circuit of Fig. 4.11(a) measures the gate capacitance of FET, and that of Fig. 4.11(b) the input capacitance of the inverter, which is the gate capacitance C_{GATE} plus the Miller capacitance C_m.

In the Miller effect capacitance measurement circuit, the capacitance C must be large enough to guarantee that $V_0 > V_{DD} - V_{THP}$, and V_0 should be as close to V_{DD} as possible. Simultaneously, the measurement accuracy must be maintained. The resistance must be properly selected to guarantee that both node X and the output node of the inverter arrive at the steady-state voltages at the measurement time. The method can be used in simulation as well as in experiments. Practically, gate capacitance is measured by integrating an electrometer and sample device in a single chip [07]. The accuracy limit is about $10^{-15} F$.

The gate capacitance and the Miller capacitance were determined by simulation using the method above and 1.75-μm CMOS with the medium current process as follows.

PFET gate capacitance (7.25 μm wide) = 0.0223 pF

NFET gate capacitance (4.00 μm wide) = 0.0113 pF

Inverter input capacitance (7.25-μm PFET and 4.00 - μm NFET) = 0.0411 pF

The input capacitance of the inverter is larger than the sum of the gate capacitances of the NFET and the PFET, and the difference,

$$0.0411\,\text{pF} - [0.0223\,\text{pF} + 0.0113\,\text{pF}] = 0.075\,\text{pF}$$

is the Miller capacitance C_m. This result shows that about 20% of the input capacitance is contributed from the Miller effect. This result is used in Section 8.4.

When Miller capacitance is determined using simulation, the result depends to a certain extent on the particular FET model used. We discussed in Section 1.4 that FET gate capacitance consists of gate-to-substrate, gate-to-channel, gate-to-source and gate-to-drain capacitances. Different FET models assign the capacitances differently.

The Miller capacitance depends on many parameters, including the node voltage slew rate, the threshold voltage of the gate, and the loading of the gate. In Figure 4.11(c), the driver of node X sees no Miller capacitance if loading of node Z by capacitance C_L is so large that the inverter is unable to switch while node X voltage is changing. Node X, after reaching nearly the final voltage, gets a continuous, low-level current from node Z until switching of node Z is finished.

4.9 FEED FORWARD

A Miller capacitance that couples the input and output of an inverter has another effect that requires attention. This is a feedforward of the input voltage. In Fig. 4.12(a) node I is originally ground potential and node O at V_{DD}. When node I makes a low-to-high transition, the voltage of node O is temporarily driven up above V_{DD}, before the NFET pulls node O down. If the NFET is very small and the input transition time is short, the feedforward voltage, V_{BK}, can be estimated from the law of charge conservation as

$$C_L V_{BK} + C_m (V_{BK} - V_{DD}) = (C_m + C_L) V_{DD} \quad \text{or} \quad V_{BK} = V_{DD} + \frac{C_m}{C_L + C_m} V_{DD}$$

The duration of the positive spike, T_{BK}, is estimated from the time the NFET requires to drain the extra charge as

$$t_{BK} = (C_L + C_m) \frac{V_{BK} - V_{DD}}{I_S}$$

Assuming that $C_L = 0.12\,\text{pF}$ (typical minimum-size inverter for 1.75 μm CMOS with three minimum-size loads), and estimating C_m as 1/8 of C_L, $V_{BK} - V_{DD} = 0.55V$. Further assuming that $I_S = 600\ \mu A$, $T_{BK} = 0.123$ ns. This estimate assumes that the input voltage switches instantly.

The estimate of the feedforward voltage is valid for small C_m / C_L. If $V_{BK} - V_{DD}$ is higher than the forward voltage drop of a PN junction, the voltage is clamped by the PN junction diode of the drain islands. An effect of the feedforward is that as the P+ diffused island is driven above the V_{DD} supply voltage, minority holes are injected into the N-tub [08].

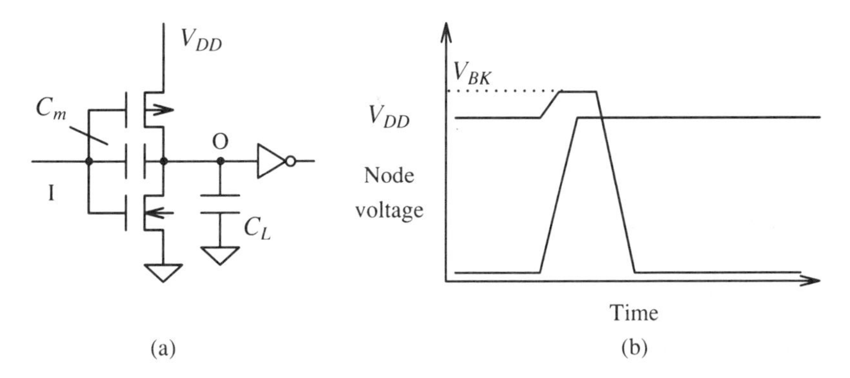

Figure 4.12 Feedforward effect.

The density of minority carriers in the substrate of an operating CMOS chip is expected to be higher than the density in the equilibrium semiconductor. Since the minority-carrier lifetime is of the order of a few microseconds, the injected carriers stay in the bulk for a long time. The excess minority carriers will influence the hold time of charge on the dynamic circuit nodes and the retention time of dynamic memories. Minority-carrier injection into substrates of MOS RAM due to potential nonuniformity in substrate was reported [09]. Feedforward aggravates the problem.

4.10 NEGATIVE RESISTANCE AND NEGATIVE CAPACITANCE

The switching of a CMOS gate is the process of charging and discharging its load capacitor by FET resistances. The delay time is reduced by decreasing the FET resistance or the load capacitance. Conventional methods of reducing delay are to optimize the logic design, adjust the FET sizes, and minimize the layout parasitics. In the following sections several techniques to reduce delay by unconventional means are discussed.

When an inverter whose equivalent internal resistance is R_D drives a capacitive load C_L as shown by Fig. 4.13(a), the delay time is $T_D = R_D C_L$. If a negative resistance $-R$ is inserted in series with the driver as shown in Fig. 4.13(b), the delay time is reduced to T_D', given by

$$T_D' = (R_D - R)C_L = T_D - (RC_L)$$

If a negative capacitance $-C$ is added in parallel with the load capacitance as shown in Fig. 4.13(c) the delay time is reduced to T_D'', given by

$$T_D'' = R_D(C_L - C) = T_D - (R_D C)$$

The problems are twofold: (1) creating negative resistance and negative capacitance, and (2) determining the characteristics of circuits that include negative components. These problems are discussed in the following sections.

4.11 NEGATIVE RESISTANCE

A number of circuits that generate negative resistance have been reported [10] [11]. The lambda diode discussed in Section 2.10 is an example. Another, more conventional technique uses a differential amplifier, as shown in Fig. 4.14(a) A pair of cross-coupled CMOS inverters are driven by voltage source V_{IN} that has internal resistance R_I. Inverters 0 and 1 are symmetrical, i.e., the switching thresholds are both at $V_{DD}/2$ and they have the same pullup and pulldown drive capability. The pair of inputs are biased at the switching threshold. A small voltage difference between terminals A_1 and A_0 is amplified and nodes A_1 and A_0 are driven by the inverters such that the voltage difference increases. This is equivalent to reduction in the internal resistance of the voltage source, R_I, and therefore the impedance looked into terminals A_1 and A_0 is equivalent to a negative resistance. At the beginning the voltages of nodes A_1 and A_0 are both $V_{DD}/2$. This bias point is stable if R_I is small, but the bias point

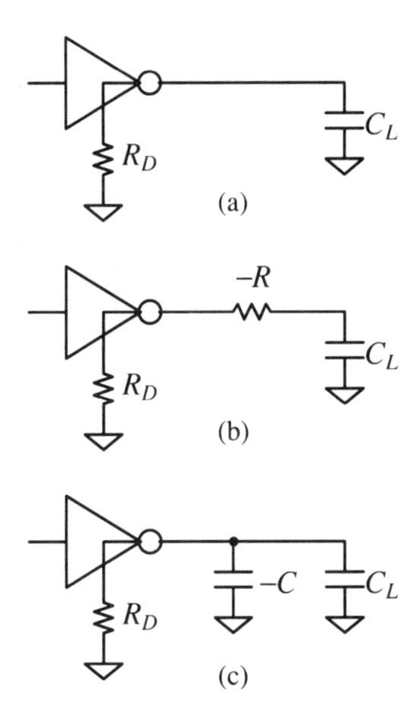

Figure 4.13 Reducing delay using negative resistance and capacitance.

becomes unstable if R_I is large. The transition from a stable to an unstable biasing point is relevant to negative resistance.

To study the stability of the bias point, the capacitors are irrelevant. The circuit of Fig. 4.14(a), without the voltage source, can then be redrawn as shown in Fig. 4.14(b) and the problem is to investigate this circuit. It is convenient to generalize the two-stage cascaded inverters to a long chain of cascaded inverters shown in Fig. 4.14(c) and to study the characteristic of the chain. All the nodes of the chain are originally biased to $V_{DD}/2$, and the small deviation of the i th node voltage, v_i, is defined by

$$v_i = V_i - \frac{1}{2}V_{DD}$$

The small currents j_i are used as variables. They satisfy

$$g_m v_{i-1} + j_i = j_{i+1}$$

where the inverters are modeled by transconductance g_m.

When the $j_i s$ are eliminated from the equations using

$$v_i - v_{i-1} = Rj_i$$

where $R = 2R_I$, since the two shunt resistors in Fig. 4.14(b) are in parallel. We obtain

$$(g_m R - 1)v_{i-1} + 2v_i - v_{i+1} = 0$$

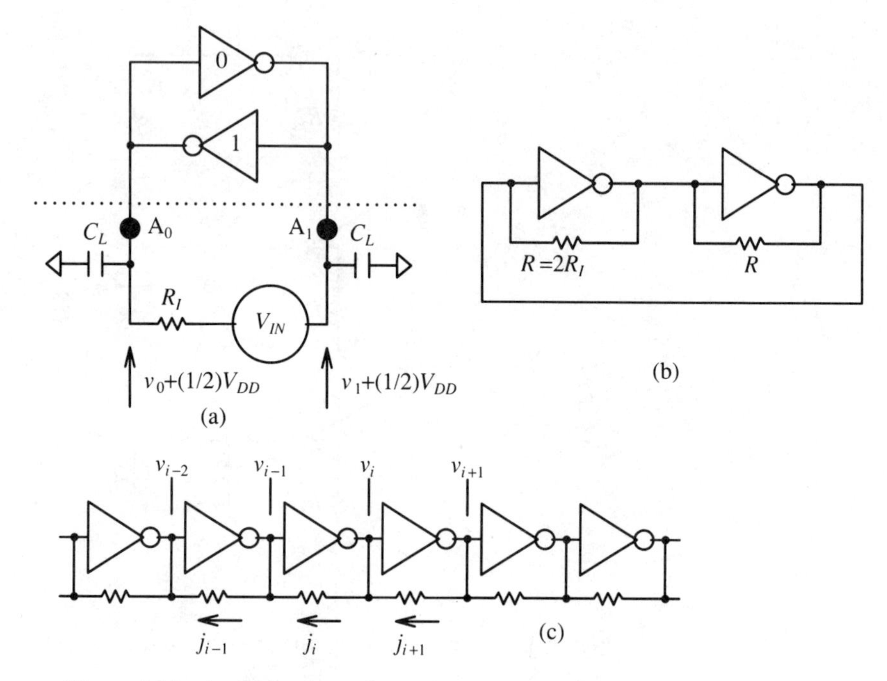

Figure 4.14 Analysis of negative resistance of a coupled inverter.

If the v_i's are related by

$$v_{i+1} = -kv_i$$

where k is the voltage gain of an inverter, gain k satisfies

$$k^2 + 2k - (g_m R - 1) = 0$$

and by solving this equation for k we obtain

$$k = -1 \pm \sqrt{g_m R}$$

The negative sign is irrelevant. We have

$$k = \sqrt{g_m R} - 1$$

and the following three cases are identified.

1. $k > 1$ or $g_m > 4/R$
2. $1 > k > 0$ or $4/R > g_m > 1/R$
3. $0 > k > -1$ or $1/R > g_m > 0$

Profiles of the node voltages of the chain are shown schematically in Figs. 4.15(a) to (c). In case 1 the inverters win over the resistors and chain switches. In case 3 the

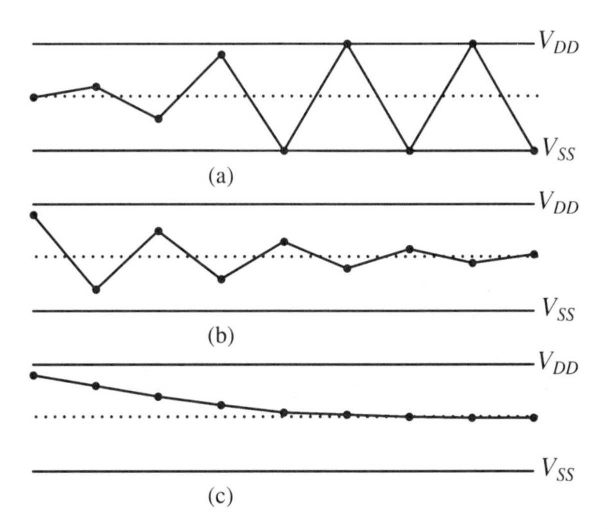

Figure 4.15 Potential profile in a long gate chain.

resistors win, and the chain does not switch. Case 2 stands between the two extremes.

The long chain of Fig. 4.14(c) is equivalent to the negative resistance circuit of Fig. 4.14(a). In Fig. 4.14(a) the electrical perturbation given to the pair of terminals circulates around the feedback loop. This same electrical phenomenon can be examined by propagating the perturbation unidirectionally along the long chain of Fig. 4.14(c). Corresponding to the three cases, the negative-resistance circuit of Fig. 4.14(a) has a stable bias point in the case of Fig. 4.15(b) and (c). The bias point is unstable in the case of Fig. 4.15(a). A negative resistance shows up in this case.

A small deviation of the voltages of nodes A_0 and A_1 of Fig. 4.14(a) satisfies the following differential equations:

$$C_L \frac{dv_0}{dt} = \frac{v_1 - v_0 - v_{IN}}{R_I} - g_m v_1 \tag{4.36}$$

$$C_L \frac{dv_1}{dt} = \frac{v_0 - v_1 + v_{IN}}{R_I} - g_m v_0$$

In Eq. (4.36), the argument of the terms $g_m v_0$ and $g_m v_1$ is time t if the inverters consist of only one stage. If the inverters consist of three or more stages, or by any other reason the circuit has an internal delay, the argument is $t - \tau$, where τ is the delay. Then one may write

$$g_m v_0(t - \tau) = g_m v_0(t) - g_m \tau \frac{dv_0(t)}{dt}$$

and therefore the equations read

$$C_L \frac{dv_0}{dt} - g_m \tau \frac{dv_1}{dt} = \frac{v_1 - v_0 - v_{IN}}{R_I} - g_m v_1 \tag{4.37}$$

$$-g_m \tau \frac{dv_0}{dt} + C_L \frac{dv_1}{dt} = \frac{v_0 - v_1 + v_{IN}}{R_I} - g_m v_0$$

Two special modes are studied. When nodes A_1 and A_0 work differentially,

$$v_0 = -v_1$$

and when they work jointly,

$$v_0 = v_1$$

In a differential operation, Eq. (4.37) simplifies to

$$(C_L + g_m \tau)\frac{dv_1}{dt} = \frac{v_{IN}}{R_I} - \left[\frac{2}{R_I} - g_m\right] v_1 \tag{4.38}$$

and in joint operation to

$$(C_L - g_m \tau)\frac{dv_1}{dt} = \frac{v_{IN}}{R_I} - g_m v_1$$

The effect of the delay of the inverters is only to modify the capacitance value. We may therefore consider the case of $\tau = 0$ for simplicity.

The case of differential operation is of special interest. The equation is rewritten as

$$\frac{dv_1}{dt} = -\frac{2 - g_m R_I}{C_L R_I}\left[v_1 - \frac{v_{IN}}{2 - g_m R_I}\right] \tag{4.39}$$

The initial condition is

$$v_1(0) = 0$$

When

$$v_{IN}(t) = 0 \ (t<0) \ \text{and} \ v_{IN}(t) = v_A \ (t>0)$$

the solution is given by

$$v_1(t) = \frac{v_A}{\chi C_L R_I}(1 - e^{-\chi t}) \rightarrow \frac{v_A}{C_L R_I} t \ \ (\chi \rightarrow 0) \tag{4.40}$$

where

$$\chi = \frac{2 - g_m R_I}{C_L R_I}$$

When the input waveform is given by

$$v_{IN}(t)=0\ (t<0) \quad \text{and} \quad v_{IN}(t)=\alpha t\ \ (t>0)$$

the solution is

$$v_1(t)=\frac{\alpha}{\chi C_L R_I}\left\{t+\frac{1}{\chi}(e^{-\chi t}-1)\right\} \rightarrow \frac{\alpha t^2}{2C_L R_I} \quad (\chi \rightarrow 0) \tag{4.41}$$

Plots of Eq. (4.40) and (4.41) versus time t are shown in Figs. 4.16(a) and (b), respectively. At the critical point $\chi=0$ the asymptotic of $v_1(t)$ when $t\rightarrow\infty$ changes drastically: In Fig. 4.16(a), $v_1(t)\rightarrow v_A/\chi C_L R_I$ if $\chi>0$, but $v_1(t)\rightarrow\infty$ exponentially if $\chi<0$. In Fig. 4.16(b), $v_1(t)\rightarrow o(t)$ when $\chi>0$, $\rightarrow o(t^2)$ when $\chi=0$, and $\rightarrow o(e^{|\chi|t})$ when $\chi<0$. $\chi<0$ means that $g_m R_I>2$ and $g_m R>4$ since $R=2R_I$. Therefore, $\chi<0$ is the case of Fig. 4.15(a).

When $t\rightarrow 0$, however, $v_1(t)\rightarrow v_A(t/C_L R_I)$ in Fig. 4.16(a) and $v_1(t)\rightarrow \alpha t^2/(2C_L R_I)$ in Fig. 4.16(b), and they are independent of the values of χ. This means that the effects of negative resistance generated by the cross-coupled inverters become effective only in the limit of large t, and not immediately after switching. At the beginning of switching, the benefit of negative resistance is not discernible. The delayed action is because the negative resistance is generated only after the signal has propagated some way along the equivalent inverter chain of Fig. 4.14(c), and that takes time since load capacitance is included in this path. This is a significant difference from negative resistance using an Esaki diode.

Negative resistance is a powerful technique to speed up nodes that are unusually slow due to heavy capacitive loading. Memory bit lines (Section 7.24), PLA bit lines (Section 5.5) and the databus of a microprocessor (Section 7.19) are a few examples. They have a common feature, that a single node is connected to many FETs, only one of which drives the node at any time, and the others add up to parasitic capacitance. Negative resistance discussed in this section is effective in speeding up this kind of node. A negative resistance generated using CMOS FETs is not able to speed up already fast CMOS nodes, whose delay time is comparable to typical CMOS gate delay, even at the expense of power. If faster (bipolar or GaAs) devices are used in the cross-coupled inverter, speed up is feasible.

An application of the negative-resistance circuit is to sense a small voltage difference developed between a pair of lines of a bus that is often found in memory devices and in microprocessors. The examples and practical details are discussed in Section 7.24. A sense amplifier is an effective technique to reduce the delay of circuit nodes that are loaded by large number of equivalent drivers.

4.12 NEGATIVE CAPACITANCE

Figure 4.17 shows a circuit that generates a negative capacitance [12]. The circuit uses the Miller effect of a noninverting amplifier M. One application of this technique is a computer databus. Suppose that the databus having capacitance C_L is precharged to V_{DD} and driver D pulls down the bus, by equivalent resistance R_D. The circuit

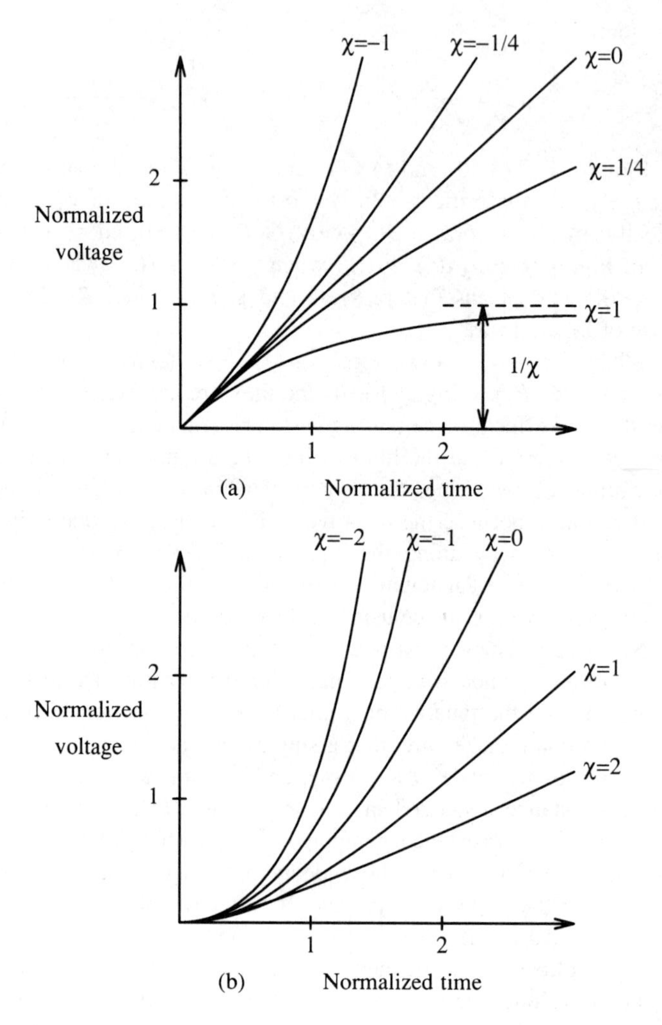

Figure 4.16 Effects of negative resistance on switching waveform.

equation satisfied by the databus voltage V_D is

$$C_L \frac{dV_D}{dt} = -\frac{V_D}{R_D} + C_0 \frac{d}{dt}(V_0 - V_D) \tag{4.42}$$

where $V_0 = AV_D$. We assume that the noninverting amplifier M has negligible delay, where A is the gain of the amplifier ($A > 0$). Equation (4.42) can be rewritten as,

$$\left\{C_L - (A-1)C_0\right\} \frac{dV_D}{dt} = -\frac{1}{R_D} V_D \tag{4.43}$$

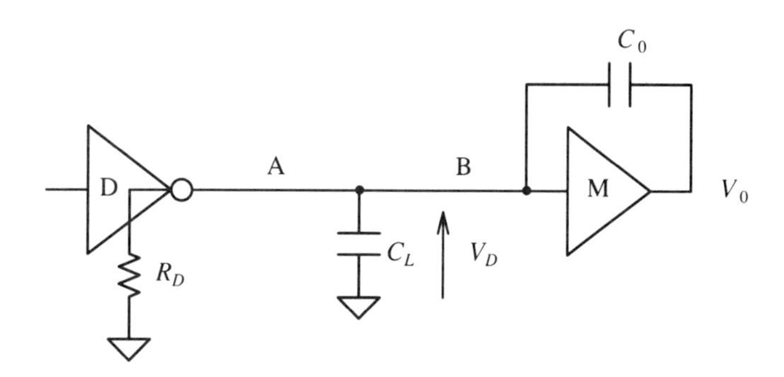

Figure 4.17 Negative capacitance generated using inverse Miller effect.

and this equation shows that the bus capacitance is reduced from C_L to $C_L-(A-1)C_0$.

Since the noninverting amplifier consists of at least two stages of cascaded inverters, the delay of the amplifier is important. If the delay τ is small and

$$\frac{V_0}{A}=V_D(t-\tau)=V_D(t)-\tau\frac{dV_D(t)}{dt} \tag{4.44}$$

is substituted into Eq. (4.42), we obtain,

$$C_0A\tau\frac{d^2V_D(t)}{dt^2}+\left\{C_L+C_0(1-A)\right\}\frac{dV_D(t)}{dt}+\frac{V_D(t)}{R_D}=0 \tag{4.45}$$

From Eq. (4.45) we observe that the amplifier delay creates a term that is a higher-order derivative of time. Assuming that $V_D \propto e^{St}$ in the equation, the natural frequency of the circuit S satisfies the following secular equation:

$$C_0A\tau S^2+\left\{C_L+C_0(1-A)\right\}S+\frac{1}{R_D}=0$$

which is solved as

$$S_\pm=\frac{1}{2C_0A\tau}\left[-\left\{C_L+C_0(1-A)\right\}\pm\sqrt{D}\right] \tag{4.46}$$

where

$$D=C_0^2A^2-2\left[C_0^2+C_LC_0+2\frac{C_0\tau}{R_D}\right]A+(C_L+C_0)^2$$

$$=C_0^2(A-A_+)(A-A_-)$$

and where

$$A_{\pm}=1+\frac{C_L}{C_0}+\frac{2\tau}{C_0R_D}\pm\left\{4\frac{\tau}{C_0R_D}\left[1+\frac{C_L}{C_0}+\frac{\tau}{C_0R_D}\right]\right\}^{1/2}$$

If τ were zero, the bus capacitance nulls when the amplifier gain is A_C, where A_C is given by

$$A_C=1+\frac{C_L}{C_0}$$

It can be shown that $A_+>A_C>A_-$. Using these quantities Eq. (4.46) becomes

$$S_{\pm}=S_R(A)\pm S_I(A)$$

where

$$S_R(A)=\frac{A-A_C}{2\tau A} \quad \text{and} \quad S_I(A)=\frac{\sqrt{(A_+-A)(A_--A)}}{2\tau A}$$

When $A_+>A>A_-$, $S_I(A)$ is imaginary. The absolute value of $S_I(A)$ in this case is given by

$$|S_I(A)|^2=\frac{A(A_++A_-)-A_+A_--A^2}{(2\tau)^2A^2}$$

This function has a maximum at

$$A=A_m=\frac{2A_+A_-}{A_++A_-}=\frac{A_C}{1+\dfrac{2\tau}{C_0R_DA_C}}<A_C$$

Using these relationships the natural frequency S can be plotted on a complex plane as a function of the amplifier gain A, as shown in Fig. 4.18. As A increases from unity two points representing S_+ and S_- move first on the real axis, and then meet when $A=A_-$. By increasing A still further the two points depart from the real axis and move on the two curves. When $A=A_C$ the real part of S changes sign. When $S_R(A)>0$, instability occurs. The real-time voltage response of the system in the stable regime can be classified as overdamped, critically damped, and underdamped, as shown in Fig. 4.18.

Using $S_{\pm}$ and assuming that the bus was originally at ground potential, the solution of Eq. (4.45) is given by

$$\frac{V(t)}{V_{DD}}=1-\frac{S_+e^{S_-t}-S_-e^{S_+t}}{S_+-S_-}-\frac{e^{S_-t}-e^{S_+t}}{(1+(C_0/C_L))(S_+-S_-)} \tag{4.47}$$

$$=1-(1-S_0t)e^{S_0t}+\frac{(t/R_DC_L)e^{S_0t}}{1+(C_0/C_L)} \quad \text{when } S_-=S_+=S_0(=S_R(A))$$

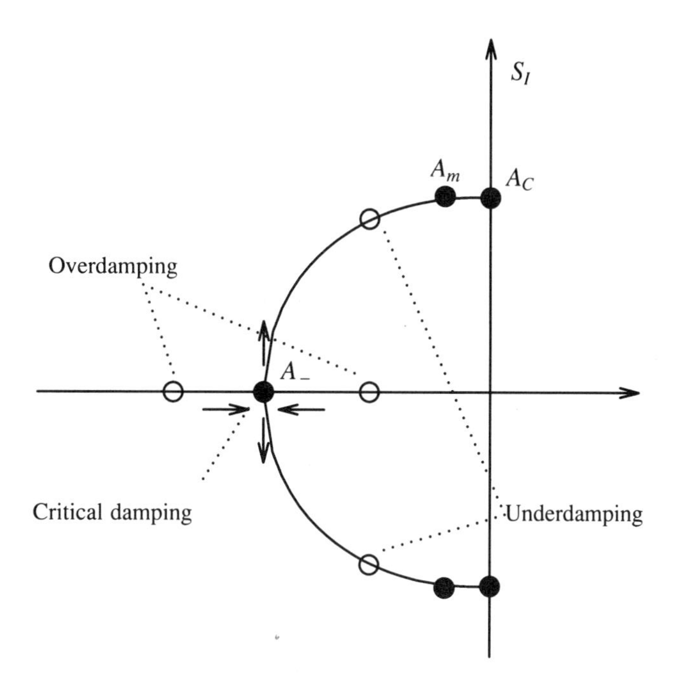

Figure 4.18 Natural frequency of negative capacitance circuit.

where S_+, S_-, and S_0 are given by Eq. (4.46). The switching time of the bus is defined as the time when the voltage reaches $V_{DD}/2$, and the time can be calculated by solving an implicit equation, $V(t) = (1/2)V_{DD}$, for t.

It is helpful to compare the switching time of the uncompensated bus with the switching time of the bus compensated to critical damping. The ratio of the switching times is plotted versus the amplifier delay τ (normalized to the time constant of the bus, $R_D C_L$), for several values of C_L/C_0 in Fig. 4.19. Figure 4.19 shows that if the switching time of the critically compensated bus is to be 70% of that of the uncompensated bus, the delay time of the amplifier should be less than 0.2 times the bus time constant $C_L R_D$. When the amplifier delay approaches the upper limit, the gain of the amplifier required to attain the critical damping approaches ∞, so that the improvement becomes unattainable. To obtain a significant improvement in bus switching time, the delay time of the amplifier must be less than 0.1 times the bus time constant.

The amplifier model of Eq. (4.44) is too simple to represent a real operational amplifier. A more realistic two-stage amplifier model was used and the circuit was simulated by a computer. Figure 4.20 shows a set of results. As the gain of the amplifier is increased the bus response changes from the overdamped to the critically damped and then to the underdamped. The underdamped bus is fast but undesirable practically because of the overshoot and because it is close to oscillation.

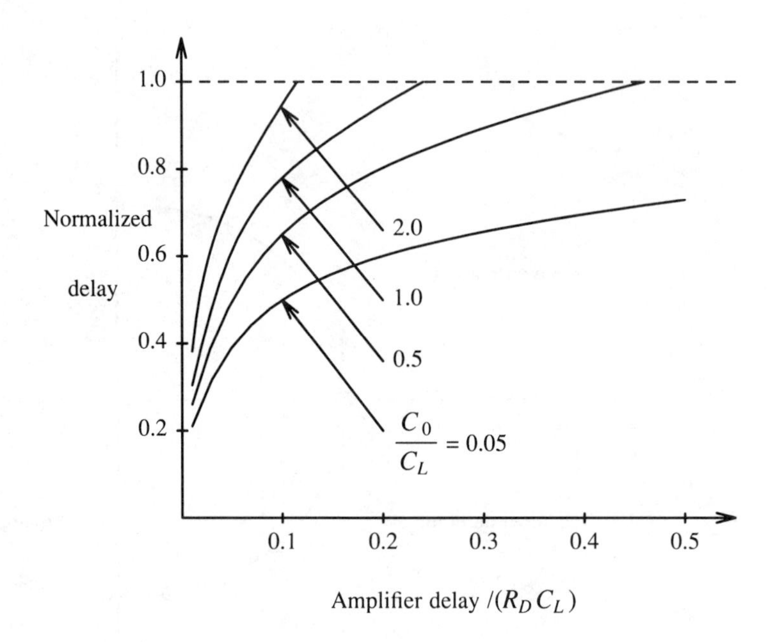

Figure 4.19 Reduction of delay by negative capacitance.

Negative capacitance has not yet been widely used, but the technique is considered important for the future. In Section 7.24 a sense amplifier that uses the negative capacitance concept is discussed. Operational amplifiers are the key element in negative resistance. CMOS operational amplifiers are too slow for this application. At present, operational amplifiers made using hybrid technology having a dynamic range wider than 10 V, and having several nanoseconds delay time are available commercially. They can be used to speed up for CMOS bus, whose amplitude is 5 V, and whose delay is 50 ns. This design is useful in special applications in a VLSI test machine interface. Progress in operational amplifier technology will allow a more attractive design than this in the near future.

4.13 CMOS D-LATCHES

A CMOS D-latch, shown in Fig. 4.21(a), is used to store 1 bit of data. When CK is high and $\overline{CK}$ is low, data D drives the latch, and node Q assumes the logic level of D, and $\overline{Q}$ the complement, $\overline{D}$. Data D comes from other logic and must make a transition and settle before CK and $\overline{CK}$ make down and up transitions, respectively. After the clock transition, input D is unable to further influence the state of the latch, and the latch is closed. The transition of the data must occur sometime before the clock transitions. This minimum time required to store the data is called a setup time. If

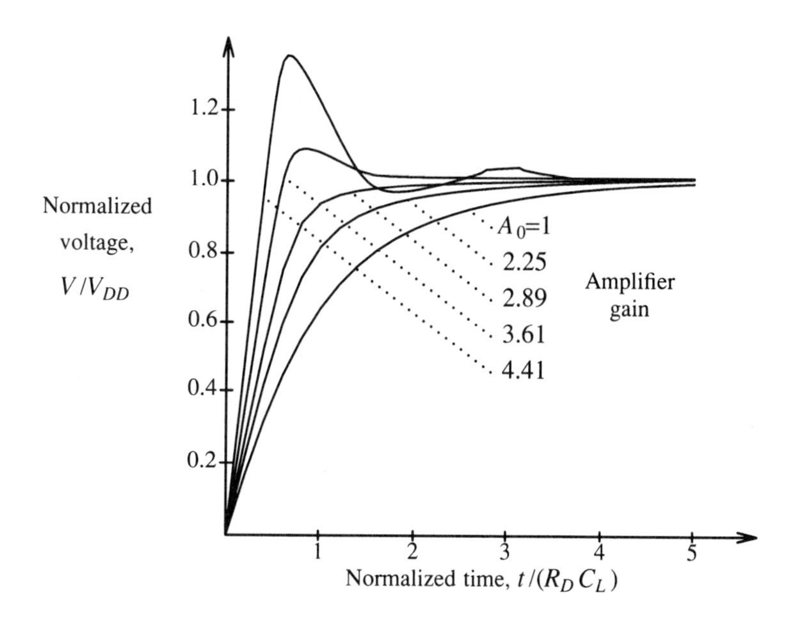

Figure 4.20 Switching waveform of a negative capatence circuit.

the data transition occurs before the setup time, the value after the transition is stored in the latch. If the transition occurs after the setup time, the data before the transition are stored. The latter case results in a logic error. The setup time may be different for up and for down data transitions.

Suppose that the input tristatable inverter (TBFI1) of Fig. 4.21(a) is disabled, and the second (feedback) tristatable inverter (TBFI2) and the inverter are both symmetrical with regard to pullup and pulldown. Then the state where the voltages of nodes Q and $\overline{Q}$ are both $V_{DD}/2$ is a metastable state. The latch stays in the state until external noise disturbs the metastability. How long the latch is able to stay in the vicinity of the metastable state can be analyzed by studying the small-signal behavior of the latch. The following is a simplified version of the theory by Flanagan [13].

The small-signal equivalent circuit of the latch in the vicinity of the metastable bias point is shown in Fig. 4.22. The circuit is analogous to that of the ring oscillator, shown in Fig. 4.6(b), but the number of cascaded stages is two. The small-signal circuit equations are

$$C\frac{dv_0}{dt}+\frac{v_0}{R}+g_m v_1=0 \tag{4.48}$$

$$C\frac{dv_1}{dt}+\frac{v_1}{R}+g_m v_0=0$$

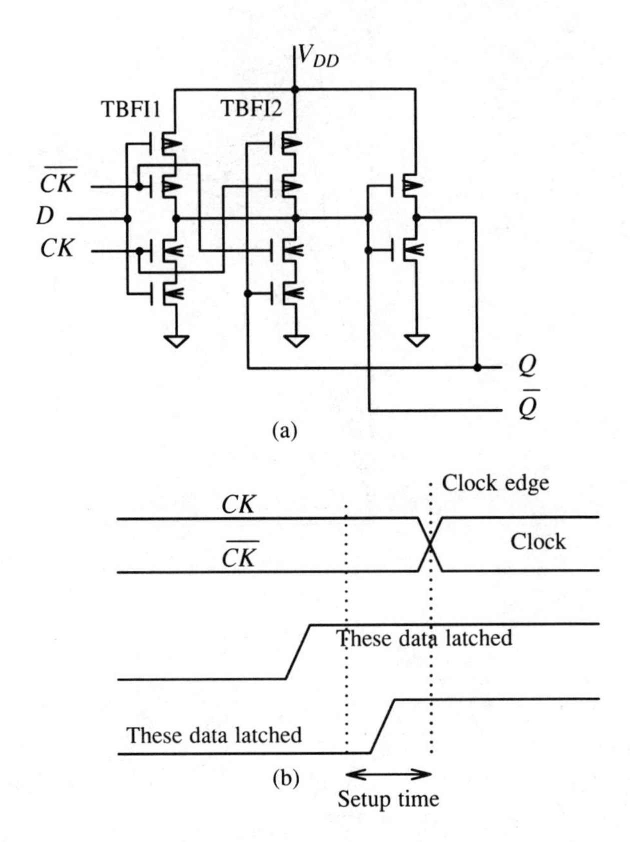

Figure 4.21 CMOS D-latch and definition of setup time.

where v_0 and v_1 are the small deviations of the voltages of nodes Q and $\overline{Q}$, respectively, from the metastable state voltage, $V_{DD}/2$. Equation (4.48) is written for the case where the Miller capacitance C_m is zero. If C_m is not zero, the following substitutions result in equations of the same form, and therefore we may consider the simplified problem with no loss in generality:

$$C \to C\left(\frac{C + 2C_m}{C + C_m}\right), \quad \frac{1}{R} \to \frac{1}{R} + \frac{g_m C_m}{C + C_m} \tag{4.49}$$

and

$$g_m \to g_m + \frac{1}{R}\frac{C_m}{C + C_m}$$

Assuming that v_0 and v_1 have time dependence like e^{St}, parameter S satisfies

$$\left(CS + \frac{1}{R}\right)^2 - g_m{}^2 = 0$$

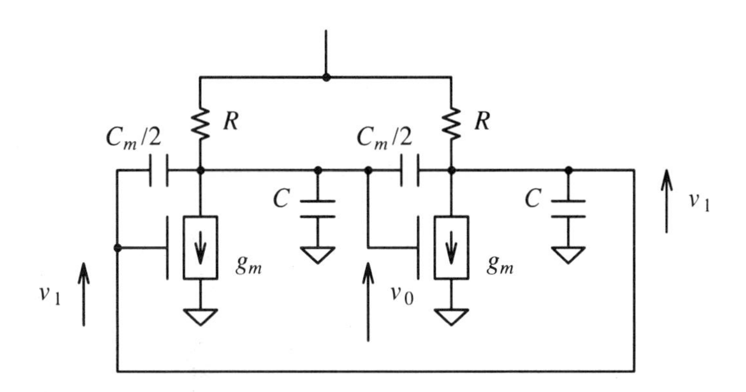

Figure 4.22 Analysis of the feedback mechanism of CMOS D-latch.

Solving this equation, we obtain

$$S = S_+ = \frac{Rg_m - 1}{CR} \quad \text{and} \quad S = S_- = -\frac{Rg_m + 1}{CR} \tag{4.50}$$

If we repeat the algebra again, assuming that $v_0 = v_1$ (the common mode) and $v_0 = -v_1$ (the differential mode), we find that $1/S_-$ is the time constant of the common-mode voltage change, and $1/S_+$ is the time constant of the differential-mode voltage change. For a latch to be able to store data, $S_+ > 0$ is required, and therefore $Rg_m > 1$.

Evolution of the node voltages can be understood by plotting v_1 versus v_0, in the $v_1 - v_0$ coordinate system shown in Fig. 4.23 [14]. By eliminating time t in Eq. (4.48) using

$$\frac{d}{dt} = \frac{dv_0}{dt}\frac{d}{dv_0} = -\frac{1}{C}\left(\frac{v_0}{R} + g_m v_1\right)\frac{d}{dv_0}$$

we obtain

$$\frac{dv_1}{dv_0} = \frac{g_m v_0 + (v_1/R)}{g_m v_1 + (v_0/R)} \tag{4.51}$$

The variables can be separated by a transform $u = v_1/v_0$ as

$$\frac{u + \gamma}{1 - u^2} du = \frac{dv_0}{v_0}$$

where $\gamma = 1/g_m R\,(<1)$. This equation can be integrated in closed form as

$$v_0 = \frac{c}{|1-u|^{\frac{1+\gamma}{2}}\,|1+u|^{\frac{1-\gamma}{2}}} \tag{4.52}$$

when $|u| < 1$. When $|u| > 1$,

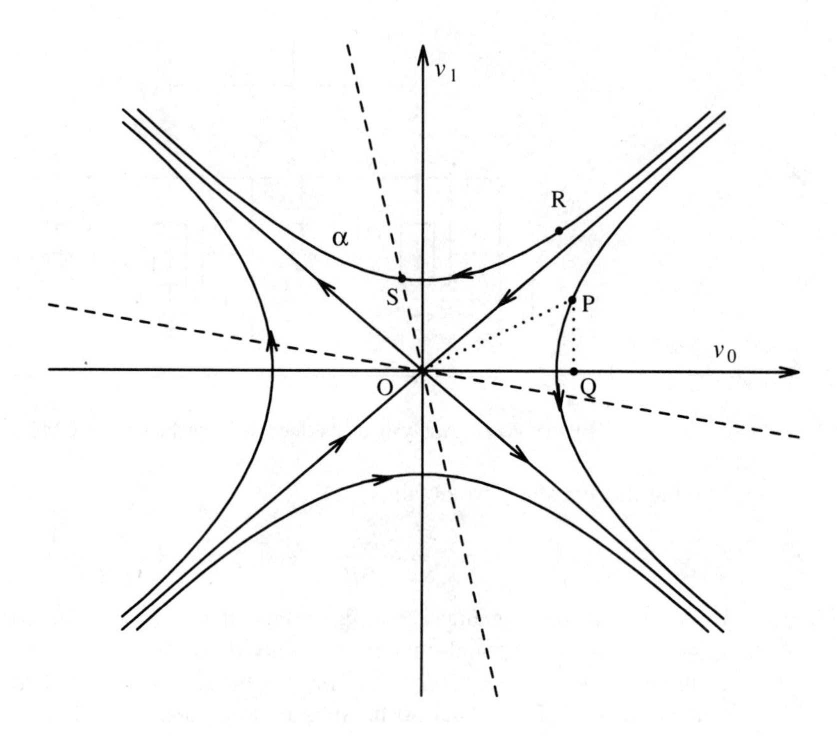

Figure 4.23 How CMOS D-latch settles.

$$v_1 = \frac{c}{\left|\frac{1}{u} - 1\right|^{\frac{1+\gamma}{2}} \left|\frac{1}{u} + 1\right|^{\frac{1-\gamma}{2}}} \tag{4.53}$$

The points on the curves move with time, in the direction indicated by the arrows. When the initial state of the latch is at point R, both v_0 and v_1 decrease with time, v_1 attains its minimum at point S, and then begins to increase, while v_0 continues to decrease. After a long time $v_0 \to -\infty$ and $v_1 \to +\infty$. This limit, however, is only qualitatively valid, since nonlinearity becomes important. This means that the latch has settled at the final state. Before v_1 reaches its minimum, the change in voltage is primarily the common mode, and the time constant is $1/S_-$. After the minimum the differential mode dominates, and the time constant changes to $1/S_+$. Point O is the metastable state [15].

What happens as the setup time varies can be observed from the simulation results shown in Fig. 4.24(a) to (c). The simulations assume 1.75-μm CMOS, a low-current process at 105 degrees centigrade and 4.75 V of supply voltage. All PFETs are 7.25 μm wide, and NFETs are 4 μm wide. When the upgoing data transition is only 3 ns before the clock transition, the latch does not capture the final data [Fig.

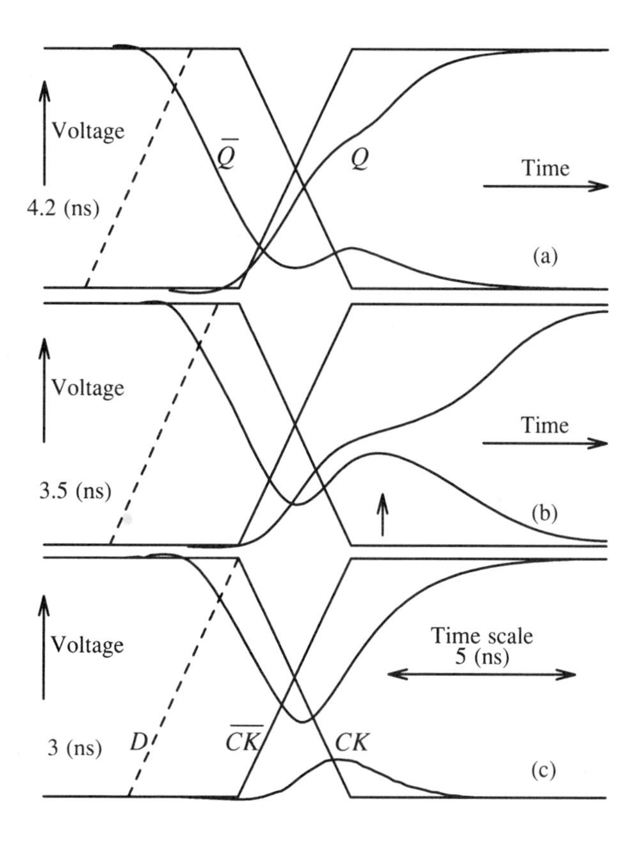

Figure 4.24 Behavior of CMOS D-latch as the data setup time is changed.

4.24(c)]. Nodes Q and $\overline{Q}$ try to make a transition, but before the state of the latch flips, the clocks turn off. When the upgoing data transition is 4.2 ns before the clock transition, the state of the latch flips, and the latch captures the data value after transition [Fig. 4.24(a)]. When the data transition occurs 3.5 ns before the clock transition, the latch captures correct data but takes a long time to settle [Fig. 4.24(b)]. When CK and $\overline{CK}$ cross over, the $\overline{Q}$ output voltage is already low. Because of the inverter delay, the Q output voltage is still low. Therefore the tristatable inverter TBFI2 of Fig. 4.23 pulls node $\overline{Q}$ up, thereby partially canceling the pulldown of TBFI1. Pullup of node Q by the inverter slows down. The voltage between nodes Q and $\overline{Q}$ remains small. A remarkable hump appears on the waveform of $\overline{Q}$ [shown by the arrow in Fig. 4.24(b)]. It takes quite a long time before the voltage difference increases and the latch settles.

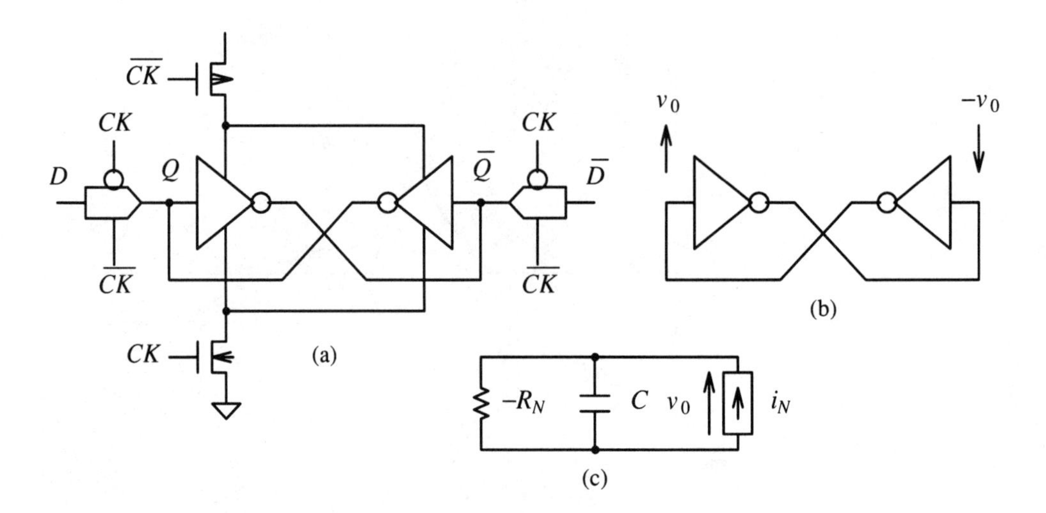

Figure 4.25 Metastable state of input latch.

4.14 METASTABLE STATE

Slow settling of the the latch discussed in the preceding section has important practical significance. Figure 4.25(a) is a schematic diagram of an input latch. D and $\overline{D}$ are input data and their complement, which come from the external circuit. If clock CK is low and $\overline{CK}$ is high, power to the cross-coupled inverters is disconnected from V_{SS} and V_{DD}. Nodes Q and $\overline{Q}$ take the signal level of D and $\overline{D}$, respectively. When CK and $\overline{CK}$ make low-to-high and high-to-low transitions, respectively, transmission gates shut off, cross-coupled inverters are powered up, and nodes Q and $\overline{Q}$ take CMOS logic levels determined by D and $\overline{D}$. Even if D and $\overline{D}$ are TTL or ECL logic signals, CMOS logic levels are recovered at nodes Q and $\overline{Q}$.

In certain system applications signal D (and $\overline{D}$) may arrive any time. This happens when the clock of the circuit that sends D and $\overline{D}$ is independent of the clock of the receiving chip. Such an arrangement is called an asynchronous system. When CK and $\overline{CK}$ make low-to-high and high-to-low transitions, respectively, D and $\overline{D}$ can be, in the worst case, at the midpoint. Then immediately after the clock transition, nodes Q and $\overline{Q}$ may take voltage levels that are very close to each other. Then the latch takes a significant time before settling at the final state. The phenomenon exists inevitably in a synchronizer circuit, which interfaces two circuits that have different clocks, and is known as metastability [16].

After a clock transition, the input latch circuit is equivalent to the cross-coupled inverters shown in Fig. 4.25(b), which is equivalent to the circuit diagram of Fig. 4.22. For simplicity, we consider here the differential-mode voltage only. From Eq. (4.48) with $v_1 = -v_0$, we have

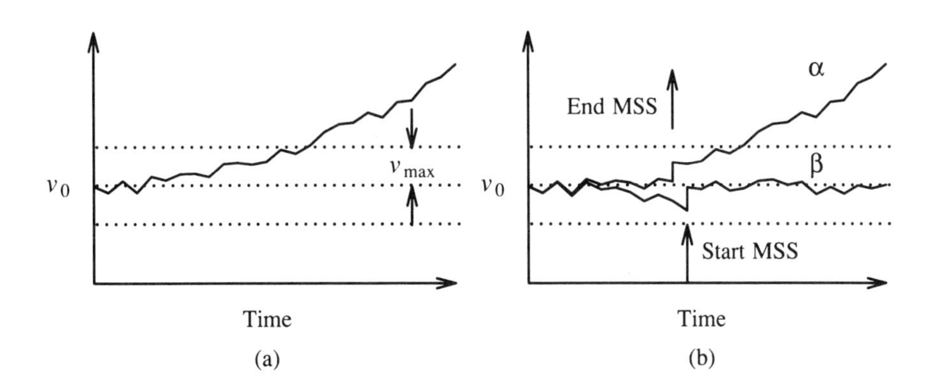

Figure 4.26 Metastable state near unstable point of equilibrium.

$$C\frac{dv_0}{dt} - \frac{1}{R_N}v_0 = i_N(t) \tag{4.54}$$

Equation (4.54) is the circuit equation of a simple equivalent circuit of Fig. 4.25(c). In the circuit a negative resistance R_N, capacitance C, and an equivalent noise generator $i_N(t)$ are included. In a metastability problem, inclusion of noise is essential [17]. Noise originates from switching of other circuits, as well as from the fundamental fluctuation mechanisms (such as thermal noise). When $|v_0(t)|$ is less than a small voltage v_{max}, the effects of noise are significant, as we see in the following. A solution of Eq. (4.54) that satisfies $v_0(0) = v_B$ is given by

$$v_0(t) = \exp(t/RC)\left\{v_B + \int_0^t i_N(\theta)\exp(-\theta/RC)d\theta\right\} \tag{4.55}$$

Equation (4.55) gives the the solution shown in Fig. 4.26(a). When v_B is small, the second term in the braces may cancel the first term v_B, and therefore the accumulated effects of amplification of the initial offset by the negative resistance is flushed out. Statistically, however, the noise effect is as shown in Fig. 4.26(b): One metastable state ends by noise (process α) while the other state goes back to the metastable state by noise (process β). If the initial offset is uniformly distributed between the limits ($-V_{DD}/2$ and $V_{DD}/2$), noise does not affect the total number of metastable states [18]. Then the time for a latch to start from offset voltage v_B and to settle at the final state, t_S, is estimated from

$$\left[\frac{V_{DD}}{2v_B}\right] = \exp\left[\frac{t}{CR_N}\right] \tag{4.56}$$

where the left-hand side equals the inverse of the probability of selecting data that fall within $-v_B$ and v_B. Using this interpretation, the probability of a metastable state that has duration longer than t, $P(t)$, is given by

$$P(t) = \exp\left[-\frac{t}{CR_N}\right] \tag{4.57}$$

where $R_N = g_m - (1/R) = (A-1)/R$, and where A is the gain of the inverter. This formula has been confirmed experimentally. The formula is useful in estimating the susceptibility of a metastable state, as well as evaluating the reliability of the circuit.

4.15 REFERENCES

01. J. A. Pretorius, A. S. Shubat and C. A. T. Salama "Analysis and design optimization of Domino CMOS logic with application to standard cells" IEEE J. Solid-State Circuits, vol. SC-20, pp. 523-530, April 1985.
02. S. Tanaka, J. Iwamura, J. Ohno, K. Maeguchi, H. Tango and K. Doi "A sub-nanosecond 8K-gate CMOS/SOS gate array" ISSCC84 Digest, pp. 260-261, February 1984.
03. A. K. Rapp, L. P. Wennick, H. Borkan and K. R. Keller "Complementary-MOS integrated binary counter" ISSCC67 Digest, pp. 52-53, February 1967.
04. R. Mueller, H-J. Pfleiderer and K-U. Stein "Energy per logic operation in integrated circuits: Definition and determination" IEEE J. Solid-State Circuits, vol. SC-11, pp. 657-661, October 1976.
05. N. Sasaki "Higher harmonic generation in CMOS/SOS ring oscillators" IEEE Trans. on Electron Devices, vol. ED-29, pp. 280-283, February 1982.
06. D. S. Babb "Pulse circuits, switching and shaping" Prentice-Hall Inc., Englewood Cliffs, N. J. 1964 C. A. Holt "Electronic circuits, digital and analog" John Wiley & Sons, Inc., New York 1978.
07. J. J. Paulos, D. A. Antoniades and Y. P. Tsividis "Measurement of intrinsic capacitance of MOS transistors" ISSCC82 Digest, pp. 238-239, February 1982 and J. J. Paulos and D. A. Antoniades "Measurement of minimum-geometry MOS transistor capacitance" IEEE J. Solid-State Circuits, vol. SC-20, pp. 277-283, February 1985.
08. J. P. Ellul, M. A. Copeland and C. H. Chan "MOS capacitor pull-up circuits for high-speed dynamic logic" IEEE J. Solid-State Circuits, vol. SC-10, pp. 298-307, October 1975.
09. B. Eitan, D. Frohman-Bentchkowsky and J. Shappir "Holding time degradation in dynamic MOS RAM by injection-induced electron currents" IEEE Trans. on Electron Devices, vol. ED-28, pp. 1515-1519, December 1981.
10. C-Y. Wu and C-Y. Wu "The new general realization theory of FET-like integrated voltage-controlled negative differential resistance devices" IEEE Trans. on Circuits and Systems, vol. CAS-28, pp. 382-390, May 1981.
11. L. O. Chua, J. Yu and Y. Yu "Bipolar-JFET-MOSFET negative resistance devices" IEEE Trans. on Circuits and Systems, vol. CAS-32, pp. 46-61, January 1985.
12. M. Shoji and R. M. Rolfe "Negative capacitance bus terminator for improving the switching speed of a microcomputer power bus" IEEE J. Solid-State Circuits, vol. SC-20, pp. 828-832, August 1985.
13. S. T. Flanagan "Synchronization reliability in CMOS technology" IEEE J. Solid-State

Circuits, vol. SC-20, pp. 880-882, August 1985.

14. R. C. Jaeger and R. M. Fox "Phase plane analysis of the upset characteristics of CMOS RAM cells" 1985 University-Government-Industry Microelectronics Symposium Digest pp. 183-187.
15. G. Lacroix, P. H. Marchegay and N. A. L. Hossri "Prediction of flip-flop behavior in metastable state" Electronics Letters, vol. 16, pp. 725-726, September 1980.
16. T. J. Chaney and C. E. Molnar "Anomalous behavior of synchronizer and arbiter circuits" IEEE Trans. on Computers, vol. C-22, pp. 421-422, April 1973.
17. G. R. Couranz and D. F. Wann "Theoretical and experimental behavior of synchronizers operating in the metastable region" IEEE Trans. on Computers, vol. C-24, pp. 604-616, June 1975.
18. H. J. M. Veendrick "The behavior of flip-flops used as synchronizers and prediction of their failure rate" IEEE J. Solid-State Circuits, vol. SC-15, pp. 169-176, April 1980.

5

CMOS Dynamic Gates

5.1 INTRODUCTION

CMOS static gates, discussed in the preceding three chapters, are the most versatile gates for CMOS VLSIs. They have limitations, however. When circuit delay is important, and where silicon area is at a premium, CMOS dynamic gates are used. In recent years more and more dynamic gates are being used to improve performance. In this chapter the properties of CMOS dynamic gates are studied, and techniques to make them work at high-performance are summarized. Dynamic gates require clocks, and involve circuit and timing design.

5.2 OPERATION OF A CMOS DYNAMIC GATE

Figure 5.1(a) shows a CMOS dynamic NAND2 gate. The gate operates as follows [01]. During precharge the clock is low, as shown in the timing diagram of Fig. 5.1(b). NFET MN1 is off and PFET MP1 is on. MN1 is called the ground switch. MP1 charges the output node O to voltage V_{DD}. The process is called precharge of the dynamic gate. During precharge, logic inputs A and B must stabilize to their final input logic level. If A is high, internal node N_A is precharged to voltage $V_{DD}-V_{TH}$, where V_{TH} is the threshold voltage of the NFET MNA, including the back-bias effect. If NFET MNA is leaky, and if the precharge time is very long, node N_A can be precharged all the way up to V_{DD}. The voltages to which internal nodes are precharged depends on the conditions and care must be exercised in circuit simulations. The problem is the same as in CMOS static gates (Section 3.8).

When the clock makes a low-to-high transition, the gate goes into the discharge phase. Then node O remains high if either of the input signals A or B is low. The logic high state is kept as long as the inputs do not change: The leakage current of the NFETs is compensated for the current from the small PFET load device MP2, which is able to supply small (but enough) current to retain the voltage of output node O at V_{DD}. If all of the input signals A and B are high, both NFETs conduct. Then output node O is pulled down, as shown in Fig. 5.1(b). PFET MP2 is designed so small that the series-connected NFETs are able to drive node O down, within several tens of mV of the ground level. The number of squares of channel of MP2 must then be more than about 30 times the number of squares of the channel of the minimum-size NFET. From this explanation the circuit obviously generates the NAND function of its two inputs.

When both input signals A and B are high, the output, O, becomes low after the pulldown delay time of the three series-connected NFETs. The delay time is shown by T_D in Fig. 5.1(b). The state of node O does not change upon clock transition if either signal A or B is at a low level. It is misleading to say, however, that the logic result in the latter case is available with no delay time. We do not know the result of the logic operation before we make sure that the output node does not pull down. Therefore, we need to wait until time T_D is reached for either a high or a low output.

The delay time T_D of a CMOS dynamic gate is, however, significantly shorter than the delay time of an equivalent CMOS static gate. This is because only fast NFETs are used to pull down the node. This point becomes clear when static and dynamic NOR gates are compared. Figures 5.2(a) and (b) show static and dynamic NOR3 gates, respectively. Static NOR gates suffer from slow pullup delay of series-connected PFETs (MPA- MPC). In the dynamic NOR gate of Fig. 5.2(b) the pullup is carried out by a single PFET, MP1, during the precharge phase. Pulldown of a dynamic NOR gates is very fast, since the number of series-connected NFETs is limited to two. In CMOS dynamic logic, NOR gates are preferred to NAND gates, as opposed to CMOS static logic, where NAND gates are preferred.

Figure 5.3 shows how much faster a dynamic gate is than the equivalent static gate. The delays of NAND(X) and NOR(X) gates are plotted versus the number of inputs, X. The delays were determined by circuit simulations, assuming 1.75-μm CMOS at the medium-current process, 105 degrees centigrade and power supply voltage of 4.75 V. The static gates were assembled from minimum-size NFETs (4-μm-wide) and 7.25-μm-wide PFETs. The dynamic gates have a 10.5-μm-wide NFET used as the ground switch, and 4-μm-wide NFETs used for the rest. The precharge PFET is 7.25-μm-wide. Both the static and dynamic gates have a minimum-size CMOS static buffer (consisting of 4-μm-wide NFET and 7.25-μm-wide PFET) as the load. For static gates both pullup and pulldown delays are plotted (indicated by arrows in Fig. 5.3).

From the figure we find that the dynamic gates (D) (shown with closed circles on the curves) are faster than the static gates (S) under all conditions. In NAND(X) gates, an extra NFET (ground switch) is in series with the rest of the NFETs. In spite

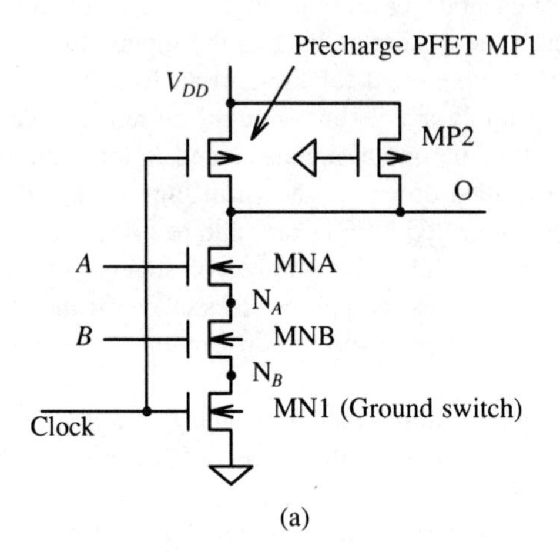

(a)

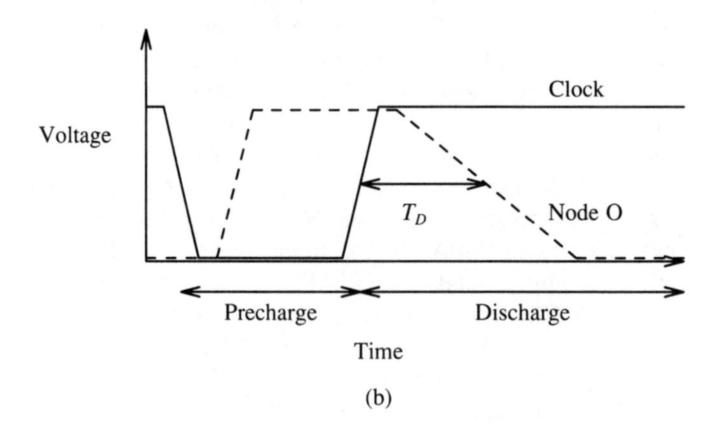

(b)

Figure 5.1 CMOS dynamic gate and clock-signal waveforms.

of the extra series NFET the dynamic gate is faster for pulldown than the static gate. This speed advantage is due to reduced capacitance at the output node and reduced overlap current. Absence of the slow pullup of the static NOR gate makes the dynamic NOR gate significantly faster. A dynamic NOR5 gate is almost 5 times faster than a static NOR5 gate. Therefore, dynamic NOR gates are the most widely used for high-speed logic operations, for example in state-vector generation in PLAs and in datapath operations.

The high-speed capability of dynamic NOR gates is made useful by proper precharge timing design, which takes no extra time slot for precharge. There is an effective FET scaling technique to attain higher speed (Sections 5.13 to 5.15), and above all, it is possible to cascade the gates with the structure of Fig. 5.1(a) over many

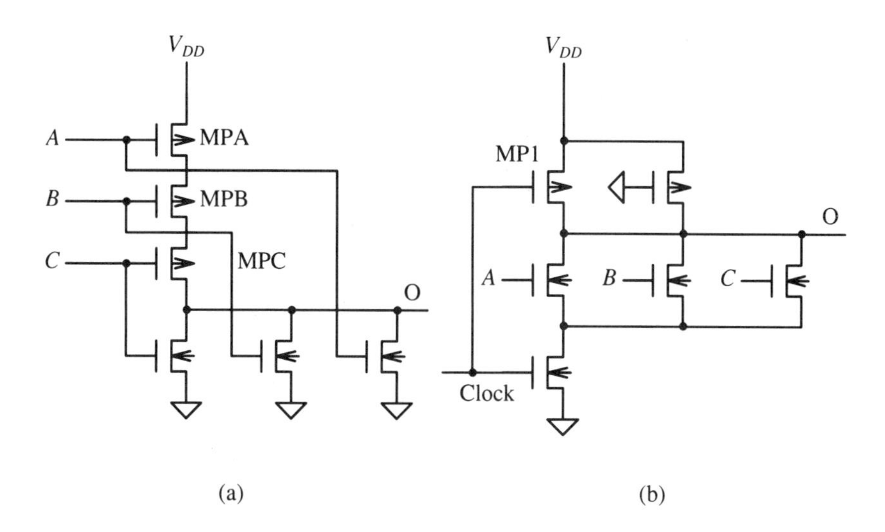

Figure 5.2 Static and dynamic NOR3 gates.

stages, in the so-called Domino CMOS configuration. Design techniques of CMOS dynamic circuit are making rapid progress now. Design of a high-performance CMOS VLSI chip, especially of a microprocessor, depends critically on the skill of applying CMOS dynamic logic in the speed-critical functions of a VLSI chip. All the relevant problems are discussed in detail in this chapter. Particular stress is given to a peculiar problem of dynamic gates: They cannot be cascaded directly as CMOS static gates can.

5.3 INPUT LATCH AND PRECHARGE TIMING

Dynamic gates always require precharge. It may be thought that the short delay of dynamic gates is greatly discounted by the additional time needed for precharge. Usually, this is not the case. In a quite general sequential circuit shown in Fig. 5.4(a) consisting of dynamic logic gates and static latches (static latch circuits are shown in Fig. 2.22) the master latch clock, MCK, and the slave latch clock, SCK, are shown in the timing diagram of Fig. 5.4(b). To avoid crowding the figure, only assertion clocks (CK of Fig. 2.22) are shown. They are nonoverlapping two-phase clocks to avoid data flow through conditions across the two latches. As soon as the slave latch closes, the positive feedback of the static slave latch settles the logic level of the latch very rapidly if node voltages have not settled yet, and the condition to begin discharge of the dynamic logic is established. Therefore, the dynamic logic circuit can be precharged by the inverted slave latch clock. In this arrangement, the dynamic logic

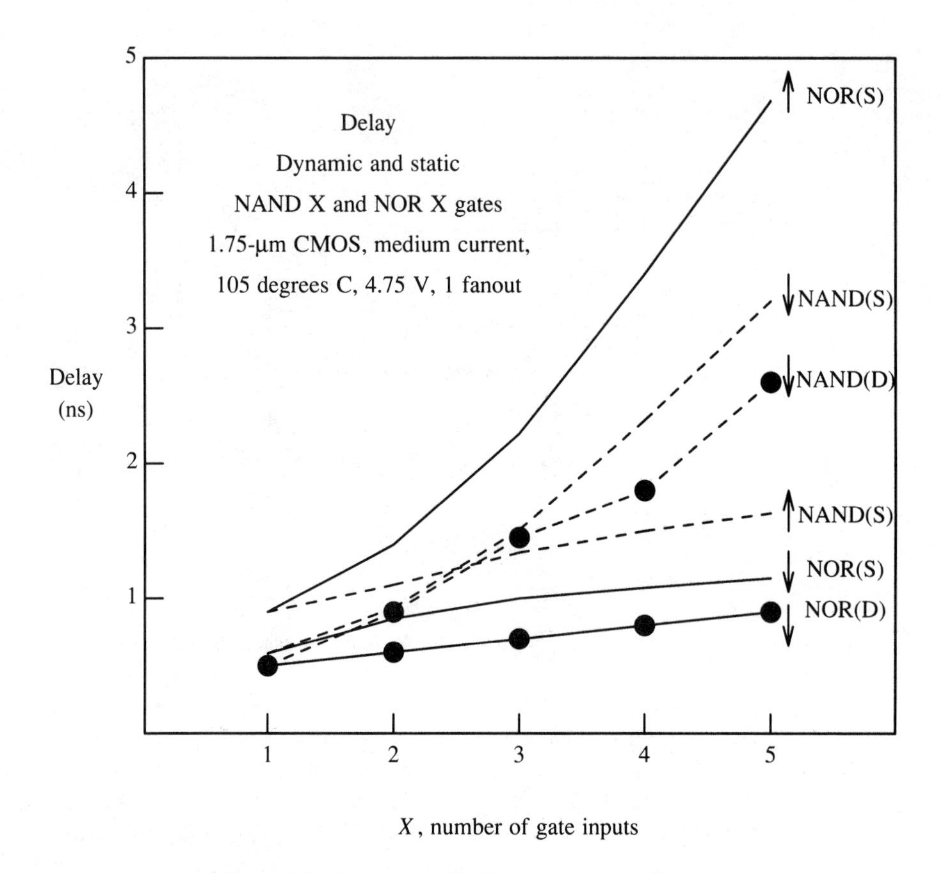

Figure 5.3 Delay of CMOS static and dynamic gates.

has less time to execute logic than the static logic, but only by the duration of the time when the slave latch is transparent. Since the time when slave latch is transparent can be made small compared with the cycle time of the system, dynamic logic never has significantly less time for logic operation than static logic. Indeed, the much shorter delay time of CMOS dynamic gates allows us to design higher-performance CMOS circuits than by using CMOS static gates alone.

In the last paragraph, we mentioned that the output voltage of a static latch settles very rapidly when the downgoing slave latch clock arrives. The voltage is securely held by a conducting FET (by PFET to V_{DD} and by NFET to ground). This is the advantage of CMOS static latch used in CMOS dynamic logic circuit, for the following reason. If the input node voltage is not held by a conducting FET, the voltage may be perturbed by noise due to capacitive coupling or by leakage current of FET. This problem occurs when input signals come from tristatable inverters and transmission gates. If the voltage of the floating input node becomes higher than the threshold voltage of NFET, the charge stored at the output node of the dynamic gate is

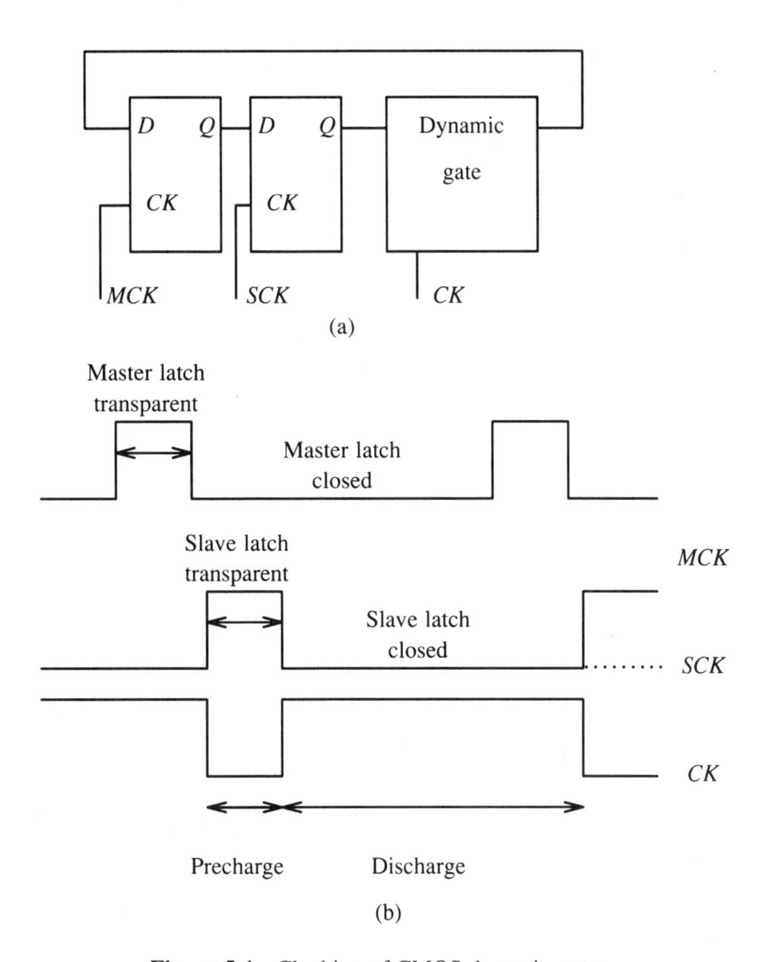

Figure 5.4 Clocking of CMOS dynamic gates.

discharged and a logic error occurs. FET threshold voltage V_{THN} is about 0.7 V. The switching threshold voltage of CMOS static gates is about $V_{DD}/2$, or 2.5 V, and noise voltage less than that does not influence logic operation. Dynamic gates are therefore about three times more noise sensitive than static gates. Noise-free input voltage is therefore an essential prerequisite to dynamic logic. For this reason, static latches are almost always used in dynamic logic circuits. In this book we assume that input signals to dynamic logic circuits are supplied from CMOS static latches.

Many latches and dynamic gates of Fig. 5.4(a) can be paralleled, interconnected together, and clocked by the same *MCK* and *SCK*. The circuit is then called a state machine, because by clocking the latches the circuit goes through a sequence of states determined by the logic and the initial conditions. A state machine used to control a microprocessor has quite complex logic circuits. Such complex logic circuits are built by cascading simpler dynamic logic gates. Cascading dynamic gates is, however, not a simple matter, as discussed in the next section.

5.4 CASCADING DYNAMIC GATES

In Fig. 5.5, gate 1 (dynamic NAND3 gate) and gate 2 (dynamic NAND2 gate) are cascaded. Suppose that $A = B = C =$ high, and $X =$ high. When the clock makes a low-to-high transition, node Y pulls down some time after the clock edge. If gate 2 is driven by a delayed clock that makes a low-to-high transition after node Y is pulled down, node Z remains high and, therefore, correct logic operation occurs. If the clock of gate 2 arrives before Y goes low, however, node Z is pulled down and a logic error results. For correct operation of the directly cascaded logic, the delay of the clock to gate 2 must be greater than a certain limit. This limit, however, depends on the complexity, and therefore delay of gate 1.

Practically, it is impossible to design an individual clock circuit for each gate, attaining the minimum and necessary clock delay for each. If more than necessary delay is used, the cumulative delay of the two stages becomes larger and the advantage of dynamic logic gates is reduced or eliminated. Furthermore, we wish to avoid complex clock generation: Practically, we wish to use a single clock edge for the entire dynamic logic circuit.

This problem, commonly called a race condition of dynamic logic gates, exists in any cascaded dynamic circuits. One way to deal with this problem is to use a dummy circuit that simulates the discharge delay of the first-stage dynamic gate. The circuit shown in Fig. 5.5(b) may replace the box "Delay" in Fig. 5.5(a). This "self-timing" scheme is convenient and is often used in memory designs. Designed properly it is relatively simple, but what is unfortunate is that there is no clock frequency for which the circuit will work if the design is not proper. Further, the circuit is rather sensitive to process variations and is too area costly to be used everywhere. This problem is discussed in detail in Section 8.12. The application is therefore limited to special circuits like memories or PLAs, where the designer's attention can be expended to make a small number of key circuits work reliably.

In addition to the self-timing scheme shown above, there are three major techniques to deal with race conditions in cascaded dynamic logic. The first is to provide a fixed delay between the first- and the second-stage clocks, which is widely used in CMOS dynamic PLAs. The second is Domino CMOS dynamic logic, and the third is NORA CMOS dynamic logic. They are discussed in the following sections.

5.5 DYNAMIC CMOS PLA

In the preceding section, with reference to Fig. 5.3, we saw that the delay of a dynamic NOR gate increases very slowly with increasing number of inputs. By cascading two stages of multi-input NOR gates, any desired Boolean function of the input

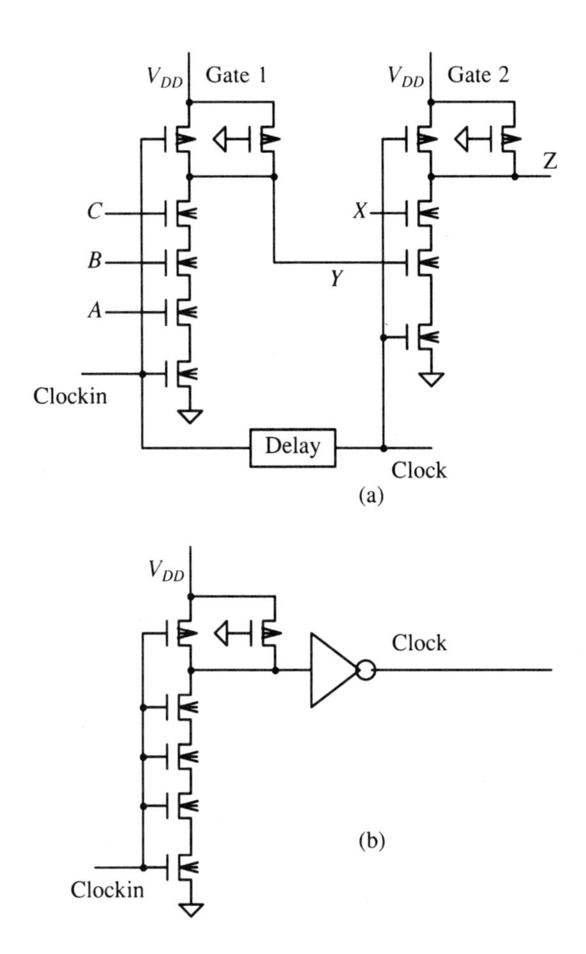

Figure 5.5 Cascading CMOS dynamic gates, and clock delay circuit.

variables can be generated. A CMOS dynamic PLA makes use of these two properties.

Logic design methods, coding methods, and structuring (folding) methods for PLAs are not discussed here. The reader is referred to references, such as [02] [03] [04]. CMOS dynamic PLAs are often used to generate state vectors for microprocessors or even datapath functions [05]. A state vector is a set of Boolean variables that control the execution units of a processor. One state vector is generated each cycle. For the PLA application, one machine cycle of the microprocessor is subdivided into four phases, 1, 2, 3 and 4, as shown in Fig. 5.6. During phase 1, the NOR gates of the first-stage dynamic logic (often called a decoder) is precharged, and the input logic variables stabilize by the end of this phase. During phase 2, the decoder NOR gates

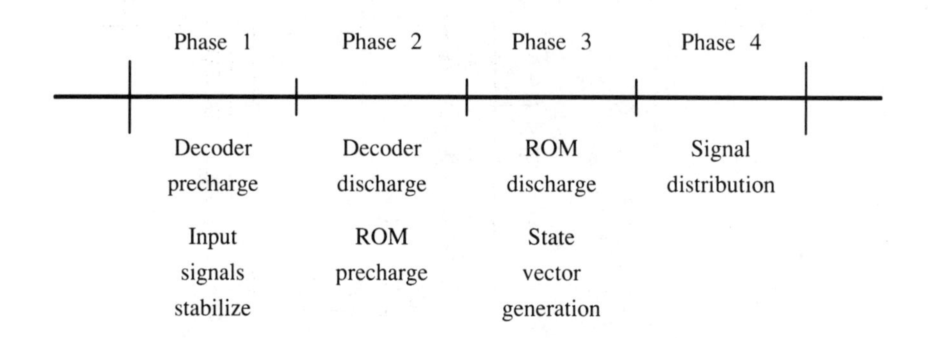

Figure 5.6 Timing of CMOS dynamic PLA.

discharge, while the second stage NOR gates (often called ROM) are precharged. During phase 3 the ROM NOR gates discharge, and by the end of phase 3 the state vector appears at the output terminals of the PLA. Phase 4 is used to route the variables to the destination. This four-phase timing is very commonly used, even in NMOS PLAs [06].

A problem of this four-phase timing scheme is that complex clocks must be generated to drive the dynamic logic circuits. A clocking scheme adapted to a single clock, however, has been reported [07]. In this scheme, static latches are provided to save the output data from the decoder. This scheme has layout problems: To provide static latches for each word line is too area consuming. If the static latches are replaced by dynamic latches, the circuit becomes noise sensitive, since input nodes to the ROM can be floating. Since PLA generates large noise spikes, noise sensitivity of floating nodes is often a fatal problem. Because of these difficulties, four-phase timing is widely adopted.

Figure 5.7(a) shows a logic diagram of a CMOS PLA, and Fig. 5.7(b) a schematic of a layout. The names of input signals ($A, B, C, \ldots$), output signals from the decoder ($\alpha, \beta, \gamma, \ldots$, called word lines), and output signals from the ROM (I, II, III, . . . , called bit lines) of Fig. 5.7(a) correspond to those in Fig. 5.7(b). In Fig. 5.7(b) solid lines represent metal, broken lines represent polysilicide, and the squares are FETs (PFETs are shaded). Input signals A, B, C and their complements $\overline{A}, \overline{B}, \overline{C}, \ldots$ come into the decoder by polysilicide wires. To generate the most general combination of the input signals, both logic polarities of the input signals are necessary. The first-stage NOR gates have output nodes α, β, $\gamma, \ldots$, which are called word lines. The word lines are precharged by the PFETs such as MP1, during phase 1 shown in the

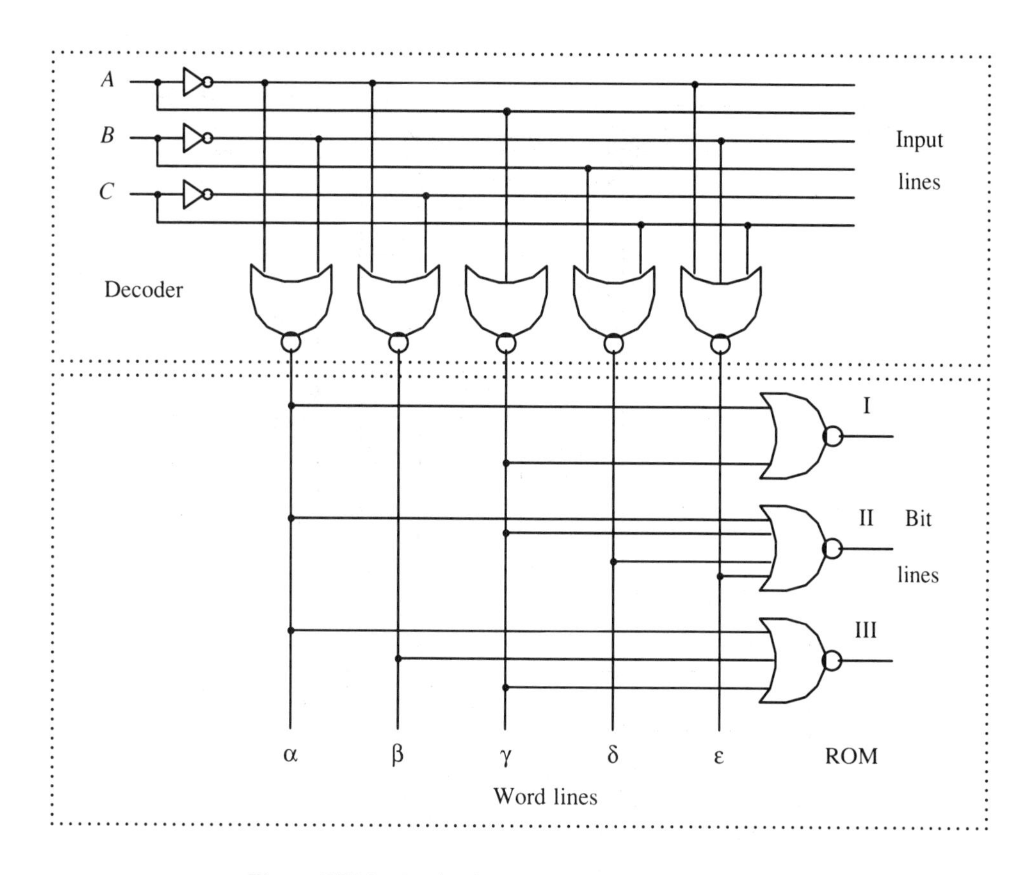

Figure 5.7(a) Logic diagram of CMOS dynamic PLA.

timing diagram of Fig. 5.6. The NFETs MN1, MN2, ..., which reflect the coding of the decoder, are connected to the word lines. The gates of these NFETs are driven by the input signals and their complements. These PFETs, NFETs, and the ground switch MN0 make the first-stage NOR gates, called a decoder.

Outputs from the decoder, the word lines α, β, γ,... go into the ROM section and change over to polysilicide lines. NFETs MNA, MNB, ... reflect the coding of the ROM. The drains of NFETs MNA, MNB, ... are the output nodes of the second-stage NOR gates called bit lines (I, II, ...). The bit lines are precharged by PFETs MPA, ... Together with the ground switch MNZ, ..., the FETs make the second-stage NOR gates (ROM).

The PLA is coded such that a word line remains high when a desired event happens (a certain combination of the input Boolean variables A, B, ... occurs). There are many desired events, and for each of them there is one word line. The specified set of word lines turn on the NFETs that discharge a bit line of the ROM, thereby

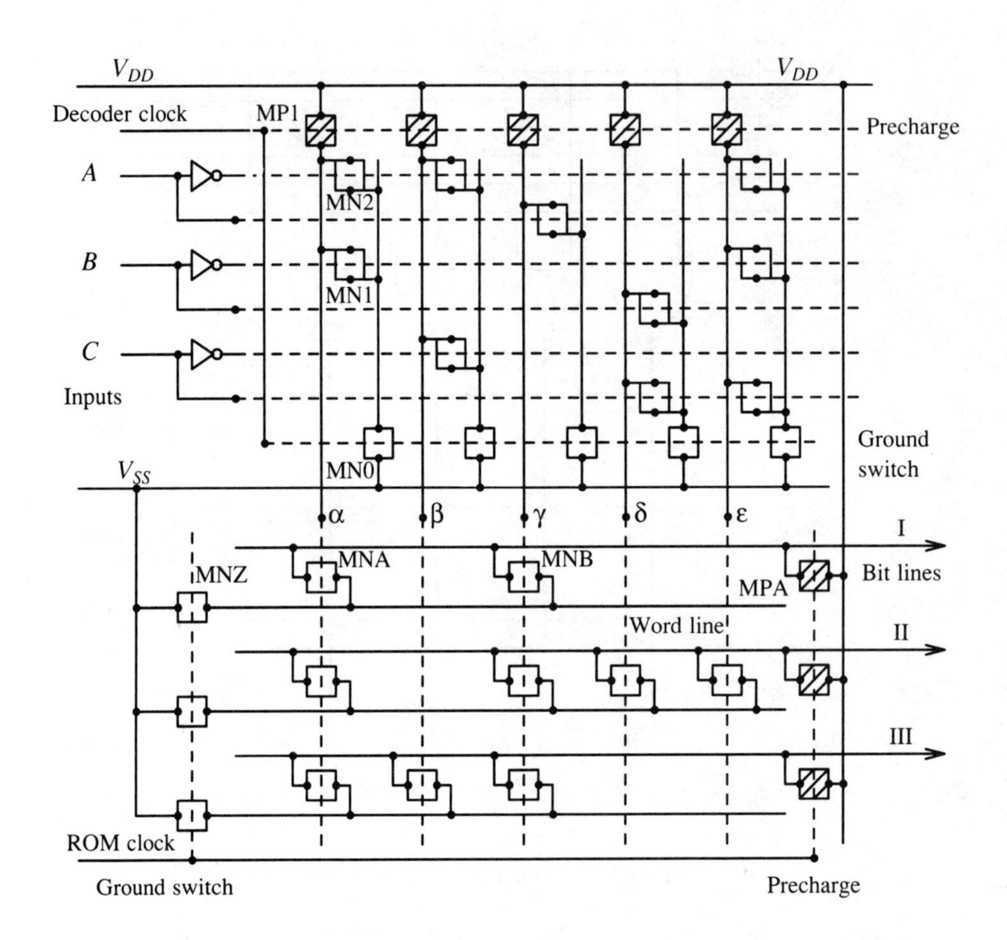

Figure 5.7(b) Structure of CMOS dynamic PLA.

indicating any one of the "desired set of the events" happens. When the bit line signal is buffered, one component of the state vector is obtained. This is a sum of the products of the input variables A, $\overline{A}$, B, $\overline{B}$, ...

In a CMOS dynamic PLA, the time available for discharging the NOR gates is limited by the duration of the phase. Therefore, there is an upper limit to the number of NFETs that can be connected to a single input line, a word line, or bit line for a given phase duration. Let the wiring capacitance of a bit line be C_W, the drain island capacitance of a single NFET (that carries PLA coding) be C_D, and the saturation current of the NFET be $I_{D\max}$. Then the pulldown delay time, T_D, is given by

$$T_D = \frac{f \cdot (C_W + NC_D)V_{DD}}{I_{D\max}} \tag{5.1}$$

where f is a numerical factor of the order of unity that includes the effects of series-connected ground switch. Since ground switches are usually very large, C_D and $I_{D\,\max}$ are proportional. Then increasing the NFET size does not decrease delay if $C_W < NC_D$. In the limit of small C_W, T_D tends to $fNC_D V_{DD}/I_{D\,\max}$, and the number of NFETs N is limited to $N_{\max}$ given by

$$N < N_{\max} = \left(\frac{1}{f}\right)\left(\frac{T_D I_{D\,\max}}{C_D V_{DD}}\right) \tag{5.2}$$

Assuming that T_D = 25 ns, $I_{D\,\max}$ = 600 μA, C_D=0.01 pF, V_{DD} = 5 V and f=2 (these are typical values for 1.75-μm CMOS at medium current process in a chip whose frequency is 10 MHz),

$$N_{\max} = \frac{2.5 \times 10^{-8} \times 6 \times 10^{-4}}{2 \times 10^{-14} \times 5} = 150$$

This estimate shows that about 100 signals can be combined within the given timing budget. If the architecture is selected properly, this capability is enough to generate the state vector of, for instance, a VLSI microprocessor.

A version of CMOS PLA that uses PFETs for coding and that operates very much like the PLA discussed above has also been used to generate the state vectors for a microprocessor [08].

5.6 DOMINO CMOS DYNAMIC LOGIC

In Section 5.4 we observed that the race condition with two cascaded dynamic gates can be eliminated if the clock of the second gate is delayed until settling of the first gate outputs. To generate such a clock using dummy gates and inverters is, however, costly. One way to eliminate the cumbersome clock circuit is to use the signal processed by the first gate to start discharging the following dynamic gates. Since an upgoing transition is necessary to clock a dynamic gate, the signal output from a dynamic gate is first inverted, and is then used to drive the input of the next gate, as shown in Fig. 5.8. This scheme is called Domino CMOS [09].

After precharge is over, all the outputs of the inverter of the Domino CMOS gates are low. As the signals are processed the outputs of some inverters become high. Once this occurs the state remains high. A low-to-high transition of a signal means that two things have happened: The answer was found, and that answer is a high level.

This type of logic is called Domino CMOS dynamic logic because the processed signal propagates like falling dominoes. We call the combination of a dynamic gate and an inverter a Domino CMOS gate. Therefore, a Domino gate is a noninverting gate. Discharge clocks to all the Domino CMOS dynamic gates make a simultaneous transition and, therefore, by the time the upcoming signal X arrives at gate 2, the ground switch of gate 2 is already turned on. In a Domino CMOS gate the only role the clock plays is to start the discharge of the first gate of the logic chain. Ground switches of the following gates may be eliminated.

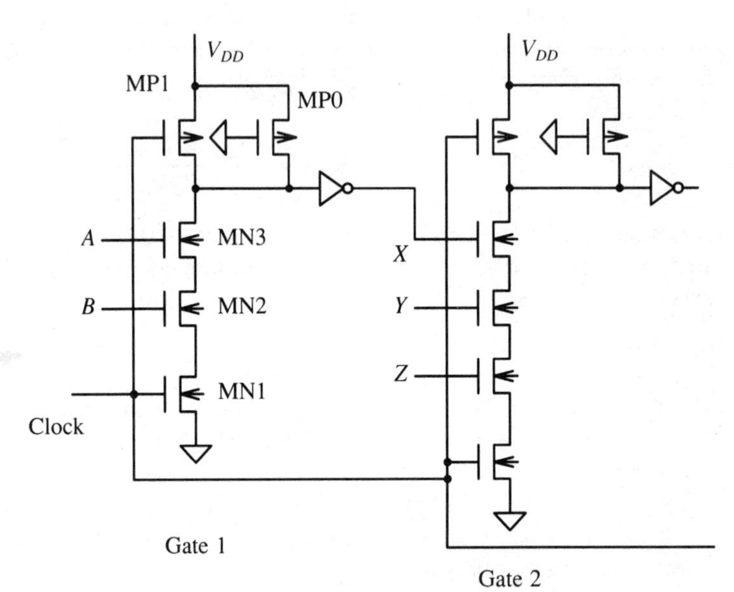

Figure 5.8 Domino CMOS gate.

The small PFET MP0 of Fig. 5.8 plays two roles. The first is that a Domino gate that has a small PFET is able to keep the logic high state indefinitely after the gate is clocked into the discharge state. The PFET replenishes the charge lost by leakage current through the NFETs. The second role is that the PFET load device keeps the output node of the dynamic gate at a reasonable impedance level when the pulldown NFET chain is off. A proper impedance level prevents noise coupling to the output node. The PFET must be designed small enough so that even a very complex and long NFET pulldown chain is able to drive the output node within several tens of mV from the ground.

Four-phase precharge-discharge logic proposed by Myers and Ivey is shown in Fig. 5.9(a) and its timing, in Fig. 5.9(b) [10]. The extra NFETs sample and hold the processed data. Therefore, the circuit can be used with two-phase, nonoverlapped timing. In the author's opinion, however, the scheme is closer to CMOS dynamic PLAs than to Domino CMOS. What is outstanding in Domino CMOS is that signals propagate over many stages, without interruption by clock.

5.7 DOMINO CMOS LOGIC DESIGN

Domino CMOS gates are noninverting. A direct consequence of this restriction is that the most general Boolean expression, the sum of products of variables, can be generated using both AND and OR gates. Since an AND gate with a large number of inputs is slow, care must be exercised to avoid using large input AND gates.

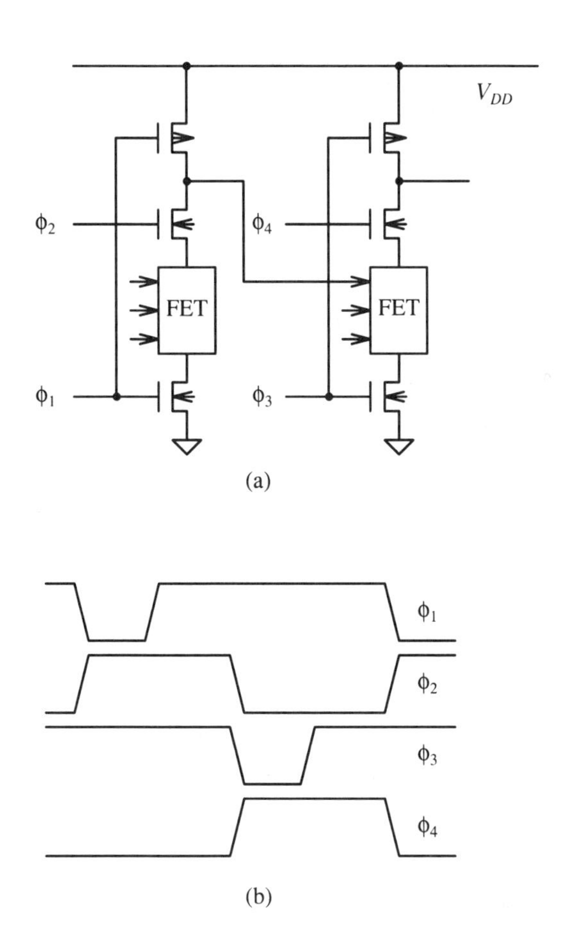

Figure 5.9 4-phase Domino CMOS.

The pulldown delay of an NFET chain is approximately a linear function of the number of series-connected FETs (N) when N is small ($N<4$). When N is large, however, the delay is proportional to N^2. This square dependence shows that the NFET chain behaves functionally in the same way as a linear RC chain (Section 5.12 and 6.8). Too complex an AND gate should be split into several cascaded stages of simpler AND gates, or the logic should be converted to negative logic that uses OR gates. The latter implementation is preferred if the number of signals combined is more than 10.

Domino CMOS allows a variety of logic implementations. The first step in Domino CMOS logic design is to redraw the given logic diagram so that the implementation with Domino logic becomes clear. Figure 5.10(a) to (c) show an example.

The given logic diagram of Fig. 5.10(a) is converted to Fig. 5.10(b) using the equivalence relations shown in Fig. 5.10(d), which are representations of De Morgan law in Boolean algebra [11]. A bubble (logic inversion) can be moved either to the left or to the right by the equivalence relations, and depending on the direction there can be different Domino CMOS logic implementations. One implementation shifts the inversion to the input, and the other to the output. One implementation can be better in some respect than the other, for example, using Domino CMOS gates that have shorter NFET chains.

The process of logic design is as follows. In Fig. 5.10(b), two nearby bubbles were collapsed, since double inversion is identity. Figure 5.10(b) can then be converted to Fig. 5.10(c), which can be implemented directly by Domino CMOS logic. In Fig. 5.10(c) the Domino gates are identified by an inverter associated with gate. The inverting dynamic gate and the static inverter cannot be separated. In the process of conversion, three gates α, β, and γ of Fig. 5.10(b) were combined together into the single AO221 Domino CMOS gate of Fig. 5.10(c). Such composition of gates and sometimes its inverse, decomposition, is commonly practiced in Domino CMOS design.

The buffer inverter of a Domino CMOS gate may be replaced by an inverting static gate and extra signals can be introduced to the gate, as shown in Fig. 5.11(a) [12]. The two NAND2 gates (marked "D") are dynamic gates, and the NOR2 (marked "S") is a static CMOS gate. When static gates are used within Domino CMOS gate chain, such as gates α and β of Fig. 5.11(b), an even number of static gates must be inserted between the Domino gates.

If not all the input signals are available simultaneously as shown in Fig. 5.11(c), the first stage of the Domino gate, which is driven by the the delayed signal, must be clocked by a delayed clock. In this way the maximum timing budget is made available for the processing. As was observed in this section, configuring Domino gates has a large degree of flexibility and freedom for new invention.

Domino CMOS is simple and is suitable for computer-aided design and layout [13]. Domino CMOS can be used in standard cell design [14]. Domino CMOS gates are silicon area economical. Layout is very similar to NMOS layout, and many ideas for compact NMOS layout [15] are applicable to Domino CMOS.

5.8 LOGIC THAT CANNOT BE IMPLEMENTED BY DOMINO CMOS

Domino CMOS gates are noninverting. This peculiar property is shared by active diode logic gates using Esaki and Josephson diodes. Domino CMOS gates and active diode gates are reset (or precharged) to one of the logic states at the beginning and then clocked into logic operation. The gates stay in the reset state, or change to the other state only once, if the result is found and it is represented by the other state. Because of the similarity of the principle of operation, many logic design techniques developed for active diode logic are applicable to Domino CMOS [16].

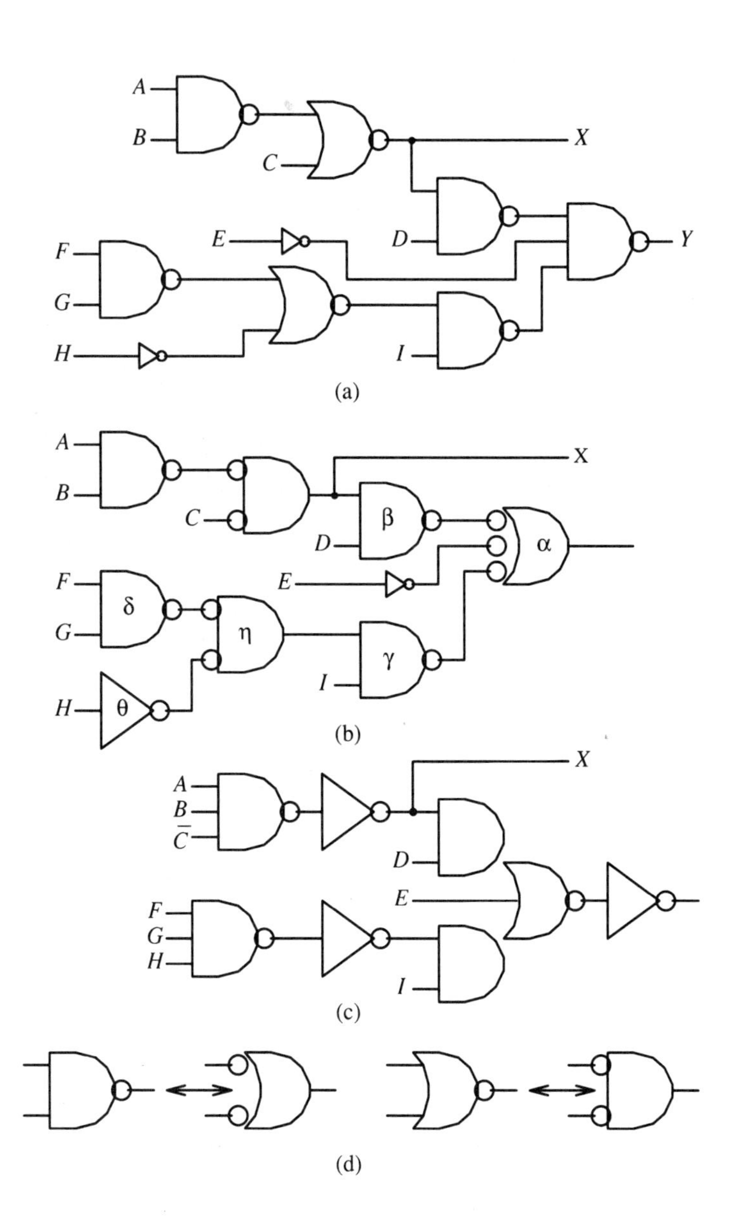

Figure 5.10 Domino CMOS logic design.

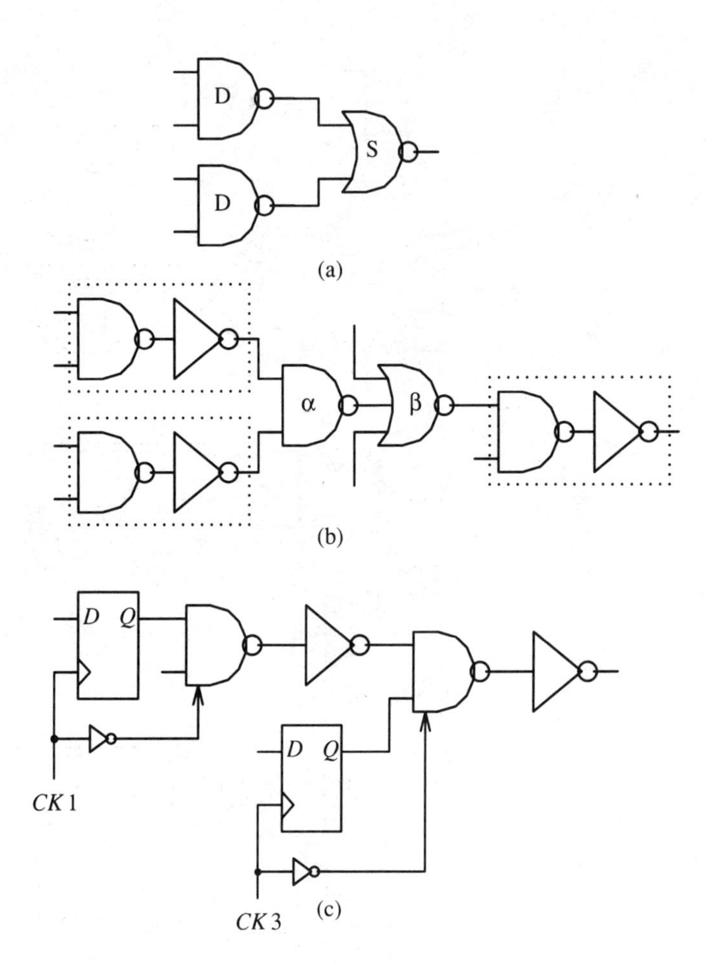

Figure 5.11 Variation of Domino CMOS logic.

CMOS logic is almost always designed using inverting logic gates. This original design must be converted to a second design using noninverting gates only. This conversion is not always feasible. We study this problem and some practical solutions in this section. When the original logic is converted to noninverting gate implementation by pairing gates as shown in Fig. 5.12(a), there may be an extra inverter (shadowed) that cannot be paired. This design is not an acceptable Domino CMOS logic. If this connection is made, the following problem occurs. Suppose that signals A, B, and C originate from latches and have values $A = H$, $B = H$, and $C = H$. Output node Z should be low when the signals are processed, but this does not happen. At the beginning of discharge, node W is low, and node X is high. Then node Y, which was

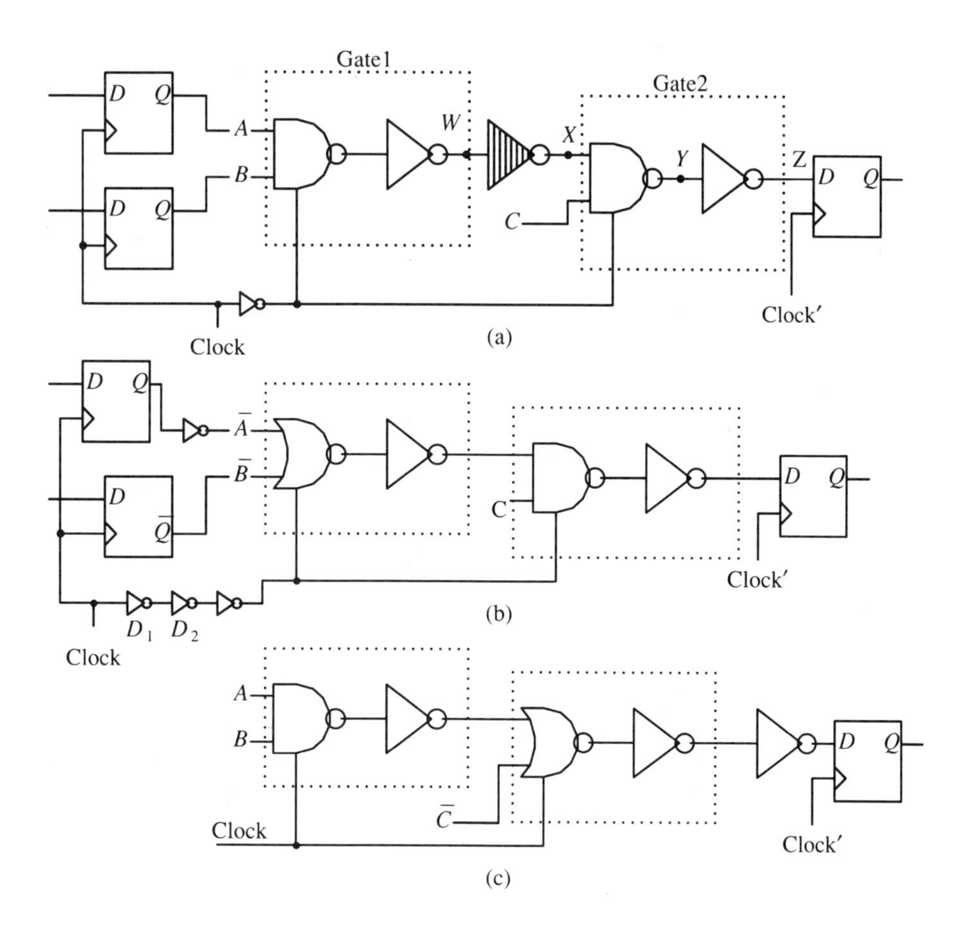

Figure 5.12 Logic that cannot be implemented by Domino CMOS.

precharged but not held at a high level, discharges. Once this happens, node Y never recovers to the high level. Node X pulls down eventually but it is too late. Output node Z remains in the high state.

It is desirable that logic is designed to avoid signal combinations of the kind described above, but this is not always possible. If the logic was originally designed for static CMOS gates, there are cases where a straightforward conversion to Domino CMOS is impossible. We wish to discuss methods to deal with this situation [17].

Since signals input to a chain of Domino gates A, B, and C always originate from static latches, and therefore $\bar{A}$, $\bar{B}$, and $\bar{C}$ are all available, an implementation as shown in Fig. 5.12(b) is possible, and this implementation is free from the problem. The circuit of Fig. 5.12(b) is practical because the discharge clock can be delayed by a

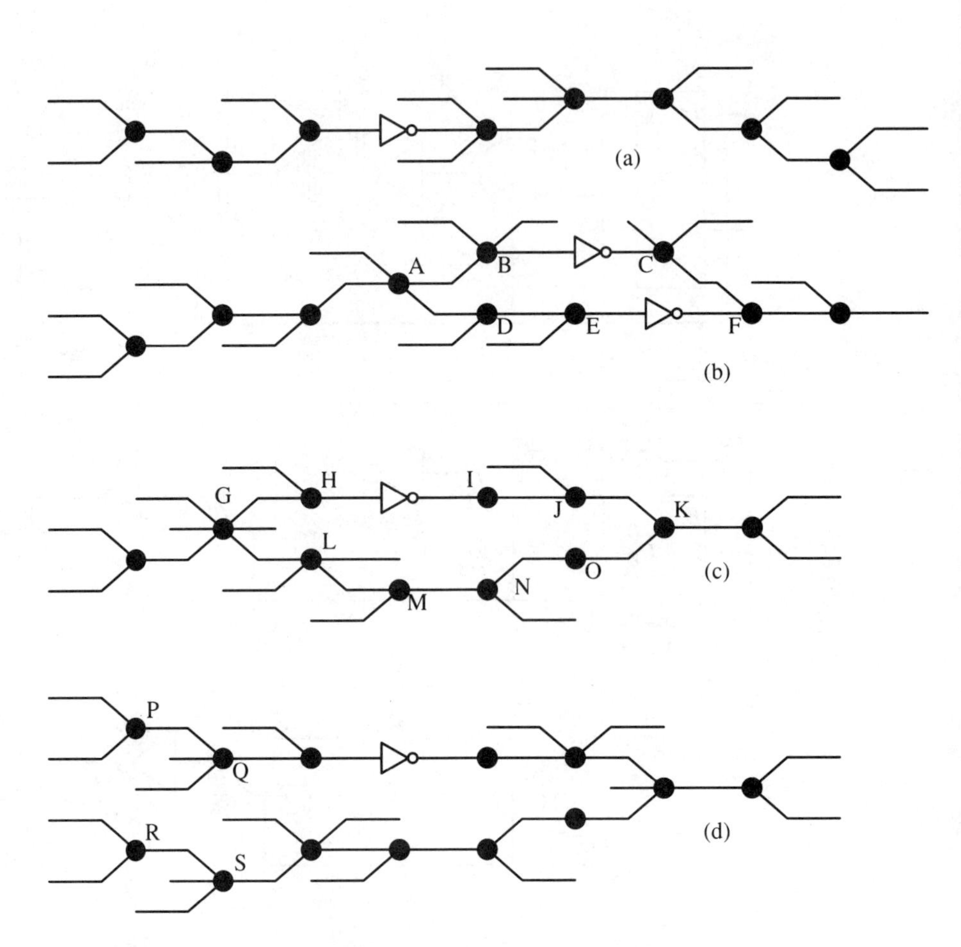

Figure 5.13(a) to (d) Extra inversion in Domino CMOS logic chain.

simple circuit until the input signals stabilize if complement of the signal is available some time later ($\bar{A}$ in this case). The clock delay circuit consists of two extra cascaded inverters, D_1 and D_2, that compensate delay of extra inversion. Similarly, the same problem can be resolved by moving the extra inversion to the right, as shown in Fig. 5.12(c). The solution that can be obtained by this kind of technique creates no serious problems.

Figures 5.13(a) to (d) are schematics of logic gate interconnections. Closed circles are noninverting, multi-input single-output gates that receive signals from the left side and deliver the processed signals to the right. If the logic chain does not contain any closed loop as shown in Fig. 5.13(a), the extra inverter can be shifted either to the left or to the right end of the chain, as we discussed in the preceding section. When the logic chain contains a loop of gates as shown in Fig. 5.13(b), and if each of the

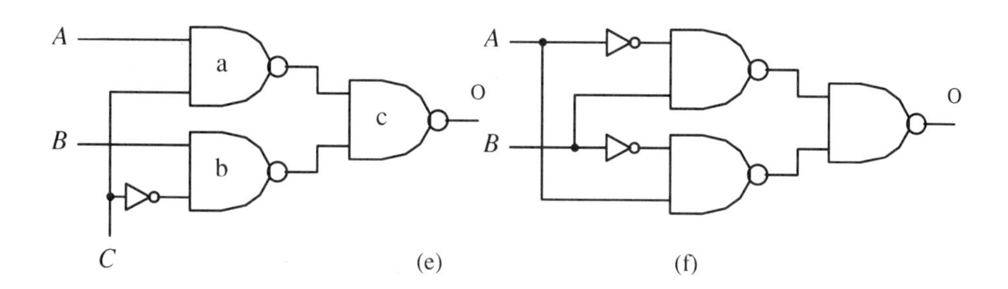

Figure 5.13(e) and (f) Multiplexer and exclusive-OR.

two branches forming the loop, (A, B, C, F) and (A, D, E, F), has an inverter, the inverters can be shifted together to the left (to gate A) and then combined together. The combined single inverter can then be shifted to the left, and the problem can be resolved as in the case of Fig. 5.13(a).

When the gate chain contains a loop shown in Fig. 5.13(c), and only one branch (G, H, I, J, K) contains an extra inverter, the inverter can be shifted to the left (to gate G), but it is impossible to shift the inverter still further to the left of gate G. In this case the original logic to the left of gate G can be duplicated, and the original loop is broken open, as shown in Fig. 5.13(d). Then the inverter can be shifted out to the left.

A logic circuit that creates difficulty of this kind is a loop of gates that contain a single inverter. Practically, logic of this kind is a variation of multiplexer, shown in Fig. 5.13(e). When control signal C is high, signal A, and when C is low, signal B is delivered to output O. Signal C branches, and one is inverted, driving NAND gates "a" and "b" separately, and then the signals from NAND gates "a" and "b" are combined by gate "c", thereby closing a loop. As shown in Fig. 5.13(f), exclusive OR is a special case of multiplexer logic. Since XOR and XNOR are frequently used we need special consideration about how to deal with them.

When a multiplexer or an XOR is required within a logic chain, often the method of shifting the exclusive-OR gate to the end of the chain works. A multibit adder with a carry-look-ahead logic is an example. In this case all the logic of the carry-look-ahead can be implemented by Domino CMOS. Full adders that consist of two exclusive-OR's are at the end of the logic chain, and therefore can be implemented by static gates.

When XOR gates must be cascaded over many stages as in an ultrafast parity generator for an error-detection/correction chip, the best way is to generate an XOR signal and an XNOR signal as a pair, as shown in Fig. 5.14(a) and (b), and to cascade the unit that has four inputs (signal and its complement for both inputs A and B) and two outputs (XOR and XNOR). This double-rail logic circuit takes more hardware, but the scheme offers superior delay characteristics of Domino CMOS logic compared to static CMOS logic.

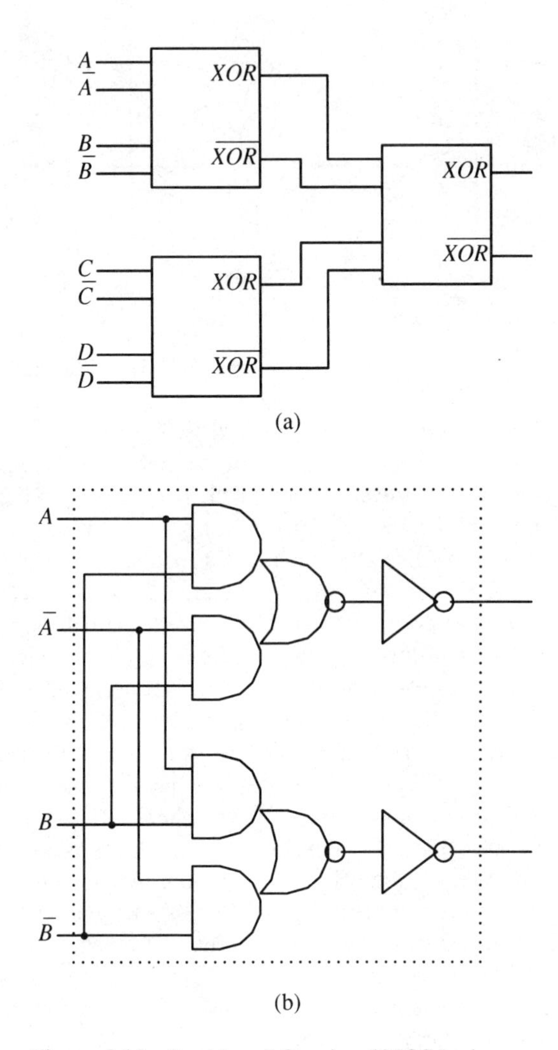

Figure 5.14 Double-rail Domino CMOS logic.

To implement the entire subsystem by double-rail Domino CMOS logic is often too costly. The latched Domino CMOS circuit shown in Fig. 5.15 generates the AND of variables A, B, and C and its complement NAND, both of which take low level before the clock makes a low-to-high transition [18]. The sense amplifier is designed asymmetrically by FET scaling, so that:

1. If node N_0 remains high (any one of A, B, or C is low), node N_1 is pulled down by the sense amplifier.

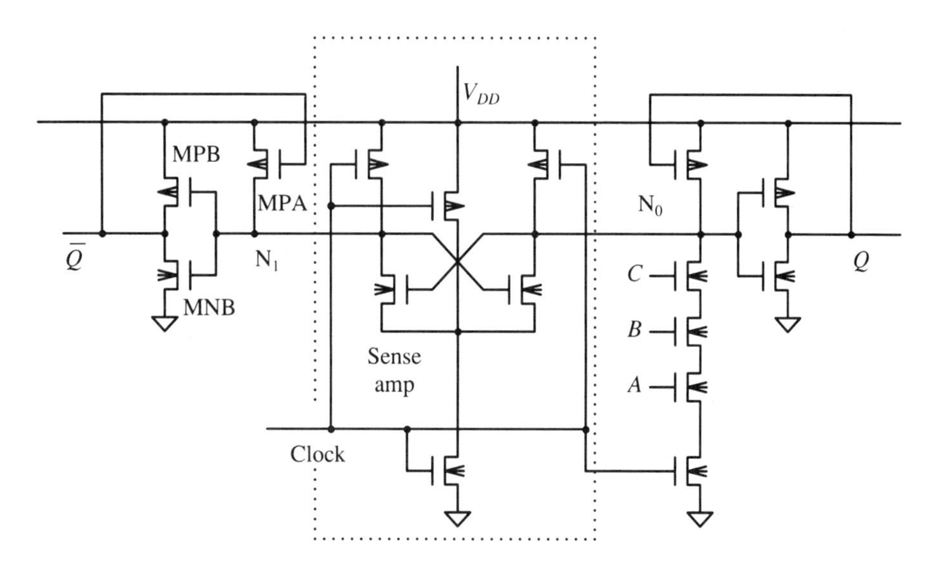

Figure 5.15 Generation of complementary signal by latched Domino CMOS.

2. If node N_0 is actively pulled down (all of A, B, and C are high), node N_1 stays high.

If node N_0 remains high for some time and then pulls down (this happens when the gate is a link of long Domino CMOS gate chain), node N_1 voltage decreases. Node N_1 pulldown stops, however, as soon as node N_0 pulls down. After that, node N_1 voltage recovers by pullup of PFET MPA. This recovery is impossible if node N_1 is pulled down below the switching threshold of the circuit consisting of MPA, MPB, and MNB. This is a clever circuit, but the circuit is not easy to design because the design requires fine adjustment of the FET sizes, to satisfy requirements (1) and (2) over all the process, temperature, and voltage variations. Latched Domino CMOS is not an all-purpose solution to the logic design problem of Domino CMOS. A similar idea using a sense amplifier with cascode voltage switch (Section 2.7) has been proposed by Grotjohn [19].

When none of the above-mentioned methods is applicable, the last technique is to introduce delayed clocks. The delayed clock can be generated by the dummy circuit as discussed in Section 5.4, or one of the chip's phase clock edges that is selected properly to balance the delays of the circuit before and after the new clocking point. However, before arrival of the clock edge, the input signals to the clocked gate must have stabilized.

As a matter of theoretical interest, how many extra clock edges are necessary to implement any given random logic? To study this question, a given logic must be reimplemented using noninverting gates and as small number of extra inverters as possible. The inverters that are either at the signal input end or at the signal output end

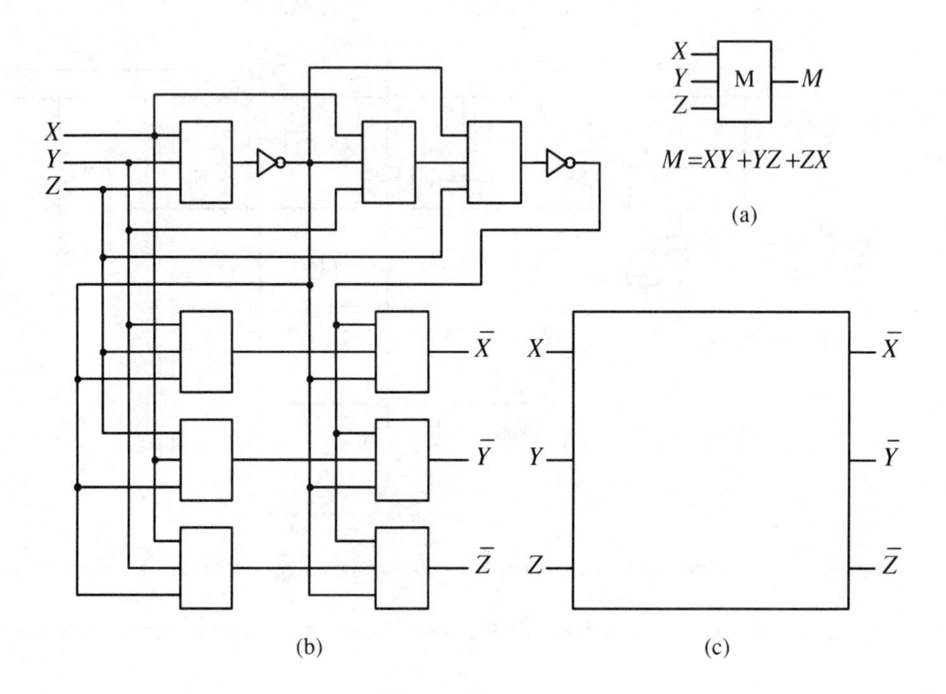

Figure 5.16 Markov's theorem.

can be dealt with by the techniques discussed before. Any (noninverting) gate that is driven by an inverter which resides inside the fabric of the logic requires a delayed clock edge. The question is now: How many inverters are necessary if any given multi-input, multi-output logic is implemented by AND gates, OR gates, and as small a number of inverters as possible?

A neat theorem proven by Markov shows that the number is only two [20]. The proof that only two inverters are sufficient is briefly summarized in the following. The proof is from Goto [21]. The rectangular box M of Fig. 5.16(a) is a majority circuit of the three input variables X, Y, and Z, that can be implemented by noninverting gates only, as

$$M(X, Y, Z) = XY + YZ + ZX \tag{5.3}$$

Using the majority box we construct the six-terminal black box of Fig. 5.16(c), which has the internal structure shown in Fig. 5.16(b). The box requires only two inverters, which are shown explicitly. It may be shown by an exhaustive check that the three outputs $\overline{X}$, $\overline{Y}$, and $\overline{Z}$ are the complements of X, Y, and Z, respectively.

If the given multi-input/multi-output circuit is redrawn (see the examples in Figs. 5.10 and 5.12) and if there are only one or two inverters, no further reduction is feasible. If three inverters are included, they may be replaced by the three pairs of

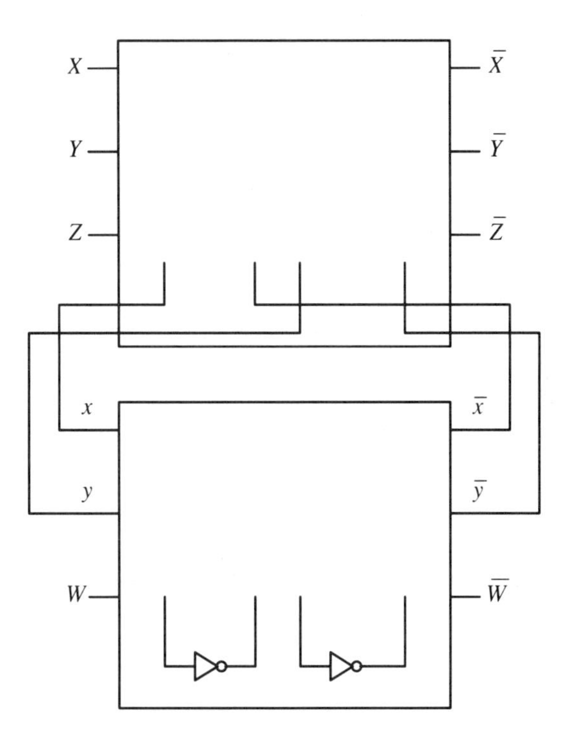

Figure 5.17 Proof of Markov's theorem.

terminals of the box of Fig. 5.16(c), and the resultant circuit has only two inverters. If the original circuit has four inverters, we use two boxes of Fig. 5.16(c) cascaded as shown in Fig. 5.17: Pairs of terminals x - $\overline{x}$ and y - $\overline{y}$ of the second box are used to replace the two inverters of the first box and the remaining four pairs are used to substitute the inverters in the original circuit. In this way any combinatorial random logic circuit can be implemented by noninverting gates and by only two inverters. Accordingly, the theoretical limit of the maximum number of extra clock edges needed by Domino CMOS implementation is only two. The theoretical circuit of Fig. 5.16(b) is, however, too complex, and the delay is too long, to be practical. The proof by Markov was given here to show that the problem of Domino logic implementation is practically never serious and there is a simple way to deal with it, by exploiting peculiarities of the cases.

5.9 NORA CMOS DYNAMIC LOGIC

In Domino CMOS logic, complex logic operations are carried out by fast NFETs. The role of PFETs is limited to precharging the gates and to buffering the logic output of the NFET pulldown chain, by pulling up and driving the fanouts. The buffer inverter

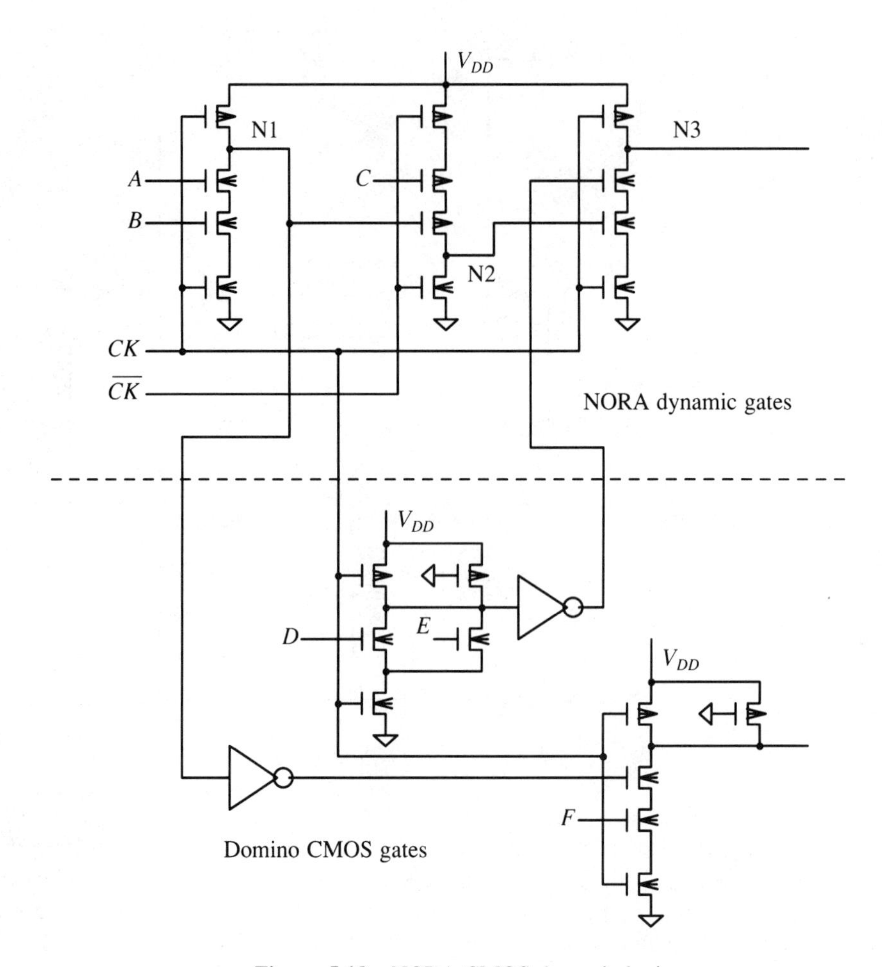

Figure 5.18 NORA CMOS dynamic logic.

of Domino CMOS can be converted into a dynamic logic gate, analogous to the static gate shown previously in Fig. 5.11(a), in which PFETs execute logic operations. An NFET in this PFET-dominated gate plays the role of predischarging the output node. This leads to a new logic circuit called NORA (standing for No Race) [22].

The NORA circuit shown in Fig. 5.18 works as follows: When the circuit is in the precharge state, clock CK is low, and clock $\overline{CK}$ is high. Then nodes N1 and N3 are precharged to V_{DD}, and node N2 is predischarged to ground. While the precharge/predischarge is in progress, input signals A, B, C, ..., settle to the final logic levels. When clock CK makes a low-to-high transition, and simultaneously $\overline{CK}$ a high-to-low transition, node N1 pulls down if A and B are both high. Suppose further that input C was logic low. Then as node N1 pulls down, node N2 pulls up.

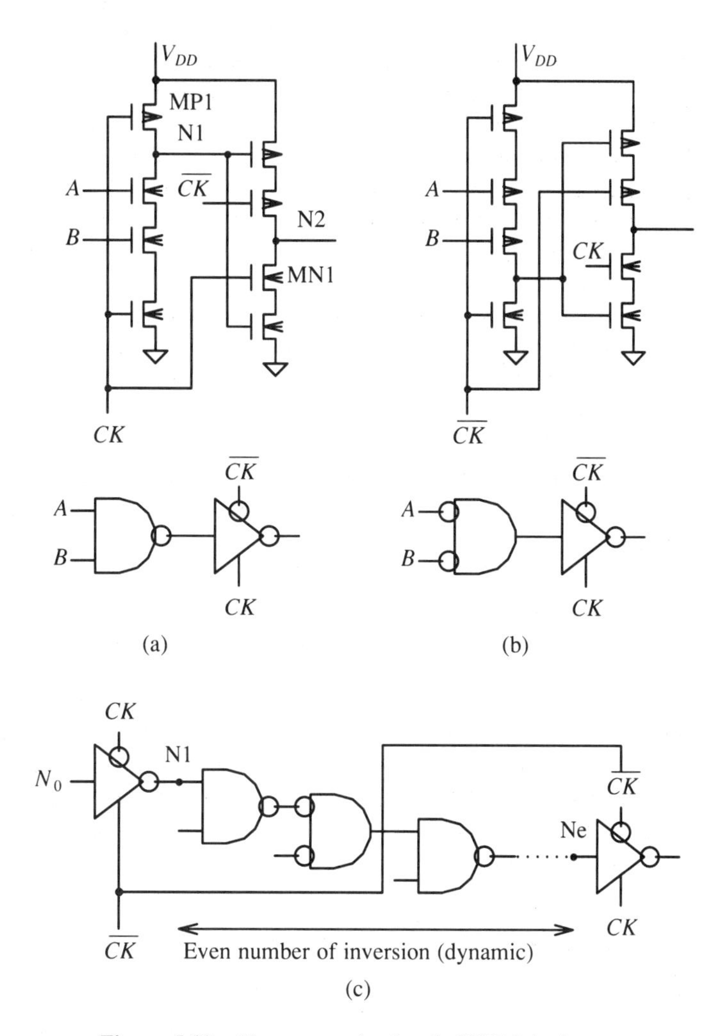

Figure 5.19 No-race mechanism in NORA logic.

If either of A and B is low, node N1 remains high, and therefore node N2 remains low. In this way logic variable $AB\overline{C}$ is generated at node N2.

NORA logic is compatible with Domino CMOS logic. As shown in the lower part of Fig. 5.18, Domino gates can be introduced into the chain of NORA logic gates. The Domino gates share the CK clock. The output node of the NFET logic is inverted by an inverter, and the output of the inverter drives the other NFET logic gates of the NORA circuits. Ground switches of NORA gates can be removed except for those gates located at the inputs of the circuit. This variation of NORA logic was proposed by Lee [23].

Original proponents of NORA logic addressed pipelined logic, in which even the data storage is carried out by dynamic latches. Some examples are shown in Fig. 5.19. In this totally dynamic design, it is possible to use the same clock for logic gates and for latches (in an IC, the two circuits that have the same clock are located nearby, and are connected by low-resistivity metal interconnects). In the circuit of Fig. 5.19(a) the pullup of node N1 at the time when the gate goes from discharge to precharge does not disturb node N2, since as NFET MP1 conducts, NFET MN1 turns off, as long as clock transition time is very short. A similar situation occurs in Fig. 5.19(b). This antirace feature does not work, however, when the clock transition is too gradual.

For convenience, we represent PFET NORA gates by the symbols shown in 19(b) (see the location of bubbles). NFET NORA gates are the same as the dynamic part of Domino CMOS gates and therefore the conventional symbols [like Fig. 5.19(a)] are used.

In the fully dynamic circuit of Fig. 5.19(c) data flow-through across the two dynamic latches does not happen if the number of inversions between the two dynamic latches are even. This is because if $\overline{CK}$ makes a low-to-high transition (assuming that CK has not arrived yet), node N1 is pulled down if N0 is high. Pulldown of node N1 may result in pulldown of node Ne. Pulldown of node Ne is prevented from influencing the state of the output dynamic latches, since the upgoing $\overline{CK}$ cuts the PFET of the dynamic latches off. NORA logic contains a number of such antirace techniques, from which the name originated.

It is the author's conservative viewpoint, however, that to construct an entire pipelined system by dynamic logic and dynamic latches only appears risky. Dynamic gates require well-established logic levels for noise immunity, and to obtain stable input voltage, static CMOS latches are far superior. Furthermore, at the present time, Domino CMOS circuits with static input latches have an established record of trouble-free performance.

5.10 NORA DYNAMIC LOGIC DESIGN PROBLEMS

When a logic diagram that was originally designed for static CMOS implementation is given as shown in Fig. 5.20(a), the diagram can be converted using the equivalent operations of Fig. 5.10(d), to Fig. 5.20(b), which can be implemented by NORA dynamic logic.

The problem of combining the signals still exists. A signal from an NFET gate cannot drive another NFET gate, and a signal from a PFET gate cannot drive another PFET gate. NFET gate and PFET gate must alternate.

A major difference between Domino CMOS logic and NORA logic is that when a chain of gates do not change logic states upon transition from precharge to discharge, all the nodes of NORA gates are held dynamically, as shown in Fig. 5.21(a). In Domino CMOS, only the output nodes of the NFET chains are held dynamically. Signals that drive the other gates originate from the static inverters, and

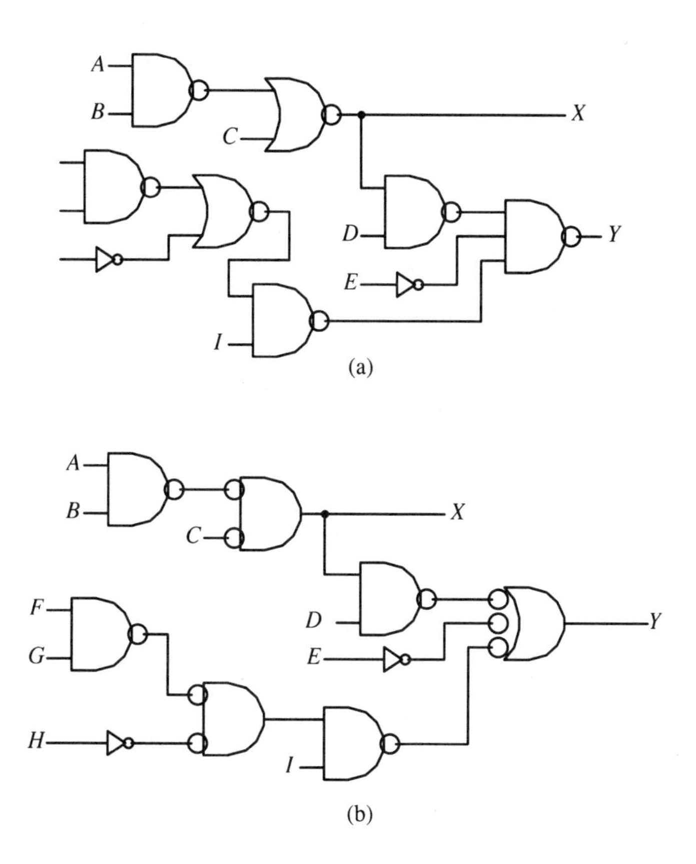

Figure 5.20 NORA logic design.

the nodes are held by conducting FETs, as shown in Fig. 5.21(b). NORA logic is, for this reason, much more susceptible to such problems as capacitive coupling of noise signals and charge sharing. A corrupted logic level that is close to the power supply or the ground, but is not exactly, has practically no effect in Domino CMOS, but may cause catastrophic failure in NORA.

In Fig. 5.22, node N1 originates from a dynamic latch that is not shown in the figure. Nodes N1 and N3 have strong capacitive coupling. Such coupling can exist if the signals are global signals of the chip. Suppose that a positive noise spike was induced on node N1. Then node N2 pulls down and node N3 pulls up. Node N1 is then pulled up more, and the circuit goes into a catastrophic failure. Such a mode of malfunction is probable, based on circuit simulations.

NORA gates suffer from charge-sharing problems. In Fig. 5.23(a), node N1 is charged up to V_{DD} while input A is low. If the gate has discharged previously, node

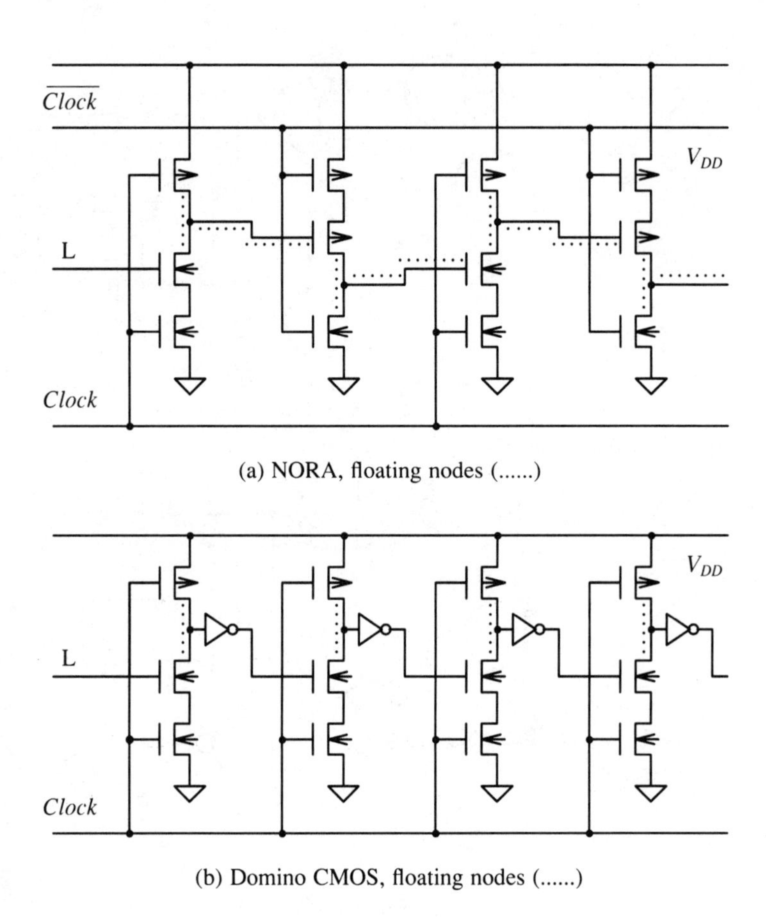

(a) NORA, floating nodes (......)

(b) Domino CMOS, floating nodes (......)

Figure 5.21 Stability of NORA and Domino logic.

N2 is at ground potential. If signal A makes a low-to-high transition, a fraction of charge of node N1 flows to node N2, and therefore the potential of node N1 drops from V_{DD} to $V_{DD}(C_1/(C_1+C_2))$ [24]. This drop, if more than the threshold voltage of the PFETs, may create a logic error or loss of noise margin in the following stages. Although Domino gates share the same problem, the effect is less, because a properly designed buffer inverter prevents propagation of unhealthy logic levels. There are several techniques for preventing charge sharing in Domino gates [25]. An example is shown in Fig. 5.23(b). The feedback PFET MP2 holds the output node even if slight charge sharing occurs. The problem of this circuit is that the size ratio of MP2 and the pulldown NFETs must be properly chosen: If MP2 is too large the circuit does not work. The delay of the circuit of Fig. 5.23(b) is significantly more than that of Fig.

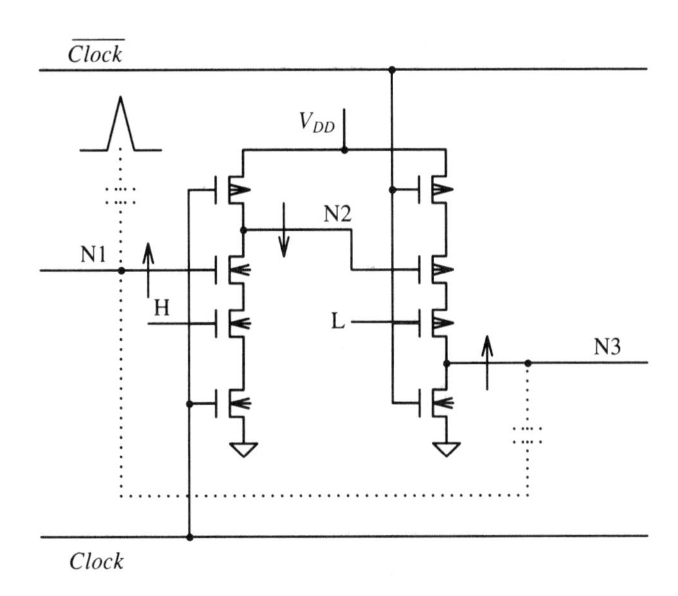

Figure 5.22 Catastrophic failure of NORA circuit.

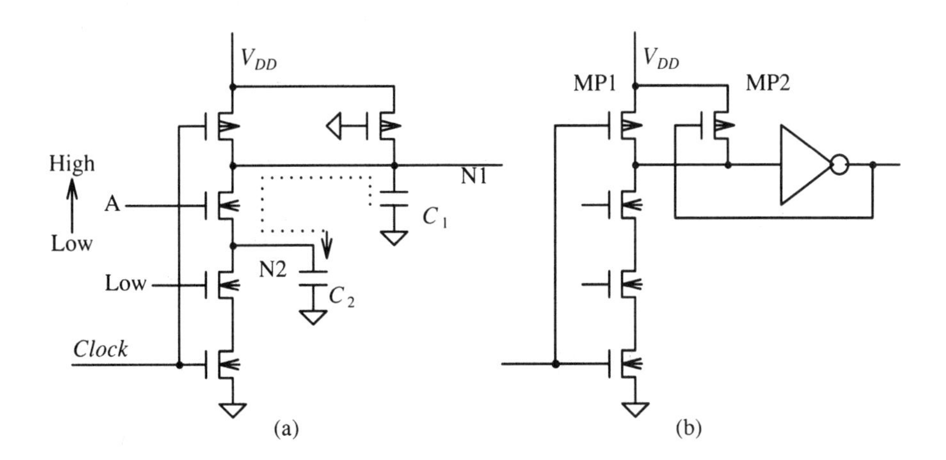

Figure 5.23 Charge sharing in dynamic logic gate.

5.23(a). Yet another way of preventing charge sharing is to precharge the internal nodes of the pulldown NFET chain [26]. This method is very reliable. Properly designed Domino CMOS gates are very robust against noise and charge-sharing problems.

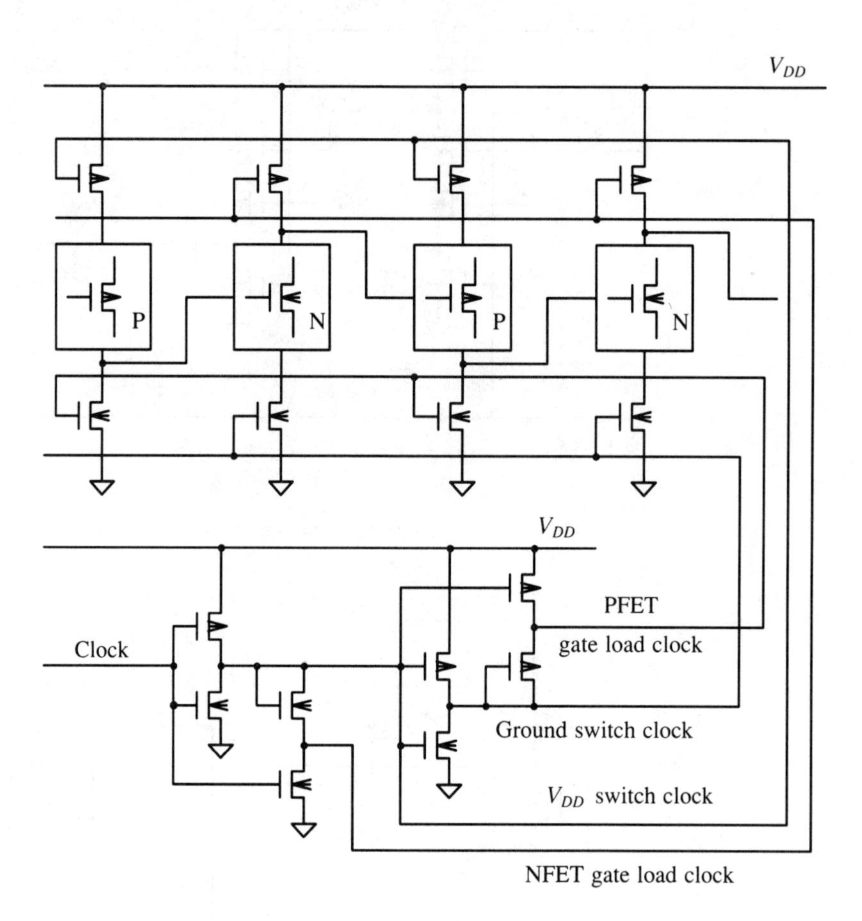

Figure 5.24 Zipper CMOS logic and clock generator.

5.11 ZIPPER CMOS

Zipper CMOS logic eases the charge sharing and the noise problems by not turning off the precharge and the predischarge FETs completely [27]. A Zipper CMOS circuit, shown in Fig. 5.24, uses a special clock generator that generates separate clock pulses for precharge PFETs and the ground switches of NFET gates (and their counterparts in PFET gates).

The ground switches (and their equivalent in PFET gates) are completely turned on during evaluation, and they are shut off completely at the time of precharge/predischarge. The precharge PFETs and the predischarge NFETs are turned on completely during precharge, but they are left slightly conducting during evaluation. Zipper CMOS gates are able to hold data indefinitely, without using conducting FET (MP2 of Fig. 5.1).

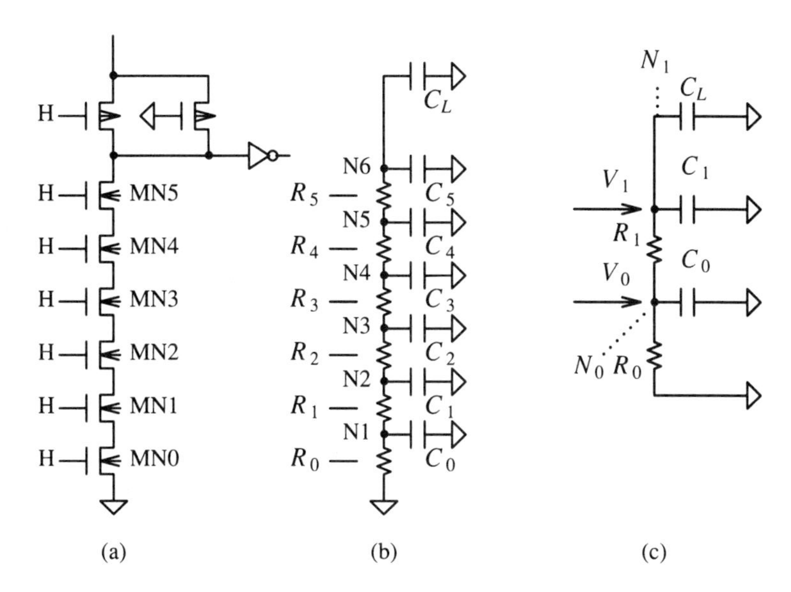

Figure 5.25 RC chain model of CMOS dynamic gates.

Clock pulses having appropriate voltage levels are generated using the threshold voltage of FETs. Since the back-gate bias effect is included, the threshold voltage is about 1.5 V, enough to keep the precharge and predischarge FETs conducting slightly.

The scheme is intended to correct the problem of NORA circuits using NMOS circuit techniques. Since PFET threshold voltage (including back-bias effect) determines the current of NFET used for predischarge (and NFET threshold voltage determines the current of precharge PFET), the standby power depends critically on process variation.

5.12 DISCHARGE DELAY OF AN FET CHAIN

In order to analyze the delay characteristics of a Domino CMOS gate, the gate must be represented by a simple and manageable circuit model that contains the essential physical mechanisms. The RC chain model used previously in CMOS static gates (Sections 3.10 to 3.14) is adequate and is the most convenient. The model is reexamined here. Figures 5.25(a) and (b) show a five-input Domino CMOS AND gate and its RC chain model, respectively. In the model, resistances $R_0, R_1, \ldots, R_5$ represent the channel resistance of the NFETs MN0, MN1, ..., MN5, respectively. Capacitance C_0 consists of the following eight components:

1. Drain-diffused island capacitance of NFET MN0
2. Source-diffused island capacitance of NFET MN1
3. 1/2 the gate-to-channel capacitance of MN0
4. Gate-to-drain overlap capacitance of MN0
5. 1/2 the channel-to-substrate capacitance of MN0
6. 1/2 the gate-to-channel capacitance of MN1
7. Gate-to-source overlap capacitance of MN1
8. 1/2 the channel-to-substrate capacitance of MN1

As is observable from this list of parasitic capacitances, it is not obvious where to credit the lumped-node capacitance. In a numerical analysis that allows more complexity in the model, the best way would be to assign the capacitances 1 to 8 to node N1 of Fig. 5.25(b). In a simplified closed-form analysis we assign the capacitances of NFET MN0 only to node N1, and so on. This latter assignment has the advantage of special simplification: Product $R_i C_i$ $(i = 0, 1, 2, \ldots 5)$ is independent of the size of the FET. This simplification is quite helpful in the later development of the theory. Capacitances 1, 2, 5, and 8 depend on the voltage of the node relative to the substrate. This dependence can be included in a numerical calculation or simulation, but is neglected in the following closed-form theory.

Resistance R_i depends on the voltage of node $i+1$, V_{i+1} (drain voltage), and the voltage of node i, V_i (source voltage), of ith NFET. We assume that the inputs switched within a negligibly short time, as was done in the theory of static gates. The gate voltage is assumed constant at V_{DD}. The current that flows through NFET MN(i), I_i, and R_i are related by

$$R_i = \left< \frac{V_{i+1} - V_i}{I_i} \right> \tag{5.4}$$

The <...> in Eq. (5.4) means an average over time during the gate switching process. As in Section 3.12, the equivalent resistance of an FET gives the gate delay time when it is used in the RC chain model of the gate. The accuracy of an RC chain model of a dynamic gate is better than static gates, due to circuit simplicity. Figure 5.26 shows gate delays calculated from the RC chain model versus delays obtained by circuit simulation. The delays are for the worst-case gate configurations (if more than one NFET is connected in parallel, only one of them is conducting and the others contributing parasitic capacitances to the nodes). A 3.5-μm CMOS at 105 degrees centigrade at $V_{DD} = 5.0$ V and the worst-case process (low current) was assumed.

The vertical axis of Fig. 5.26 is the results from the RC chain model with the delay times normalized to the time constant of a 3.5-μm CMOS minimum-size NFET (0.68 ns). A linear relationship holds between the normalized delays determined by the RC chain model and the delays determined by simulations within an accuracy of about 5%. This result demonstrates the validity of the RC chain model. Reflecting the simpler structure of a dynamic gate than a static gate, the agreement is better than that obtained for the static gates.

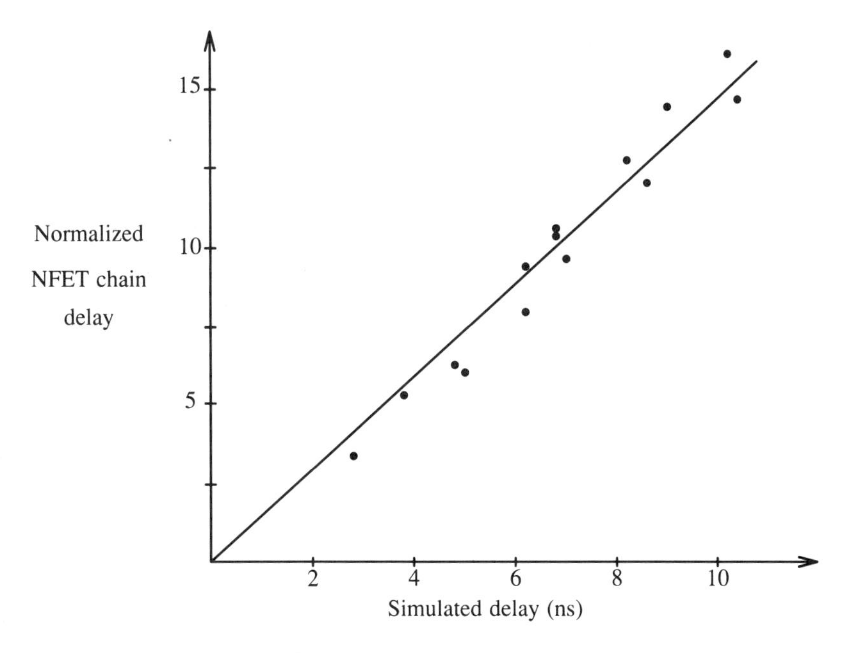

Figure 5.26 Accuracy of RC chain model.

5.13 SCALING OF THE NFET CHAIN

It is common practice of CMOS circuit designers to adjust the FET sizes for optimum delay, subject to other design constraints. When they try to reduce the delay of a Domino CMOS gate, they see a curious effect. The effect is discussed in this section and can be explained by the RC chain model.

Figure 5.27(a) shows a Domino CMOS six input NAND gate. The chain of seven NFETs is usually laid out as shown by layout A of Fig. 5.27(b). We find that layouts B to G, in spite of the smaller FET sizes, give shorter delay. Delays of the gates of layouts A to G determined by circuit simulations are also shown in the figure. The simulations were carried out assuming the 2.5-μm CMOS technology and the medium current process. The width of the FETs of the unscaled chain (layout A) was 44 μm, and a minimum-size buffer was used. The delay decreased by more than 10% by reducing the total area of the NFETs by as much as 30%. This is quite a surprising result [28].

When the size of the FET MN6 of Fig. 5.27(a) is made smaller, two effects occur. First, the resistance of the FET increases, and therefore the charge on the output node takes more time to be drained to the ground. This mechanism increases the delay. If the length of the NFET chain is long, the increase in resistance of only one FET becomes less important, and therefore the effect becomes less significant. Second, the parasitic capacitance of FET MN6 decreases. Since the charge stored in the capacitance decreases, the delay time decreases.

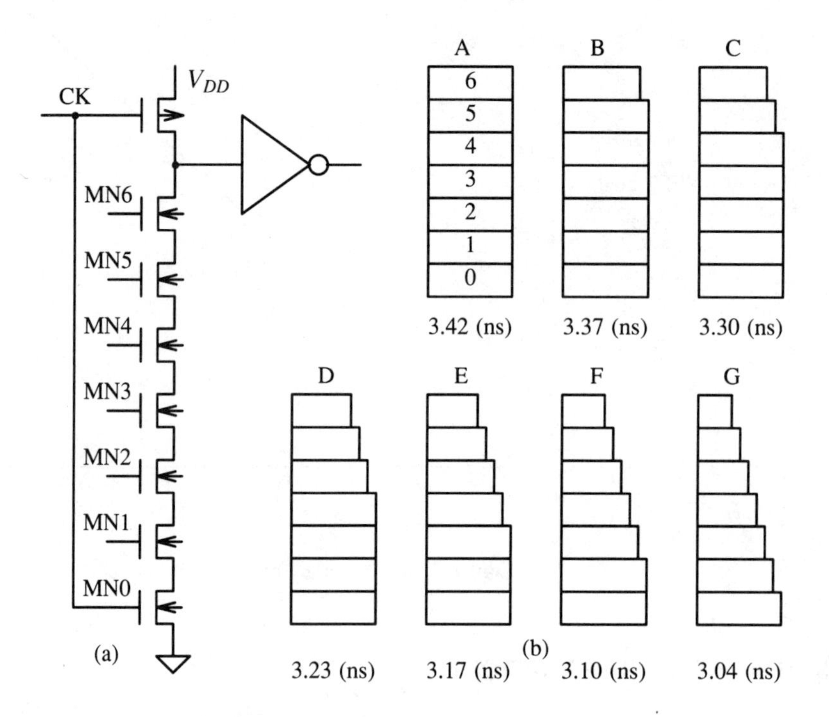

Figure 5.27 Trapezoidal scaling of FETs in Domino CMOS.

Since the charge drains through the summed resistance of FETs MN0 to MN6 the effect is more significant when the length of the NFET chain is longer. Since the first effect is minimized in a long NFET chain while the second is maximized, we expect that the second effect predominates if the NFET chain is longer than a certain minimum, and in a long NFET chain, the delay time can be decreased by decreasing the size of FET MN6.

The physical mechanism discussed above is independent of the details of the channel resistance of the FET. Therefore, we can get a deeper insight into the problem by studying the RC chain model. In this model all the components are assumed to be linear. By this simplification we are able to derive closed-form results for how long an NFET chain needs to be in order to have a decrease in delay when the FET nearest the output node shrinks.

In the most drastically simplified RC chain model of Fig. 5.25(c), all the resistances and the capacitances of the RC chain except for the top link are combined into single components R_0 and C_0. R_1 and C_1 are the resistance and the capacitance of the last link, and C_L is the load capacitance of the output node consisting of the gate capacitance of the fanouts, and so on. Node 0 and node 1 are originally charged to V_{DD}. The voltage of node 1 decays to $V_{DD}e^{-1}=0.36V_{DD}$ after time T_1 given by the Elmore's formula (Section 6.8),

$$T_1=R_0(C_0+C_1+C_L)+R_1(C_1+C_L) \tag{5.5}$$

When the size of the last FET is increased by a fraction, Δk, in the equivalent circuit C_1 is replaced by $C_1(1+\Delta k)$ and R_1 by $R_1/(1+\Delta k)$. Then we obtain,

$$T_1=R_0(C_0+C_1+C_L)+R_1(C_1+C_L)+R_0\left[C_1-\frac{R_1}{R_0}C_L\right]\Delta k \tag{5.6}$$

if Δk is small. If

$$C_L<\left[\frac{R_0}{R_1}\right]C_1 \tag{5.7}$$

T_1 decreases by decreasing the size of the top FET. R_0/R_1 is the number of series-connected FET minus one, times a factor γ that takes into account the many effects that make a real FET different from a linear resistor. We expect $\gamma < 1/2$ (see reference [28]). If we use an approximation $\gamma = 1/2$ we obtain

$$C_L < (N-1)\frac{C_1}{2} \tag{5.8}$$

If this condition is satisfied, the delay time decreases by decreasing the size of the NFET closest to the output node. Table 5.1 shows a comparison between the theory and an ADVICE simulation, carried out with 44-μm-wide NFETs of a 2.5-μm CMOS process, at a medium current process. Equation (5.7) holds with reasonable accuracy, and the small disagreement can be explained by the uncertainty in choosing γ. This result establishes the validity of the physical model and shows that the effect of decreasing delay with decreasing transistor size can occur in a real CMOS gate.

TABLE 5.1 CRITICAL VALUE OF C_L

Number of FET	C_L by theory	C_L by simul.
5	0.94(pF)	0.80(pF)
10	2.11	1.50

5.14 DRIFT OF POTENTIAL PROFILE IN A DISCHARGING FET CHAIN

Further insight into this phenomenon is gained by studying the evolution of the potential profile within the FET chain. We consider a very long equivalent RC chain that can be analyzed by a differential equation. We will show that the potential profile effectively drifts from the grounded end of the FET chain to the output node, with a velocity that can be enhanced by FET scaling. The series resistance $R(x)$ and the parallel capacitance $C(x)$, where x is a nondimensional number specifying the location of a scaled FET, satisfy

$$R(x)C(x)=\frac{1}{D} \tag{5.9}$$

where D has the dimension of s^{-1}. In 2.5-μm CMOS technology for the medium current process, $D = 4.2 \times 10^{9} s^{-1}$. The time-dependent voltage of the RC chain $V(x,t)$ satisfies the equation

$$D\frac{\partial^2 V}{\partial x^2}+D\frac{C'(x)}{C(x)}\frac{\partial V}{\partial x}=\frac{\partial V}{\partial t} \tag{5.10}$$

When $C'(x)/C(x)=-\sigma$ (constant), we have $C(x)=C(0)e^{-\sigma x}$. This is the case where the FET size $W(x)$ decreases exponentially, as $W(x)=W(0)e^{-\sigma x}$. For convenience we define the uniform scale-down ratio, $S=e^{\sigma}$. If we substitute

$$V(x,t)=\Phi(x-vt,t) \tag{5.11}$$

into Eq. (5.10) we obtain

$$D\frac{\partial^2 \Phi}{\partial \xi^2}=\frac{\partial \Phi}{\partial t} \tag{5.12}$$

provided that

$$\xi=x-vt \quad \text{and} \quad v=D\sigma \tag{5.13}$$

Equations (5.9) to (5.12) lead to the following conclusion: If $\sigma=0$ ($S=1$, unscaled chain) the voltage profile of the NFET chain will vary with time as

$$V(x,t)=V_{DD}\, erf\left(\frac{x}{2\sqrt{Dt}}\right) \tag{5.14}$$

where

$$erf(\Theta)=\left(\frac{2}{\sqrt{Dt}}\right)\int_0^{\Theta} e^{-u^2}du$$

Equation (5.14) is plotted by curve A of Fig. 5.28. When $\sigma \neq 0$ the equation needs to be solved numerically. Curves B and C are the solutions when S (= e^{σ}) is greater than 1. Observation of the two curves indicates that they are concave upward in the vicinity of the origin (grounded end) and the potential profile moves to the right without changing shape while the chain discharges. The movement is speeded up by transistor scaling ($\sigma > 0$). This velocity of motion is given by Eq. (5.13). When σ is negative this motion slows discharge, as is shown by curve D of Fig. 5.28.

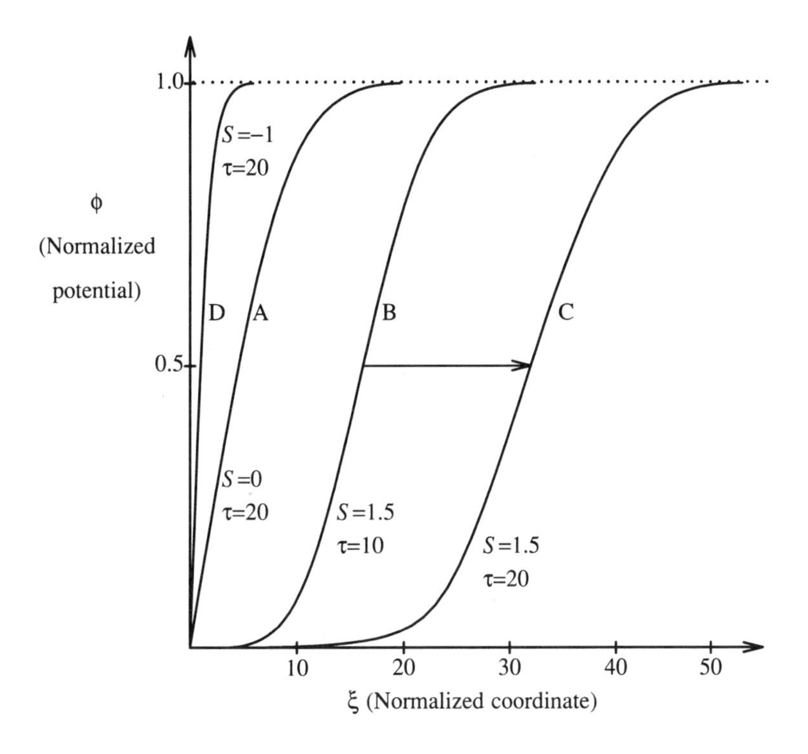

Figure 5.28 Potential profile drift in scaled RC chain.

We now study a long NFET chain, whose resistance and capacitance are voltage dependent. Evolution of the potential profile in a FET chain consisting of 20 NFETs was simulated using ADVICE circuit simulator, and the results are shown in Fig. 5.29. 2.5-μm CMOS at the medium current process was used. Figure 5.29(b) shows the potential profile in an unscaled chain. The sudden voltage increase from 3.2 V to 5 V (arrow mark) is the voltage drop due to the threshold voltage (including the back-bias effect) of the topmost FET. Figure 5.29(a) shows evolution of the potential profile of a scaled NFET chain. The potential profile moves toward the right, approximately maintaining the same shape, and the NFET chain discharges significantly faster than the unscaled chain. The potential profile becomes concave upward in the vicinity of the grounded end, which is an inevitable consequence of the moving potential profile in a chain with one end grounded. Figure 5.29(c) shows the potential profile within an inversely scaled chain. The discharge is very significantly slowed, and the potential profile is convex upward in the vicinity of the grounded end, a sign of absence of drift motion. Figure 5.30 shows the motion of the 4 V point on the moving potential profile [refer to Figs. 29(a) to (c)]. The slope of the plot is the velocity of the drifting motion. The velocities are summarized in Table 5.2.

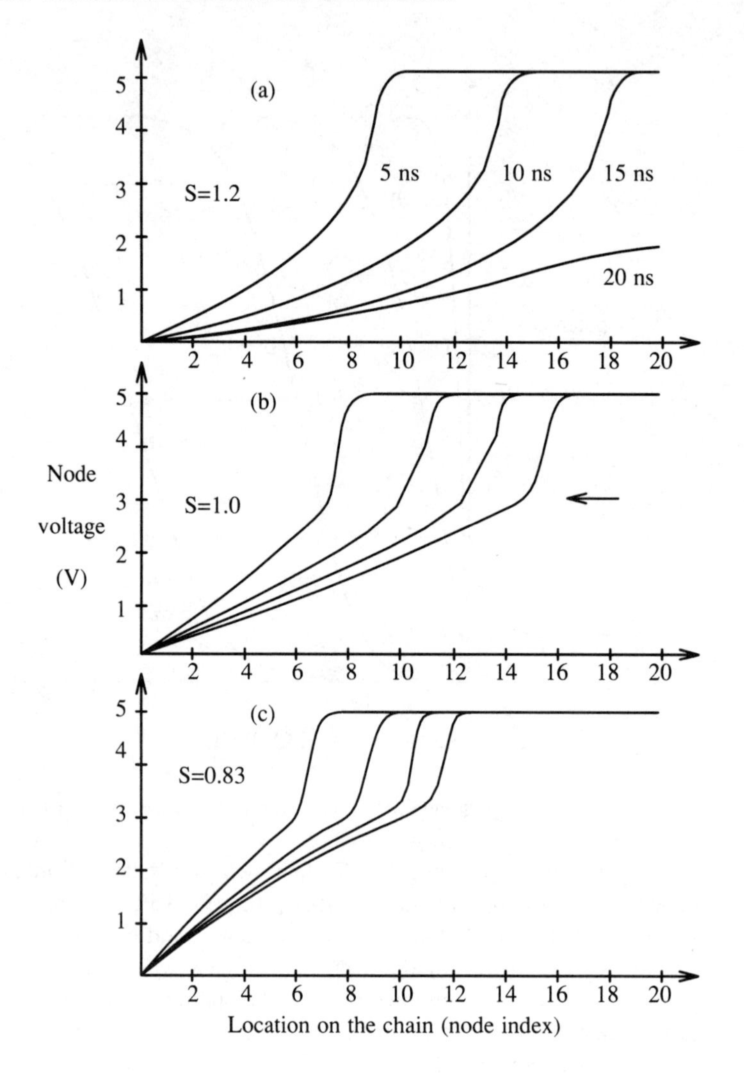

Figure 5.29 Potential profile drift in scaled FET chain.

In the table, the velocities in excess of the velocity for $S = 1.0$ are tabulated. Using Eq. (5.13), D was determined, in the range 2.3 to $2.6 \times 10^9\ \mathrm{s}^{-1}$. The value of D does not depend much on S or σ. This result shows that the analysis based on the linear RC chain model is reasonable. The value of D determined from the FET discharge characteristics, $D = 4.2 \times 10^9 \, s^{-1}$ is about 60% larger than the value determined from the motion of the potential profile. This discrepancy can be explained from the difference between a linear resistor and a real nonlinear, three-terminal NFET, and also by the way the velocity of motion was determined from Fig. 5.30 and therefore the discrepancy is not detrimental to the theory.

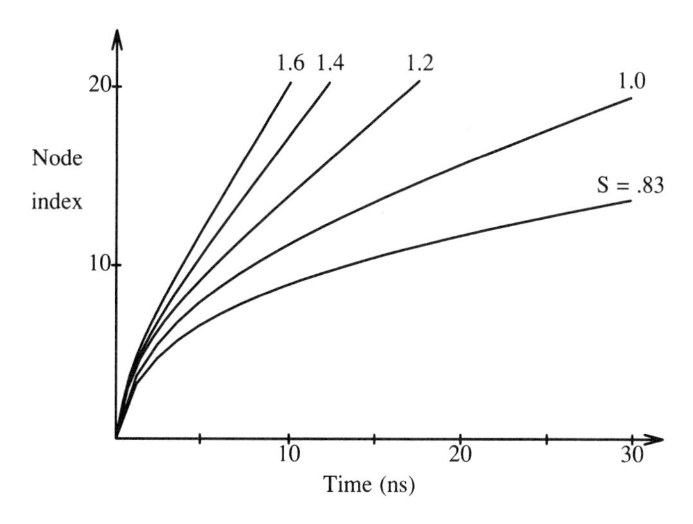

Figure 5.30 Drift velocity of potential profile in FET chain.

TABLE 5.2 VALUES OF σ AND D

S	σ	velocity (stage/ns)	D (from the velocity) (s^{-1})
1.6	0.470	1.21	2.57×10^9
1.4	0.336	0.86	2.56×10^9
1.2	0.182	0.42	2.30×10^9

The technique of graded scaling of NFETs is particularly useful in very complex OAI gates such as that shown in Fig. 5.31. Domino CMOS gates are usually scaled so that the switching delays from all the logically equivalent inputs are the same. Then all the FETs within the same level have the same size. The worst-case switching delay occurs when only one of the equivalent FETs in each level is conducting, and we assume the condition. The size of a single level 3 FET is fixed at $W(3) = 44$ μm and the size of the other FETs is scaled following

$$W(i) = W(3)\left\{1 - \left[\frac{\alpha}{100}\right](i-3)\right\}$$

where α is the gradient of the FET size in percent. This linear scaling is not the best scaling, but it is convenient to see the effects of scaling. The delay time versus α determined by simulation is shown in the upper curve of Fig. 5.32. An improvement of over 30% can be achieved by scaling the size of the FETs. When the same scaling technique is applied to a NAND6 gate, the delay versus α is shown in the lower curve of Fig. 5.32. An improvement of more than 20% can be achieved.

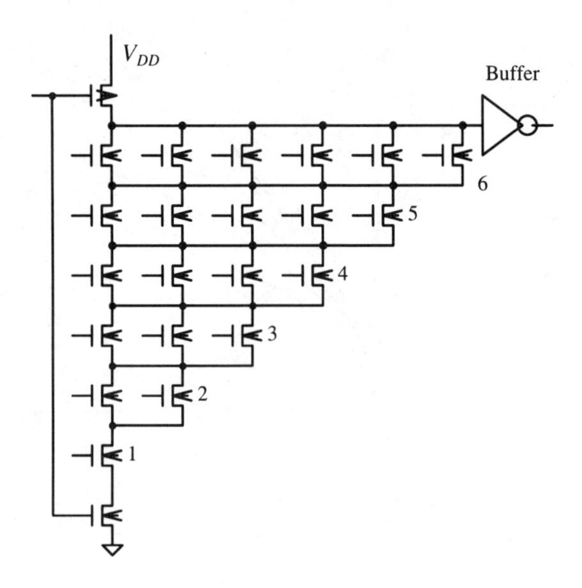

Figure 5.31 Complex dynamic OAI gate.

5.15 OPTIMUM NFET SCALING

The graded FET scaling technique in Domino CMOS is now widely practiced, but how much scaling is required is determined by circuit simulation. To determine the optimum scaling by random search using simulation is costly. An RC chain model supplies some insight into the optimum scaling. We consider the RC chain model shown in Fig. 5.25(b), which has cascaded $(N+1)$ RC links. If capacitance $C_0, \ldots, C_N$ and C_L are charged to V_{DD} at time $t=0$, time T_D required for the voltage of the output node to drop to $V_{DD}/2$ is given by, with good approximation (Section 6.8),

$$T_D = f\left\{R_0(C_0+C_1+\ldots+C_N+C_L)+R_1(C_1+\ldots+C_L)+\ldots+R_N(C_N+C_L)\right\} \quad (5.15)$$

where f is a numerical factor of the order of unity.

Capacitance $C_0, C_1, \ldots, C_N$ and C_L consists of components such as gate-to-channel capacitance, source-to-substrate and drain-to-substrate capacitance, and so on, as before. If the FET sizes are given, these capacitances can be computed from the oxide thickness and the two capacitance coefficients characterizing diffusion islands: C_A (capacitance per unit drain/source area) and C_P (capacitance per unit peripheral length of the drain/source). The equivalent resistance per unit width of an NFET can then be determined using these capacitance values, the delays obtained by simulations, and the delay computed from Eq. (5.15), by seeking the best overall agreement. We discussed this problem in Section 5.12.

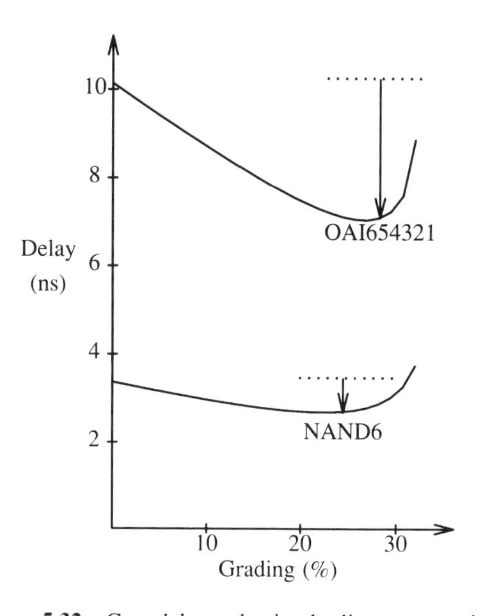

Figure 5.32 Gate delay reduction by linear trapezoidal scaling.

The problem is now stated as follows: When the circuit of a Domino CMOS gate is given, practical restrictions on the NFET size are set, and delay of the pull-down NFET chain is approximated by Eq. (5.15), what size NFETs in the chain give minimum delay? Practical restrictions on the FET size are, for example, the maximum FET size is fixed, and the total area of the FET is fixed. This problem can be solved using the following algorithm and the Monte Carlo technique.

The algorithm is as follows. Using physical insight into the problem, the initial width of the Xth FET from the grounded end, $W_0(X)$, is set as a function of X [or equivalently a list of $W_0(X)$ versus X, for integer X in the range from $X=0$ to $X=N$, is given]. The set of widths $W_0(X)$ should satisfy the requirements set by the practical restrictions. This is the "first guess" of FET sizes. A positive number, B ($B<1$) is chosen. Then the following steps are repeated.

1. Delay of the chain having FET size $W_0(X)$ is computed. The delay is T_D.
2. Using parameter B, the size of the Xth NFET, $W(X)$ $(1 \le X \le N)$, is randomly generated within the range

$$BW_0(x) \le W(x) \le W_0(x)/B$$

3. The set of sizes, $W(X)$ is checked for conformity with the restrictions. If not in conformance, new $W(X)$ are generated until an acceptable $W(X)$ is generated.
4. Delay of the chain having FET size $W(X)$ is computed using Eq. (5.15). If the new delay, T_D', is larger than T_D, the steps 2 to 4 are repeated.
5. If $T'_D < T_D$, $W(X)$ is a better guess than $W_0(X)$, and therefore $W_0(X)$ is replaced by $W(X)$ and T_D by T'_D.

When steps 1 to 5 are repeated many times (say, 10,000 times), the set of FET sizes that give the minimum delay are found. For faster convergence the value of B is gradually increased toward 1, as steps 1 to 5 are repeated.

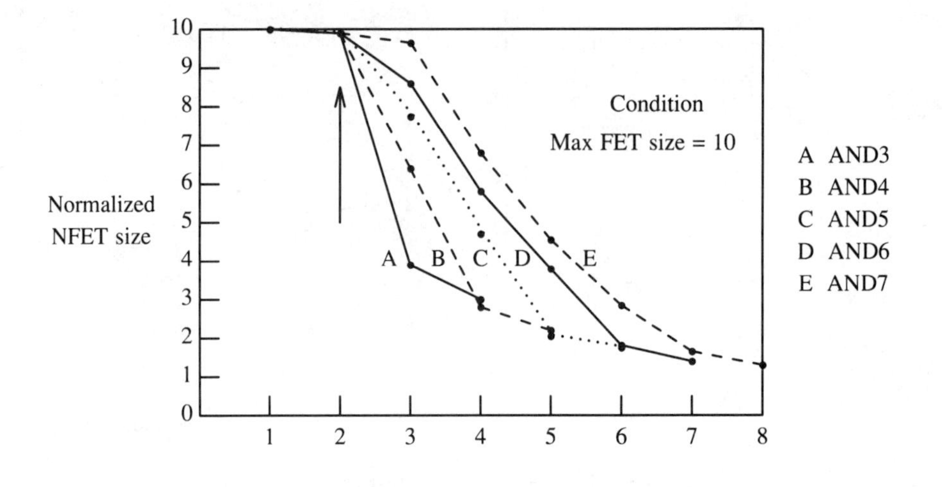

Figure 5.33 Optimum FET scaling in Domino CMOS.

The delay of the static inverter that buffers the NFET chain is an integral part of the Domino CMOS gate delay. The delay can be computed as the charging process of the load capacitor by the resistance of the PFET. In a Domino CMOS circuit the rate of voltage change at the input of the buffer varies over a wide range. A correction due to the rate variation is required, as discussed in Section 3.17.

Figure 5.33 shows the FET sizes versus location from the grounded end that give the best delay for AND3 to AND7 Domino CMOS gates. The restriction was that the NFET size should be less than 10 times the minimum-size. Minimum-size inverters are used to buffer the output of the NFET chain.

Optimum delay is attained when the FET size decreases from the grounded end, but a very significant "shoulder" (shown by the arrow) exists at the ground end. Obviously, the optimum scaling law with practical restrictions is not a linear law or an exponential law. To find the optimum scaling has therefore been a difficult problem. For this type of extremum problem the Monte Carlo method is quite powerful.

One important design consideration is that the chain should be scaled less if the capacitive load at the output node of the chain is more. Figure 5.34 shows the results of optimization for 1 ×, 3 ×, 5 ×, and 20 × load (unit of minimum-size inverter).

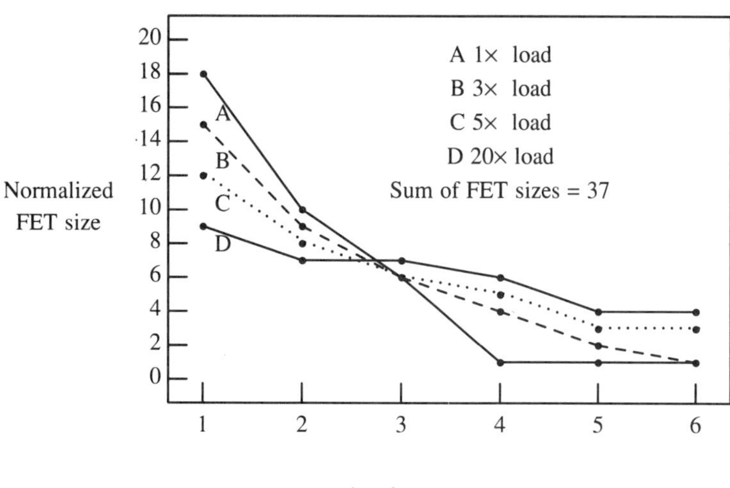

Figure 5.34 Effects of load capacitance on FET scaling.

The layout constraint is that the total area occupied by the NFETs should be less than 37 times the minimum-size NFET. As is observed, the heavier the capacitive load, the smaller the scaling. The conclusions of Figs. 5.33 and 5.34, originally found from design experience, are now confirmed by the Monte Carlo analysis.

High-performance Domino CMOS gates should be designed using large NFETs, so that capacitive loading to the output node is minimized. A buffer of Domino CMOS protects output node from capacitive loading, thereby helping reduce delay. A NORA gate cannot claim unconditional delay improvement by removing the inverter.

5.16 FAILURE MODE OF A DYNAMIC GATE

A dynamic gate fails in the high-frequency limit in a peculiar way. Figure 5.35(a) shows a dynamic inverter that is precharged during interval t_p of Fig. 5.35(b), and is discharged subsequently: The output data should be available time t_d after the upgoing clock edge, when the receiving latch (not shown in the figure) closes. In order to analyze the precharge/discharge characteristic of the gate, PFET MP1 is substituted by an equivalent resistor R_P, and NFETs MN1 and MN2 by R_N. If the load capacitor is completely discharged at the beginning of the precharge, the capacitor is charged to voltage V_0 after precharge time t_p, which is given by

$$V_0 = V_{DD} \left\{ 1 - \exp(-t_p / t_{p0}) \right\}$$

where $t_{p0} = R_P C_L$. The precharge voltage level should be high enough to provide a logic 1 to the load.

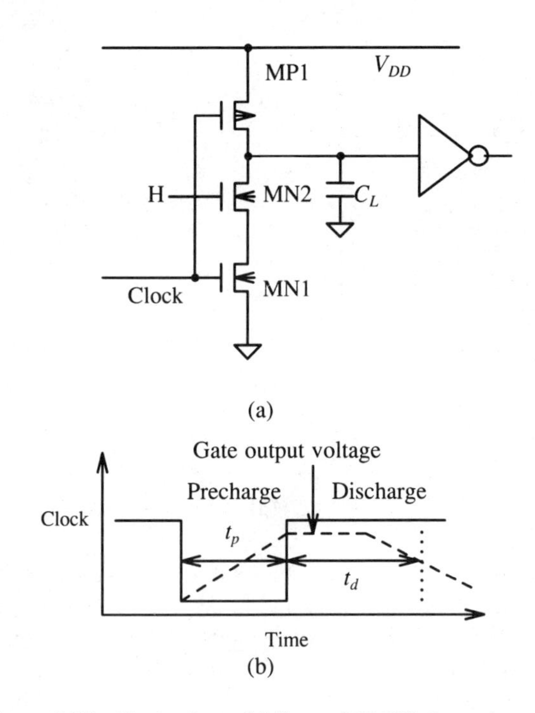

Figure 5.35 Mechanism of failure of CMOS dynamic gate.

Therefore

$$V_0 > \frac{1}{2} V_{DD} \tag{5.16}$$

is required. This leads to the condition

$$t_p / t_{p0} > \log 2 = 0.693$$

This result was derived not including margin for noise. If the dynamic gate is influenced from noise, Eq. (5.16) is modified by adding effective noise voltage to the right-hand side, and the following discussions remain unchanged. If MN2 is on during discharge, however, the output node voltage drops from V_0 to V_1 after time t_d as

$$V_1 = V_0 exp(-t_d/t_{d0})$$

where $t_{d0} = 2R_N C_L$. Since this voltage should be less than $V_{DD}/2$,

$$V_1 < \frac{1}{2}V_{DD} \tag{5.17}$$

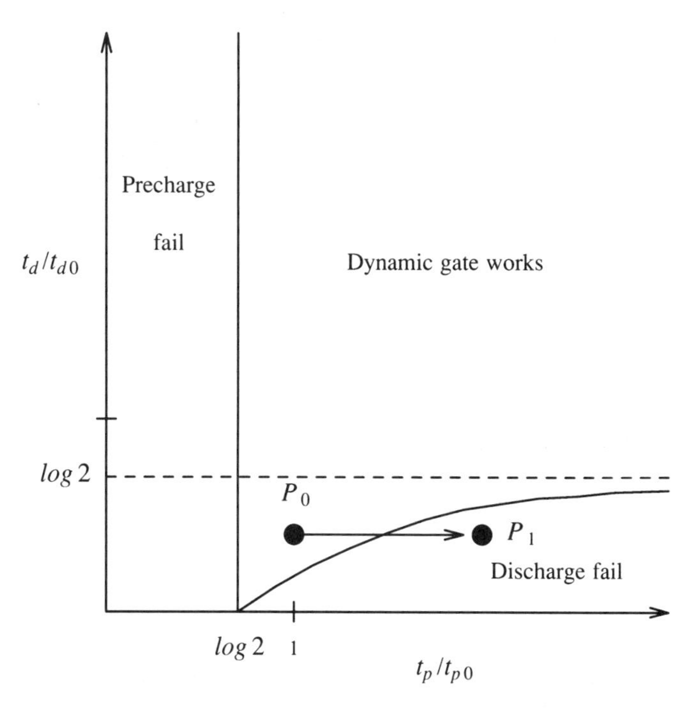

Figure 5.36 Failure of dynamic gate-precharge and discharge failures.

is required. Equations (5.16) and (5.17) amount to the following: For the dynamic gate to operate correctly,

$$\frac{t_p}{t_{p0}} \geq \log 2 \quad \text{and} \quad \frac{t_d}{t_{d0}} \geq \log 2 + \log\left\{1 - exp(-t_p/t_{p0})\right\} \tag{5.18}$$

Figure 5.36 shows the plots of the two unequalities. The dynamic gate works correctly in the wedge-shaped region. If R_N is kept unchanged and R_P is reduced, t_{p0} decreases, and therefore point P_0 (that represents correct operation of the dynamic gate) moves into point P_1, which represents a discharge failure of the gate. Since dynamic gates are usually complex, the slower discharge determines the operation of the dynamic gate in the limit of the highest frequency. Then, improving the precharge circuit of the dynamic gate results in a discharge failure and reduction in the maximum frequency of the circuit. A designer needs to pay attention to this subtle problem, since a design change or process improvement intended to increase maximum frequency of operation.

5.17 REFERENCES

01. J. A. Cooper, J. A. Copeland and R. H. Krambeck "A CMOS microporcessor for telecommunication applications" ISSCC77 Digest, pp. 138-139, February 1977.
02. H. Fleisher and L. I. Maissel "An introduction to array logic" IBM J. Res. and Develop., vol. 19, pp. 98-109, March 1975.
03. S. Y. Hwang, R. W. Dutton and T. Blank "A best-first search algorithm for optimum PLA folding" IEEE Trans. on Computer-Aided Design, vol. CAD-5, pp. 433-442, July 1986.
04. G. H. Mah and R. Newton "PANDA: a PLA generator for multiply-folded PLAs" ICCAD84 Digest, pp. 122-124, 1984.
05. M. S. Schmookler "Design of large ALUs using multiple PLA macros" IBM J. Res and Develop., vol. 24, pp. 2-14, January 1980.
06. R. A. Wood "High-speed dynamic programmable logic array chip" IBM J. Res. and Develop., vol. 19, pp. 379-383, July 1975.
07. E. Hebenstreit and K. Horninger "High-speed programmable logic arrays in ESFI SOS technology" IEEE J. Solid-State Circuits, vol. SC-11, pp. 370-374, June 1976.
08. K. Manabe, N. Someya, M. Ueno, H. Neishi, M. Imai, S. Okamoto and K. Suzuki "A C^2MOS 16-bit parallel microprocessor" ISSCC76 Digest, pp. 14-15, February 1976.
09. R. H. Krambeck, C. M. Lee and H. S. Law "High-speed compact circuits with CMOS" IEEE J. Solid-State Circuits, vol. SC-17, pp. 614-619, June 1982.
10. D. J. Myers and P. A. Ivey "A design style for VLSI CMOS" IEEE J. Solid-State Circuits, vol. SC-20, pp. 741-745, June 1985.
11. H. T. Nagle, Jr., B. D. Carroll and J. D. Irwin "An introduction to computer logic" Prentice-Hall, Inc., Englewood Cliffs, New Jersey, 1975.
12. J. A. Pretorius, A. S. Shubat and C. A. T. Salama "Charge redistribution and noise margins in Domino CMOS logic" IEEE Trans. on Circuits and Systems, vol. CAS-33, pp. 786-793, August 1986.
13. A. F. Murray and P. B. Denyer "A CMOS design strategy for bit-serial signal processing" IEEE J. Solid-State Circuits, vol. SC-20, pp. 746-753, June 1985.
14. J. A. Pretorius, A. S. Shubat and C. A. T. Salama "Optimization of Domino CMOS logic and its applications to standard cells" CICC84 Digest, pp. 150-153, 1984.
15. D. A. Pucknell and K. Eshraghian "Basic VLSI design, principles and applications" Prentice-Hall of Australia, Sydney, 1985.
16. E. Goto, K. Murata, K. Nakazawa, K. Nakagawa, T. Motooka, Y. Matsuoka, Y. Ishibashi, H. Ishida, T. Soma and E. Wada "Esaki diode high-speed logic circuits" IRE Trans. on Electronic Computers, vol. EC-9, pp. 25-29, March 1960.
17. R. K. Brayton, C-L. Chen, C. T. McMullen, R. H. J. M. Otten and Y. J. Yamour "Automated implementation of switching functions as dynamic CMOS circuits" CICC84 Digest, pp. 346-350 1984.
18. J. A. Pretorius, A. S. Shubat and C. A. T. Salama "Latched Domino CMOS logic" IEEE J.

Solid-State Circuits, vol. SC-21, pp. 514-522, August 1986.
19. T. A. Grotjohn and B. Hoefflinger "Sample-set differential logic (SSDL) for complex high-speed VLSI" IEEE J. Solid-State Circuits, vol. SC-21, pp. 367-369, April 1986.
20. A. A. Markov "On the inversion complexity of a system of functions" Journ. A. C. M. vol. 5, pp. 331-334, October 1958.
21. E. Goto, "On the puzzles of automata", in "The way to information science", vol. A. 1. 1, pp. 67-92, Kyoritsu Publishing Co., Tokyo Japan (in Japanese).
22. N. P. Goncalves and H. J. de Man "NORA: a racefree dynamic CMOS technique for pipelined logic structures" IEEE J. Solid-State Circuits, vol. SC-18, pp. 261-268, June 1983.
23. C. M. Lee, unpublished work.
24. Y. T. Yen "Intermittent failure problems of four-phase MOS circuits" IEEE J. Solid-State Circuits, vol. SC-4, pp. 107-110, June 1969.
25. V. G. Oklobdzija and R. K. Montoye "Design-performance trade-offs in CMOS-Domino" IEEE J. Solid-State Circuits, vol. SC-21, pp. 304-306, April 1986.
26. J. A. Pretorius, A. S. Shubat and C. A. T. Salama "Analysis and design optimization of Domino CMOS logic with application to standard cells" IEEE J. Solid-State Circuits, vol. SC-20, pp. 523-530, April 1985.
27. C. M. Lee and E. W. Szeto "Zipper CMOS" IEEE Circuits and Devices Magazine, pp. 10-17, May 1986.
28. M. Shoji "FET scaling in Domino CMOS gates" IEEE J. Solid-State Circuits, vol. SC-20, pp. 1067-1071, October 1985.

6

Circuit Performance Evaluation

6.1 INTRODUCTION

Circuit simulation techniques are the basis of CMOS VLSI chip electrical design. In a traditional, discrete-component circuit design, the standard approach was selection of the proper circuit followed by intuitive optimization of parameters such as delay and frequency response. The designed circuit was built, and then the performance was evaluated, using test equipment like a pulse- or sine-wave generator and a cathode-ray oscilloscope. If performance was not satisfactory, component substitution might be used to try to improve performance.

An MOS VLSI chip has one essential impossibility. Direct and accurate observation of the node voltages by a device like an oscilloscope is, for the most part, impossible. This point is clear from an estimate of the node capacitance of a typical circuit. The output node capacitance of a minimum-size inverter with single fanout is about 0.075 pF. To observe a switching waveform of a very lightly loaded node like this, a capacitive-divider probe is used. The input capacitance of a high-speed oscilloscope probe is about 15 pF. To observe the node without loading, significantly more than a factor of

$$\frac{15\text{pF}}{0.075\text{pF}} \times 10 = 2000\,(\text{times})$$

attenuation of the voltage is necessary. If a 5 V signal is attenuated that much, the signal is buried under noise. Experimentally, the absolute minimum dividing capacitor is about 0.1 pF. If the internal node waveforms must be observed, the nodes must be buffered and the buffered output must be taken out to the probing pad. Again, the output waveform of a minimum-size gate is not observable directly and undisturbed, since the buffering adds extra capacitance and delay.

Still another problem relates to the small dimensions of the interconnects. Probing an unprepared node usually destroys the chip by scratching the metal features with the probe. Scanning electron microscope observation does not destroy samples, but the expensive test equipment is often unavailable to VLSI designers, even in a large design organization. Furthermore, breadboard model of a VLSI chip may take more time to build than the chip itself. This observation leads to the conclusion that a CMOS VLSI chip must be built blindly, from the viewpoint of traditional circuit designers. Circuit simulation techniques provide a guide to help designers.

Design failure of VLSI chips is very costly. It takes several months to complete diagnostics and redesign, and this delay seriously affects the marketing position of the chip. A VLSI chip must be practically perfect (usable for prototype design) by the first mask. This difficult design task is accomplished with the help of circuit and logic simulation.

Another application of circuit simulation is to CMOS circuit research. When designers simulate circuits they often find new and interesting electrical phenomena, or new ideas for circuit design or optimization. It is therefore important that accurate information be derived from circuit simulation. To get accurate information, however, the simulators must be used carefully. Understanding the working mechanisms of a simulator helps a designer to carry out correct simulations. Simulated results must be interpreted before useful information is extracted from the results. There is a limit to the accuracy of simulations. Gate-level and circuit-level simulations have both advantages and disadvantages. Further, it is often desired to estimate delays of a large number of circuits without time-consuming simulations. These problems are discussed in this chapter.

6.2 GATE-LEVEL AND CIRCUIT-LEVEL TIMING SIMULATIONS

There are two different approaches to simulating the delays of CMOS digital circuits: by circuit-level simulation and by gate-level simulation. The difference of the two approaches lies in the level where the model based on the basic physics is introduced into the simulator.

In a circuit-level simulator, individual devices and individual interconnects are modeled using device physics. In a gate-level simulator, delays of individual gates and trees of the interconnects are modeled using circuit theory. A circuit-level simulator is therefore based on the more fundamental level of physics, and it is often taken for granted that the accuracy of a circuit level simulation is better than the accuracy of a gate-level simulation.

In reality the circuit delay problem is never so simple: Circuit delay depends on a number of conditions to which the circuit is subjected, such as waveforms of the input signals, the relative delay of more than one input signal, and details of the loading gates. These conditions are introduced to the simulation program by the input data. If any of these conditions are set up differently, the delays may be different, as much as several tens of percent. Therefore, it is appropriate to say that a circuit-level simulator is more accurate when the conditions are well specified. But this requirement creates a problem.

To set the conditions to a circuit accurately, the other circuits that exchange signals must be accurately included in the simulation. Then the number of elements that are included in the simulator code can be huge. Since a circuit-level simulator contains more specifications than a gate level simulator, the upper limit of the size of the circuit that can be simulated is severely limited.

An entire VLSI chip simulation by a circuit-level simulator appears impracticable. A gate-level simulator is more economical for this problem. Already an entire VLSI chip timing simulation is routinely run for varieties of input data setup. From this point of view the delay simulation by a good gate-level simulator can be potentially more accurate and the results more realistic than the simulation results obtained from simplified circuit-level simulation.

At present, the required level of sophistication of a gate-level simulator has not yet been achieved. Early gate-level simulators were often too restrictive, the ideosyncrasies were unnatural to designers, and they used inaccurate gate-delay algorithms. These problems have been rectified and the new generation of gate-level simulators are significantly improved.

6.3 CIRCUIT SIMULATORS

A circuit simulator is a program that accepts the connectivity information of the given circuit, the types and values of the components, and information about the sources of excitation, such as the power supply voltages and the input waveforms, and that determines the node voltages, the currents, and the states of charge as functions of time. A historical review was given by Pederson [01]. CMOS circuit simulators, first written in the 1960's [02], underwent generations of development. Although the latest circuit simulators are very complex and sophisticated, their function is transparent to the user. A designer is able to use a circuit simulator like a breadboard model. In case there is a doubt he is always able to check sanity of the simulator by running a simulation that gives known results. This is one distinct advantage of a circuit simulator over a gate-level simulator.

A circuit simulator determines the DC initial conditions of the given circuit and then numerically integrates the circuit equations, which are simultaneous nonlinear integro-differential equations satisfied by the node voltages and currents. A circuit simulator stands directly on the basic principles of physics. As long as an integrated circuit is represented accurately by a lumped equivalent circuit, and if individual circuit components are accurately described by the respective physical model, there is not much room for error to creep into the numerical analysis.

Major errors in circuit simulation originate from the input data and the modeling of the devices. Accuracy of a circuit simulation depends strongly on how well the active, nonlinear components like FETs are described in the simulator software. A FET model used in a circuit simulator is more complete than the simple FET model discussed in Sections 1.6, 1.7, 1.9, 1.10 and 1.11.

One of the most widely used MOSFET models, the CSIM (Compact Short-channel IGFET Model) [03] describes the drain current, I_D, of a FET by the following formula: Definitions of the major parameters such as V_{DS} (drain voltage relative to source), V_{GS} (gate voltage relative to source), V_{SB} (source voltage relative to substrate), V_{TH} (threshold voltage), and ϕ_F (Fermi level) are given in Chapter 1, and the definitions of the new parameters are given in the following. Drain current I_D for the three regions is given as follows.

1. Triode region. If $V_{GS} > V_{TH}$ and $0 < V_{DS} < (V_{GS} - V_{TH0})/(a - \eta)$,

$$I_D = \frac{1}{1-(\Delta L/L)} \frac{B}{1+U_0(V_{GS}-V_{TH})} \left\{ (V_{GS}-V_{TH})V_{DS} - (a/2){V_{DS}}^2 \right\} \tag{6.1}$$

where L and ΔL are channel length and its variation (by drain voltage, Section 1.6) and where

$$V_{TH} = V_{TH0} + K_1 \left\{ (2\phi_F + V_{SB})^{1/2} - (2\phi_F)^{1/2} \right\} - \eta V_{DS} \tag{6.2}$$

is the threshold voltage. V_{TH0} is the threshold voltage at $V_{SB} = V_{DS} = 0$. Parameters B and K_1 are given by (Section 1.6)

$$B = \frac{\varepsilon_{OX} \mu W}{T_{OX} L} \quad \text{and} \quad K_1 (= H \text{ in Section 1.6}) = \frac{(2\varepsilon_S q N_A)^{1/2}}{C_{OX}}$$

respectively. Parameters a and η are given by

$$a = \left\{ 1 + \frac{0.5 K_1 g(2\phi_F + V_{SB})}{(2\phi_F + V_{SB})^{1/2}} \right\} \left\{ 1 + U_1(V_{GS} - V_{TH}) \right\} \tag{6.3}$$

where

$$g(x) = 1 - \frac{1}{1.744 + 0.8364x} \tag{6.4}$$

and

$$\eta = \frac{\eta_0}{1 + G_{DC}/\sqrt{2\phi_F + V_{SB}}} \quad \text{where} \quad \eta_0 = \frac{\varepsilon_S}{\varepsilon_{OX}} \frac{T_{OX}}{L}$$

2. Saturation region. If $V_{DS} > (V_{GS} - V_{TH0})/(a - \eta)$ and $V_{GS} > V_{TH}$,

$$I_D = \frac{1}{1-(\Delta L/L)} \frac{B}{2a\left\{ 1 + U_0(V_{GS} - V_{TH}) \right\}} (V_{GS} - V_{TH})^2 \tag{6.5}$$

3. Cutoff region. If $V_{GS} < V_{TH}$,

$$I_D = 0 \tag{6.6}$$

Subthreshold current (Section 1.11) is not included in this FET model used for digital circuit simulation. This omission is not a problem, since the analysis of circuits that are influenced by subthreshold current [e.g., initial condition of gates (Section 3.8) and PLA bit line leakage problem (Section 1.11)] is carried out using a simplified closed-form circuit models.

Equations (6.1) to (6.6) and the model parameters $B, U_0, U_1, a, \eta, G_{DC}, V_{TH0}$, and K_1 are interpreted as follows. If $\Delta L = 0$, $K_1 = 0$, $U_0 = U_1 = 0$, $a = 1$, and $\eta = 0$, Eqs. (6.1), (6.5), and (6.6) reduce to Eqs. (1.10a) and (1.10b) of Section 1.6. These are the characteristics of a long-channel FET at low-field operating conditions, with negligible back-bias effect.

Equation (6.2) describes the back-bias effect. V_{TH0} is the threshold voltage of the FET when the source is tied to the substrate. By comparing Eq. (6.2) with Eq. (1.7), we find that $K_1 = H$. Notation K_1 is commonly used in CSIM model. Therefore, from Eq. (6.3), a is a parameter that describes the effects of the charge that exists in the depletion layer behind the channel. The charge brings the FET in the saturation region at lower drain to source voltages than $(V_{GS} - V_{TH0})$. This effect, and the back-bias effect of the threshold voltage, are both related to parameter N_A (substrate doping) through K_1. The effects of substrate doping when $V_{DS} = 0$ can be included in the theory by adjusting the threshold voltage by Eq. (6.2). The substrate doping effects in saturation cannot be included by adjusting the threshold voltage alone. Therefore, factor a is included in the denominator of Eq. (6.5), but not of Eq. (6.1).

In Section 1.6 we saw that drain voltage V_{DS} affects channel current through two different mechanisms. The first is by reducing the effective channel length from L to $L - \Delta L$, and the second by inducing extra charge on the channel. The first effect is included in the CSIM model by a factor $1/(1-(\Delta L/L))$ in Eqs. (6.1) and (6.5). The effect is associated with the channel shortening with increasing drain-source voltage. Since the effective channel length decreases with increasing drain voltage, the drain current increases, and therefore the drain conductance in the saturation region is not zero. The second effect is included through parameter η. Parameter η describes the degree of distortion of the electric field of the silicon-silicon dioxide interface from the assumed one-dimensional field. The drain voltage influences the channel charge through the bulk silicon that is on the back side of the channel. Therefore, parameter η depends on the aspect ratio T_{OX}/L. Through parameter η, threshold voltage V_{TH} decreases with increasing drain voltage as shown in Eq. (6.2).

Parameter U_0 introduces the effects of channel carrier mobility reduction by the transverse field ($V_{GS} - V_{TH}$ term) and by the lateral field (V_{DS}). Both effects make the saturation current of a FET closer to proportional to the gate voltage rather than to the square of the gate voltage.

The effect of the drain-source field is the drift velocity saturation, but in this formulation no velocity-field characteristic of the surface carriers of silicon is involved.

The effect of drift velocity saturation is included in the theory using the mobility deterioration factor, which depends on the combined lateral and transverse fields. The effect of the source-drain field is proportional to V_{DS}/L, where L is the length of the channel of the FET. An experimental check of the mobility deterioration factor is reported [04]. In the CSIM model the combined effects are approximated by the mobility deterioration factor,

$$U_0 = \text{constant} + \frac{\text{constant}}{L}$$

This is not an exact formula. The const/L term applies to the V_{DS} only, and the first term to V_{GS} only. For simplicity, the mobility degradation factor contains the terms as a sum. The effects of drift velocity saturation is included through parameter U_1 also. Parameter a increases with increasing U_1. Then saturation voltage $(V_{GS} - V_{TH})/(a - \eta)$ decreases, and the FET reaches saturation at the smaller drain voltage. Parameter η in the denominator reflects the channel charge induced directly by drain; the extra charge increases the saturation voltage.

All the parameters involved in the CSIM model were explained in terms of the simpler FET model. Practically, these parameters are determined by fitting the experimental data obtained from test FETs with the characteristics predicted from the model.

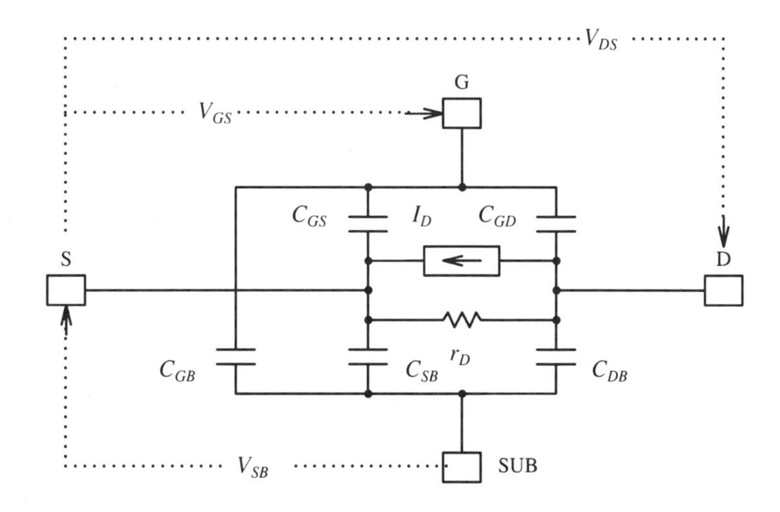

Figure 6.1(a) Equivalent circuit of FET.

Equations (6.1) to (6.6) describe the DC characteristics of FET well. The equivalent circuit of an NFET is shown in Fig. 6.1(a). In the equivalent circuit, I_D is given by Eqs. (6.1), (6.5), and (6.6). Resistance r_D is a part of I_D model representing incomplete saturation of FET. A simple way of computing r_D is to use the Early

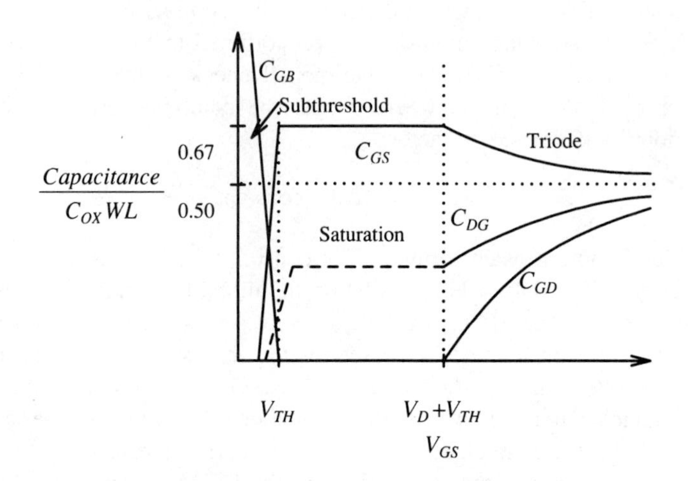

Figure 6.1(b) Gate capacitance versus gate voltage V_{GS}.

voltage, V_E defined by

$$V_E = \left| \frac{I_D}{\partial I_D / \partial V_D} \right|_{\text{satur}} - V_D = r_D I_D - V_D$$

and is approximately independent of gate and drain voltages. If V_E is given from I_D model, r_D can be determined using I_D and V_D.

Capacitances C_{SB} and C_{DB} are primarily PN junction capacitance of the source- and drain-diffused islands, respectively (Section 1.16). Gate capacitances C_{GB}, C_{GS} and C_{GD} consist of overlap capacitance (the capacitance between the gate and N+ diffused islands) and the capacitance between the gate and the channel. When gate voltage is less than the threshold voltage V_{TH}, C_{GS}, and C_{GD} equal their respective overlap capacitances. The rest of the gate is directly coupled to the substrate, and

$$C_{GB} = C_{OX} (WL)$$

where $C_{OX} = \varepsilon_{OX} / T_{OX}$, and (WL) is the area of the gate directly over the channel. C_{GB} decreases with increasing gate voltage, and as the channel is formed at $V_{GS} = V_{TH}$, C_{GB} vanishes as shown in Fig. 6.1(b).

Suppose that the NFET sustains positive drain-source voltage V_{DS}. When V_{GS} is in the range from V_{TH} to $V_{DS} + V_{TH}$, the NFET is in saturation. The drain side of the channel is pinched off. The gate is capacitively coupled to the source through the channel but not to the drain. The charge on the channel is calculated from Eq. (1.6) as

$$Q_{\text{channel}} = \int_0^L Q_I(x) dx = C_{OX} W \int_0^L \left\{ V_{GS} - V_{TH} - V(x) \right\} dx \tag{6.7}$$

Using

$$I_D = C_{OX} W \mu (V_{GS} - V_{TH} - V) \frac{dV}{dx}$$

$$Q_{\text{channel}} = C_{OX}{}^2 W^2 \frac{\mu}{I_D} \int_0^{V_{DS}} (V_{GS} - V_{TH} - V)^2 dV$$

$$= \frac{C_{OX}(WL)}{3} \frac{(V_{GS} - V_{TH})^3 - (V_{GS} - V_{TH} - V_{DS})^3}{(V_{GS} - V_{TH} - (1/2)V_{DS})V_{DS}}$$

where we assumed, for simplicity, lightly doped substrate [$H = 0$ in Eq. (1.7)] and low-field, long-channel FET. In saturation, $V_{DS} = V_{GS} - V_{TH}$ and $I_D = (1/2)C_{OX}\mu(W/L)(V_{GS} - V_{TH})^2$. We then have

$$Q = \frac{2}{3} C_{OX}(WL)(V_{GS} - V_{TH})$$

Therefore the gate-to-source capacitance is $(2/3)(WL)C_{OX}$ plus the overlap capacitance [03]. The gate-to-drain capacitance, according to this simplified model, is just the overlap capacitance. C_{GS} and C_{GD} shown in Fig. 6.1(b) [03] do not include the overlap capacitance, and therefore

$$C_{GS} = \frac{2}{3} C_{OX}(WL) \quad \text{and} \quad C_{GD} = 0 \tag{6.8a}$$

in the saturation region. This theory assumes long-channel, low-field FET model (Section 1.6). Capacitance calculation including the effects of carrier velocity saturation and short channel has been reported [05] [06].

When $V_{GS} > V_{DS} + V_{TH}$, the channel exists all the way. Then the gate to channel capacitance, $C_{OX}(WL)$, is shared by the source and the drain. When $V_{GS} \gg V_{DS}$, the effects of nonuniform potential along the channel is negligible, and from the symmetry [we assume the FETs shown in Fig. 1.10(a) or (b), not Fig. 1.10(c), of Chapter 1],

$$C_{GS} = C_{GD} = \frac{1}{2} C_{OX}(WL) \tag{6.8b}$$

Equations (6.8a) and (6.8b) show that the capacitances assigned to the source and the drain nodes depend significantly on the biasing condition of the FET. This dependence has an impact on the RC chain model of static and dynamic gates discussed in Sections 3.9 to 3.14, 5.12 to 5.15 and 8.2 to 8.5, and on the theory of Miller effect in Section 4.8. Since the voltage dependence brings complexity in the theory, we use averaged capacitance. As for the Miller capacitance in Section 4.8, however, the coupling capacitance C_m originates practically from gate-drain overlap capacitance and wiring capacitance that are voltage independent. Therefore, Miller capacitance has clear definition. As for the uncertainty of where to assign the capacitance of the RC chain model, there is a problem that is more fundamental than voltage dependence. Some capacitances appearing in the FET model cannot be represented by simple, two-terminal capacitances. Furthermore, the simple model based on the equivalent circuit of Fig. 6.1(a) violates charge conservation in numerical simulation, and the two problems are related.

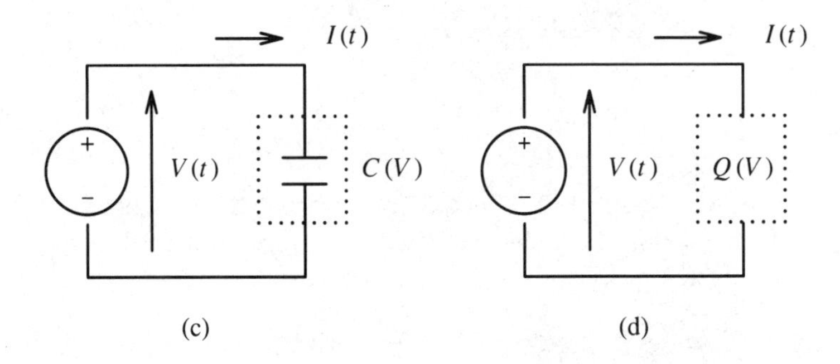

Figure 6.1(c) and (d) Conservation of charge in voltage-dependent capacitors.

Violation of charge conservation in voltage-dependent capacitance is understood from Figs. 6.1(c) and (d). In the figures, time-dependent voltage source $V(t)$ drives the voltage-dependent capacitor shown in dotted squares. The capacitor is modeled in two different ways: In Fig. 6.1(c) as a voltage-dependent capacitor $C(V)$, and in Fig. 6.1(d) as a functional relationship between stored charge Q and terminal voltage V, $Q(V)$. In Fig. 6.1(c), the charge the voltage source supplies to the capacitor between time $t = 0$ and t, $Q(t)$, is determined from

$$Q(t) = \int_0^t I(t)dt = \int_0^t C(V(t))\frac{dV(t)}{dt}dt$$

Numerical integration involves quantization error. If $V(t)$ is a periodic function, it is easy to see that the error accumulates and after a long time the voltage source loses or gains significant charge. This is an artifact of numerical analysis.

The same problem is solved in Fig. 6.1(d) as follows. Current $I(t)$ is given by charge $Q(V(t))$

$$I(t) = \frac{d}{dt}Q(V(t)) = Q'(V)\frac{dV(t)}{dt}$$

and therefore

$$Q(t) = \int_0^t I(t)dt = \int_0^t Q'(V)\frac{dV}{dt}dt = Q(V(t)) - Q(V(0))$$

This result is not influenced by accumulating integration error.

A recent trend is to develop an accurate FET model based on the charge model [05] [06] [07]. In this model, charges stored in gate, channel (source and drain), and substrate, Q_G, Q_C (Q_S and Q_D), and Q_B, are functions of voltages V_{GS}, V_{DS}, and V_{SB}, and the capacitance coefficients

$$C_{XY} = \frac{\partial Q_X}{\partial V_Y}$$

where Q_X is any of Q_G, Q_C, Q_S, Q_D, and Q_B, and V_Y is any of V_{GS}, V_{DS}, and V_{SB}, are used as the equivalent-circuit parameters. An immediate consequence of this approach is that $C_{XY} \neq C_{YX}$, which means that capacitances cannot be described as simple two-terminal elements. An example is that C_{GD} and C_{DG} shown in Fig. 6.1(b) are significantly different in the saturation region. Drain voltage is unable to influence gate charge ($C_{GD} = 0$), but gate voltage is able to control drain charge ($C_{DG} \neq 0$). The advanced FET modeling is necessary in precision analog design. Tsividis gave a review of this problem [08]. As for digital design the simple equivalent circuit of Fig. 6.2(a) is still adequate for many problems, but from a theoretical viewpoint it is desired that the RC chain model be reformulated on the charge model of the FET. This is very desirable, since some digital circuit design is now no less precise than analog design (Section 8.12). Analysis of digital circuit delay based on the more precise FET model will be the important new area of research.

6.4 SIMPLIFICATION OF THE EQUIVALENT CIRCUIT

Figure 6.2(a) shows a typical CMOS circuit laid out using the Gate-Matrix symbolic layout style [09]. Before writing circuit simulator code, the layout is represented by an equivalent circuit, shown in Fig. 6.2(b). The circuit is described by the circuit simulator code. SPICE (or its upgraded version, ADVICE) is a very popular circuit simulator. As for SPICE codes the reader is referred to the original reference [10]. The simulator code, the device specification file, and the CMOS process specification file are run by a computer. The results are plotted on a graph paper or on CRT screen for the designer's examination.

Although a circuit simulator is able to handle a large number of components, it is not a wise practice to represent all the components that exist on the chip, in the equivalent circuit. The circuit simplification task is, at present, carried out by the designers who understand the circuit. Efforts to computerize the input, including the resistance parasitics, have not yet been successful. It is the author's opinion that entirely automated input generation is less desirable than interactive code extraction; the software should display the relevant mask patterns and accept instructions from the designers. In this section a short summary of methods for drawing an adequate equivalent circuit is given.

When an equivalent circuit of the layout of Fig. 6.2(a) is drawn as shown in Fig. 6.2(b), the coupling circuit between the two stages contains four resistors and five capacitors. Except for the most precise simulations the equivalent circuit is too complex. The circuit should be simplified to the circuit of either Fig. 6.2(c) or Fig. 6.2(d), but the problem is how to determine the resistance and the capacitance values.

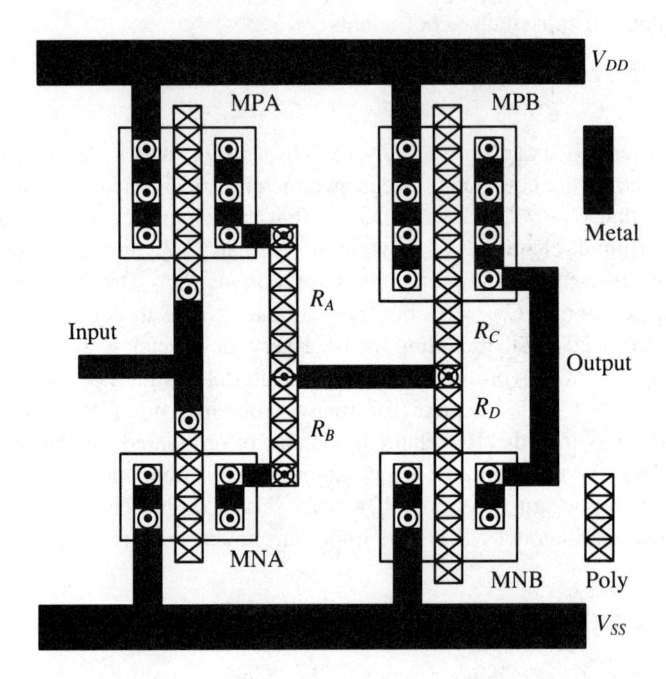

Figure 6.2(a) Layout of cascaded two-stage inverters.

A comparison with the equivalent circuit of a gate-level simulator which takes only the capacitive parasitics into account gives a hint: The sum of the capacitances of the nodes connected by the wiring resistances must be the same. From this requirement the drivers see an equal load capacitance for all three cases (b), (c), and (d), and therefore the three circuits have, to the first approximation in the limiting case of small parasitic resistance, equal delays.

The next problem is how to include resistances and how to split the capacitances between the nodes separated by the resistances. This approximation cannot be good for all the possible layout configurations. We assume a typical layout. If the circuit is to be simplified to Fig. 6.2(c), C_1 and C_2 can be set,

$$C_1 = C_A + C_B + \frac{1}{2}C_E, \quad C_2 = C_C + C_D + \frac{1}{2}C_E \tag{6.9}$$

and in Fig. 6.2(c),

$$C_1 = C_A + C_B + C_E, \quad C_2 = C_C, \text{ and } C_3 = C_D \tag{6.10}$$

Let us assume that the input of the first-stage inverter is pulled up within a negligibly short time. Then the effects of pulldown by NFET MNA is transmitted to PFET MPB through resistance $R_B + R_C$. In the gates of the FETs of the first-stage

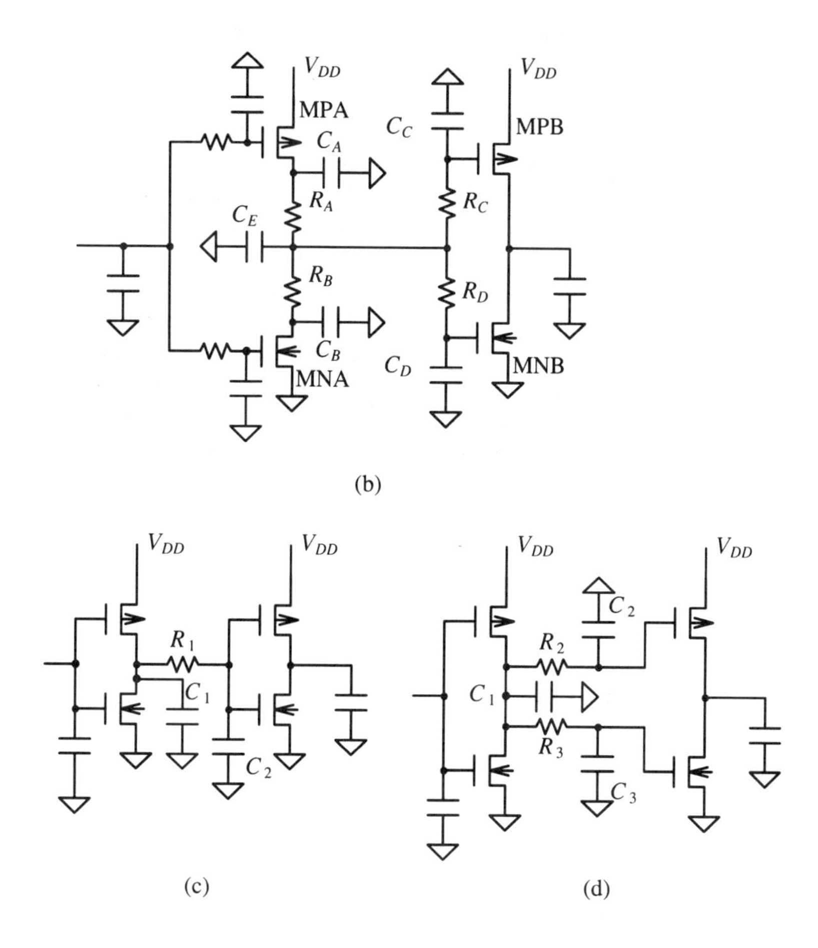

Figure 6.2(b) to (d) Equivalent circuit of the cascaded inverters.

pull down, the effects are transmitted from PFET MPA to NFET MNB through resistance $R_A + R_D$. In circuit (d) if we choose

$$R_2 = R_B + R_C, \quad R_3 = R_A + R_D \tag{6.11}$$

circuit (d) becomes faster than circuit (b) since C_C was separated from C_D, but becomes slower than circuit (b) because C_A, C_B, and C_E were merged. The two errors are of opposite polarity. Then circuit (b) and circuit (d) should have approximately the same delay.

In circuit (c), resistor R_1 is chosen as the average,

$$R_1 = \frac{1}{2}(R_A + R_D + R_B + R_C) \tag{6.12}$$

As is observed from the crude nature of the approximation, to attain a very high accuracy is difficult. In spite of that, circuits (c) and (d) have fewer resistors and capacitors, and therefore are more desirable in a large-scale simulation. These simplification techniques do not work if the values of R_A and R_B differ significantly or if the values of R_C and R_D differ significantly. Such an asymmetrical layout is usually avoided. The simplification technique must be used with caution. A difficult requirement is that the simplification guideline cannot be very complex, since a designer must deal with hundreds of circuits if such simplifications are needed.

As was observed from this example, the inputs to a circuit simulation involve subjective judgments of the designer who selects the equivalent circuit. An experiment conducted by one of the author's associates and the author himself showed, however, that the subjective difference is surprisingly small. Practically, the two simulated waveforms, when superimposed, had no visible gap between the plots. This experience showed that a simple and rational guideline to convert a silicon layout to an equivalent circuit is of great value. This section is the first attempt in this direction. Better and simpler prescriptions are awaited.

6.5 ERRORS IN CIRCUIT SIMULATIONS

Since circuit simulation and analysis are practically the only means of gaining knowledge of CMOS VLSI circuits, we need to understand various causes that make simulation result errors. Here we list various mechanisms that create simulation errors.

(A) Parasitics That Are Not Modeled

Real CMOS circuits never have ideal structures, as represented by the circuit diagram. In generating the capacitance parasitics of polysilicide wiring, the capacitance of the wire to the silicon substrate and to that of the neighboring polysilicide wiring can be accurately modeled using a conventional capacitance modeling program. The capacitance to the metal conductor structures that exist above, however, cannot be modeled accurately. As shown in Fig. 6.2(a), the metal wires are conventionally perpendicular to the polysilicide wires. The three-dimensional structure, with complex edge shapes of the reflown intermediate oxide, is difficult to model. Because of the modeling difficulty the extra parasitics are usually not included. This omission creates simulation error.

Inclusion of wire-to-wire coupling capacitance creates a fundamental problem: The circuit simulator code becomes unmanageably complex and large if they are included. Especially important is the Miller capacitance that exists between the drain and the gate of an inverting gate. This capacitance creates waveform distortion and thereby, large delay uncertainty.

(B) Initial Conditions and the Past History of a Gate

A CMOS gate stores past history as charge (see Section 3.8). The effect is especially significant if not all the FETs of a gate are located closely and therefore wiring capacitance within a single gate is large. Charge stored in the internal node capacitance depends on the past history of the gate. Uncertainty in the node charge results in uncertainty in delay. It is bad layout practice not to pack all the FETs of a single CMOS gate within a small area.

(C) Simultaneous Switching

In a simple delay simulation, only one input signal changes state, and all others have steady logic levels that enable the gate. This is not always true. If two signals change simultaneously in the same direction, the switching delay of a gate is different from that determined by a simulation with a single transition (only one input switching).

(D) Miller effect of loading

To simplify the equivalent circuit of a complex CMOS layout, loading of fanout gates that are not on a direct signal path is often substituted by FETs or by equivalent inverters. This simplification creates uncertainties in capacitive loading by the Miller effect, and the uncertainty amounts to 20 to 30% (Section 4.8).

(E) Variation in the Device Parameters within a Single Wafer

A ring oscillator study showed that variation in gate polysilicon width amounts to 10 to 15% within a single wafer [11]. The variation results in a significant simulation error.

(F) Simulation of Periodic Phenomena over a Long Period of Time

Because of charge conservation problems of device models (Section 6.3) and accumulation of integration error, it is difficult to simulate a periodic or quasi-periodic phenomena over a long period of time. An example is dynamic response of a self-excited oscillator.

6.6 GATE-LEVEL-CIRCUIT DELAY SIMULATION

Timing problems in a CMOS VLSI chip can occur anywhere in the chip. Even for an experienced and careful designer, it is impossible to weed out all the critical paths and the timing problems by pencil-and-paper design. An entire VLSI chip contains several tens of thousands of gates.

It is impossible to simulate the chip by a circuit simulator. In the author's experience, a circuit simulation that contains more than 1000 components is infrequent, and the cost is high. A gate-level delay simulator is used to detect the critical paths of a VLSI chip, when the first-cut design is finished. The gate-level simulator is also used to detect potential race conditions.

A gate-level delay simulator works with gates as fundamental units. It uses simplified algorithms to compute gate delays, and then the simulator sums the delays along the logic path to determine the total delay. Depending on the degree of sophistication of the delay computing algorithm, the accuracy of a gate-level simulator varies over a wide range.

The most fundamental constraint of a gate-level simulator is the type of circuit nodes that can be used for simulation. Two nodes connected by a resistive polysilicon wire are certainly distinct nodes for a circuit simulation. If the two nodes have to be distinct in a gate-level simulator, there are a number of fundamental problems.

1. The number of nodes increases drastically. Since the database of a gate-level simulator is already large, further increase in the number of node variables creates serious computing time and storage problems.

2. In an entire-chip timing simulation, it is often necessary to examine whether or not the designed chip is executing correct logic operations. Logic specification of a VLSI chip is customarily described by a language that is unable to distinguish two nodes connected by a resistor. Therefore, an interfacing problem with a logic description language occurs.

3. Many simple gate-delay computation algorithms are unilateral. Signals must flow from input to output. A bidirectional element such as a resistor causes confusion.

4. In a distributed RC network such as a resistive interconnect there is arbitrariness in deriving an equivalent circuit for simulation (Section 6.4). Since a gate-level timing simulation is often carried out simultaneously to verify logic integrity of the laid-out circuit, extraction of the circuit connectivity information as well as the parasitic values is carried out by a computer. Extraction of a reasonable equivalent circuit, including series resistance parasitics from a layout that does not contain fragmented parasitics, has not yet been successful.

For reasons 1 to 4 most gate-level timing simulators that are able to simulate an entire chip from data extracted from the layout do not include the effects of delay by resistive interconnects. Because of this simplification, a gate-level simulator is not very accurate in predicting delays of the circuits that include extensive polysilicon wiring, like the datapath, PLAs, RAMs and ROMs of a microprocessor. Simulation of such special subsystems must be carried out by a circuit simulator.

To design a custom chip by polycell or standard cell technique, or to design a gate-array chip, gate-level timing simulators are used almost exclusively. This is because of a tight design schedule that does not allow extensive circuit simulation. To improve the performance of such a chip and to improve the accuracy of the gate-level timing simulation, the majority of polycell, standard cell, or gate array chips use second-level metal interconnect to avoid resistive interconnect delay.

We often need to correlate the results of a gate-level simulator to those of a circuit simulator. The design strategy is as follows: When the first-cut design is finished, the circuits of the entire chip are extracted from the layout, and a gate-level timing simulation is run. In this simulation, logic integrity is first checked, and then we let the timing simulator point out the critical paths between the internal master and slave latches, in ascending order of the limiting frequencies. The timing simulator is set up for the worst-case process, temperature, and power supply voltage. The designer then examines the layout using the circuit simulator and either rescales the FETs or simplifies the logic of the critical paths and produces an upgraded layout, the circuit is extracted, and the entire chip simulation is run again. VLSI chip timing design is an iterative process. Special circuits (e.g., I/O buffers and sense amplifiers) that are not accurately simulated by gate-level simulator are analyzed using circuit-level simulation. After several iterations the limiting frequency determined by the gate-level simulator becomes higher than the specified frequency by a safety margin, say by 10%. At this point the timing design is considered finished.

In detecting critical paths between the chip's internal master-slave latches, a timing simulator often assumes the existence of perfectly synchronized clock edges. This assumption is often not satisfied: When a critical path delay cannot be reduced by logic optimization and FET scaling, clock to the latch that receives the processed data is delayed. This kind of circuit must be verified using circuit-level simulation. Although perfectly synchronized clocks are not the best clocks to obtain the highest possible performance, certainly they are from the timing simulation viewpoint.

6.7 MECHANISM OF A GATE-LEVEL TIMING SIMULATOR

A gate-level timing simulator is the most complex software used in a CMOS VLSI design. Since a gate-level timing simulator does not yet have specifications that are agreed upon, various "natural" assumptions are made by CAD writers that do not appear "natural" to designers. Furthermore, to make the complex software generation feasible, and to reduce computing time, many heuristic approaches are used. Many of them, although convincing at first sight, are not really valid for all the cases. This section is intended to review some of the techniques used in gate-level simulators that are relevant to CMOS designers for interpreting the simulation results.

(A) Gate Delay Computation Algorithm

Elmore's RC delay algorithm, which is discussed in the next section, is most frequently used [12] [13]. The equivalent resistances of FETs are determined so that the delays become consistent with the circuit simulation results. In a cascaded gate chain, pullup and pulldown delays of the gates are separately computed and then added together, taking proper transition polarities of each stage. This algorithm gives the delay of a long chain of logic gates reasonably well. This is understandable, since the equivalent resistance is determined by averaging over many different gates. The only case where the agreement suffers is in a short chain of gates where the averaging fails to work. If a gate-level simulator is used to detect a critical path consisting of a long chain of gates, the accuracy is high.

In the other gate-level simulator, gate delays are determined by interpolating the data read from the tables [14]. In the MOTIS simulator [15], FET current values are stored in device tables, and the software computes the discharge waveforms of nodes using various simplifying techniques. The algorithm is in between a gate-level and a circuit-level simulator.

(B) Problems of Parasitics

PFET-diffused island capacitance and NFET-diffused island capacitance depend on the node voltage. In a gate-level simulator the voltage dependence is neglected. The way the capacitance is averaged affects the delay computation precision. Capacitance appropriate for pullup delay computation is not the same as the capacitance appropriate for pulldown delay computation.

(C) Dependence on the Input Switching Time

Gate delay depends on the rate of change of the input voltage. Accurate inclusion of this dependence into a gate-delay computation is difficult. In "Crystal" the effects of reduced drive capability of FET for slow input switching is included using equivalent resistance of FET dependent on the input slew rate. This solution is not valid under all the conditions. A designer should watch nodes that switch very slowly and should reexamine the node using a circuit-level simulator.

(D) Data-Independent Timing Analysis

Some gate-level timing simulators do not have the capability of a logic simulation for an entire chip. From the simplification a curious problem occurs. The gate-level simulator searches for a gate or a latch that is clocked, and a fictitious signal starts from there. The fictitious signal stops at another clocked gate or a latch. The simulator computes the delay of the path assuming that all the gates in the path are enabled. In practice this may not be true. Furthermore, the data that are latched into the second latch may not be used at all. Therefore, a critical path that is not relevant to the operation of the chip could be reported. This is an example of the assumptions that may fool circuit designers. The user of any gate-level simulator should be aware of the peculiarities of the particular simulator.

(E) Bidirectional Path

In a circuit that contains many bidirectional components, the direction of signal flow is difficult to establish by software. In a data-independent timing analysis this may create a huge number of irrelevant critical paths, especially in a structure like a barrel shifter. Very complicated criteria were used to determine the direction of signal flow by "TV". Some of them (to use FET size ratios as guides) are too restrictive. Some gate-level timing simulators assume prescreened design data. In this problem, some help from the designer supplying the design intention is required. This information could be provided in the layout file or in a separate logic circuit description file. The author thinks that many designers are willing to supply such information to make software generation easier.

The problems of a gate-level timing simulator are numerous. A gate-level simulator must resolve the most difficult problems of CMOS VLSI, understanding the circuits from the layouts, identifying the critical paths and assigning the timing budgets, selecting proper operating conditions of the circuit, and determining delays within a reasonable computing time. A gate-level simulator is nevertheless an absolute necessity in future CMOS VLSI design, and therefore the degree of sophistication and flexibility of this type simulator will be the measure of sophistication in CAD engineering. The quality of a gate-level simulator determines the organizational aspects of a VLSI design project. Different from a logic simulator (which is used at the beginning of the project) and from a design rule checker (which is used at the end), a gate-level timing simulator is used in the middle (chip assembly phase) of the project, when the work load is heavy and coordination is difficult. New generations of gate-level simulators have a number of improvements over earlier simulators [16].

The author believes that a gate-level timing simulator should be developed to ease communication between the designer and the software, even to the extent that some algorithms can be controlled by the designer. An effort to automate every operation is difficult and almost certainly not very useful to experienced designers.

6.8 METHOD OF DELAY COMPUTATION FOR AN RC CHAIN

In this section we discuss a useful method of estimating delays of an RC chain (that includes CMOS circuits) in closed form. The formula allows us to compute the delay of a complex RC chain with reasonable accuracy. We give a detailed proof of the formula for the simplest case of the two-stage cascaded RC chain circuit shown in Fig. 6.3(a), and from that we gain experience with the approximations in the formula, which was first derived by Elmore [17].

With reference to Fig. 6.3(a), the problem can be stated as follows. At $t=0$, capacitors C_0 and C_1 are both charged to V_{DD}, and the circuit begins to discharge. The objective is to find a simple closed-form formula that includes R_0, R_1, C_0, and C_1 and that gives the time needed for node 1 voltage $V_1(t)$ to drop from V_{DD} to V_{DD}/e, where e is the base of the natural logarithm (2.718). The result is

$$T_D = R_0(C_0+C_1)+R_1C_1 = R_0C_0+(R_0+R_1)C_1 \tag{6.13}$$

and the proof follows. The circuit equations satisfied by node 0 voltage $V_0(t)$ and node 1 voltage $V_1(t)$ are

$$C_0 \frac{dV_0}{dt} = \frac{V_1 - V_0}{R_1} - \frac{V_0}{R_0} \tag{6.14}$$

$$C_1 \frac{dV_1}{dt} = -\frac{V_1 - V_0}{R_1}$$

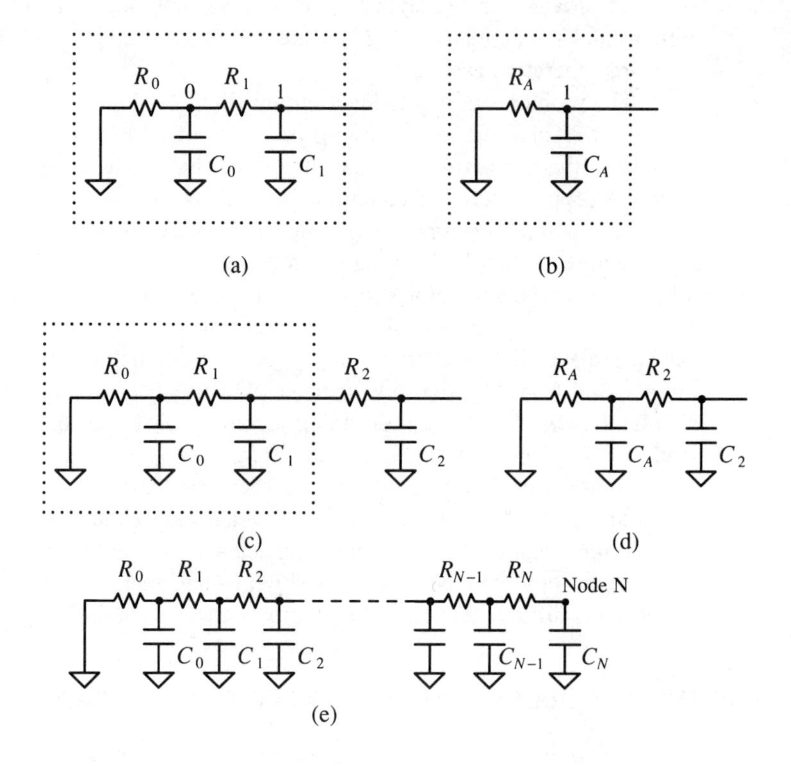

Figure 6.3 RC chain circuits used to prove Elmore's delay formula.

These simultaneous equations can be solved by the standard method of circuit theory: The node voltages are given as the sum of terms of the form const.e^{-St}, where constant S is positive. We replace the operator d/dt by $-S$ and we seek the equation satisfied by S.

We have

$$S^2 - \frac{C_0 + C_1 + (R_1/R_0)C_1}{R_1 C_0 C_1} S + \frac{1}{R_0 R_1 C_0 C_1} = 0 \tag{6.15}$$

If α and β are the roots, they satisfy

$$\alpha + \beta = \frac{C_0 + C_1 + (R_1/R_0)C_1}{R_1 C_0 C_1} \qquad \alpha\beta = \frac{1}{R_0 R_1 C_0 C_1} \tag{6.16}$$

Node voltages $V_0(t)$ and $V_1(t)$ are written as

$$V_0(t) = V_{00}e^{-\alpha t} + V_{01}e^{-\beta t} \tag{6.17}$$

$$V_1(t) = V_{10}e^{-\alpha t} + V_{11}e^{-\beta t}$$

and they satisfy the following four initial conditions at $t = 0$.

$$V_0(0) = V_1(0) = V_{DD}$$

$$\left[\frac{dV_1(t)}{dt}\right]_{t=0} = 0 \qquad \left[\frac{dV_0(t)}{dt}\right]_{t=0} = -\frac{V_{DD}}{C_0R_0}$$

We obtain

$$V_0(t) = V_{DD}\frac{(C_0R_0\beta - 1)e^{-\alpha t} - (C_0R_0\alpha - 1)e^{-\beta t}}{C_0R_0(\beta - \alpha)} \tag{6.18}$$

$$V_1(t) = V_{DD}\frac{\beta e^{-\alpha t} - \alpha e^{-\beta t}}{\beta - \alpha}$$

When $R_0 \gg R_1$, Eq. (6.15) has one solution that is small: The S^2 term in Eq. (6.15) is neglected and we obtain

$$S \approx \frac{1}{R_0(C_0 + C_1)} \qquad (= \alpha)$$

and another solution that is large: The term in Eq. (6.15) that does not contain S is neglected and we obtain

$$S \approx \frac{C_0 + C_1}{R_1C_0C_1} \qquad (= \beta)$$

We define α and β in this way, so that $\beta \gg \alpha$. When t is positive, the terms in Eq. (6.18) that involve $e^{-\beta t}$ may be approximated by zero. Then

$$V_1(t) = V_{DD}\frac{\beta}{\beta - \alpha}e^{-\alpha t} = V_{DD}\,e^{-\alpha t + \log\frac{\beta}{\beta - \alpha}}$$

and therefore by requiring

$$-\alpha T_D + \log\frac{\beta}{\beta - \alpha} = -1$$

we get

$$T_D = \frac{\alpha + \beta}{\alpha\beta} \tag{6.19}$$

when $\beta \gg \alpha$. Using Eq. (6.16), we obtain Eq. (6.13).

When $R_1 \gg R_0$, Eq. (6.15) has the solution

$$s \approx \frac{1}{R_1C_1} \qquad (= \alpha)$$

and another solution,

$$s \approx \frac{1}{R_0 C_0} \qquad (=\beta)$$

and again $\beta \gg \alpha$. Then following the same manipulation we arrive at Eq. (6.13). When neither $R_0 \gg R_1$ nor $R_1 \gg R_0$ is satisfied, the simplification does not work, but the formula is still quite accurate.

The error when $R_0 = R_1 = R$ and $C_0 = C_1 = C$ is estimated as follows. Since

$$\alpha = \frac{3-\sqrt{5}}{2}\left(\frac{1}{RC}\right) = \frac{0.3819}{RC} \quad \text{and} \quad \beta = \frac{3+\sqrt{5}}{2}\left(\frac{1}{RC}\right) = \frac{2.6180}{RC}$$

the delay time is determined from

$$\frac{1}{e} = \frac{1}{2\sqrt{5}}\left\{(3+\sqrt{5})e^{-\frac{3-\sqrt{5}}{2}\left(\frac{t}{RC}\right)} - (3-\sqrt{5})e^{-\frac{3+\sqrt{5}}{2}\left(\frac{t}{RC}\right)}\right\}$$

as $t = 3.03(RC)$. Equation (6.13) gives $t = 3(RC)$. Therefore, even in this case the agreement is good.

The analysis shows that the two-stage RC chain shown in Fig. 6.3(a) can be approximated by the single-stage RC chain shown in Fig. 6.3(b), consisting only of R_A and C_A. We showed that

$$R_A C_A = T_D = (R_0 + R_1)C_1 + R_0 C_0$$

The two circuits can be made quite similar in their charactersitics by further requiring that the low-frequency impedance looking into the single port is the same. This requirement means that CMOS gates and interconnects work from DC to a maximum frequency: Therefore, two circuits, one of which is a simple approximation of the other, must give the same resistive impedance at DC. This requires that

$$R_A = R_0 + R_1$$

When the three-stage circuit shown in Fig. 6.3(c) is considered, the first two stages can be substituted for by Fig. 6.3(b). Then the circuit of Fig. 6.3(d) has a delay

$$\begin{aligned} T_D &= (R_A + R_2)C_2 + R_A C_A \\ &= (R_0 + R_1 + R_2)C_2 + (R_0 + R_1)C_1 + R_0 C_0 \end{aligned}$$

As a generalization, the discharge delay of the last node of the cascaded N-stage RC chain circuit is given by

$$T_D = \sum_{i=0}^{N} R_i \sum_{j=i}^{N} C_i = \sum_{i=0}^{N} C_i \sum_{j=0}^{i} R_i \tag{6.20}$$

When R_i and C_i are the wiring resistance and capacitance, respectively, the formula is used to estimate wiring delay of a CMOS VLSI chip. When R_i is the equivalent resistance of a FET channel and C_i is the drain/source island parasitic capacitance, the formula estimates gate delay. Many gate-level timing simulators use this formula.

When N identical polysilicon segments are cascaded, the delay given by Eq. (6.20) is $(RC/2)N(N+1)$. If r and c are the series resistance and the parallel capacitance per unit length of the polysilicon wire, respectively, and if L is the total length of the wire, $R = (Lr/N)$ and $C = (Lc/N)$. The delay is then $T_D = (L^2/2)(rc)(N+1)/N \rightarrow (L^2/2)(rc)$ when $N \rightarrow \infty$. Therefore, the polysilicon wiring delay is proportional to the square of the wire length.

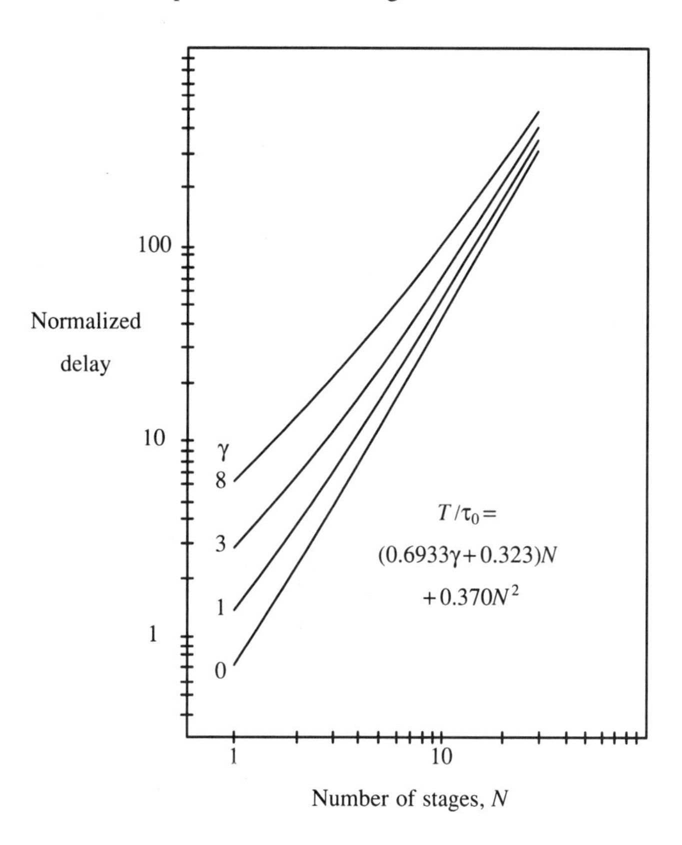

Figure 6.4 Delay of a capacitively loaded, uniform RC chain.

A special case where N identical RC stages are cascaded and the last stage has extra capacitance C_L approximates a loaded polysilicon line delay and the delay of a conventional, uniformly scaled CMOS static gate. The delay is given by

$$T_D = (C + C_L)NR + C(N-1)R + \ldots + C \cdot 2 \cdot R + C \cdot 1 \cdot R$$

$$= CR\left\{(\gamma + \frac{1}{2})N + \frac{1}{2}N^2\right\} \tag{6.21}$$

where $\gamma = C_L / C$. T_D is the time required for the last node voltage to drop $1/e = 0.367$ times the voltage at the beginning.

If we wish to find the delay time for the voltage to drop to half, $\log 2 = 0.693$ is multiplied by Eq. (6.21). We have

$$\frac{T_D(50\%)}{CR} = (0.6931\gamma + 0.3466)N + 0.3466N^2 \tag{6.22a}$$

A numerical analysis shows that slightly modified numerical coefficients give the best over all agreement. Figure 6.4 shows the delay time determined from the numerical analysis, versus N, for several different values of γ, which can be approximated by

$$\frac{T_D(50\%)}{CR} = (0.693\gamma + 0.323)N + 0.370N^2 \tag{6.22b}$$

When an infinitely long polysilicon line is charged to V_{DD} and one end of the line is pulled to ground by a driver whose internal resistance is negligible, the potential profile as a function of position, x, and time, t ($x=0$ at the driving end and $t=0$ at the beginning of the discharge) is given by

$$\phi(x,t) = \frac{2}{\sqrt{\pi}} V_{DD} \int_0^{\frac{x}{2}\sqrt{\frac{RC}{t}}} e^{-\xi^2} d\xi \tag{6.23}$$

$$\rightarrow \frac{V_{DD}}{\sqrt{\pi}} \left[\frac{RC}{t} \right]^{1/2} x \quad (x \rightarrow 0). \tag{6.24}$$

The current that flows through the driver is

$$I(t) = \frac{1}{R} \frac{\partial}{\partial x} \phi(x,t) \Big|_{x=0} = \frac{V_{DD}}{\sqrt{\pi}} \left[\frac{C}{Rt} \right]^{1/2} \tag{6.25}$$

From Eqs. (6.23), (6.24) and (6.25) we conclude the following. During time interval 0 and t, a section of the RC line of length

$$\frac{\sqrt{\pi}}{2} \left[\frac{t}{RC} \right]^{1/2}$$

or approximately $\sqrt{t/RC}$, discharges. For the part of line of length L from the grounded end to discharge, time

$$t_L = \frac{4}{\pi} RCL^2$$

is required and the time increases as the square of the length of the line.

The problems discussed in this section are special cases of a nonuniform RC chain or a tree [18]. A simple formula to estimate the upper and lower bounds of the node voltages were given by Rubinstein et al. [19] [20] [21]. An approximate closed-form analysis is given by Sakurai [22]. A very general algorithm for computing the delays of an RC network, including effects of the initial voltage, has been reported [23].

6.9 CONCEPTUAL DESIGN AND CIRCUIT SIMULATION

In many cases when a state-of-the-art CMOS VLSI chip is designed, it is not clear whether the designed logic circuit meets timing specifications after circuit optimization. Although larger FETs and duplicated hardware help shorten the delay, there is a practical limit.

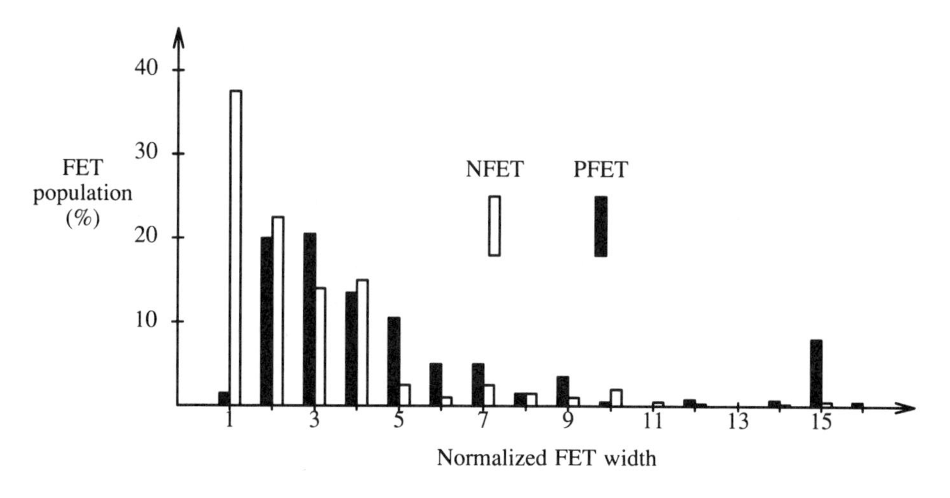

Figure 6.5 Statistics of FET size (static gates).

It is a general practice to establish a standard of design trade-offs. It is very costly in silicon area to implement unusually long logic chains using large FETs to meet timing specifications. In order to examine whether or not a certain logic design can be implemented using reasonable silicon area to meet timing specifications, we need a standard of selecting FET sizes and of estimating parasitic wiring capacitances. If we have such a set of standards, we make initial estimates before conceptually designing and simulating the logic circuit. If the design estimates meet the timing specification, we study the timing yield of the chip. If not, we consider a different logic implementation.

In order to obtain a guideline we need past experience. Studies of successful chips provide what is state-of-the-art practice in CMOS circuit design. Figure 6.5 shows statistics of the size of FETs used in a random logic functional block in a

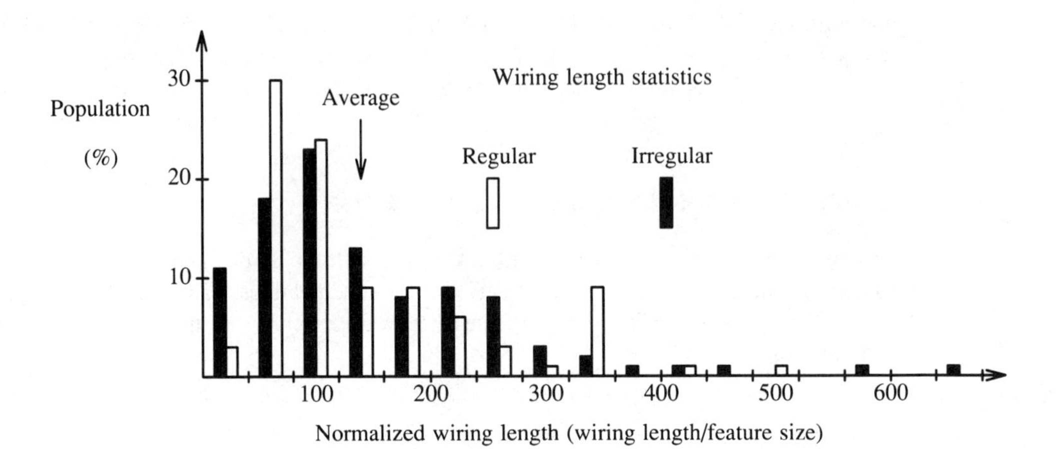

Figure 6.6 Statistics of wiring length.

successful 1.75-μm CMOS design.

In present-day layout technique, a symbolic layout method is often used which allows only quantized sizes of FETs. The minimum-size FET is defined by design rules derived from the photolithographic requirements. Larger FETs are integer multiples of the minimum-size FET. In Fig. 6.5, the horizontal axis is the FET size relative to the minimum-size FET. The histogram shows that minimum-size NFETs are the most frequently used. As the size of FET increases, the population drops off rapidly. A small number of especially large FETs are used for clock drivers and for powerful buffers.

The PFET size statistics closely reflect the NFET statistics. It is our customary practice to choose PFETs about twice as large as NFETs, except when there is some reason not to do so. The two statistics match if the horizontal scale of the PFET statistics is shrunk by a factor of 2. From the statistics, an ordinary logic circuit is implemented using one to two times the minimum-size NFET, and two to four times the minimum-size PFETs. In exceptional cases the FET sizes can be doubled or tripled, but if the timing specification is still not met, it is better to consider a different logic implementation.

In order to carry out circuit simulations we need to know typical wiring parasitics of circuits. Figure 6.6 shows distribution of the normalized wiring length for regular, datapath-like structures and for irregular, random logic-like structures. Wiring within a single gate is not included in the statistics. The parasitics of the signals confined entirely within a logic block are dependent on the block size. The average wire length follows $\sqrt{area}$ dependence [24], except for very small blocks and blocks having an odd aspect ratio. There is a statistical formula derived by Donath [25] that gives the population of the wire having length λ, $f(\lambda)$, by

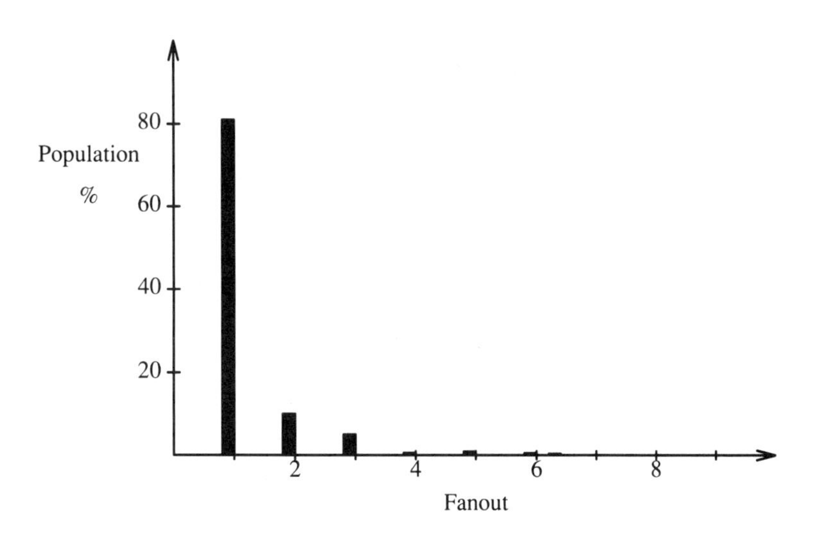

Figure 6.7 Statistics of fanout.

$$f(\lambda) = \frac{g}{\lambda^{\gamma}} \quad (L_0 < \lambda < L) \tag{6.26}$$

$$= 0 \quad (L < \lambda)$$

where L_0 is the size of a single gate and g is the normalization factor. The exponent γ is related to Rent's exponent p by $\gamma = 3 - 2p$. Rent's rule is discussed in Section 7.3. A similar result was derived by Feuer [26]. The relationship appears to hold well for many layouts.

Parasitics within a gate are usually limited by the capacitance of the diffused islands. A gate layout having extensive internal wiring can usually be improved.

The statistics of the number of destination of signals are shown in Fig. 6.7. The majority of signals have only one destination. In practice it is always necessary to check the delays of signals having a large number of destinations.

The FET size and interconnect statistics of Figs. 6.5, 6.6, and 6.7 refer to manual layout. FET sizes used in polycell or standard cell layout are larger than those used in manual layout. Four to six times the minimum-size NFETs and correspondingly larger PFETs are used, because the lengths of the interconnects, and consequently the load capacitances, are larger.

Conceptual designs and simulations are carried out using typical FET sizes, wiring parasitics, and fanouts. The data shown here are only examples. Different standards apply for different design facilities and for special devices. It is, however, important for a designer to maintain such design standards and to evaluate circuit designs by carrying out conceptual designs and by simulation at the earliest phase of the VLSI project.

6.10 METHOD OF ESTIMATING LOGIC CIRCUIT DELAYS WITHOUT SIMULATION

Circuit simulation is tedious and expensive. In a VLSI development project the number of circuits that need simulation should be kept to the minimum. A technique to achieve this objective is the subject of this section.

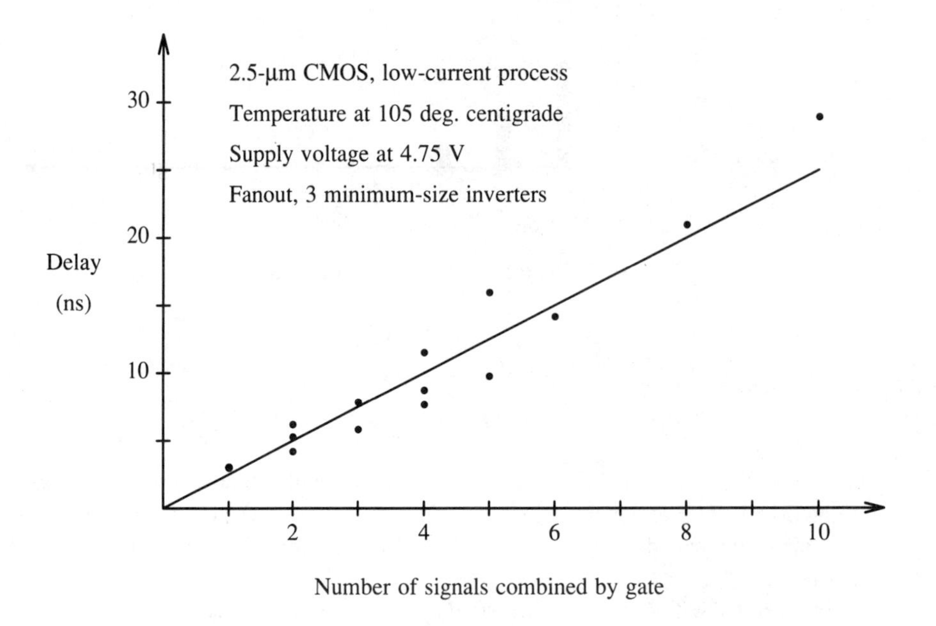

Figure 6.8 Gate delay time versus the number of signals combined.

The need for circuit simulation arises when the logic circuit to be implemented using CMOS gates consists of too long a chain of gates or consists of too complex gates. If the delay of the gate chain cannot be reduced below the specification, even if the FET sizes are optimized and privileges are given in routing of the signals, the logic implementation must be changed. Many such evaluations on a logic design can be carried out without conceptual design or simulations. Figure 6.8 shows the delays of commonly used 2.5-μm CMOS NAND, NOR, and AOI gates (averaged over pullup and pulldown transitions) versus the number of inputs. The FETs used were minimum-size NFET (5 μm wide) and the second minimum-size PFET (8.75 μm wide) of 2.5-μm CMOS. The simulation conditions are summarized in the figure. Gate delays are approximately proportional to the number of signals combined: $\tau_0 = 2.5$ ns per signal. If a gate that has N_1 inputs is cascaded with a gate that has N_2 inputs, delay of the two-stage cascaded circuit is approximated by

$$T_D = \tau_0(N_1 + N_2) \tag{6.27}$$

This formula is valid, after the FETs are scaled, so that each gate has reasonable loading. When the circuit is optimized by adjusting the FET sizes, the fanout ratio of all the gates should fall within the reasonable range (3 to 6). If the delay per signal, τ_0, is defined using the central fanout ratio, the delay given by Eq. (6.27) is a good approximation of the delay of the optimized circuit [27].

It is interesting to note that the delay of an optimized circuit which is the final product is determined only by the number of signals combined by the logic chain. This relationship is useful to identify impossible logic designs: If τ_0 is 2.5 ns and the timing budget is 25 ns, it is impossible to combine more than 10 signals. It is important to redesign the logic or to change the basic circuit techniques (such as conversion to dynamic logic), rather than to dwell on the impossible logic design.

6.11 REFERENCES

01. D. O. Pederson "A historical review of circuit simulation" IEEE Trans. on Circuits and Systems, vol. CAS-31, pp. 103-111, January 1984.
02. A. Feller "Computer analysis and simulation of MOS circuits" ISSCC69 Digest, pp. 134-135, February 1969 and D. Frohman-Bentchkowsky and L. Vadasz "Computer-aided design and characterization of digital MOS integrated circuits" IEEE J. Solid-State Circuits, vol. SC-4, pp. 57-64, April 1969.
03. S. Liu and L. W. Nagel "Small-signal MOSFET models for analog circuit design" IEEE J. Solid-State Circuits, vol. SC-17, pp. 983-998, December 1982.
04. L. A. Akers, M. Holly and J. M. Ford "Transconductance degradation in VLSI devices" Solid-State Electronics, vol. 28, pp. 605-609, June 1985.
05. B. J. Sheu, D. L. Scharfetter, C. Hu, and D. O. Pederson "A compact IGFET charge model" IEEE Trans. on Circuits and Systems, vol. CAS-31, pp. 745-748, August 1984 and B. J. Sheu and P. K. Ko "An analytical model for intrinsic capacitances of short-channel MOSFETs" IEDM84 Digest, pp. 300-303, December 1984.
06. G. W. Taylor and W. Fichtner "An analytic MOSFET model including internodal capacitances: Results on device scaling and parasitic limitations" IEDM81 Digest, pp. 215-218, December 1981.
07. D. E. Ward and R. W. Dutton "A charge-oriented model for MOS transistor capacitances" IEEE J. Solid-State Circuits, vol. SC-13, pp. 703-708, October 1978 and S. Y. Oh, D. E. Ward and R. W. Dutton "Transient analysis of MOS transistors" IEEE J. Solid-State Circuits, vol. SC-15, pp. 636-643, August 1980.
08. Y. Tsividis "Problems with precision modeling of analog MOS LSI" IEDM82 Digest, pp. 274-277, December 1982.
09. A. D. Lopez and H. F. S. Law "A dense Gate-Matrix layout style for MOS LSI" ISSCC80 Digest, pp. 212-213, February 1980.

10. L. W. Nagel "Spice 2: A computer program to simulate semiconductor circuits" Memo ERI-M520, University of California, Berkeley, California, May 1975.
11. J. M. Cassard "A sensitivity analysis of SPICE parameters using an eleven-stage ring oscillator" IEEE Trans. on Electron Devices, vol. ED-31, pp. 264-269, February 1984.
12. J. K. Ousterhout "Crystal: A timing analyzer for n-MOS VLSI circuits" in Caltech Conference on Very Large Scale Integration, edited by R. Bryant, pp. 57-69, Computer Science Press, 1983.
13. N. P. Jouppi "TV: An nMOS timing analyzer" in Caltech Conference on Very Large Scale Integration, edited by R. Bryant, pp. 71-85, Computer Science Press, 1983.
14. P. Subramanian "Modeling MOS VLSI circuits for transient analysis" IEEE J. Solid-State Circuits, vol. SC-21, pp. 276-285, April 1986.
15. B. R. Chawla, H. K. Gummel and P. Kozak "MOTIS-an MOS timing simulator" IEEE Trans. on Circuits and Systems, vol. CAS-22, pp. 901-910, December 1975.
16. T. G. Szymanski "Leadout: A timing analyzer for MOS circuits" ICCAD86 Digest, pp. 130-133, 1986.
17. W. C. Elmore "The transient response of damped linear networks with particular regard to wideband amplifiers" Journal of Applied Physics, vol. 19, pp. 55-63, January 1948.
18. E. N. Protonotarios and O. Wing "Theory of nonuniform RC lines part II: Analytic properties in the time domain" IEEE Trans. on Circuit Theory, vol. 14, pp. 13-20, March 1967 and "Computation of the step response of a general nonuniform RC distributed network" IEEE Trans. on Circuit Theory, vol. 14, pp. 219-221, June 1967.
19. J. Rubinstein, P. Penfield, Jr., and M. A. Horowitz "Signal delay in RC tree networks" IEEE Trans. on Computer-Aided Design, vol. CAD-2, pp. 202-211, July 1983.
20. J. L. Wyatt and Q. Yu "Signal delay in RC meshes, trees and lines" ICCAD84 Digest, pp. 15-17, 1984.
21. C. A. Zukowski "Relaxing bounds for linear RC mesh circuits" IEEE Trans. on Computer-Aided Design, vol. CAD-5, pp. 305-312, April 1986.
22. T. Sakurai "Approximation of wiring delay in MOSFET LSI" IEEE J. Solid-State Circuits, vol. SC-18, pp. 418-426, August 1983.
23. T. M Lin and C. A. Mead "Signal delay in general RC networks" IEEE trans. on Computer-Aided Design, vol. CAD-3, pp. 331-349, October 1984.
24. K. C. Saraswat and F. Mohammadi "Effect of scaling of interconnections on the time delay of VLSI circuits" IEEE Trans. on Electron Devices, vol. ED-29, pp. 645-650, April 1982.
25. W. E. Donath "Wire length distribution for placements of computer logic" IBM J. Res. Develop., vol. 25, pp. 152-155, May 1981.
26. M. Feuer "Connectivity of random logic" IEEE Trans. on Computers, vol. C-31, pp. 29-33, January 1982.
27. M. Shoji "Reliable chip design method in high-performance CMOS VLSI" ICCD86 Digest, pp. 389-392, October 1986.

7

CMOS VLSI Circuits

7.1 INTRODUCTION

There is no universally accepted definition of a VLSI chip. At the time of writing a reasonable definition is a single chip that integrates over 100,000 FETs and whose major area is not occupied by repetitive primitive cells. There are a number of chips reported that satisfy this criteria. A 32-bit CMOS microprocessor including 375,000 FETs has been reported [01].

At present, the density of fatal processing defects is at least 3 per square centimeter. Then the maximum chip area of practical designs is about 1 square centimeter (Section 9.16). The average area occupied by a FET is then equal to or less than

$$(10^4)^2/10^5 = 1000\,\mu\text{m}^2$$

In order to produce a VLSI chip, the processing technology, jointly with the design methods, must attain the FET density. Statistics show that the area per FET of a chip of 1.75-μm CMOS designed by the standard cell method is about 3000 μm^2. In order to gain a factor of 3 improvement in the FET density, the feature size must shrink down to about 1 μm. Therefore, an automatically designed VLSI using standard cell method is just about to emerge.

Manual layout is able to produce more densely packed chips. In a manually laid-out logic circuit, a density of 400 square μm per FET has been achieved, using 1.75-μm CMOS. Therefore, CMOS VLSI chips that exist now are exclusively laid out manually. Figure 7.1 shows the WE32000 CPU chip. The chip contains about 100,000 FETs and is processed by 1.75-μm CMOS technology. The top row consists of six PLAs that generate state vectors of the CPU. The logic circuits of the middle row process the state vector further and deliver control signals to the bottom row, which is the datapath. The chip is surrounded by the I/O frame, and the predominantly white area is the metal power and ground bus. In this chip only PLAs, ROMs, and the wiring channels are automatically generated (these regular functional blocks have high FET density). The rest of the chip was laid out manually, using a Gate-Matrix symbolic layout [02].

7.2 GATE-ARRAY AND STANDARD CELL DESIGNS

Design of a VLSI chip is very time consuming. Gate-array and standard cell design methods are intended to reduce the designer's work by unloading most of the design and the layout job to the computer [03]. Both design methods produce a chip of the form shown schematically in Fig. 7.2(a). In the logic column, gates are stacked, and they are connected together using the wires in the wiring channel. In a gate-array device, wafers are processed up to metallization step and stockpiled. Wafers are later *personalized* by the metal level processing. Since polysilicon or polysilicide levels are not available for personalization, two level metal interconnects are prerequisite to gate-array devices. Sometimes three levels of metal interconnects are used to reduce the area occupied by power buses. Standard cell devices can be designed using only one level of metal and one level of polysilicon/polysilicide. For performance reasons two levels of metal interconnects are usually available.

A typical gate-array cell consists of two PFETs and two NFETs, as shown in Fig. 7.2(b) [04] [05]. There are designs that include three pairs of FETs [06] [07]. Some new designs use two stacked columns of FETs [08]. A special, high performance design uses eight-FET cells; four of the FETs are normal size, and the other four are smaller FETs. The structure is suitable for implementing static memory cells [09].

When a tristatable inverter is required, for instance, four FETs are connected as shown in Fig. 7.2(c) [10]. When a more complex gate is required, FETs from many cells are connected together. Subsystem functions such as registers, adders, and multipliers are designed using libraries of connectivity worked out previously.

A recent trend in gate-array technology is to make the structure uniform to improve device usage. The first attempt was to remove isolation between devices by using a gate isolation technique [11]. More recently the wiring channel was removed, in the so-called "sea of gates" design [12]. Some sea-of-gate devices have no dedicated output drivers [13].

A standard cell design facility offers a number of logic gate standard cells, enough to implement any CMOS static logic. The cells are already designed, laid out, tested, and characterized. All the cells have equal widths, and the cell boundaries can be joined without conflict. The designer needs to place the cells in columns, to arrange the columns in rows, and then to wire the gates together, using place and route software.

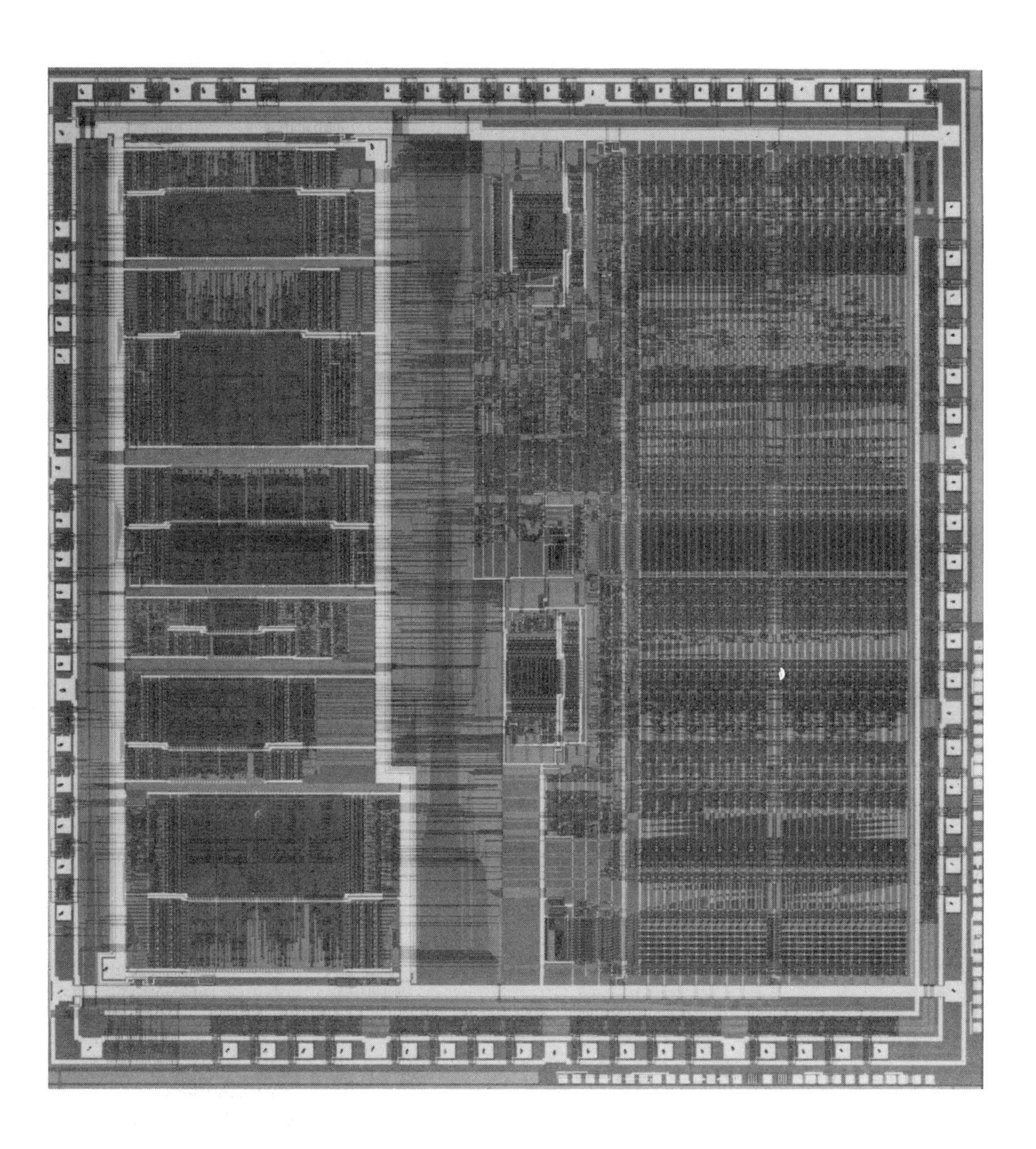

Figure 7.1 WE32000 CPU chip.

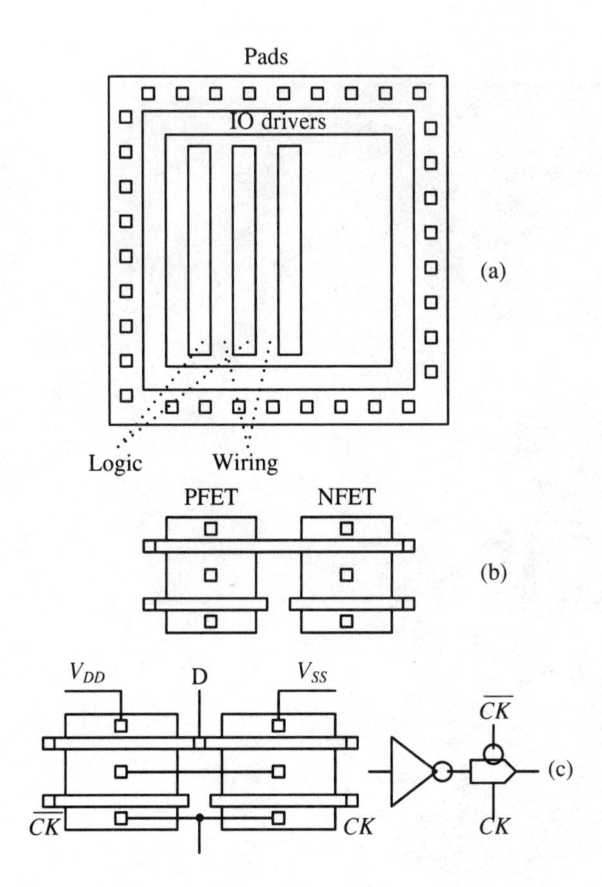

Figure 7.2 Gate-array device, FETs and gate assembly.

Standard cell and gate-array devices use relatively large FETs. Conventionally, from 4 to 7 times the minimum-size FETs are used. Several FETs can be paralleled together in a gate-array device to obtain the effect of even wider FETs. In standard cell design, high-power cells that are made of large FETs are provided. The varieties are limited, however, and a fine optimization of the circuit delays is not possible. The PFET/NFET size ratio is often close to unity. In a very complex gate the fixed size ratio may result in an unbalance of the pullup/pulldown delay times. Use of very complex gates is often discouraged to avoid timing problems.

Timing verification of a gate-array or standard cell device is usually carried out using a gate-level timing simulator (Section 6.6 and 6.7). Two levels of high-conductivity (metal) interconnects are required to improve the accuracy of the gate-level timing simulation, by reducing wiring resistance delays.

Standard cell and gate-array devices may include functional cells that consist of many gates, such as memories [04] [06], ROMs and PLAs. Analog functions can be mixed with digital functions. For analog circuits, FETs having longer channel lengths are used. Precision analog circuit requires that certain FETs are matched closely. It is difficult to achieve precision matching if gate length is close to the limit of optical resolution.

Although layout of a standard cell device is carried out quickly by computer, the layout style does not provide a very high FET density. The small FET density results in long delays due to the wiring parasitics. Unless the designer takes special precautions, priority signals such as clocks are not routed properly. It appears to be extremely difficult at present to design a very high performance chip without frequent intervention of an experienced designer.

7.3 PLANNING A VLSI CHIP I: ARCHITECTURE

Design methodology of VLSI chips is an active current subject of research. Many textbooks have been published on the subject [14] [15] [16] [17] [18]. Here only the most relevant subjects are reviewed. Before starting a VLSI development project, several basic decisions must be made.

(A) How to Split the System into VLSI Chips

The system must be split into chips satisfying the following two requirements.

1. Proper number of FETs must be assigned to each chip by taking FET density into account [19] such that the absolute maximum chip size is about 1 cm^2. FET density depends, above all, on how much memory is included in the chip.
2. Each chip has a reasonable number of I/O terminals.

The chip area is estimated from the logic design specifications of the chip, by separately estimating the area occupied by the gates and by the interconnects, and then adding them together. Each area estimate is carried out using a semiempirical algorithm [20].

The maximum number of I/O terminals is determined from the number of bonding pads that can be placed on the periphery of the chip. At present, the bonding pads are about 120 μm square, and the minimum center-to-center distance is about 250 μm. Then a 1 cm × 1 cm chip can accommodate 160 terminals if the chip's edges are occupied only by the pads. This limit can be lifted only if an advanced packaging technology that allows placement of pads inside the chip is available.

There is an empirical relationship between the number of terminals, the number of gates, and the average number of stages of the cascaded logic gates in a subsystem, called Rent's rule. A large system is conventionally designed using the top-down design method. At first the global structure of the system is drawn in the block diagram shown in Fig. 7.3(a), in which the function of each block (nodes of the graph shown by the circles) is specified by words and the interconnecting signals (edges of the graph) are specified only by bus names, such as data, address, and status. As the design progresses one step, a single functional block is replaced by several subblocks (nodes) as shown in Fig. 7.3(b), and a more precise number of interconnects are specified, like a databus for the first byte, second byte, and so on. In a system designed following this methodology, Rent's rule holds.

Let the design procedure be iterated l times, by which a single node in the first schematic is expanded into N_C descendent nodes. If a single node is replaced on the average by q nodes in a single design iteration, we have

$$N_C = q^l$$

The group of nodes originating from the single node of the first schematic is connected to the other nodes by N_I interconnects. Let the number of interconnects of the first schematic be A, and in a single design iteration a single interconnect is replaced on the average by S interconnects. Then we have

$$N_I = AS^l$$

If each node is replaced, on the average, by λ series-connected nodes, the number of stages a signal must travel from the input to the output, N_D, is given by

$$N_D = B\lambda^l$$

where B is the number of stages in the first schematic.

Since $q = \lambda S$, the following results are derived:

$$N_I = AN_C{}^p \tag{7.1}$$

$$N_D = BN_C{}^{\alpha}$$

where

$$p = \left(\frac{\log S}{\log q}\right) \quad \text{and} \quad \alpha = \left(\frac{\log \lambda}{\log q}\right)$$

and further we have

$$\alpha = 1 - p$$

The parameter p is called Rent's exponent, and from experience p falls in the range from 0.47 to 0.75 [21]. The first line of Eq. (7.1) is often called Rent's rule. If Rent's exponent is calibrated by past design experiences, the formula is accurate enough to be helpful in estimating the number of I/O terminals on a chip or to a subblock within the chip. The second line of Eq. (7.1) is useful to make a crude estimate of the signal delay. As shown in Section 6.10, the optimized delay is approximately proportional to $N_D \overline{f_I}$, where $\overline{f_I}$ is the averaged fan-in of a gate [22].

Some terminals must be assigned for power and ground connections. The number depends on the number of output drivers, the capacitive load of the drivers,

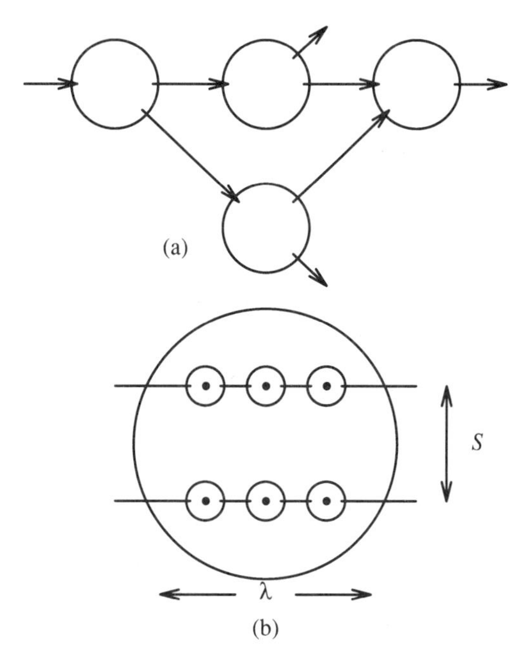

Figure 7.3 Top-down design method leading to Rent's rule.

and the time available to drive the capacitive load. In a device such as a microprocessor that has multibit data and address buses from which data are driven out simultaneously, about 10% or more of the I/O pads must be allocated for power and ground connections to minimize switching noise.

(B) Signal Transfer Timing

VLSI chips must have reasonable chip-to-chip communication time. To attain this objective, the most time-critical logic paths must be kept within a single chip, or as small a number of chips as possible. An example of this problem is the address translation delay of a VLSI microprocessor system. If a CPU and an address translator (a memory management unit or MMU) are integrated into a single chip like the INTEL 286 CPU chip, no significant time is lost to transfer the virtual address from the CPU section to the MMU section. It is impossible, however, to integrate extensive memory management hardware into a CPU chip at present. If the MMU is a separate chip, time is lost just to transfer the address from the CPU to the MMU.

(C) Timing Design Options

The most conservative and reliable timing scheme for a sequential device is to clock the master and slave latches in the chip by two-phase, nonoverlapping clock signals.

Adoption of this timing scheme allows easy circuit design and dependable chip operation, and is therefore definitely the best choice for many cases. In the other timing scheme using edge-triggered master-slave latches, clock generation and distribution are more involved. Tight control of clock delay and skew are necessary.

(D) System Clocks and Synchronization of Chips

This problem is fundamental to chip-to-chip signal transfer timing. If an asynchronous hand shaking scheme is used, chips need not be synchronized, but the processing efficiency is not the best. Further, there is a problem of metastability of the input latches (Section 4.13 and 4.14). Signal transfer between chips is not guaranteed to be successful, although the chance of failure can be made small, as discussed in Section 4.14. Synchronized signal transfer between the chips has no such disadvantages, and therefore the scheme is increasingly used in CMOS VLSI systems.

The design problem of this scheme is that the internal clocks of all the chips must be well synchronized. To attain this objective, the best idea is to produce all the necessary clock signals by a dedicated clock chip, and distribute the primary clocks with minimum skew, by a strong CMOS logic level driver. Each chip receiving the primary clock does only buffering, or minimum decoding. Reducing the clock delay from the clock input terminal to the internal clock is essential: Variation of this delay by processing conditions, temperature, and power supply voltage may create serious clock skew.

(E) Power Consumption

The power of a CMOS VLSI chip is the hardest parameter to estimate (Section 9.7). The author's experience is that the accuracy of chip power estimates based entirely on the logic design specifications is only $\pm 40\%$. There is no way to estimate accurately how often the internal circuits charge or discharge. It is easier and is more accurate to estimate the power consumption of a chip from a similar chip designed earlier, by applying a modification factor.

The chip power must be known, however, to evaluate the heat dissipation requirements of the package. The thermal resistance of an ordinary chip carrier package is about 20 degrees centigrade per watt of power. It is an advantage of CMOS that a chip does not dissipate very significant standby power.

(F) Architecture of a Chip

An advantage of any MOS technology is availability of a high level of integration that provides a lot of hardware that allows execution of system operations in parallel. However, CMOS, and generally any MOS circuits, cannot operate any particular part of the chip at a significantly higher speed than the rest, even if power and silicon area are expended at the particular part. Such operation, available in mixed-technology systems consisting of MOS, bipolar TTL, and ECL devices, is unavailable in pure

MOS systems. The chip architecture must make the best use of the characteristics of MOS technology.

When the chips exchange commands, the command words must be chosen so that parallel execution is easy and efficient. This matter is especially important in selecting the instruction set of a CMOS microprocessor system.

(G) Testability

VLSI chips having a huge number of internal sequential circuits and a limited number of I/O terminals have severe testability problems. If the problem is not attended to, a huge number of test vectors become necessary. This point is clear even from a simple binary counter: An N-stage cascaded counter requires 2^N clock cycles, or clock vectors to complete a test. When N is the modest number of 16, the number is already 65,000.

At present, some testability features, such as parity check, internal register scan, internal register access, or self-testing, became a more or less essential part of the design. Testing cost is a significant fraction of the VLSI chip cost, and testability features help reduce the chip cost.

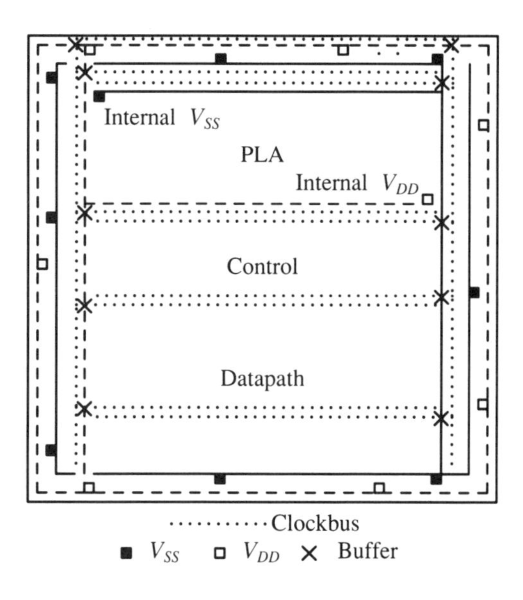

Figure 7.4 Chip structure of WE32000 microprocessor.

7.4 PLANNING A VLSI CHIP II: STRUCTURING CMOS VLSI

To lay out a CMOS VLSI chip, a rational chip structure is important. Four often conflicting requirements must be met. They are:

1. Low impedance, noise-free power, and ground buses
2. Primary clock distribution by uninterrupted metal wires
3. Critical circuit contained in a small area
4. Efficient silicon area usage

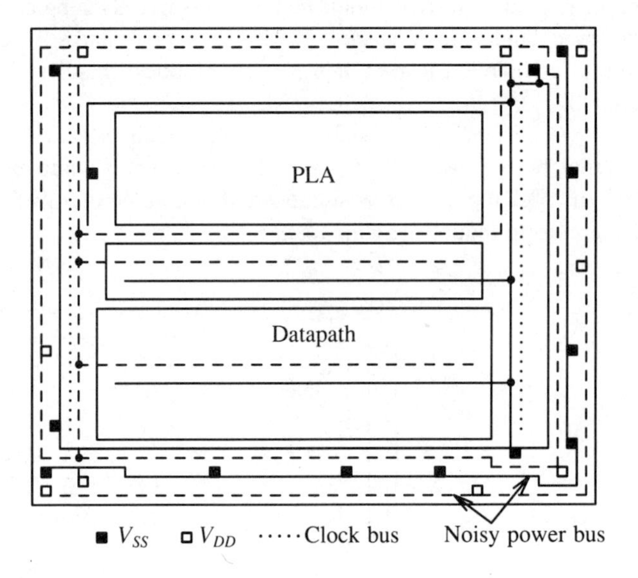

Figure 7.5 Chip structure of WE32100 microprocessor.

Structuring a VLSI chip begins with allocating chip area for functional components such as datapath, controls, memories, random logic, PLAs, and so on. The I/O pin locations are also assigned. After the functional blocks are allocated to their respective areas, means to distribute the power, ground, and clocks to all blocks are worked out. Output pins and I/O pins are grouped together since they require separate power and ground bus. When the chip drives heavy capacitive loads, the power and the ground buses of all the output drivers are separated from the rest, and the noisy power and the ground buses are provided with many bonding pads. It is customary to group the input pins together, because they require noise-free power and ground buses. Noisy and quiet power/ground buses are connected by resistance of silicon substrate, but they are not connected by metal interconnects on the chip. Input buffers that are part of I/O ports must be supplied by quiet power and ground buses. Noisy and quiet power buses are separately bonded to the package power terminal. The result of this design effort is a chip floor plan such as that shown in Figs. 7.4 and 7.5.

In Fig. 7.4 the power, ground, and clock bus structures of the WE32000 CPU are shown. The vertical power buses feeding the interior of the chip have bonding pads on both the top and bottom ends. For PLAs that generate large current spikes, extra power and ground pads are placed in the interior of the chip. The clock buses are designed using continuous metal wires wherever possible. There are a pair of clock buses carrying two primary clocks. When the clock bus crosses over the power bus, a buffer is provided to eliminate the delay. Until recently, most of CMOS technology had only one level of metal interconnects. The polysilicide clock bus crossunder creates a large clock skew if all the clock loads beyond the crossunder are directly connected to the clock bus. A clock buffer provided at the far side of crossunder (from the clock source) reduces clock skew to the minimum.

Figure 7.5 shows the structure of the WE32100 CPU chip. One set of independent noisy power and ground buses feeds the output drivers of the address and the data buses. Power and ground pads are located so that the voltage drop across the quiet power bus is minimized. Recently, even some gate-array devices have a separate power bus for the output drivers.

7.6 PLANNING A VLSI CHIP III: SELECTING CMOS TECHNOLOGY

Selection of the wafer processing technology creates subtle but significant differences in the chip performance. If a VLSI chip has a powerful, multibit datapath, a way of propagating control signals quickly across the structure is required. A two-level metal technology or a silicided polysilicon technology is required. If the chip contains a lot of memory, two levels of polysilicon helps to reduce the size of the static memory cell to a quite significant degree. If a lot of random logic exists, two level metal interconnects reduce the area per FET quite drastically. A PLA requires silicided polysilicon for rapid propagation of the internal signals. If a lot of dynamic logic is required, the NFET must have good characteristics. If a dynamic PLA is used, the subthreshold leakage current cannot exceed a certain maximum, as discussed in Section 1.11. These requirements are often conflicting [e.g., high channel current and small subthreshold leakage], but they must all be met simultaneously. It appears that no single technology meets the requirements for all the chips designed at one time. Flexibility in modifying the details of the processing technology is essential in modern VLSI technology.

7.6 CHIPSET SYNCHRONIZATION AND SIGNAL TRANSFER

The most difficult design decision is whether or not a certain extra function should be integrated into a single chip or should be split into two chips. When more than one VLSI chips exchange signals in synchronized steps, strict timing requirements must be met. When the clock of chip A ticks earlier than the clock of chip B, chip A-to-chip B signal transfer is easy, but chip B-to-chip A transfer is hard. The latter transfer becomes the system's critical path. To prevent this difficulty, the internal clocks of the chips should be closely synchronized [23].

Since the two chips are processed subject to different conditions, the delay of the internal clock from the clock input is not the same. The best way to resolve the problem is to minimize the delay of the internal clock from the input clock. If the worst-case delay from the input clock edge to the internal clock edge is T_D nanoseconds, the worst-case internal clock skew is T_D times the process variation, which typically amounts to $T_D/2$ if the fastest chip A and the slowest chip B communicate. By reducing T_D the range of variation can be narrowed.

If a system consists of more than one chip, the clock buffer and the decoder circuits of all the chips should have the same delay specification; the delay, as well as process dependence of the delay of the clock circuits, should be designed to be the same. When the chips are processed by the same process line, the statistical distributions of internal clock delays are usually quite similar and a selection procedure can be used to obtain a matched chip set. The technique is a practical and reliable solution to the chip-to-chip clock skew problem. More sophisticated techniques using a feedback control of the clock edges have been proposed, but the techniques have not yet been widely practiced [23].

If, by some means, a signal can be sent out from one chip and then received by the other chip without seriously disrupting the master-slave transfer timing, a VLSI chip can be split into two smaller chips. The combined yield of the two smaller chips is higher than the yield of a single large chip, and therefore the technique provides a useful design alternative.

If a signal is sent out from a chip, the output buffer must drive the capacitance of the package and the printed circuit board (PCB) wiring. If a conventional package is mounted on a PCB the capacitance including PCB traces is typically about 10 pF per package. Then a minimum capacitive load of 20 pF is expected for a signal going out from a chip. Within a VLSI chip the signal starts at the output node of a minimum-size buffer. The signal is buffered by a chain of successively larger buffers to drive the 20 pF capacitive load. The input capacitance of a minimum-size buffer is about 0.033 pF. If fanout of 3 is assumed (this is the optimum scaleup ratio, see Section 8.8), five stages of buffering are necessary. The total delay is estimated to be 6 ns. The size of the output buffer is 243 times the minimum size.

This is an example of an exercise called *conceptual design*. This conceptual design shows that the scheme can work, as far as the timing is concerned, for a chip timed using nonoverlapping two-phase timing (see the next section) up to 10 MHz. The average duration of one phase (phases 1 to 4 of Fig. 7.6) is 25 ns. A 6 ns loss in a timing budget is still tolerable. The scheme, however, is unworkable since the output driver (the last stage of the buffer chain) is too large. The output driver is comparable to the databus driver in size. A design providing 100 such output drivers and sending the state vector out to the datapath chip is very difficult for I/O design and for power bus noise control. This is one reason why MOS VLSI microprocessors are implemented as a single chip.

The capacitance between the chips can, however, be reduced at least by 10 times if advanced packaging technology such as IBM's flip-chip technology is used [24]. This advanced packaging technology application is still not for the highest possible frequency, but allows implementation of large and fast cache and very complex and extensive CPU control logic systems that operate at 10 to 25MHz clock frequency, with very high yield.

7.7 CHIP TIMING DESIGN

A crucial decision a designer must make at a very early phase of a VLSI development project is how to clock the chip. Two basic questions must be answered: How to clock the data storage devices within the chip, and how to generate necessary clock edges. We discuss here the first problem.

To execute a logic operation on the data stored in a set of latches and to deposit the answer into a second set of latches, the input latch data must be held stable until the answers are deposited into the output latches and the latches are closed. The input latches cannot change the data to the output latches before the output latches close. If the two latches are transparent simultaneously, the data flow through the both latches uninterrupted. The circuit may settle in a wrong state, or it may even oscillate. Anyway logic error is inevitable. This data flow-through condition must be avoided completely.

Two schemes of clocking the internal latches are widely used. One is to use a pair of two-phase nonoverlapping clocks and the other is to use a single clock in an edge-triggered master-slave latch configuration.

The cleanest way of achieving this objective is to use a pair of two-phase, nonoverlapping clocks as shown in Fig. 7.6. In Fig. 7.6 only assertion clocks are shown. CMOS latches require yet another antiassertion clock, which is complementary to the assertion clock. The latter is not shown to avoid confusion. For convenience the interval when the master latch is transparent is called phase 1, and the interval when the slave latch is transparent is called phase 3. There are two "dead" phases, phase 2 and phase 4, between phases 1 and 3. Logic (M) operates on the master latch data during phases 1, 2, and 3, and at the end of phase 3 the processed data are latched into the slave latch. Logic (S) operates on the slave latch data during phases 3, 4, and 1, and at the end of phase 1 the processed data are latched into the master latch. Let the duration of phase 1, 2, 3, and 4 be T_1, T_2, T_3, and T_4, respectively. Then the time available for logic (M), T_M, is given by

$$T_M \leq T_1 + T_2 + T_3 = T_P - T_4$$

and for logic (S),

$$T_S \leq T_3 + T_4 + T_1 = T_P - T_2$$

where T_P is the clock period. Therefore, more time is available for logic processing if

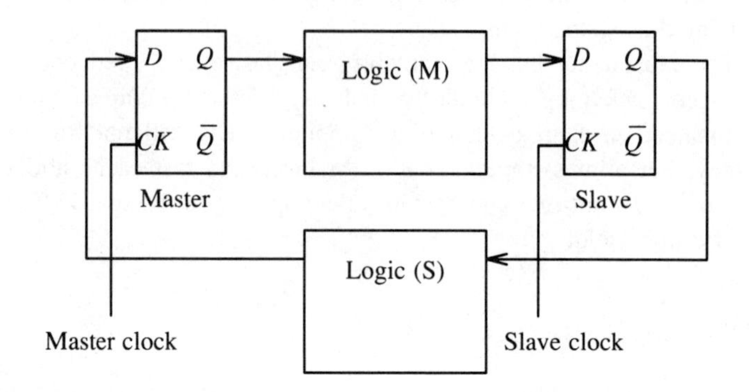

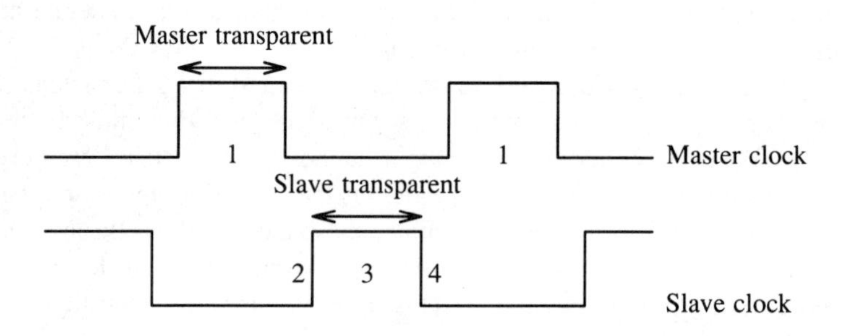

Figure 7.6 Nonoverlapping two-phase clocking of master and slave latches.

the dead phases 2 and 4 are made shorter. The dead phases, however, cannot be made shorter than a certain practical limit. The master and slave latches of Fig. 7.6 can be any pair of latches in a VLSI chip. The distance between their locations can be as much as 1 to 2 cm. It is very difficult to synchronize clocks over an entire chip very accurately. If the accuracy of the clock edge is $\pm\Delta T$, the dead phase cannot be shorter than about $2\Delta T$. If ΔT is 3.5 ns in a 10 MHz microprocessor, the dead phase cannot be made shorter than about 7 ns.

It is difficult to generate a clock signal that has accurate duration of dead phases. Four equally-spaced phases are most often used [25] because of convenience in generating such a clock waveform. In this case the duration of the dead phases is sufficient and the system is trouble-free. The cost of adopting the design is a loss of about 13% of processing time (about 1/2 of one phase), which is usually tolerable.

Logic (M) and (S) in Fig. 7.6 can be dynamic logic. The clock edge to start discharge of the dynamic logic may be chosen to be the same as the clock edge that shuts off the latches that supply data to the dynamic logic (Section 5.3).

The problem of using two-phase nonoverlapping clocks lies in generation and distribution of the clocks. The edge-triggered master-slave data transfer scheme

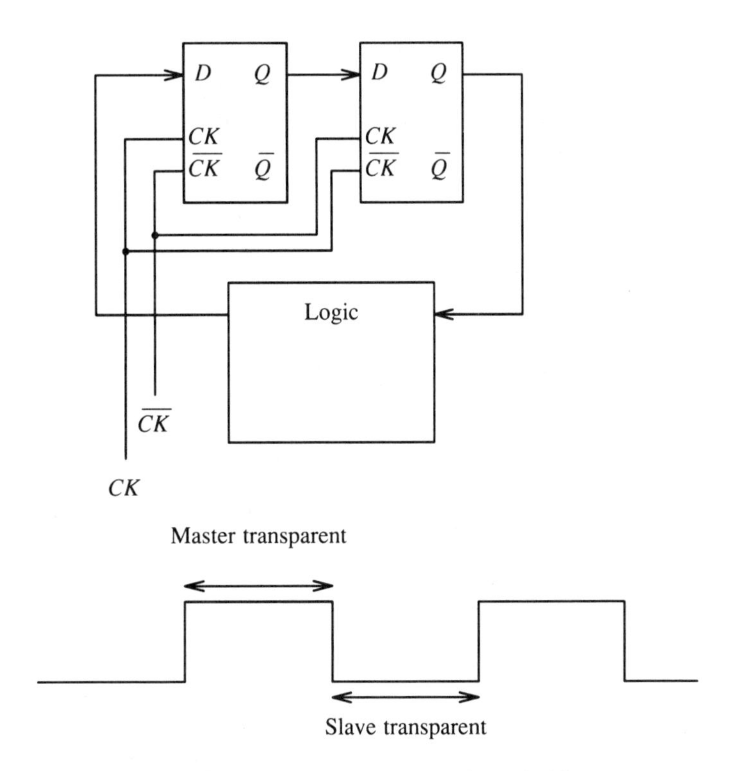

Figure 7.7 Edge-triggered master-slave clocking.

shown in Fig. 7.7 requires only one clock signal. This is equivalent to two-phase nonoverlapping clocking with zero dead time. In this design the clock inputs of the master and the slave latches are cross-connected by high-conductivity interconnects so that the clocks of the pair of M/S latches are complementary and well synchronized. Since some delay is involved in transferring data from the master to the slave latches, this layout practice effectively prevents data flow-through across the two latches. The problem is, however, that the data can still flow through the M/S pair if the master and the slave latches are at different locations. If delay is inserted between the master and the slave latches, the requirement for strict synchronization of the clocks is relieved. It is usually enough to insert two to four stages of inverters between the master and the slave latches.

Clocking schemes that use more complex clock signals have been reported. In the C^2MOS by Manabe et al. [26] [27] the cascaded clocked CMOS logic gates (discussed in Section 2.9) were driven by a clock whose frequency is eight times higher than the basic machine frequency, and the circuit is strongly pipelined. The general trend is, however, that high-performance CMOS VLSI chips of the 1980s use simple clocking schemes.

7.8 CLOCK DISTRIBUTION

Two realistic conditions to which a CMOS VLSI chip is subjected determine the baseline of the clock distribution problem. The first is that the maximum size of a CMOS VLSI chip is about 1 cm × 1 cm. A chip having a larger area will not have an economical yield. Then the maximum length of a clock bus is about 2 cm. The second condition is that approximately 10 to 15% of the random logic FETs (not including datapath, memories, and I/O) on the chip accept clock signals. A 1.75-μm CMOS microprocessor has 30,000 random logic FETs, and 3,000 FETs must be clocked. This is a huge capacitive load. We consider the problem of clock distribution keeping the two conditions in mind.

Resistance of a 2-cm-long metal wire in 1.75-μm CMOS is given by, using a 0.04 Ω/square sheet resistance for aluminum,

$$R(W)=\frac{800}{W\,(\mu\text{m})}(\Omega)$$

and therefore R=400 Ω for W=2 μm and R=80 Ω for W=10 μm. The capacitance of the wire can be estimated by the formula given in Section 1.14, as

$$C(W)=2(\text{cm})\times\left\{2.5+0.345(W-2)\right\}\quad(W>2)$$

and C=5 pF for W=2 μm and C=10.5 pF for W=10 μm. The time constant $T_D(W)$ of the wire is given by

$$T_D(2)=\frac{1}{2}R(2)C(2)=1\ (\text{ns})$$

and

$$T_D(10)=\frac{1}{2}R(10)C(10)=0.42\ (\text{ns})$$

where the delay due to the inductance of the wire (light-velocity delay) is neglected. Therefore, a signal takes at least 1 ns to reach the end of 2-μm-wide and 2-cm-long *unloaded* metal wire. This delay is practically too long to be ignored in precision timing design (delay comparable to gate delay). In a CMOS VLSI chip of about 1 to 2 μm feature size, a wire having W=10 μm is used to distribute the primary clocks.

Let the gate capacitance of the minimum-size inverter be $C_{G\min}$. For 1.75-μm CMOS,

$$C_{G\min}=0.035\,\text{pF}$$

and if there are N gates driven by the clock bus, the delay is given by

$$T_D(W,N)=\frac{1}{2}R(W)\left\{C(W)+NC_{G\min}\right\}\tag{7.2}$$

$$=T_D(W)\left\{1+\frac{NC_{G\min}}{C(W)}\right\}$$

This formula assumes that the clocked gates are in the immediate vicinity of the clock bus.

If we require $T_D(W,N) < fT_D(W)$ where f is a factor allowed for deteriorating the delay of unloaded clock bus by gate loading,

$$N \le \frac{C(W)}{C_{G\min}}(f-1) = 143(f-1) \text{ for } W=2$$

and

$$= 300(f-1) \text{ for } W=10$$

If the intrinsic delay of the wire is 0.42 ns, and if a maximum clock delay of 1 ns is allowed, $f=2.38$, then 414 minimum-size inverters can be driven by the 10-μm-wide clock bus. This number is, however, an order of magnitude less than the 3,000 clocked FETs in a realistic chip. In a CMOS VLSI of 1 to 2 μm feature size, 10-μm-wide metal wires are considered very wide. If a pair of clock bus, each wider than 10 μm, meander through a VLSI chip, many signals and power/ground buses are interrupted, and many problems happen in the chip layout. Therefore, increasing the width of the clock bus more than that is not a rational solution to the problem. Furthermore, a clock bus cannot be laid out using conductive metal wire only. When a clock bus crosses over a power or a ground bus, the part of the clock bus must be a polysilicon or polysilicide. Series resistance of the crossovers add to the resistance of the bus, thereby aggravating the problem. Two metal-level CMOS technology is desirable, but it is not available for cost-sensitive design. We conclude that a global clock bus cannot drive all the clocked gates in a CMOS VLSI directly without causing serious clock skew and clock delay.

The structure shown schematically in Fig. 7.8(a) is a solution to this problem. A clock buffer is provided for each logic block [25] [28]. The buffered clock drives all the gates in the block. If more than one primary clock is required, the scheme shown in Fig. 7.8(b) can be used. When the primary clock has to cross over a power bus, a buffer is provided there. If the primary clock must be decoded to generate the phase clocks, the decoder circuit as shown in Fig. 7.8(c) is used. The gates that are driven by the primary clock bus are designed as small as the delay specifications permit, so that the delay across the length of the primary clock bus is minimized.

The local clock decoders are designed so that all the clock edges at the destinations are synchronized within a close tolerance limit. Figure 7.9 shows a histogram of the clock edges in the WE32000 CPU, designed using 2.5 μm CMOS (low-current process, 105 degrees centigrade and supply voltage of 4.75 V). When clock signals of both polarity are required, one of them is the buffered output of the other. Then the pair of clock edges are separated by one-stage buffer delay. The two broad peaks observed at ±1.5 ns are the populations of such paired clock edges. When only one polarity clock signal is required, it can be adjusted exactly, by scaling FETs. The narrow population peak at 0 ns represents such clock edges.

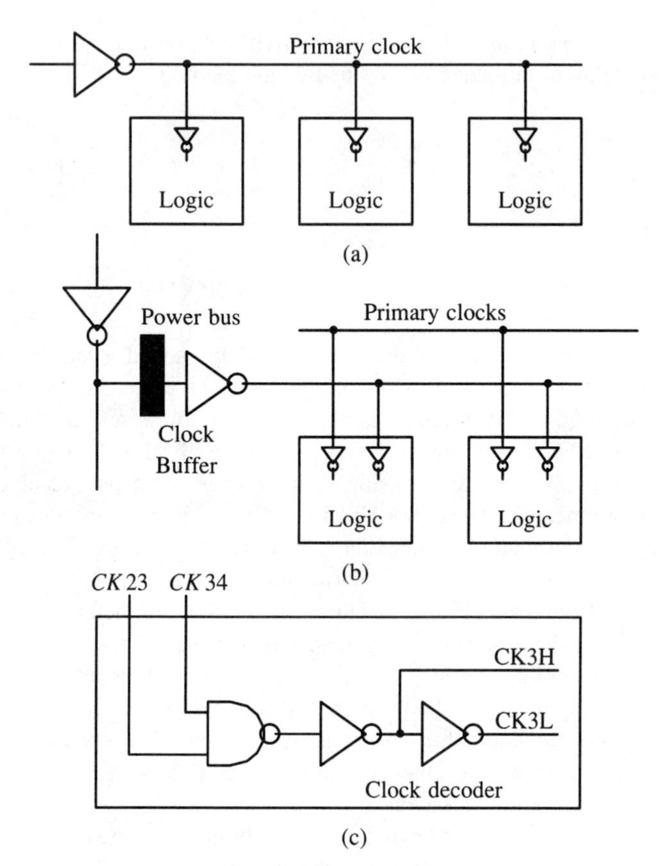

Figure 7.8 Distributed clock system.

It is also possible to adjust the local clock decoding/buffering delay so that the delay of the primary clock on the clock bus is compensated. This technique will be used in very high speed CMOS IC, whose clock frequency is higher than 50 MHz.

7.9 LOGIC PARTITION

CMOS VLSI circuits are usually laid out by group efforts. When the logic design is completed and the FETs are sized, the logic gates are partitioned into a number of smaller blocks, each of which is given to an individual member of the layout team. There is an important consideration in how this partition is carried out. If the sizes of the blocks are greatly different, some of the electrical designs, like local clock delay adjustment or sizing of the gates, cannot be carried out with a uniform standard. From this viewpoint it is convenient to determine the block size based on the following criteria [25].

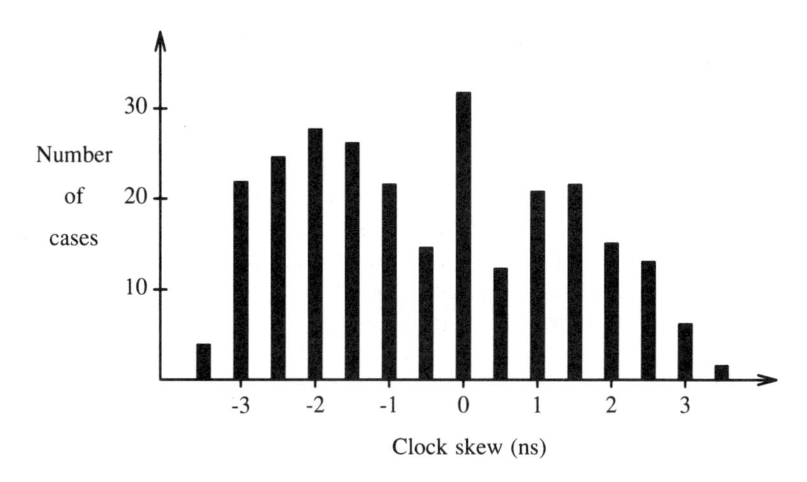

Figure 7.9 Statistics of clock edges in distributed clock system.

Suppose that the block is laid out in a $L\,\text{cm} \times L\,\text{cm}$ square area. Many logic blocks may be considered approximately square shape. Logic blocks that are either very narrow or very wide should be considered separately. A minimum-size inverter at the lower left hand corner drives L cm of polysilicon and then L cm of metal wire, and three minimum-size fanouts. Let the gate capacitance of the minimum-size inverter be $C_{G\,\min}$pF, the polysilicon series resistance be R_P (Ω/cm), the capacitances of the polysilicide and metal wires be C_P and C_M pF/cm, respectively, and the equivalent resistance of the minimum-size inverter be R_D. The delay is given by Elmore's formula (Section 6.8),

$$T_D(L) = R_D\left\{(C_{DP} + C_{DN}) + (C_P + C_M)L + 3C_{G\min}\right\} + R_P L\left\{\frac{1}{2}C_P L + C_M L + 3C_{G\min}\right\}$$

If $L = 0$,

$$T_D(0) = R_D\left\{(C_{DP} + C_{DN}) + 3C_{G\min}\right\}$$

We then set a factor $f > 1$, and determine block size L by requiring that

$$T_D(L) = fT_D(0)$$

In this design criteria we allow the wiring to deteriorate the circuit performance by the factor f. If the logic circuit is partitioned so that each block has $L\,\text{cm} \times L\,\text{cm}$ area, the local delays of gates (parasitics included) are the same over the entire chip, which is desirable from the ease of project coordination for the design uniformity.

7.10 CLOCK GENERATION

Clock signals that drive the chips must meet the specifications on frequency, duty cycle, and mutual phase relationships. Modern VLSI systems are driven by crystal stabilized clocks. A crystal oscillator is built using bipolar devices, because they have large transconductance at small-signal amplitude, which is favorable to start oscillation (Section 2.18). Even using fast bipolar devices it is difficult to control the waveforms of a crystal oscillator whose frequency is above 30 MHz. One expects nothing except a stable frequency from a bare crystal oscillator.

A conventional design strategy to generate a desired clock waveform from the more or less uncontrolled clock signal from a crystal clock oscillator is to generate the crystal clock signal at several times the desired frequency, and then to count down to obtain accurately spaced clock edges. The circuit shown in Fig. 7.10(a) is called a Moebius counter, since $\overline{Q}$ of the second D-latch is connected to input D of the first D-latch. This is a twisted connection, and from this comes the name.

When the circuit is clocked, a signal at half the input frequency emerges from Q and $\overline{Q}$ outputs of the latches, as shown in Fig. 7.10(b). The two waveforms, observed at nodes N_1 and N_2 have exactly 50% duty-cycle, and the delay between the two signals is T_H, the width of the input clock signal. This circuit is adequate to produce a single 50% duty-cycle clock signal that is required to drive a CMOS VLSI chip designed using the edge-triggered master-slave latches.

Since the clock edges available from nodes N_1 and N_2 are skewed by the width of the input clock pulse, the circuit is not adequate to generate a pair of two-phase, 50% duty-cycle clocks with equally spaced phases [$CK23$ and $CK34$ of Fig. 7.8(c)]. It is possible, however, to generate a four-times-the-basic-frequency crystal clock, and count down once by the circuit of Fig. 7.10(a). The 50% duty cycle clock signal generated in this way is used to drive another countdown circuit. In this way an ideal pair of skew-free, 50% duty cycle two-phase clocks can be generated.

When the clock generator countdown circuit starts from a quiescent state and is then clocked, the clock signals from the circuit depend on the initial state of the counter. If the circuit should start from a specified clock phase, initialization capability must be added to the counter.

As the clock frequency of a CMOS VLSI chip increases, crystal generation of 4 times the base frequency becomes increasingly difficult. It is desirable to use 2 times the base frequency clock signal and to correct the duty cycle to 50%. The simple analog feedback circuit shown in Fig. 7.11 keeps the duty cycle of the output clock at 50%. The output clock duty cycle is determined by averaging the output clock by an RC circuit with a long time constant. The averaged voltage, proportional to the duty cycle, is compared with a reference voltage, $V_{DD}/2$. If there is any difference, the operational amplifier adjusts a control voltage. When the control voltage is high, the control gate pulls up slowly but pulls down quickly. An upgoing input pulse comes out as a wider downgoing pulse (Section 2.15). When the control voltage is low, an upgoing input pulse comes out as a narrower downgoing pulse. A high-gain operational amplifier keeps the duty cycle set at 50% [29].

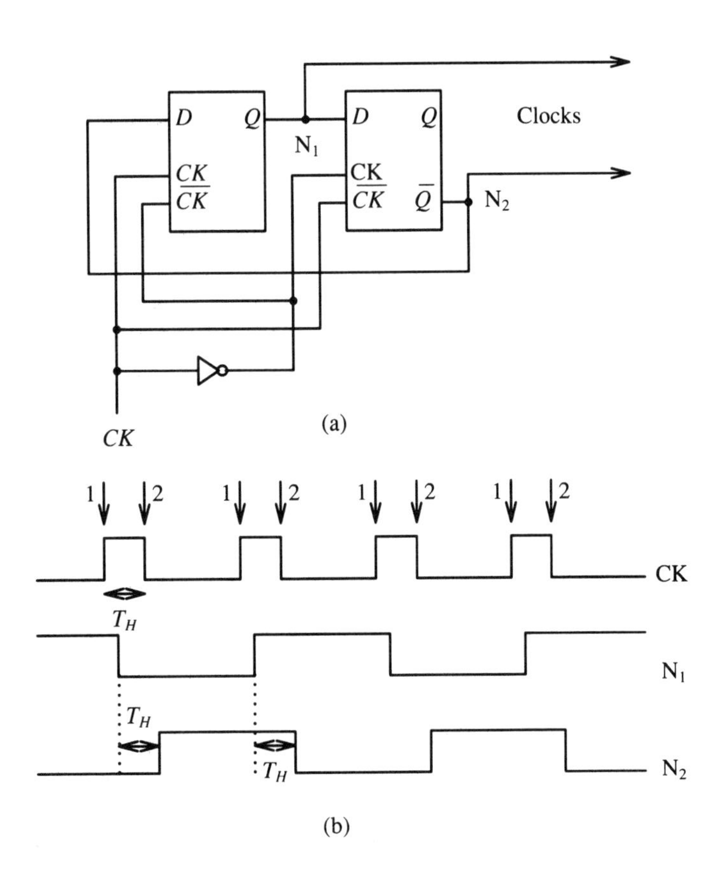

Figure 7.10 Clock generation by countdown.

7.11 CMOS VLSI SUBSYSTEMS

In the following sections CMOS VLSI subsystems that are assemblies of CMOS gates and FETs and that execute specific, well-defined functions are discussed. There is a new trend in CMOS VLSI design methodology, which is an attempt to assemble a VLSI chip from subsystems rather than from individual gates. Some of CMOS VLSI subsystems can be complex. Often, two subsystems that have identical functions may have different circuit-level or even gate-level implementations.

7.12 OUTPUT DRIVERS

A large CMOS inverter is used to send out a signal from a CMOS chip. When the output logic level is CMOS level, the PFET/NFET size ratio is chosen at about $\sqrt{\beta} = 1.8$ or more, to attain delay minimum and pullup-pulldown delay symmetry. The

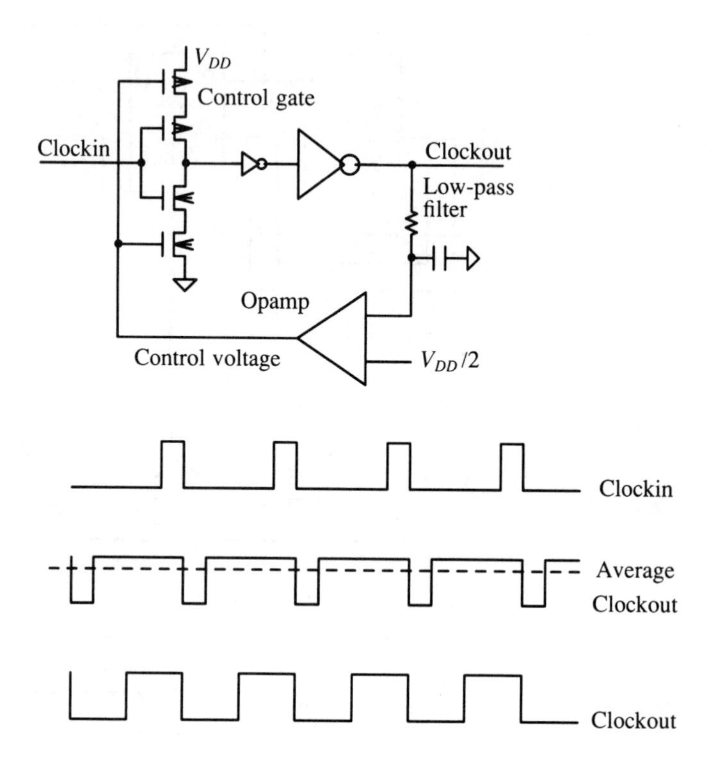

Figure 7.11 Clock duty cycle correction circuit.

detail is discussed in Sections 8.8 and 8.9. If the logic level is a TTL level, the PFET/NFET size ratio is smaller than $\sqrt{\beta}$, usually about 1. This smaller ratio is necessary to make pulldown delay (from 5 V to about 1.4 V) and pullup delay (from 0 V to 1.4 V) matched. Except for large FET sizes and latchup protection features, output driver design is conventional.

If an output driver is required to have a capability of interrupting the signal by tristating, the circuits shown in Fig. 7.12(a) are commonly used. The CMOS tristatable driver of Fig. 7.12(a) works as a noninverting buffer if $\overline{TRI}$ is high. When $\overline{TRI}$ is low, however, the gate of the NFET is low and the gate of the PFET is high, independent of data D, and the output terminal is open circuited. A conventional tristatable driver shown in Fig. 2.1 is not suitable for this application. Since a conventional tristatable inverter has two FETs connected in series, the sizes of the FETs must be doubled to get the same drive capability and cause intolerable area penalty.

The NMOS tristatable driver of Fig. 7.12(b) works in the same way. Since the pullup is carried out by an NFET, it is impossible to drive the output terminal above voltage $V_{DD} - V_{TH}$ (V_{TH} is the FET threshold voltage with back-bias effect). This

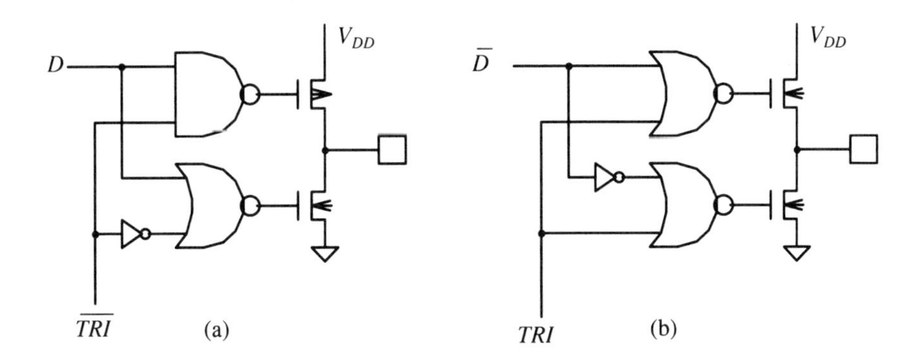

Figure 7.12 Output drivers.

output driver is used for TTL level loads. Insufficient pullup of the source follower circuit is not a problem, since TTL logic high level is only 2.4 V above the ground, and up to that voltage the NFET source-follower pullup is fast.

The output driver requires special design consideration. An output driver must drive the external capacitance strongly, enough to transmit the signal within a short time. If the time is too short, however, the current spike generated by the driver can upset normal operation of the chip's circuits that share the power/ground bus.

Let us consider this problem for the case of CMOS logic level signals. CMOS-level communication between chips is widely used, since logic level conversion is associated with delay as discussed in Section 3.3, and the delay is often intolerable. In a CMOS signal distribution network, the load is capacitive; no DC path exists from the output driver to the power supply levels. Let the load capacitance be C_L and the capacitance be initially charged to power supply voltage level V_{DD}. When a CMOS output driver discharges the capacitance, the current $I(t)$ is zero at $t=0$, increases with time, and attains a maximum at $t=t_{\max}$. Then $I(t)$ decreases to zero with time. The current waveform $I(t)$ is shown schematically in Fig. 7.13(a).

When the discharge current flows through inductance, L, of the package and on the chip power bus, a noise voltage $LdI(t)/dt$ is generated. The noise generated by the driver is proportional to the slope of curve $I(t)$ of Fig. 7.13(a). The slope is less if the time required to complete switching is more. The time t_S of Fig. 7.13(a) is the time required to complete discharge of the load capacitor, and it is called a flight time. Flight time indicates how long a gate is in transition; practically, the time is defined as the interval required for output node voltage of the gate to change from 5% to 95% of the total logic swing. If flight time t_S is longer, the noise is less, but the signal transmission takes more time. There is a complementary relationship between the flight time, t_S, and the maximum noise. The area enclosed by curve $I(t)$ of Fig. 7.13(a) is the charge stored in capacitor C_L, $Q_L = C_L V_{DD}$. Then the problem is to find

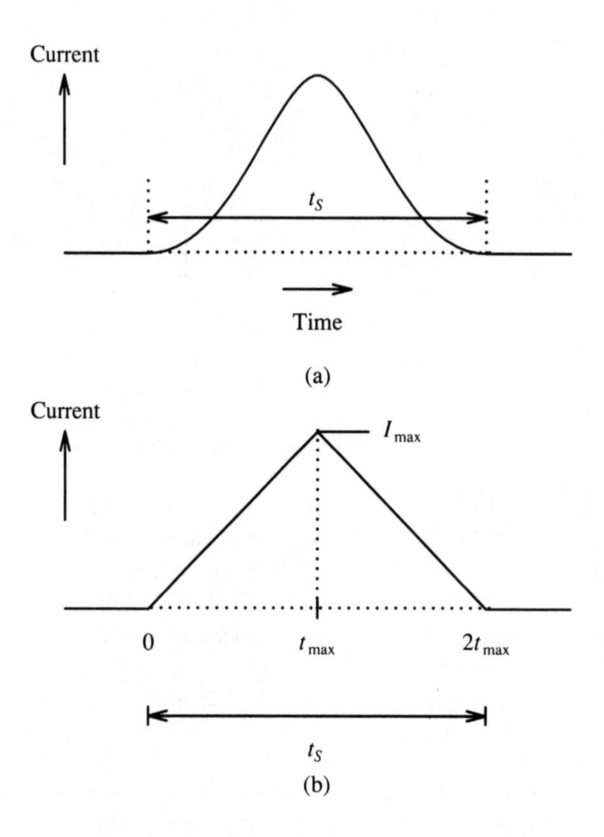

Figure 7.13 Discharge current waveform of output driver.

the curve that surrounds area $C_L V_{DD}$ that has base width t_S and that has minimum slope. By inspection the desired curve is an equilateral triangle shown in Fig. 7.13(b).

Then

$$\left| \frac{dI}{dt} \right|_{max} = \frac{I_{max}}{t_{max}} \tag{7.3}$$

$$I_{max} t_{max} = Q_L = C_L V_{DD}$$

and

$$t_S = 2t_{max}$$

By eliminating I_{max} and t_{max} among the equations, we have

$$\left| \frac{dI}{dt} \right|_{max} t_S^2 \geq 4Q_L = 4C_L V_{DD} \tag{7.4}$$

and this is the theoretical limit of $|dI/dt|_{max} t_S^2$ of any output driver [30]. If power supply voltage V_{DD}, load capacitor C_L, and flight time t_S are given, $|dI/dt|$ cannot be less than $4C_L V_{DD}/t_S^2$. Experience shows that a well-designed output driver comes close to this limit. If $C_L = 150\,\text{pF}$ and $V_{DD} = 5$ V, $Q_L = 750\,\text{pC}$.

Then

$$\left| \frac{dI}{dt} \right|_{max} \geq 3 \times \frac{10^{-9}}{t_S^2} \tag{7.5}$$

If $t_S = 2\times 10^{-9}$ s, $|dI/dt|_{max} \geq 0.75 \times 10^9 = 750\,\text{mA/ns}$. If the lead inductance of a power connection (bond wire) is 2 nH, and if the driver discharges through the bond wire, 1.5 V of noise spike is expected. This numerical estimate shows clearly that power bus noise control is one of the fundamental problems in high-speed CMOS VLSI design.

7.13 CONTROL OF THE PARALLEL OUTPUT DRIVERS

In CMOS VLSI microprocessors, the data and the address come out as 8-, 16- or 32-bit buses. If that many drivers charge or discharge their capacitive loads simultaneously, the power-bus noise becomes too large. The fundamental solution to this problem is advanced packaging technology. Another solution to this problem is to stagger the output drive. In some cases the lower bits of the address of a microprocessor are required earlier than the higher bits of the address. In this case the higher bits of address are delayed relative to the lower bits of address, thereby avoiding overlap of the current spikes. This solution makes sense only when special relief in timing requirements exists from the system's architecture and is not a general-purpose solution.

In Fig. 7.14(a) current pulses from output drivers are staggered, so that the cumulative current pulse looks like a low trapezoid rather than a high triangle. $|dI/dt|_{max} = 2I_{max}/t_S$ and the total charge is $4I_{max}t_S/2 = 2I_{max}t_S$. If we design a driver that has a flight time of $2.5t_S$, however, the maximum current I'_{max} will be $1.6I_{max}$ and therefore $|dI/dt|_{max} = 1.3I_{max}/t_S$, as shown in Fig. 7.14(b). This design is simpler than a staggered design for general-purpose applications. The height of the noise spike is lower, but the width is wider than the staggered design.

An interesting technique of reducing dI/dt of an output driver is to reduce the charge of the load capacitor as shown in Fig. 7.15. Before the driver (consisting of FETs MP1 and MN1) drives the capacitive load, preset signal *PS* is pulled high. NFETs MN2 and MN3 conduct, and output node Z, along with nodes P and N, settle to the same voltage. If the beta ratio of FETs MP1 and MN1 is unity, the voltage is $V_{DD}/2$. After the voltage has settled, *PS* is pulled down and the tristate signal *TRI* is driven low, thereby enabling the output driver. Since the initial output node voltage was half, dI/dt can be reduced to half [31]. Some additional circuits to control the DC current (the overlap current of the output stage and conflicting drives of the NAND and NOR gates) to a reasonable limit are desired.

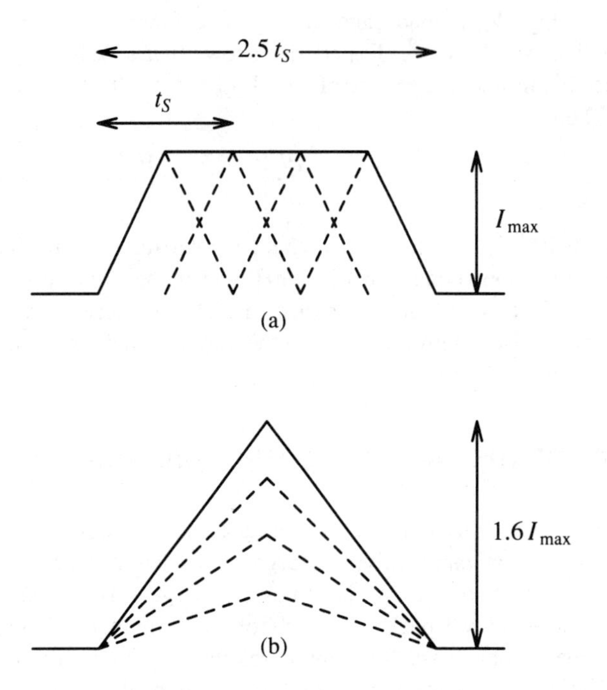

Figure 7.14 Staggered output driver, discharge waveform.

An analog feedback technique can be used to reduce the temperature sensitivity of the output buffer delay, thereby easing the noise problem [32]. Noise control is difficult because of square dependence of noise on delay time and a wide range of variation of delay time with process and temperature. If either process variation or temperature variation can be reduced by a circuit technique, noise control becomes significantly easier.

7.14 ADDERS

An adder for two N-digit numbers consists of N full adders that generate N digits of sum, and a carry generation circuit that produces carries to all the N bit positions. The $(N+1)$st digit sum is generated by the carry generator. Here we discuss carry-generation circuits. Full adders were discussed in Section 2.14.

The carry to the $(i+1)$st bit is defined by the Boolean equation

$$C_{i+1} = (A_i + B_i)C_i + A_i B_i \tag{7.6}$$

and a logic circuit that implements the logic equation is shown in Fig. 2.31. The unit circuit is cascaded over N stages, and it has two gates of delay per stage. Therefore,

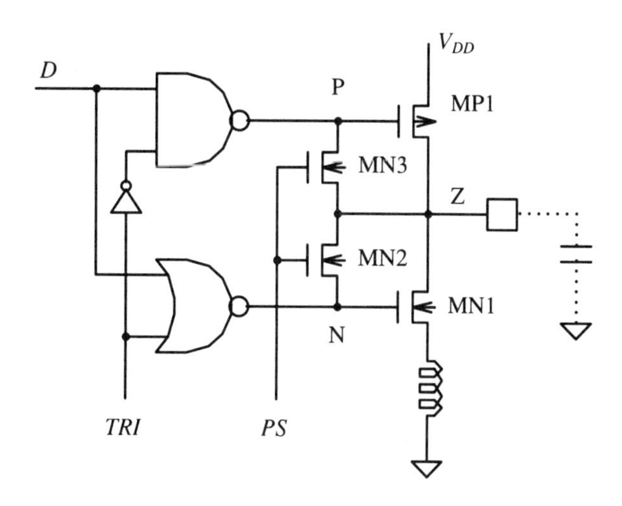

Figure 7.15 Method of reducing dI/dt of output driver.

in a 32-bit adder, 62 gate delays are required to generate the thirty-first bit carry. This delay time is usually too long to be tolerable.

A technique to shorten a logic chain consists of looking ahead the carry. Figure 7.16 shows the block diagram of a 9-bit, two-stage carry-lookahead adder, intended to show the general organization. Full adders (labeled "FA") add A_i, B_i, and C_i, thereby generating sums S_i (not shown, for simplicity). In addition, the box generates variables "generation," G_i, and variables "propagation," P_i, defined by

$$G_i = A_i B_i \qquad P_i = A_i + B_i \tag{7.7}$$

$G_i = high$ means that the ith bit position of the adder generates a carry. $P_i = high$ means that if $C_i = high$, then $C_{i+1} = high$.

The logic in the CL boxes works as follows. Carry C_1 is generated if A_0 and B_0 are both high, and therefore the 0th bit of the adder generates a carry, or if at least one of A_0 or B_0 is high, and carry input C_0 is high. Carry C_2 is generated, in a similar manner, if any one of the following conditions are satisfied.

1. A_1 and B_1 are both high ($G_1 = A_1 B_1$).
2. A_0 and B_0 are both high, and at least one of A_1 or B_1 is high [$P_1 G_0 = (A_1 + B_1) A_0 B_0$].
3. C_0 is high, at least one of A_0 or B_0 is high, and at least one of A_1 or B_1 is high [$P_1 P_0 C_0 = (A_1 + B_1)(A_0 + B_0) C_0$].

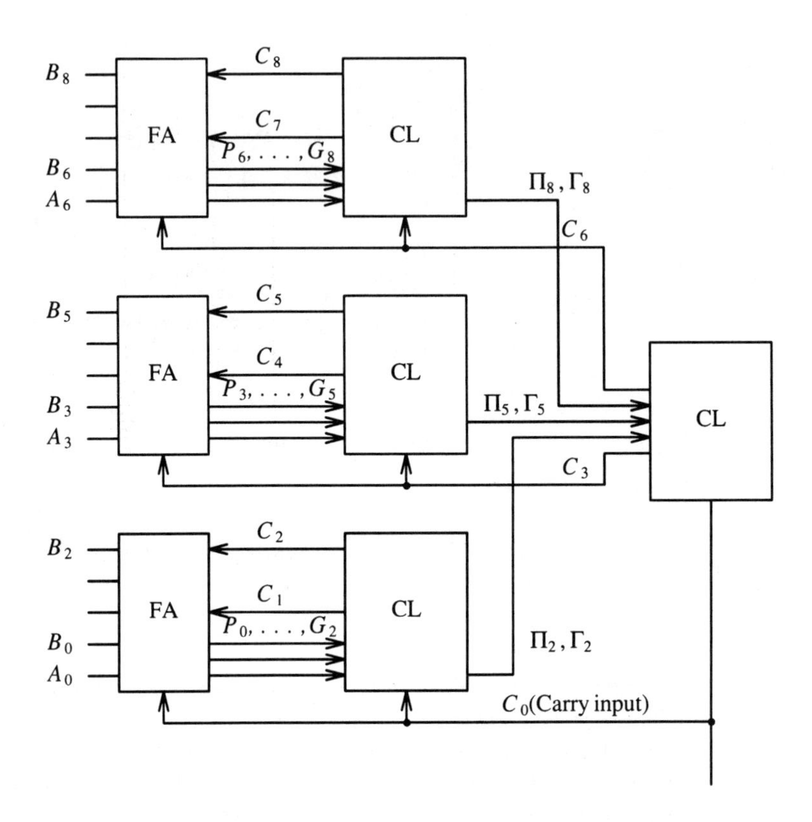

Figure 7.16 Cascaded carry lookahead circuit.

We have

$$C_2 = G_1 + P_1 G_0 + P_1 P_0 C_0$$

The logic circuit is shown in Fig. 7.17.

The carry-lookahead circuit can be cascaded in the configuration of multiple-level carry-lookahead of Fig. 7.16. If

$$\Gamma = G_2 + P_2 G_1 + P_1 P_2 G_0 \tag{7.8}$$

is high, the box CL generates carry. Further if

$$\Pi = P_0 P_1 P_2 \tag{7.9}$$

is high, the carry input to the lowest bit of the box propagates across the box. Π 's and Γ 's from a CL box can be used in exactly the same way as P_i and G_i generated directly from the input data to the adder. Π and Γ are called group propagate and group generate, respectively. Therefore, the carry-lookahead box, CL, can be cascaded

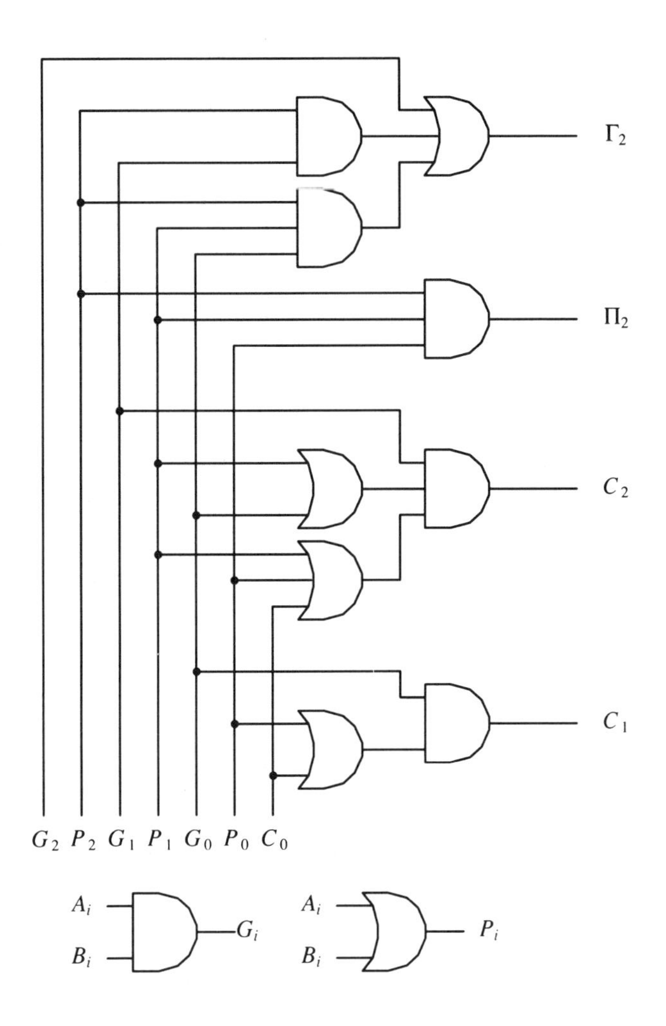

Figure 7.17 Carry-lookahead logic.

in a multistage lookahead, as shown in Fig. 7.16. Using a carry-lookahead circuit, the length of the carry generation chain can be reduced by a very significant factor.

Carry-lookahead circuits have logic that can be implemented by noninverting gates only. Therefore, the speed-sensitive circuit can be implemented by the fast Domino CMOS dynamic logic [33].

Generating carries for higher bit positions takes time, even by carry lookahead logic. An effective way to reduce the delay is to use two adders for the high-bit positions; one has a fixed input carry 0, and the other 1. The adders generate answers before the carry from the lower-bit adder arrives. As soon as the carry arrives, the carry selects one of the two higher-bit results. Adders with this architecture can have very high performance [34].

7.15 CARRY LOOKAHEAD AND CARRY BYPASS

One practical problem of a cascaded carry-lookahead adder is that the layout of the adder is rather irregular and silicon area usage is inefficient. This increases the effort for layout and reduces logic speed because of long connections. A 32-bit carry-lookahead adder can be structured so that carries of every 4-bit section $(A_0B_0...A_3B_3)(A_4B_4\ ...A_7B_7)...$ are first computed, and the carries of four such sections $(A_0B_0...A_{15}B_{15})$ and $(A_{16}B_{16}...A_{31}B_{31})$ are then computed. The FET count of the second-level lookahead circuit is approximately 1/4 of that of the first-level lookahead circuit. Therefore the second-level circuit is very sparse if laid out in a regular 32-bit structure. Further, the circuit has many long and irregular interconnections. The layout of such a circuit requires a lot of effort.

A widely used method of resolving this problem is to produce the generation and propagation variables for every N-bit (N is usually 2, 4, or 8) section and then bypass the carry, using the logic circuit shown in Fig. 7.18. The structure has an advantage that the circuit that belongs to every N-bit position is exactly the same; if only one of them is laid out, the pattern can be repeated for the other locations.

The delay of the combined carry-lookahead and carry-bypass circuit can be very short. An AND-OR21 gate used at every N-bit position is a relatively simple gate and its delay is small. The circuit shown in Fig. 7.18 is for Domino CMOS implementation. If CMOS static logic is preferred, alternating AND-OR-INVERT and OR-AND-INVERT gates can be used to save an extra inversion delay, with alternating carry-lookahead circuits generating opposite polarity P and G signals.

When the bit width of the adder, B, is given, the problem of choosing the best N and M (M is the number of repetitions of the basic units, where $B=MN$) can be solved as follows.

In an optimized CMOS circuit, the delay is approximately proportional to the number of signals combined, as discussed in Section 6.10. The AND-OR gate of Fig. 7.18 combines three signals per stage, and therefore $(M-1)$ cascaded stages of the carry bypass circuit combine $3(M-1)$ signals. A justification for estimating the circuit delay by this simple algorithm is given in Section 6.10. The number of signals combined to generate the propagation and the generation variables of an N-bit group is determined as follows. To generate P and G at each bit position requires a combination of two signals. To generate the group propagation Π, all N $P's$ ("propagation" variable) are ANDed together, and therefore Π is the result of combining $(N+2)$ signals. How to count the number of signals combined was discussed in Section 6.10. Group generation, Γ, is a generalization of the formula of the preceding section,

$$G = \sum_{i=1}^{N} G_i \left[\prod_{j=i+1}^{N} P_j \right] \tag{7.10}$$

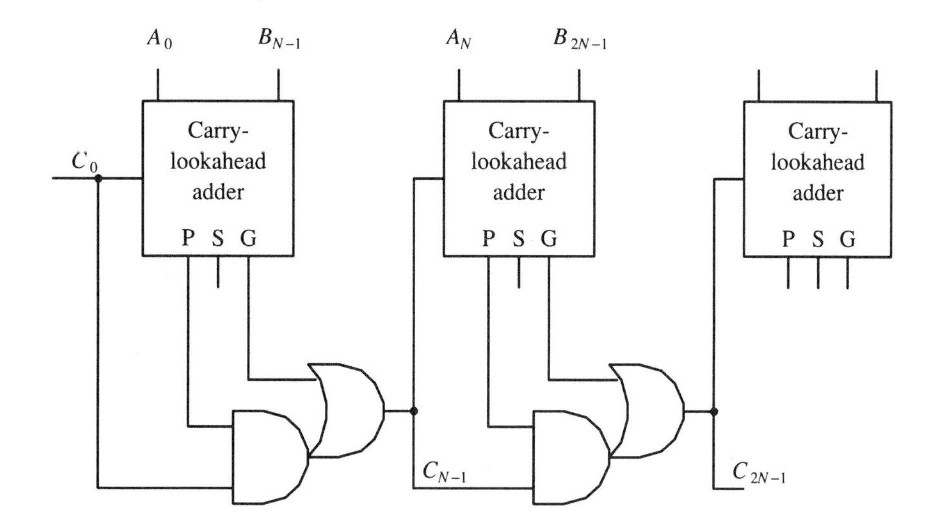

Figure 7.18 Carry-lookahead and carry-bypass combination.

If the combination is implemented exactly as indicated in the formula, $N(N+1)/2$ generate and propagate signals must be combined. In the formula, however, the same signals are combined many times. If the duplication is not counted, $2N-1$ signals are combined. Depending on the implementation, the delay to produce Γ is $N(N+1)/2+1$ or $2N+1$ (2 extra combinations are to generate P and G). The former is a pessimistic estimate and the latter is an optimistic estimate. The carry at the $(B-N)$th bit must be further processed to generate the last carry bit. This combines $N(N+1)/2$ or $2N-1$ signals within the first-level carry circuit associated with the last N bits, depending on the implementation. Therefore, the delay of the B-th bit carry is given by

$$T_D(O) = (2N-1)+2+(2N-1)+3(M-1) = 4N+3M-3 \tag{7.11}$$

for the optimistic estimate and

$$T_D(P) = N(N+1)+2+3(M-1) = N(N+1)+3M-1 \tag{7.12}$$

for the pessimistic estimate, where N and M are related by

$$NM = B$$

The problem of minimizing $T_D(O)$ or $T_D(P)$ is straightforward. $T_D(O)$ has its minimum when

$$N = N_{\min} = \left(\frac{3B}{4}\right)^{1/2} \tag{7.13}$$

is satisfied, and $T_D(P)$ has its minimum determined by solving the equation

$$2N_{min}^3 + N_{min}^2 - 3B = 0 \tag{7.14}$$

B versus N_{min} is shown in Table 7.1.

TABLE 7.1 OPTIMUM N

B	N(optimistic)	N(pessimistic)	N(realistic)
8	2.4	2.1	2
12	3.0	2.5	3
16	3.5	2.7	3
20	3.9	2.9	3
24	4.2	3.2	4
28	4.6	3.3	4
32	4.9	3.5	4

The best estimate of the integral number N is also given. For a 32-bit adder, N=4 is the optimum. $T_D(P)$ for this optimum value is 40. The number of signals combined to generate the last carry of a full, three-stage cascaded, 32-bit carry-lookahead adder is 45 (This number depends on logic implementation). Therefore, following the argument of Section 6.10, the delay of the two schemes can be made comparable by proper FET scaling. Then it is unnecessary to use multistage carry lookahead that is fast but complex. A mixed carry-generation scheme has the additional advantage of layout simplicity [35]. Evaluation of various carry-lookahead circuits from a viewpoint similar to the theory of this section was reported by Rhyne [36].

A carry-lookahead adder can be implemented using a PLA. Large adders can be implemented sequentially, computing the answer for a few bits every cycle. A PLA adder that generates result in one cycle is more involved. A straightforward implementation results in a design having unpractically many word-lines. If the input data to the decoder are pre-processed and if exclusive-OR gates are provided at the ROM outputs, the number of word-lines can be reduced to a practical number [37] [38].

7.16 MULTIPLIERS

Multipliers are classified into two types. The first generates all the bits of answer simultaneously. In the second the answer comes out one bit at a time sequentially. We discuss them in the following two subsections.

(A) Conventional Multipliers

Multiplication of two N-bit numbers,

$$A = (A_{N-1} \ldots A_3 A_2 A_1 A_0)$$

$$B = (B_{N-1} \ldots B_3 B_2 B_1 B_0)$$

by the conventional algorithm is carried out as follows. Two $2N$-bit storage registers are allocated, and number A is stored in bits $N-1$ to 0 of one of them (Register A), and bit locations from $2N-1$ to N are cleared (0's are stored). The second $2N$-bit storage register is cleared (Register C). The following steps are then repeated:

1. For integer i from 0 to $N-1$,
(a) if $B_i = 0$, $RegC \rightarrow RegC$ (no operation, partial product = 0).
(b) if $B_i = 1$, $RegC + RegA \rightarrow RegC$ (partial product $= A \times 2^i$).

2. Shift Register A left by one, and set bit 0 to 0.

This simple method of multiplication can be expedited using two sets of adders and registers, by which the even and the odd partial products are added in parallel. After all the partial products are exhausted, the results from the two adders are added together to generate the product. By this scheme $N \times N$-bit multiplication can be carried out by $(N/2)+1$ adder delays [39].

This procedure is a direct analogy of the method used in pencil-and-paper arithmetic. Booth's method is an improvement of this basic algorithm [40]. There are several versions of Booth's algorithm, but the following is the most interesting. If the number B consists of a section,

$$B = (B_{N-1}....01111110)$$

that is, if B contains a long sequence of 1's, B may be rewritten as

$$B = (B_{N-1}....10000000) - (0....00000010)$$

If this relationship is used, the first eight steps of multiplication can be simplified to the following:

1. Shift A by one position and

$$RegC - RegA \rightarrow RegC$$

2. Shift A further by six positions,

$$RegC + RegA \rightarrow RegC$$

In other words, reading the number B from the LSB end, and simultaneously shifting the number A to the left, if a transition $0 \rightarrow 1$ is found in the digit sequence of the number B, Reg A is subtracted from Reg C, and if a transition $1 \rightarrow 0$ is found, Reg A is added to Reg C. Since an ordinary number often contains a sequence of 1's, the method speeds up multiplication quite significantly.

To execute Booth's algorithm an adder must have the capability of subtraction and also the capability of shifting by an arbitrary number of positions at high speed. Booth's algorithm essentially reduces the number of required partial products. The reduced number of partial products can then be added by several adders in parallel, thereby further speeding up the multiplication.

$N \times N$-bit multiplication requires $f \cdot N$ ($f \approx 0.5$–1) steps of addition. If each addition is carried out using a carry-lookahead adder, the delay time is significant. A Wallace tree multiplier uses internal representation of numbers that consists of two components S (Sum) and C (Carry). If this special representation is used, a carry-lookahead adder is necessary only to produce the final answer of multiplication: The multiplier's internal operation can be carried out without propagating carries [41]. Wallace tree multipliers require a lot of gates and silicon area. The scheme is now used in some VLSI designs [42].

(B) Carry-Save Multipliers

A multiplier of this type operates on the following principle. When two numbers $(A_3A_2A_1A_0)$ (multiplicand) and $(B_3B_2B_1B_0)$ (multiplier) are given, the product of the LSB of the multiplicand and the entire multiplier,

$$(A_0B_3 \;\; A_0B_2 \;\; A_0B_1 \;\; A_0B_0)$$

is generated, and this is added in parallel to the initial content of the answer register [which was (0 0 0 0)] at each of the four bit positions. Then the data of the answer register are shifted by one bit to the right and added to the product of the second bit of the multiplicand and the entire multiplier,

$$(A_1B_3 \;\; A_1B_2 \;\; A_1B_1 \;\; A_1B_0)$$

at each of the four bit positions. This addition may generate carries. The carries are saved in latches at the same bit position. Then the sum is shifted to the right by one bit position while keeping the carry at the same position, and the sum, the carry (which is not shifted), and the product

$$(A_2B_3 \;\; A_2B_2 \;\; A_2B_1 \;\; A_2B_0)$$

are added together. The carry is saved, the sum is shifted, and the same operations are repeated until all the bits of the multiplicand are exhausted.

This operation requires N cycles of the clock, where N is the number of digits of the multiplicand. The answer can be automatically truncated to the number of digits of the multiplier, and this feature is sometimes convenient for digital signal processing applications. The multiplier can be assembled from the simple cell shown in Fig. 7.19 [43] [44]. The AND gate generates the partial product A_iB_j. The M/S latch C saves the carry of the current addition, and the M/S latch S sends the sum to the next location.

7.17 SHIFTERS

To shift an N-bit array of data one bit at a time, an ordinary shift register is used. The shift register can be assembled from cascaded CMOS D-latches in a master-slave configuration. The data are shifted by one bit position every time the M/S latches are

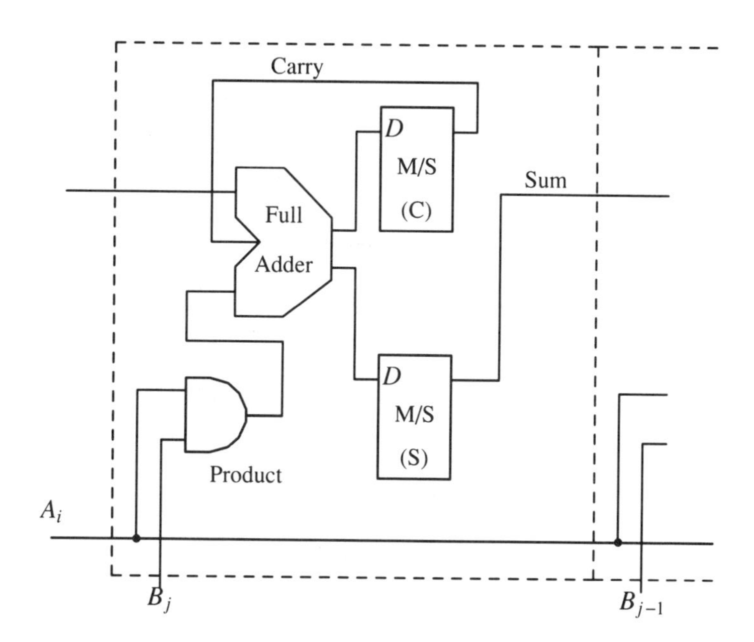

Figure 7.19 Carry-save multiplier.

clocked. If the shifter is a component of an ALU (Arithmetic-Logic-Unit), either the input or the output latches of the adder can be converted to a shift register and the shifting operation can be carried out there.

If there is a need for shifting an arbitrary number of bit positions within one clock cycle, a dedicated shifter/rotator called a barrel switch is used. Figure 7.20 shows the principle. In the figure, 6-bit data arrive from the bottom, and the rotated data exit from the right side. At each intersection of the input lines I_i and the output lines D_j there is a transmission gate that is controlled by the diagonal control line. All the transmission gates on a single control line (indicated by the dashed lines) open and close together. If control line S_1 is activated, data I_0 come out from port D_5, I_1 from D_0, and so on. Therefore, the data were rotated by one bit position. If the rotated data must be returned to the input port (this is the case where $I_0 - I_6$ is a data bus), the rotated data that emerge from port $D_0 - D_5$ are temporarily latched, and then the data drive back, by activating the control line S_0 (no rotation). A shifter of this structure can be advantageously located at the point of the databus where the signals branch off to feed the control part of a microprocessor. Then properly aligned data are available at the control area destination.

A rotator can be implemented using CMOS dynamic logic. Transmission gates are replaced by pass transistors, and there is no need to distribute shift control signals of both polarities. The output nodes are precharged by PFETs.

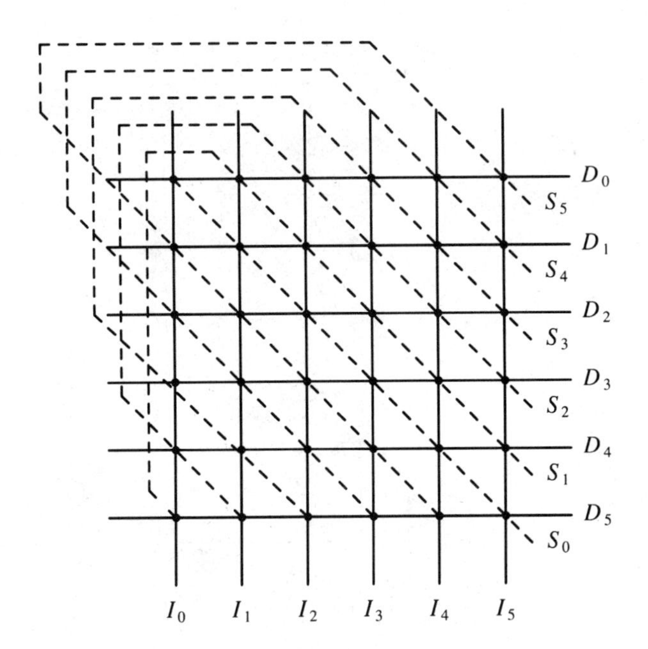

Figure 7.20 Barrel switch schematic.

Shift control signals are usually available in a binary-coded format. The coded signals must be decoded to activate one of the shift control lines, which costs time. The shifter shown in Fig. 7.21 uses the coded shift control signals directly. The circuit consists of complex Domino CMOS gates that steer the input data first by one bit position, then by two bit positions, and so on. In this way the coded shift control signal can be used directly to control each stage of the shifter. The circuit has a relatively small number of FETs but rather extensive wiring [45].

7.18 ERROR DETECTION AND CORRECTION

As the scale of integration increases and therefore the feature size of the technology decreases, the charge stored in a circuit node decreases. Already in 1.75-μm-feature-size CMOS, the charge stored in the simplest node is as small as an order of magnitude less than the charge deposited by the most powerful cosmic-ray ions. Under this condition, the only way to secure the integrity of a system working in a hazardous environment (e.g., exposure to strong electromagnetic interference and ionizing radiation) is to provide an architectural protection measure. At present, there is no way to provide error-detection/correction capability to a processing unit, except to provide more than one chip running concurrently and to take a vote on the results. Many processors now designed have the capability to operate under such an arrangement. In a data storage system such as a mass memory, however, there is a way to detect and

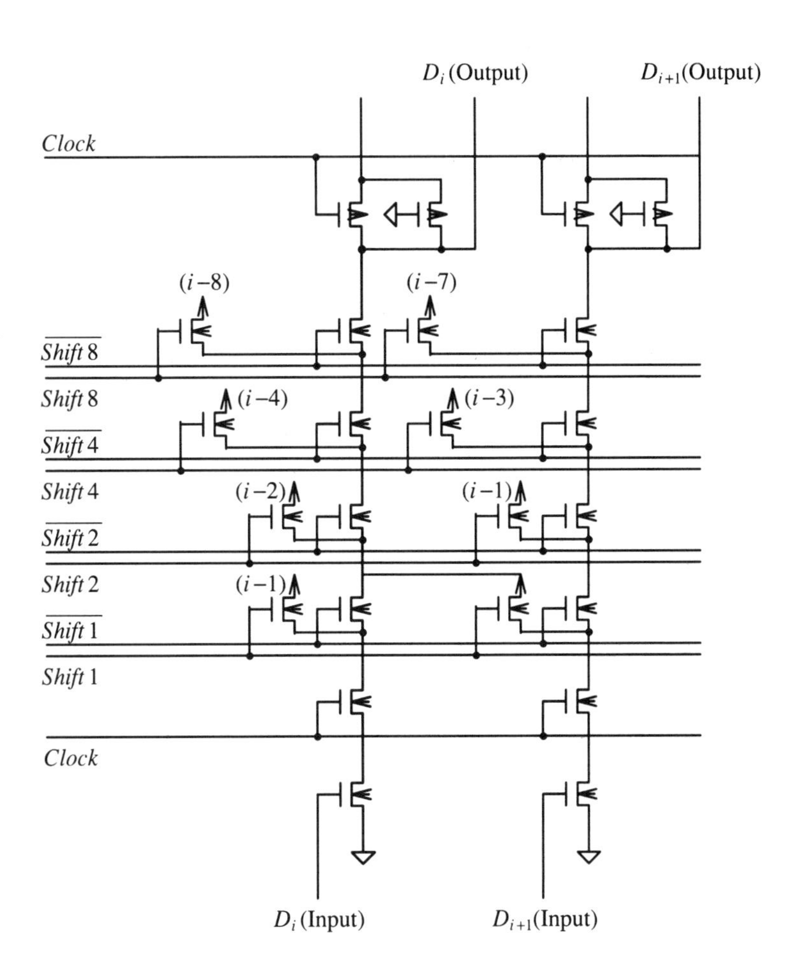

Figure 7.21 Domino CMOS barrel switch.

correct a small number of errors without duplicating the entire hardware. The technique is known as error-detecting/correcting codes (ECC) [46].

When a set of 32-bit data is stored in a memory for an extended period, the probability of one bit out of a set of 32 bits suffering from environmental influence, P, is small but significant. The probability that 2 bits of data within the same set of 32-bit data are in error is of the order of P^2: If P is a small probability, P^2 is insignificantly small. Ordinary ECC provides the capability to detect and correct single-bit errors and to detect and report 2-bit errors.

When a set of 32-bit data ($D_0 - D_{31}$) is stored in a memory, 7 extra bits, called check bits ($C_0 - C_6$), are produced from ($D_0 - D_{31}$), and they are stored together.

TABLE 7.2 PARITY TABLE OF ERROR-CORRECTING CODE

	Checkbit								Checkbit						
Bit	0	1	2	3	4	5	6	Bit	0	1	2	3	4	5	6
0	x	x	x	x	—	—	x	16	—	x	x	x	—	—	—
1	—	x	—	x	—	—	x	17	x	x	—	x	—	—	—
2	—	x	—	—	x	—	x	18	x	x	—	—	x	—	—
3	—	—	x	—	x	—	x	19	x	—	x	—	x	—	—
4	x	x	x	—	x	—	x	20	—	x	x	—	x	—	—
5	—	—	—	x	x	—	x	21	x	—	—	x	x	—	—
6	x	x	—	x	x	—	x	22	—	x	—	x	x	—	—
7	x	—	x	x	x	—	x	23	—	—	x	x	x	—	—
8	x	x	—	—	—	x	—	24	—	x	—	—	—	x	x
9	x	—	x	—	—	x	—	25	—	—	x	—	—	x	x
10	—	x	x	—	—	x	—	26	x	x	x	—	—	x	x
11	x	—	—	x	—	x	—	27	—	—	—	x	—	x	x
12	—	x	—	x	—	x	—	28	x	x	—	x	—	x	x
13	—	—	x	x	—	x	—	29	x	—	x	x	—	x	x
14	x	—	—	—	x	x	—	30	—	—	—	—	x	x	x
15	—	—	x	—	x	x	—	31	x	—	x	—	x	x	x

$C_0 - C_6$ are the parity bits of selected set of data bits. Table 7.2 shows how the parity is taken. Check bit C_0 is, for example, the parity of D_0, D_4, $D_6 - D_9$, $D_{11}, D_{14}, D_{17} - D_{19}, D_{21}, D_{26}, D_{28}, D_{29}$, and D_{31}. $C_0 - C_6$ and $D_0 - D_{31}$ constitute a total of 39 bits of data.

Suppose that the data were stored in the memory and that any one bit suffered from ionizing radiation. When $D_0 - D_{31}$ and $C_0 - C_6$, which contain one error are read from the memory and a new $C_0' - C_6'$ are generated from D_0–D_{31}, the following possibilities occur.

1. If any one bit in $D_0 - D_{31}$ suffered, either three or five disagreements between $C_0 - C_6$ and $C_0' - C_6'$ are found.
2. If any one bit in $C_0 - C_6$ suffered, one disagreement between $C_0 - C_6$, and $C_0' - C_6'$ is found.
3. If a 2 bit error occurred in $D_0 - D_{31}$ or $C_0 - C_6$, an even number of disagreement is found between $C_0 - C_6$ and $C_0' - C_6'$.

In case 1, by reading Table 7.2 backwards it is possible to identify the bit that suffered from the error. Using this technique, one bit error can be corrected.

The circuit that executes the error-detecting/correcting operation is a parity generator that can be implemented efficiently by Domino CMOS (Section 5.8). In the future such systems will be important in space and military systems [46].

ECC codes are now used in 1-Mbit dynamic memory. The ECC code used by Yamada et al. consists of partitioning a memory in $m \times k$ cells, and adding m horizontal and k vertical parity bits. The scheme has the advantage of easy error correction [47]. ECC are often used with a system maintenance strategy, thereby increasing the capability of detecting both hard and soft memory errors [48]. Error-correcting code functions are included in one disk controller chip [49].

7.19 DATABUS

A databus is a common medium for communication within microprocessors. The bus should be designed to allow signal transfer from any source to any destination with equal delay time. This requirement sets the objective for databus design. The simplest databus structure is shown in Fig. 7.22(a). The bus is driven by N tristatable drivers $D_1, D_2, \ldots, D_N$, of which only one drives the bus at any moment. If the equivalent resistance of FET of the driver is r_D, the equivalent internal resistance of the tristatable driver is $2r_D$, since two FETs are connected in series (Section 2.2). The driver has output capacitance C_D, each of N receivers has capacitance C_R, and the wiring of the bus has capacitance C_B. Capacitance C_B is practically capacitance of metal wire that stretches across the width of VLSI chip, and therefore it is independent of the number of devices communicating to the bus. Then the delay of a tristatable inverter to drive the bus, T_D, is given by

$$T_D = 2r_D \left\{ N(C_D + C_R) + C_B \right\} \rightarrow t_D N \quad \text{when } N \rightarrow \infty \tag{7.15}$$

where

$$t_D = 2r_D (C_D + C_R)$$

The delay increases linearly with the number of logic blocks that communicate through the bus. This dependence cannot be changed, whatever technique may be used. The problem is how to reduce the coefficient of N in Eq. (7.15). There are several ways to achieve the objective.

1. A driver that is tristatable but that has only one FET in the charge/discharge path. Delay t_D will decrease by a factor of 2. Such a driver is shown in Fig. 7.12. A bipolar-CMOS (BiMOS) driver is an excellent bus driver. A bus driver can be bootstrapped to increase the current (for details, Section 2.9).

2. Use a dynamic databus. The bus structure shown in Fig. 7.22(b) works as follows. During phase 1 the precharge PFET is turned on, while all the pulldown NFETs are turned off. The bus is precharged to V_{DD}. In phase 2, only one of the pulldown NFETs, say the ith NFET, is activated. If D_i is high, the NFET is turned on and the bus is pulled down. If D_i is low, the bus is left high. At the end of phase 2 the data are strobed into the destination. This scheme is quite effective since there is only one large PFET for pullup rather than N large PFETs for each driver, connected

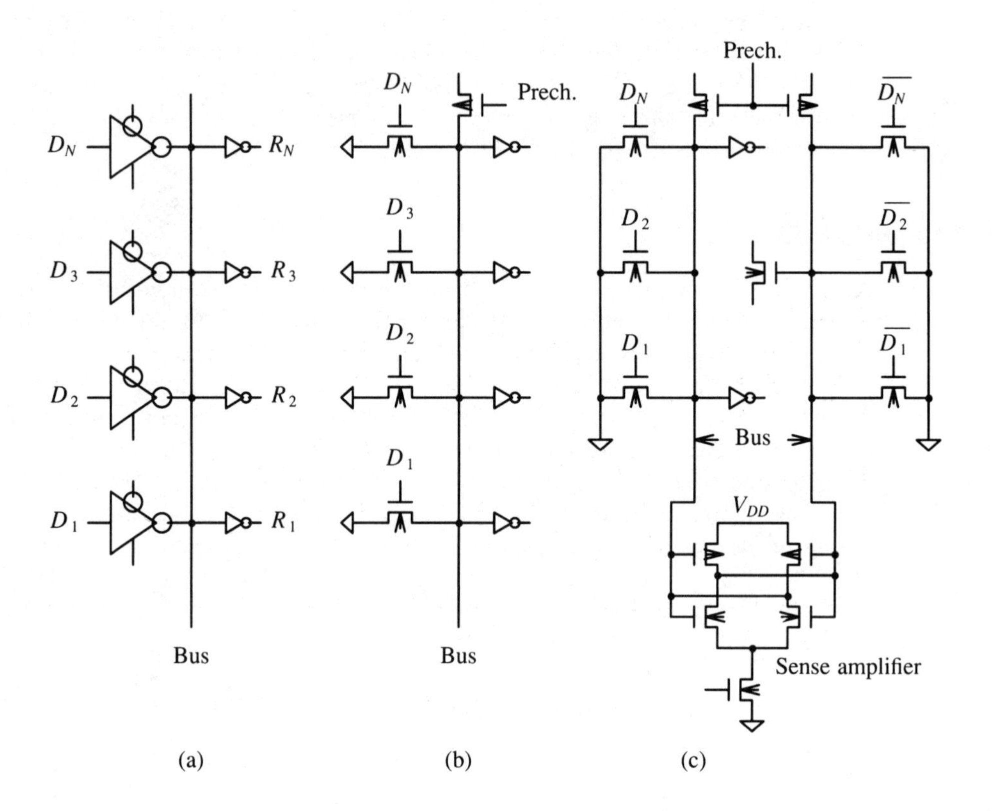

Figure 7.22 Structure of databus.

to the bus. Therefore, the bus capacitance contributed from the driver output capacitance is reduced from C_D to, approximately,

$$C_D \frac{1}{\beta+1} + C_D \frac{\beta}{\beta+1} \frac{1}{N} = C_D \frac{N+\beta}{N(\beta+1)} \rightarrow C_D \frac{1}{\beta+1} \; (N \rightarrow \infty) \tag{7.16}$$

Since $\beta = 2$ to 3 this is a very significant reduction in the bus capacitance. Most state-of-the-art CMOS microprocessors use a dynamic databus. However, a dynamic bus can be susceptible to noise. If many static latches strobe data in synchronously, the voltage of the precharged bus may be perturbed significantly. In such a case it is possible to design a static bus, with a greatly reduced PFET size in the tristatable drivers. Since the tristatable drivers cannot pull up fast enough, a central precharge is provided. Such a hybrid of a static and a dynamic bus has the advantages of stability against noise and fast drive capability.

3. Use a double-rail bus with sense amplifier. In a memory-intensive microprocessor architecture, a pair of wires that carry a data value and its complement can be used as a data bus, as shown in Fig. 7.22(c). In this design, a sense amplifier can be used to speed up the bus response. The bus is precharged while the sense amplifier is turned off. Some time later the sense amplifier is turned on, and thereafter the data level on the pair of wires settles very rapidly. The databus is often charged to a voltage lower than V_{DD} to accelerate the feedback action of the sense amplifier.

4. Use a negative capacitance generator. The details were discussed in Chapter 4.

5. Split a bus into several sections. Separating a databus into several sections is desirable for reducing bus transaction delay as well as for architectural flexibility to allow pipelining. A single pass transistor can be used to separate a dynamic bus into sections. Splitting a static bus requires a bidirectional bus driver, which is a cross-coupled tristatable inverter, only one of which is driving. Transmission gates are not desirable to connect two buses, since the heavy capacitive load of a static bus cannot be driven by the passive device.

The number of bus drivers can be reduced by multiplexing the input-output of more than two logic blocks, thereby reducing the number of devices connected directly to the bus.

7.20 MEMORY DEVICES

Circuit techniques originating from memory device development are expected to play an ever-increasing role in digital CMOS VLSI technology for two reasons: State-of-the-art VLSI chips often contain a significant amount of memory, and many circuit techniques invented to improve memory device performance have direct application to logic circuits in digital VLSI chips. The following four sections are intended to review memory circuit techniques that are considered useful in CMOS VLSI design, not as a comprehensive or a balanced presentation of memory circuit technology. An excellent review of MOS memory circuits is given by Hunt [50].

The performance of a memory device is measured by the silicon die area per bit of storage and the access time. Figure 7.23 shows access time versus storage cell area of static and dynamic memories, including the most recent designs. NMOS was the predominant memory technology of the 1970s. In the past 5 years, however, CMOS has entered the mainstream. CMOS design has the advantage of small power consumption for the peripheral circuits. Since CMOS is more area consuming than NMOS, the storage cells developed for NMOS are still used. It is customary to use mixed NMOS/CMOS circuits in the present-day high-performance MOS memories.

MOS read/write memories [random access memories-(RAM)] are classified into static and dynamic memories, as shown in Fig. 7.23. A dynamic memory requires periodic refresh to keep the data from deteriorating in the cell. A dynamic random access memory (DRAM) storage cell is a small area capacitor. Current design has 4-

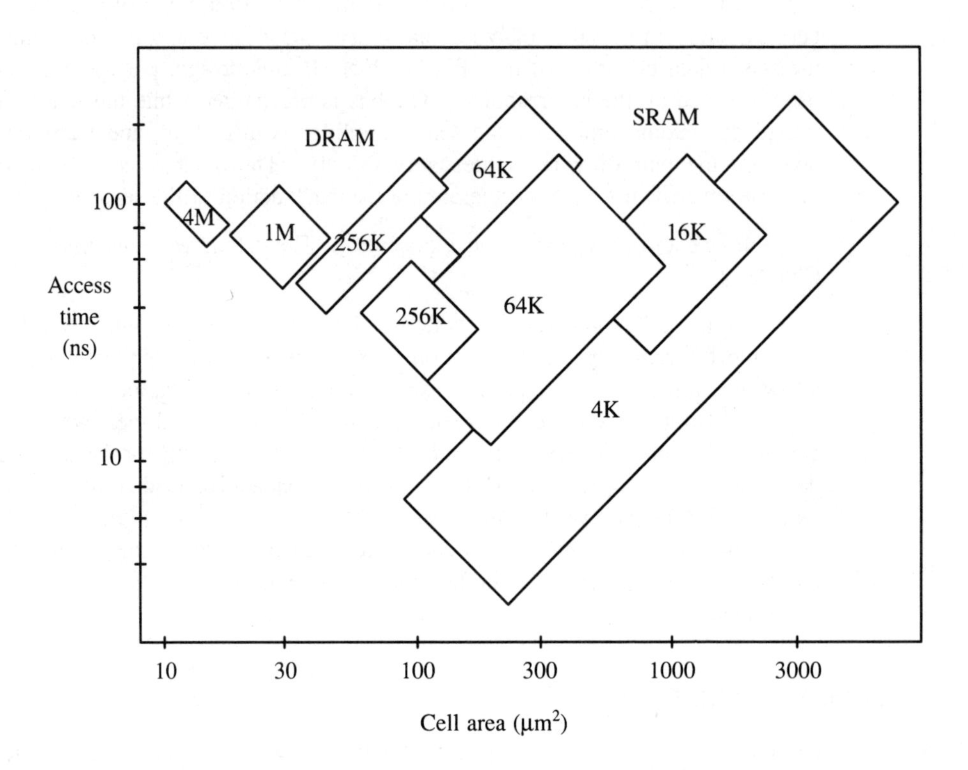

Figure 7.23 Cell area versus access time of MOS memories.

Mbit capacity. The DRAM is considered a milestone of VLSI technology. A review of DRAM technology has been given by Lewyn [51]. A static random access memory (SRAM) keeps the data as long as the chip power is on. A SRAM storage cell, a flip-flop, requires more area than the capacitor used in DRAMs, but its access time is short. A 4-K bit CMOS memory with a 5-nanoseconds access time has been reported [52].

A read-only memory (ROM) and electrically programmable read-only memory (EPROM) retain data even if the power supply is off. ROMs and EPROMs are static memories, but the difference between a ROM and a RAM is more fundamental: The information is stored as a permanent (or semipermanent in case of EPROM) current path in a ROM and EPROM that cannot be destroyed by accessing the cell, but in a RAM the information is stored as a voltage level that can be destroyed by access. A ROM can be read out reliably, by a simple current sensing circuit. To read a RAM cell, a voltage sensing must be exercised carefully so that the cell data are not destroyed, or a mechanism to rewrite the read data back into the cell must be provided.

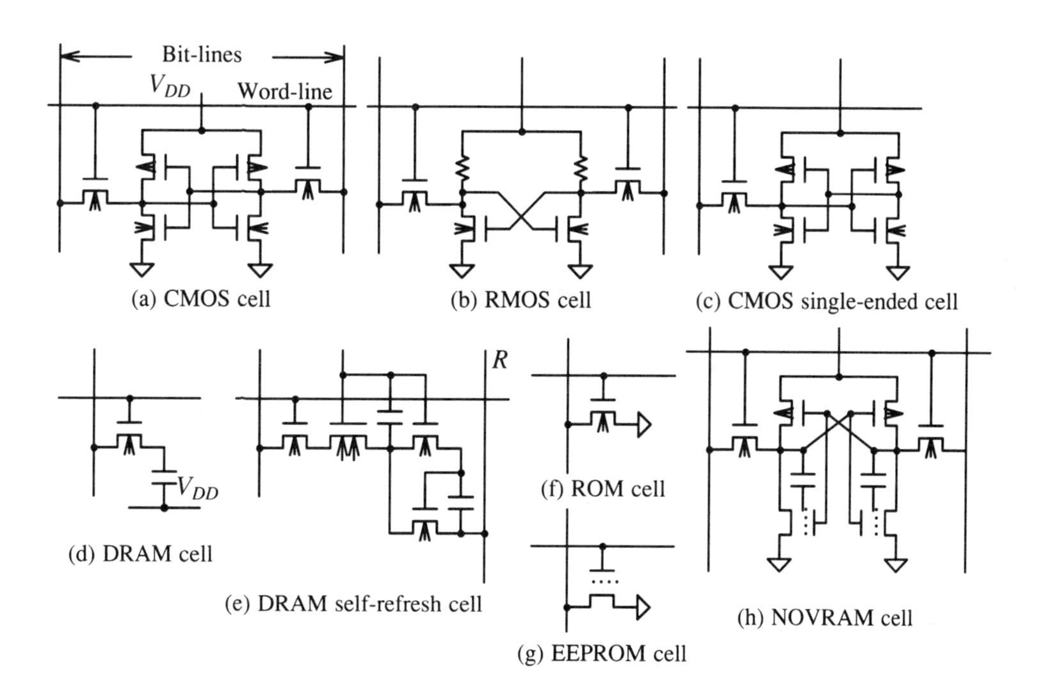

Figure 7.24 Static, dynamic, and read-only memory cells.

7.21 MEMORY CELLS

Figure 7.24 shows various MOS memory cells. Figure 7.24(a) is a standard, six-transistor CMOS static memory cell. To read data, the pair of bit lines are first precharged, while the word line is held low. As soon as the word line is driven high, the pair of pass transistors connect the bit lines to the cell. Depending on the state of the cell, one of the bit lines is pulled down and the other is pulled up, thereby placing the stored data on the bit lines. The readout process does not disturb the state of the memory cell if the ratio of the pass FET size to the pulldown FET is small enough.

To write data to the cell, the word line is pulled up, and complementary values are placed on the bit lines. The write process works if the ratio of the size of the pass FET and the FET in the cell is more than a certain minimum. FET sizes must be optimized for reliability of read/write operation, and for performance.

The CMOS memory cell has no standby power. If the PFETs in the cell are replaced by depletion-mode NFETs silicon area is saved but the cell dissipates standby power. A modern NMOS cell uses a pair of undoped polysilicon resistors, typically 10 $g\,\Omega$ each for pullup, as shown in Fig. 7.24(b) [53]. The resistor's characteristic is very complex [54]. The size ratio of the FETs must still be observed, and generally the design to obtain a wide parameter margin is difficult if very little standby power is required.

The CMOS cell area can be reduced by removing one of the bit lines as shown in Fig. 7.24(c) [55], and by contacting polysilicon directly onto the diffusion islands (polycon, or buried contact [56]). The cell design is more difficult to secure margin of operation than the design of a conventional cell.

Figure 7.24(d) shows a dynamic memory cell [57]. The far end of the storage capacitor is usually held at V_{DD}. The pass transistor was exclusively an NFET in earlier designs, but PFETs are now used frequently because of their tolerance to the high-field reliability problem. If NFETs are used, they are surrounded by a grounded P-tub. Then the floating drain node of the NFET discharges to ground potential. If the far end of the storage capacitor is at V_{DD}, the storage capacitor charges up after a long time. The charged state is stable, but the discharged state is not. The time required for a discharged capacitor to charge up to a significant level is of the order of several tens of milliseconds, or more. Before such a point is reached, the correct potential of the cell must be restored. The operation is called refresh. The cell shown in Fig. 7.24(e) does the refresh every time the refresh line, R, is driven up and down [58]. The refresh operation is transparent to the cell access.

Modern high-capacity DRAM cells use a trenched capacitor [59] which is formed on the wall of a vertical hole cut into the silicon substrate.

Figure 7.24(f) shows a ROM cell. The presence or absence of a transistor at the word-line/bit-line intersection determines the stored value. Figure 7.24(g) shows an EPROM cell. The FET has a floating polysilicon gate between the channel and the control gate. An EPROM cell can be written by biasing both the word and the bit-lines significantly above V_{DD}, thereby inducing avalanche breakdown of the drain PN junction, and by steering the hot electrons to the floating gate. A similar FET is used in a standard six-transistor cell of Fig. 7.24(h) to design a nonvolatile RAM cell (which retains information even if power is turned off, but otherwise works in the same way as SRAM) [60]. The cell stores the data permanently by pulsing V_{DD} momentarily above the normal level.

A serial-parallel ROM combines very high storage density and modestly fast access time. In Fig. 7.25, PFET MP1 precharges the output node of the block of series-connected cell strings to V_{DD}. As the precharge ends one of the vertical column select lines B_i is pulled up and one of the FET columns is connected to the output node. Simultaneously, all but one horizontal select lines are pulled up. If the FET on the selected column under the selected line (whose gate was not pulled up) is a depletion-mode NFET, the output line is pulled down. If the FET is an enhancement-mode FET the output node remains at V_{DD}. In this way the content of the ROM is read [61]. CMOS technology must be supplemented with a depletion-mode NFET made by using a threshold adjust implant. The number of columns, N, must be large enough to provide sufficient capacitance on the output node to hold the potential of the output node against charge sharing. The length of the column cannot be too long, to prevent the charge sharing as well as to avoid a long delay for readout. Since one bit

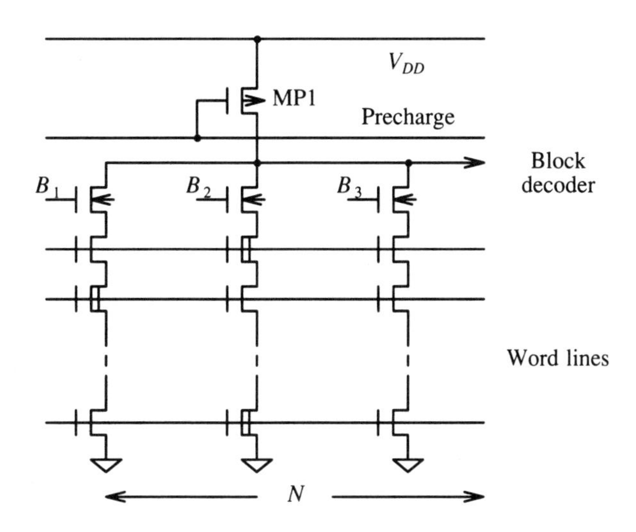

Figure 7.25 Series ROM.

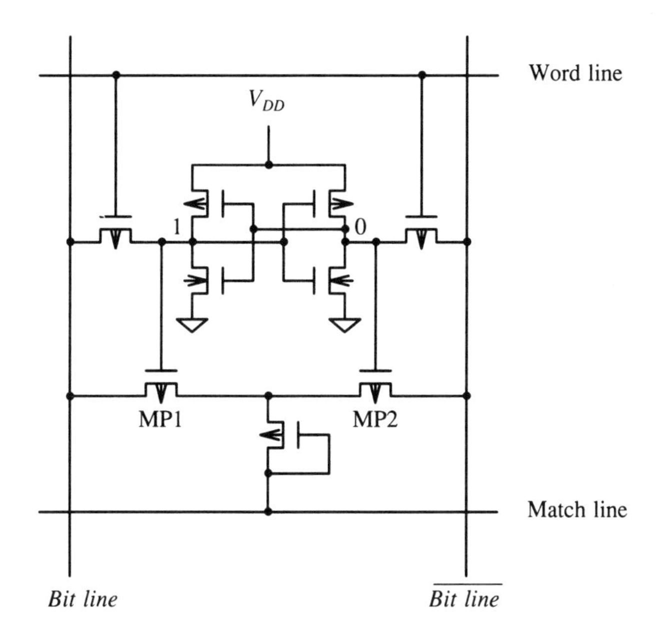

Figure 7.26 Content-addressable memory cell.

of data can be stored at every intersection of the horizontal polysilicide lines and the vertical diffusion lines, the ROM attains the maximum data storage density the processing technology offers.

A content-access memory (CAM) stores a data item consisting of a tag and a value. Instead of giving an address, a data pattern is given to the tag section of the CAM and whether or not an item that has the specified tag exists in the memory is determined. If such an item exists, the memory sends out the data. A CAM cell must also be readable and writable just as an ordinary memory cell.

The tag part of a CAM cell is shown in Fig. 7.26. Identical cells are repeated in a horizontal array that has a common *Matchline*, to obtain the necessary number of tag bits for each item. Suppose that the stored tag bit is "1" if node 1 is high. Before operation, the *Matchline* is predischarged to ground. To check whether or not a tag bit "1" is stored in this cell, *Bitline* is held high, and $\overline{\mathit{Bitline}}$ is held low. If node 1 is high, PFET MP1 is off, and therefore *Matchline* remains low, indicating coincidence at this bit location. If the stored data in the cell are low, however, *Matchline* is pulled up by MP1 and only one such pullup indicates that coincidence was not found. To check whether or not a tag bit "0" is stored in the cell, $\overline{\mathit{Bitline}}$ is held high and *Bitline* is held low. If "0" is stored in the cell, node 0 is high, and therefore *Matchline* stays low, indicating a hit. CAMs are most frequently used in virtual/physical address translation [62]. Dynamic CAM cell was reported recently [63].

7.22 MOS STATIC MEMORY

Figure 7.27 shows a block diagram of an MOS static memory [50]. The chip has 2^{2N} storage locations, any of which can be accessed by $2N$ bits of address. The address is given conventionally in TTL logic levels. TTL-to-CMOS level converters, discussed in Sections 2.9 and 8.7, are used to interface the signals. N bits of the address data are the ROW address, and the rest are the COLUMN address, where the ROW address selects one word-line from 2^N word lines and the COLUMN address selects one bit-line pair from 2^N bit-line pairs. At each intersection of the word-line and the bit-line pair, a static memory cell is placed, of the type shown in Figs. 7.24(a) to (c).

Decoding the N-bit address into 2^N word-lines is carried out using static or dynamic gates. NMOS memory chips use exclusively NMOS static NOR gates, shown in Fig. 7.28 [52]. CMOS memories use modified NMOS NOR gates, whose depletion load devices are replaced by PFETs. CMOS static NAND or NOR gates are too slow and too bulky for this application. Dynamic row decoders are used in recent designs. Dynamic decoders have advantages of speed, low power, and easy design.

Decoding of the column address to select a bit-line pair is carried out by static gates. Once the row address is entered into the cell array, a set of 2^N data from the selected row appears at the cell array outputs. Cell array outputs are 2^N pairs of bit-lines, all of them are precharged and the voltages of the two lines of each pair equalized before the word-line is pulled up. One line of the bit-line pair is pulled down and the other is pulled up by the accessed cell, and the differential signal appears at the

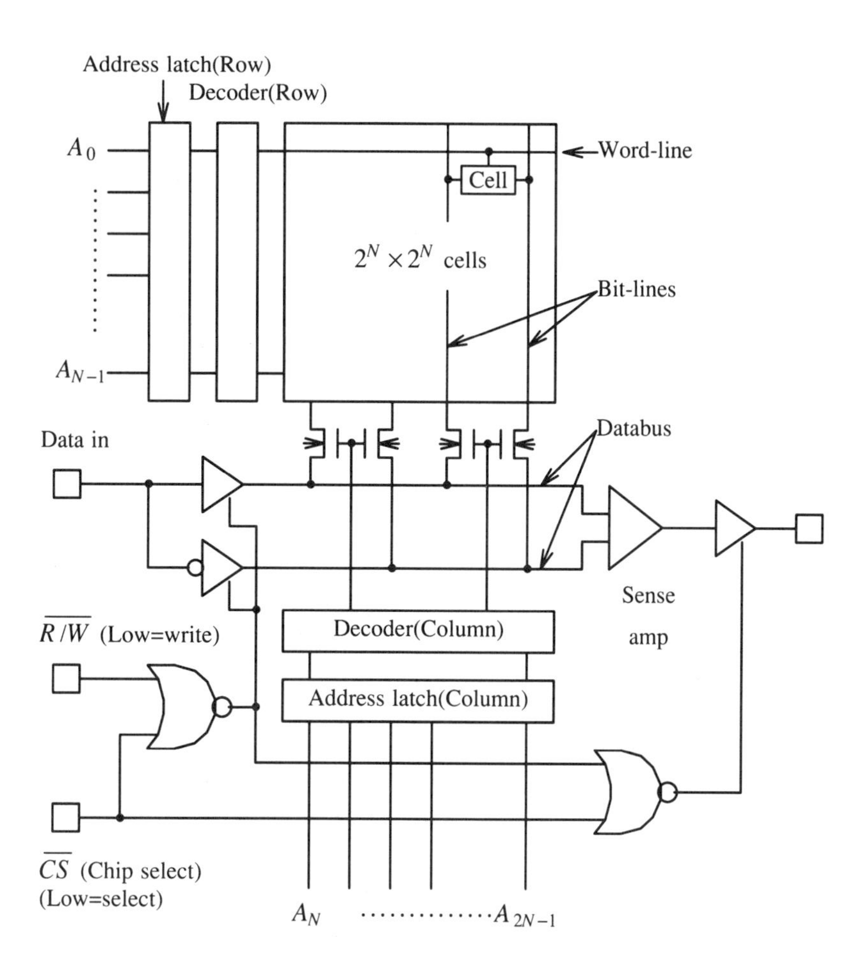

Figure 7.27 MOS static memory, block diagram. (Reprinted by permission of John Wiley & Sons, Inc.)

cell array output. Low-capacity MOS memories were practically all $2^{2N} \times 1$ bit architecture. One bit out of the 2^N column bit is selected by the column address. This is not an efficient way of using the data read by the time-consuming row access. One way of using the set of 2^N column data efficiently is by rapidly switching the column address while maintaining the row address unchanged. To make the best use of this mode of operation, it is convenient not to use dynamic column address decoders.

The cell data connected to the databus pair is amplified by a differential amplifier, often called a sense amplifier. The sense amplifier is a cascaded one-to-three stage differential amplifier. Sophisticated techniques are used to improve

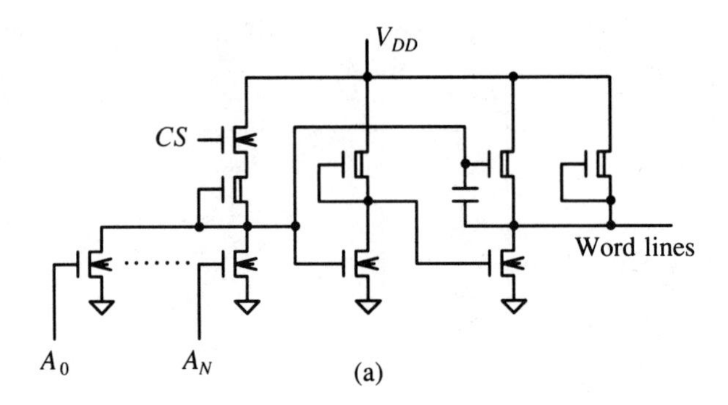

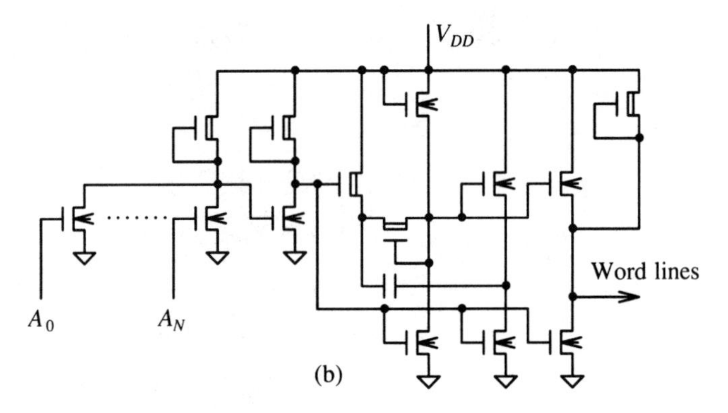

Figure 7.28 Address decoder and word-line driver.

sensitivity, delay and stability (Section 7.24). In some designs a differential preamplifier is provided for each bit-line pair.

The data to be written in the memory cell are driven into the databus pair by a pair of tristatable drivers that are controlled by the read/write signal $R/\overline{W}$. The write data drives the bit-line pair through the selected pass transistors.

Figure 7.27 shows the block diagram of a simple MOS static RAM of the 1970s. Recent static RAMs are very sophisticated. Some features of recent static RAMs that are different are summarized in the following. Modern, high-capacity static or dynamic memories are never organized in a single $2^N \times 2^N$ cell array. The memory is subdivided into $x = 2^{(N_R + N_C)}$ blocks of $2^{N - N_R} \times 2^{N - N_C}$ cells each, where x is the number of the blocks and N_R and N_C are small integers. In 1-Mbit-4-Mbit DRAMs, x is about 16. Since a word line goes over the channel of the pass transistors of the memory cells, a word line must be a polysilicon or a polysilicide wire. Unless the high-resistivity interconnect is strapped by a second level of metal wiring, the resistance-capacitance delay of the word line is quite significant. Second-level metal is area consuming. Division into many small arrays minimizes the delay.

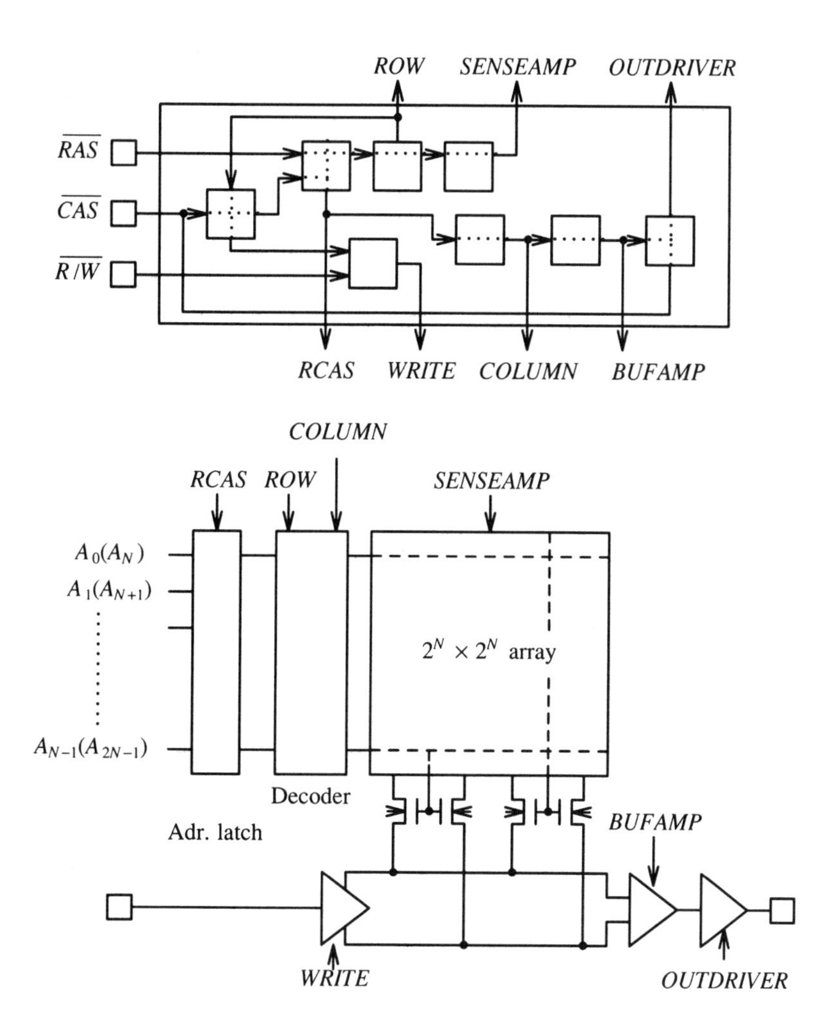

Figure 7.29 MOS dynamic memory, block diagram. (Reprinted by permission of John Wiley & Sons, Inc.)

When the array is split, some predecoding of the address is carried out. Address predecoding has the second objective of reducing the size of the address decoder, which must fit in the unit memory cell width. The low power advantage of divided memory arrays is exploited in recent CMOS designs [64] [65]. Only the address decoders and the peripheral circuits of the accessed subarray are activated by the clock and, unless clocked, the CMOS peripheral circuits consume no power.

The clocks that drive the recent memory circuit are generated by a circuit that detects a change of the address [66]. The circuit consists of an exclusive-OR gate at each address bit position. One input of the exclusive-OR gate is the address bit and the other is the delayed address bit. When the address bit changes, the exclusive-OR gate generates a pulse whose width equals the delay. When all the outputs from the exclusive-OR gates are-ORed, an address transition pulse is generated, which initiates a sequence of activities in the chip; bit line precharge, dynamic row address decoding, and pullup of the selected word line.

7.23 MOS DYNAMIC MEMORY

Dynamic memory is always considered as the front line of the largest scale of integration [67] [68] [69]. Figure 7.29 shows a block diagram of an MOS dynamic memory [50]. A capacitive memory cell connected to the bit line is able to change the bit line voltage only very slightly, typically only 100 to 200 mV in memories of 1-Mbit capacity. This small voltage must be amplified to recover a full logic level, and this process takes a long time. Therefore, it is unnecessary to input the column address at the same time as the row address. The row and the column adress are multiplexed, and they are input from the same address port, as shown in Fig. 7.29. The address decoders are also shared by the row and column addresses. An example of address decoder and word line driver are shown in Fig. 7.28.

The sequential input of addresses requires two address strobes, the Row Address Strobe $\overline{RAS}$ and the Column Address Strobe $\overline{CAS}$. The address buffer, the address decoder, and the rest of the circuits are operated by a complex sequence of pulses generated from $\overline{RAS}$ and $\overline{CAS}$. The time relationship of the pulses is maintained against process variations by the signal flow relationship in the clock generator circuit: The circuit is a chain of logic in which one clock edge generates the others, as shown in Fig. 7.29. Such a clock generation scheme is called an interlock [70].

In a dynamic memory, the charge stored in the capacitor determines the overall performance of the memory. Since the capacitor is charged by a source follower of the pass transistor, the maximum voltage stored in the capacitor is $V_{DD} - V_{TH}$ if the gate of the pass transistor is driven by a regular MOS level signal, where V_{TH} is the threshold voltage of the FET. Approximately 30% less charge is stored in the cell than the maximum possible charge (capacitance times V_{DD}). To avoid this problem, the word lines are driven above V_{DD} when the data are written.

The signal propagation delay along a word line is a significant problem in DRAM design. A solution to this problem is to split the word line into several segments, as shown in Fig. 7.30(a) [71]. The global word line is aluminum, local word line sections are polysilicon, and the structure is called a divided word line [72].

The idea of segmentation can also be applied to a pair of bit lines as shown in Fig. 7.30(b). The delay of a bit line is limited by the time constant of the devices that are connected to the bit line, and no delay reduction is expected by scaling the device

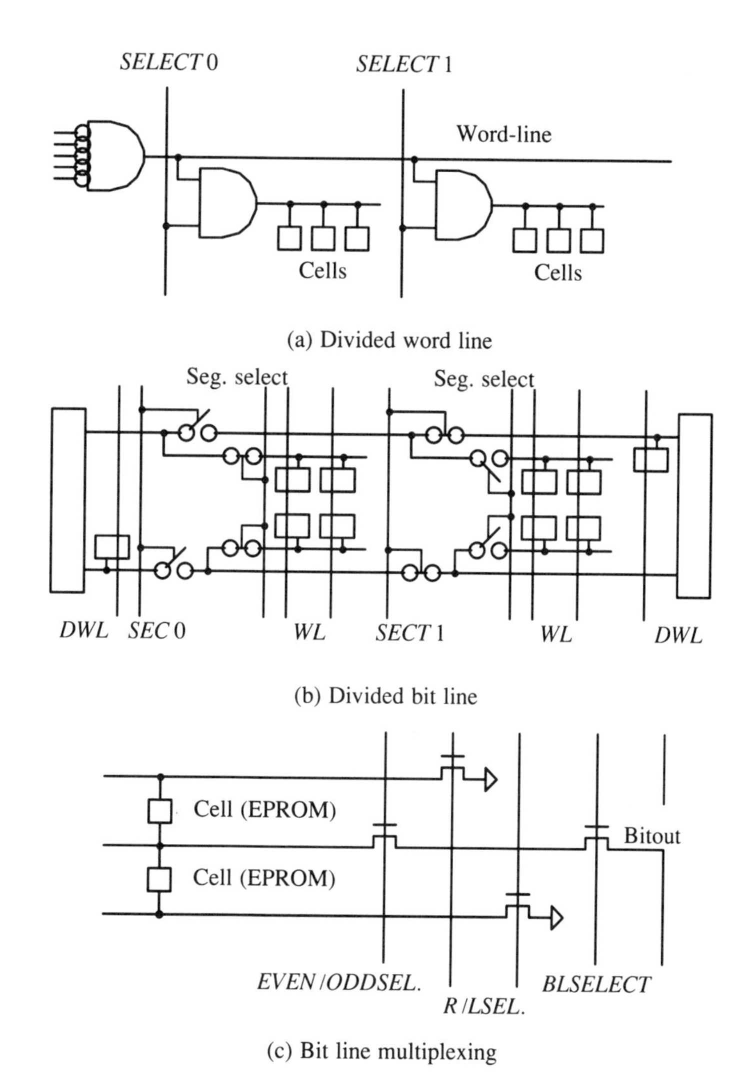

Figure 7.30 Various memory speedup techniques.

(Section 7.19). Segmentation is very effective in reducing the delay [73]. A similar scheme of bit line multiplexing is shown in Fig. 7.30(c).

7.24 SENSE AMPLIFIER

Figure 7.31 shows the circuit of a sense amplifier and bit line of a DRAM. Storage cells C(L) and C(R) (there are many of them in parallel) have capacitance C. Before access, precharge signal *PRE* is pulled up. The signal is bootstrapped (driven above

V_{DD}) and therefore bit lines BL(L) and BL(R) are both charged to V_{DD}, and the lower plates of the two dummy cells whose capacitance is half, (C/2), are connected to ground. The two bit line potentials are equalized by NFET MN1.

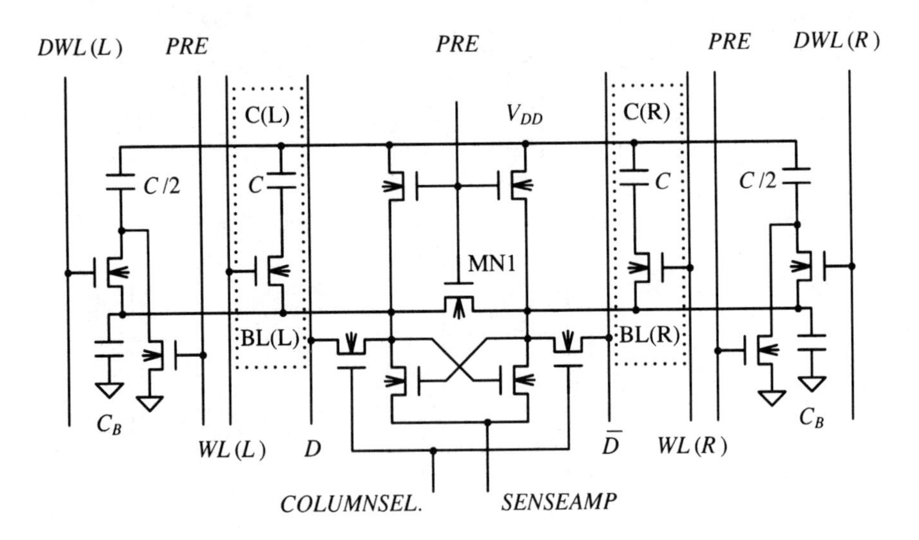

Figure 7.31 Sense amplifier, bit lines, and cells.

The capacitances of bit lines BL(L) and BL(R) are equal at C_B. If cell C(L) is accessed, $WL(L)$ and right dummy word line $DWL(R)$ are pulled up. Then the potential of the right bit line BL(R), $V(R)$, is

$$V(R) = \frac{C_B}{C_B + (C/2)} V_{DD}$$

and the potential of the left bit line BL(L) when the cell stores data 1 (potential V_{DD}), $V(L1)$, is

$$V(L1) = V_{DD}$$

and the bit line potential when the cell stores data 0 (the ground potential), $V(L0)$, is

$$V(L0) = \frac{C_B}{C_B + C} V_{DD}$$

Then

$$V(L0) - V(R) = \frac{-(C/2)C_B}{\Delta} V_{DD}$$

is negative and

$$V(L1) - V(R) = \frac{(C/2)}{C_B + (C/2)} V_{DD}$$

is positive, where $\Delta = (C_B + C)(C_B + (C/2))$. The difference is amplified by the differential amplifier to restore the digital voltage level. This half-capacitance dummy-cell method automatically provides the reference voltage to the amplifier and equalizes capacitive coupling to the bit lines. When the bit line potential settles to the final voltage, reflecting the logic level, the storage cell potential is also set at that level, by bootstrapping the word line WL(L) above V_{DD} potential. If the original cell potential was not exactly 0 or V_{DD}, the cell voltage is restored to 0 or V_{DD} by the read operation. Every time the cell is read, the cell potential is *refreshed*.

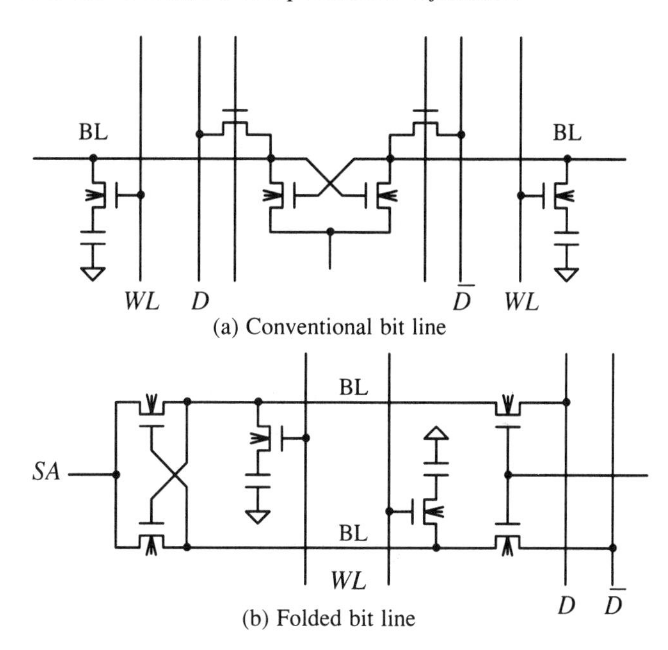

Figure 7.32 Conventional and folded bit line structure.

In the earlier DRAM layouts, bit lines on the left and on the right sides were placed physically to the left and to the right side of the sense amplifier, as shown in Fig. 7.32(a). In recent designs the bit lines are folded and are placed side by side, as shown in Fig. 7.32(b) [74] [75]. This configuration allows better balancing and less capacitance coupling to the sense amplifier. The sense amplifier and the pass transistors to the databus can then be placed at the opposite end of the bit line pair.

A number of techniques are used to speed up sensing. Bit lines are precharged to half V_{DD} voltage to improve sense amplifier response, and simultaneously to avoid current spikes [76]. Figure 7.33(a) shows application of a small voltage spike when the sensing begins. The small pulse, applied through the capacitors to both inputs of the differential amplifier, brings the sense amplifier to a region of higher transconductance, thereby increasing the gain-bandwidth and reducing the time needed to generate valid logic levels [50]. The circuit of Fig. 7.33(b) uses a negative capacitance that appears at the input of the differential amplifier to reduce the capacitance of the input circuit [77].

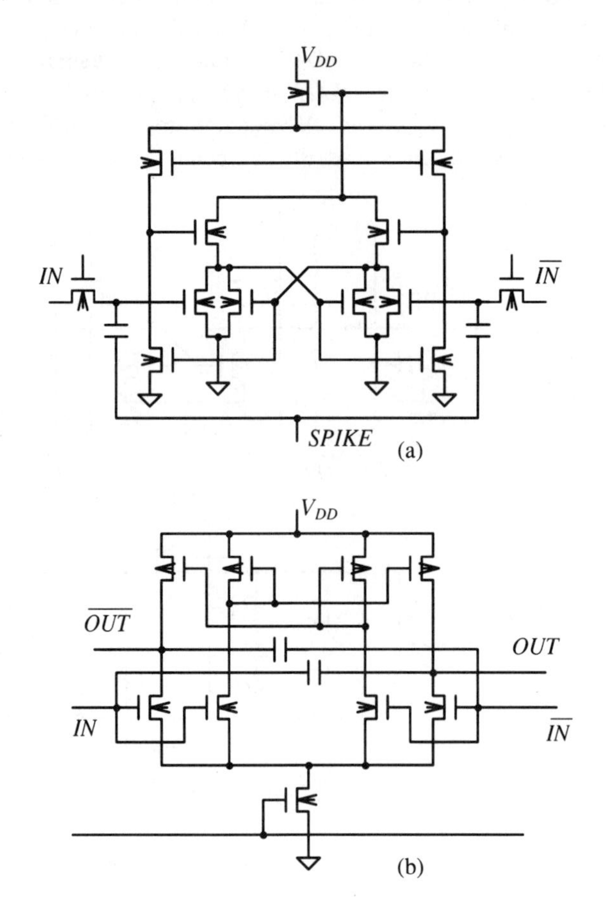

Figure 7.33 Sense amplifier speed-up techniques.

Bipolar sense amplifiers have better characteristics than CMOS amplifiers [65]. More than one stage of sense amplifier is often used [78].

Sense amplifier sensitivity is limited by NFET mismatch. Mismatch in the threshold voltage of nearby FETs amounts to 10 to 20 mV. The circuit shown in Fig. 7.34 compensates for this mismatch. First clock ϕ_P is pulled up, and nodes N_1 and N_2 are precharged to the *same* voltage, close to V_{CC}. Clock ϕ_T is low, and therefore BL1 and BL2 are precharged through NFET M1 and M2, respectively, to voltages that

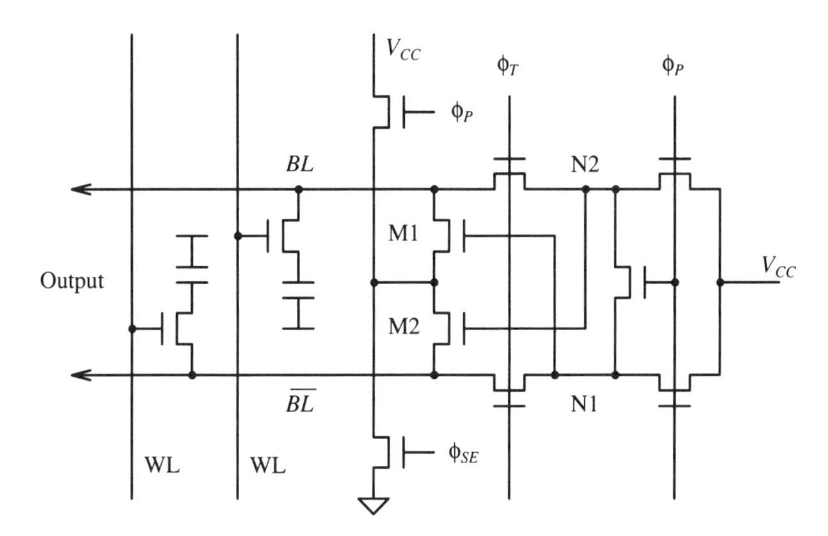

Figure 7.34 Compensation of FET threshold voltage.

reflect this *difference* in the threshold voltages of these NFETs. If M1 has a higher threshold voltage than M2, BL1 is precharged to a lower voltage than $BL2$. When clock ϕ_T is pulled up, the sense amplifier starts driving the bit lines from the voltage that reflects the threshold voltage difference. The circuit therefore automatically compensates for the threshold voltage difference [79].

Figure 7.35 shows simple sense amplifiers used in ROMs and EPROMs [80]. The amplifiers use either simple positive feedback, or a negative feedback technique to bias the single-ended amplifier at the most sensitive biasing condition. If read timing is not critical ROM can be read without using a sense amplifier. The bit line is precharged, and a word line is pulled up. When the bit line voltage is settled, it is strobed into a CMOS latch.

7.25 LOGIC BLOCK-TO-LOGIC BLOCK SIGNAL TRANSFER

A CMOS VLSI chip consists of logic blocks that execute their assigned functions. Their terminals are wired together when the chip is assembled. Signal propagation delay of the wiring channels is usually quite significant, and if a long polysilicon or polysilicide interconnect is used as a crossover, the delay cannot be reduced significantly, even by increasing the driver size. It is therefore necessary to investigate schemes that are least dependent on the inter-logic-block signal delay [30].

In the top figure of Fig. 7.36, a signal originates from block A and goes to block B, is combined with signals originating in block B, transferred to block C, and further combined with signals from block C. The logic operation requires two sequential

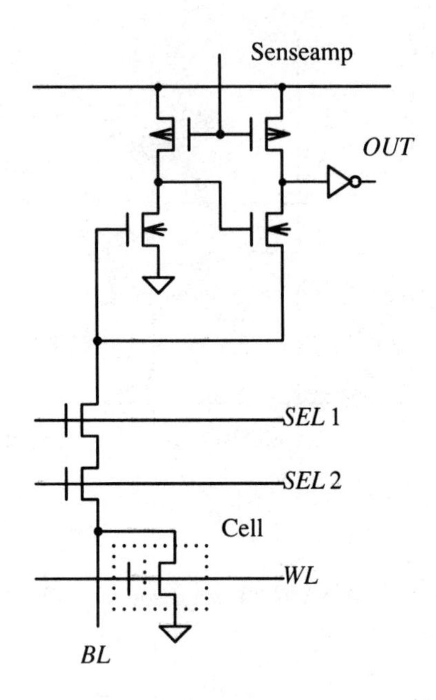

Figure 7.35 Simple sense amplifier.

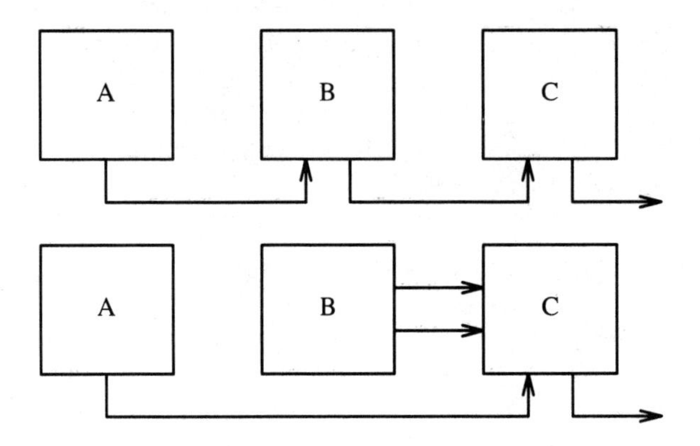

Figure 7.36 Method of reducing signal transfer delay.

block-to-block signal transfers. If some logic circuits are moved from block B to block C and if necessary, some logic circuits are duplicated in block B, the reconfigured logic shown in Fig. 7.36 (bottom) executes the same logic much faster, since only one (simultaneous) block-to-block signal transfer is involved. Generally, the designer is not concerned about increased power consumption by logic duplication in the process of optimization of the critical path delay. It is practical to spend a little more power or silicon area to attain much higher speed.

The signal of Fig. 7.37 has a large fanout. Such a signal is almost always an enabled/disabled clock signal. If the signal is wired in a straightforward way, as shown by the top figure, the delay is significant because of large fanout. An examination usually shows that some destination like B requires the signal immediately while the rest C,..., Y, Z can wait for some time. Then it makes sense to separate the signals into two, and to buffer the signals that have large fanouts and that are not immediately needed, thereby reducing the delay to block B.

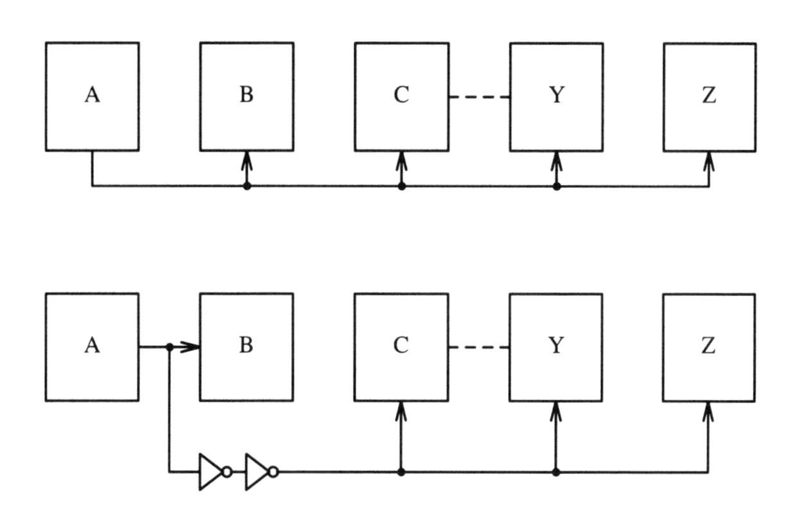

Figure 7.37 Selective buffering of high fanout signals.

7.26 CHIP PERFORMANCE IMPROVEMENTS AND ESTIMATING THE PERFORMANCE

One of the hardest parameters for a VLSI designer to predict is the clock frequency of a VLSI chip. It is a common practice of the microprocessor industry to introduce a new generation device at a significantly lower clock frequency than the ultimate frequency in the datasheet. It usually takes some time to improve the chip speed to the ultimate frequency in the datasheet. This is, however, an expensive development process; since any layout change of a VLSI chip requires that all the design verification and some chip reliability tests be redone. If the first design achieves the performance objective, considerable cost can be saved. This is especially true when the operating frequency of chip is specified by the user. Following are the key points of the project management.

The baseline requirement is the existence of well-synchronized clocks. In a large project such as a VLSI development, a designer usually does not have time to accurately tailor the clock delays and the gate delays. The most practical way is to synchronize clocks as much as possible and to make only a small number of manageable exceptions for delayed or advanced clocks. If well-synchronized clocks exist, it is possible to design a VLSI chip at the desired frequency if the following conditions are satisfied [30].

1. The functional subsystems, such as memories, adders, shifters, PLAs, ROMs, and I/O, meet the basic timing specification.
2. The random logic does not have excessively long chains of logic.

If the two conditions are strictly met, the major design task is to scale the gates carefully and be cautious not to cause the gate's performance to deteriorate by excessive interconnect resistance. This design method has the advantage of easy design check by simple CAD tools.

It is usually safe to do simulations for the worst case: The low-current process, high temperature, and the lowest power supply voltage. If only one type of product is required to meet the timing specification, this is the way.

However, if the design objective is to produce as many high-frequency chips as possible, it should be kept in mind that not all critical paths speed up as the process speeds up. A critical path that contains a lot of resistive wiring parasitics does not speed up at the same rate that FET limited paths speed up. It is important that the critical paths of this type be thoroughly cleaned up before a design is committed to silicon.

7.27 REFERENCES

01. Y. Yano, J. Iwasaki, Y. Sato, T. Iwata, K. Nakagawa and M. Ueda "A 32b CMOS microprocessor with on-chip virtual memory managemrent" ISSCC86 Digest, pp. 36-37, February 1986.
02. A. D. Lopez and H. S. Law "A dense Gate-Matrix layout method for MOS VLSI" IEEE J. Solid-State Circuits, vol. SC-15, pp. 736-740, August 1980.
03. S. L. Hurst "Custom-specific integrated circuits" Marcel Dekker Inc, New York, 1985.
04. T. Sano, S. Matsukura, K. Hashimoto, Y. Ohuchi, O. Kudo and H. Yamamoto "A 20ns CMOS functional gate array with a configurable memory" ISSCC83 Digest, pp. 146-147, February 1983.
05. T. Saigo, H. Tago, M. Shiochi, T. Hiwatashi, S. Shima and T. Moriya "A 20K-gate CMOS gate array" ISSCC83 Digest, pp. 156-157, February 1983.
06. M. Takechi, K. Ikuzaki, T. Itoh, M. Fujita, M. Asano, A. Masaki, and T. Matsunaga "A CMOS 12K-gate array with flexible 10Kb memory" ISSCC84 Digest, pp. 258-259, February 1984.

07. T. Itoh, M. Takechi, M. Fujita, K. Ikuzaki "A 6, 000-gate CMOS gate array" ISSCC82 Digest, pp. 176-177, February 1982.
08. Y. Takayama, S. Fujii, T. Tanabe, K. Kawauchi, T. Yoshida and K. Yamashita "A 1ns 20K CMOS gate array series with configurable 15ns 12K memory" ISSCC85 Digest, pp. 196-197, February 1985.
09. H. Takahashi, S. Sato, G. Goto, T. Nakamura, H. Kikuchi and T Shirato "A 240K transistor CMOS array with flexible allocation of memory and channels" ISSCC85 Digest, pp. 124-125, February 1985.
10. T. Kobayashi, H. Tago, T. Moriya and S. Yamamoto "A 6K-gate CMOS gate array" ISSCC82 Digest, pp. 174-175, February 1982.
11. K. Sakashita, M. Ueda, T. Arakawa, S. Asai, T. Fujimura and I Ohkura "A 10Kgate CMOS gate array based on a gate isolation structure" IEEE J. Solid-State Circuits, vol. SC-20, pp. 413-417, February 1985.
12. A. Hui, A. Wong, C. Delloca, D. Wond and R. Szeto "A 4. 1K gates double metal HCMOS sea of gate array" CICC85 Digest, pp. 15-17, 1985.
13. T. Arakawa, M. Ueda, Y. Saito, T. Fujimura and S. Asai "A basic cell buffer 440K transistor CMOS masterslice" ISSCC86 Digest, pp. 78-79, February 1986.
14. C. A. Mead and L. A. Conway "Introduction to VLSI systems" Addison-Wesley Publishing Co., Reading, Mass., 1980.
15. A. Mukherjee "Introduction to nMOS and CMOS VLSI systems design" Prentice Hall Inc., Englewood Cliffs, N. J., 1986.
16. N. Weste and K. Eshraghian "CMOS VLSI design" Addison-Wesley, Reading, Mass. 1985.
17. L. A. Glasser and D. W. Dobberpuhl "The design and analysis of VLSI circuits" Addison-Wesley, Reading, Mass. 1985.
18. M. Annaratone "Digital CMOS circuit design" Kluwer Academic Publishers, Boston, 1986.
19. W. A. Clapp and A. Feller "CMOS custom LSI computer design and fabrication" ISSCC73 Digest, pp. 166-167, February 1973.
20. H. Kitazawa and K. Ueda "Chip area estimation method for VLSI chip floor plan" Electronics letters, vol. 20, pp. 137-138, February 1984.
21. W. V. Vilkelis "Lead reduction among combinatorial logic circuits" IBM J. Res. Develop., vol. 26, pp. 342-348, May 1982.
22. W. E. Donath "Equivalence of memory to random logic" IBM J. Res. Develop., vol. 18, pp. 401-407, September 1974.
23. E. Berndlmaier, J. Dorler, J. Mosley and S. Weitzel "Delay regulation, a performance concept" ICCC80 Digest, pp. 701-704, October 1980.
24. R. W. Johnson, J. L. Davidson, R. C. Jaeger and D. V. Kerns, Jr. "Hybrid silicon wafer-scale packaging technology" ISSCC86 Digest, pp. 166-167, February 1986.
25. M. Shoji "Electrical design of BellMac32A microprocessor" ICCC82 Digest, pp. 112-115, October 1982.
26. K. Manabe, N. Someya, M. Ueno, H. Neishi, M. Imai, S. Okamoto and K. Suzuki "A C_2MOS 16-bit parallel microprocessor" ISSCC76 Digest, pp. 14-15, February 1976.
27. T. Shigematsu, Y. Suzuki, N. Kokado, Y. Kudo, S. Makino, Y. Nagashima and H. Tajiri "An electronic auto-channel selection system using C^2MOS LSI in a TV receiver" ISSCC76 Digest, pp. 72-73, February 1976.
28. E. G. Friedman and S. Powell "Design and analysis of a hierarchical clock distribution system for synchronous standard cell/macrocell VLSI IEEE J. Solid-State Circuits, vol. SC-21, pp. 240-246, April 1986.

29. R. H. Krambeck and M. Shoji "Skew-free clock circuit for integrated circuit chip" U. S. patent 4, 479, 216, October 23, 1984.
30. M. Shoji "Reliable chip design method in high performance CMOS VLSI" ICCD86 Digest, pp. 389-392, October 1986.
31. T. Wada, H. Shinohara, and S. Kayano "A low noise data-out buffer in CMOS SRAM" Japanese Telecommunication Society Meeting Digest, pp. 2–229, 1986.
32. J. Pathak, H. Kurowski, R. Pugh, R. Shrivastava and F. Jenne "A 19ns 250mW programmable logic device" ISSCC86 Digest, pp. 246-247, February 1986.
33. B. T. Murphy, R. Edwards, L. C. Thomas and J. J. Molinelli "A CMOS 32b single chip microprocessor" ISSCC81 Digest, pp. 230-231, February 1981.
34. Y. Kaji, N. Sugiyama, Y. Kitamura, S. Ohya and M. Kikuchi "45ns 16x16 CMOS multiplier" ISSCC84 Digest, pp. 84-85, February 1984.
35. K. Yoshida, I. Yamazaki, K. Doi, H. Nozawa, T. Shibata and M. Ueno "A 16-bit LSI minicomputer" ISSCC76 Digest, pp. 12-13, February 1976.
36. T. Rhyne "Limitations on carry lookahead networks" IEEE Trans. on Computers, vol. C-33, pp. 373-374, April 1984.
37. A. Weinberger "High speed programmable logic array adders" IBM J. Res. Develop., vol. 23, pp. 163-178, March 1979.
38. M. S. Schmookler "Design of large ALUs using multiple PLA macros" IBM J. Res and develop., vol. 24, pp. 2-14, January 1980.
39. J. Iwamura, K. Suganuma, M. Kimura and S. Taguchi "A CMOS/SOS multiplier" ISSCC84 Digest, pp. 92-93, February 1984.
40. T. Kohonen "Digital circuits and devices" Prentice Hall Inc., Englewood Cliffs, New Jersey, 1972.
41. C. S. Wallace "A suggestion for a fast multiplier" IEEE Trans. on Electronic Computers, vol. EC-13, pp. 14-17, February 1964.
42. A. E. Gamal, D. Gluss, P. H. Ang, J. Greene and J. Reyneri "A CMOS 32bit Wallace tree multiplier-accumulator" ISSCC86 Digest, pp. 194-195, February 1986.
43. N. Ohwada, T. Kimura and M. Doken "LSI's for digital signal processing" IEEE J. Solid-State Circuits, vol. SC-14, pp. 214-220, April 1979.
44. D. Hampel, K. E. McGuire and K. J. Prost "CMOS/SOS serial-parallel multiplier" IEEE J. Solid-State Circuits, vol. SC-10, pp. 307-314, October 1975.
45. S. M. Kang and R. H. Krambeck "Data shifting and rotating apparatus" U. S. patent 4, 396, 994, August 2, 1983.
46. R. W. Hamming "Error detecting and error correcting codes" Bell System Technical Journal, vol. 26, pp. 147-160, April 1950 and W. T. Greer "32-bit EDAC chips fix single-bit errors efficiently" Electronic Design, pp. 269-274, January 6, 1983.
47. J. Yamada, T. Mano, J. Inoue, S. Nakajima and T. Matsuda "A submicron VLSI memory with a 4b-at-a-time built-in ECC circuit" ISSCC84 Digest, pp. 104-105, February 1984.
48. "A system solution to the memory soft error problem" IBM J. Res. and Develop. vol. 24, pp. 390-397, May 1980.
49. A. Chan, D. Deschene and S. Yuen "A 25Mb/s CMOS disk data controller" ISSCC86 Digest, pp. 120-121, February 1986.
50. R. W. Hunt "Memory design and technology" Large Scale Integration, edited by M. J. Howes and D. V. Morgan, John Wiley & Sons, Chichester, 1981.
51. L. L. Lewyn and J. D. Meindl "Physical limits of VLSI DRAM's" IEEE J. Solid-State Circuits, vol. SC-20, pp. 231-241, February 1985.

52. K. J. O'Connor "A 5ns 4K x 1 NMOS static RAM" ISSCC83 Digest, pp. 104-105, February 1983.
53. V. G. McKenny "A 5V-only 4K static RAM" ISSCC77 Digest, pp. 16-17, February 1977.
54. J. E. Mahan, D. S. Newman and M. R. Gulett "Gigaohm range polycrystalline silicon resistors for microelectronic applications" IEEE Trans. on Electron Devices, vol. ED-30, pp. 45-51, January 1983.
55. K. Goser and M. Pomper "Five-transistor memory cells in ESFI MOS technology" IEEE J. Solid-State Circuits, vol. SC-8, pp. 324-326, October 1973.
56. R. G. Stewart and A. G. F. Dingwall "16K CMOS/SOS asynchronous static RAM" ISSCC79 Digest, pp. 104-105, February 1979.
57. V. L. Rideout "One-device cells for dynamic random-access memories: A tutorial" IEEE Trans. on Electron Devices, vol. ED-26, pp. 839-852, June 1979.
58. H. J. Boll "Automatic refresh dynamic memory" ISSCC76 Digest, pp. 132-133, February 1976 and J. M. Caywood, J. C. Pathak, G. L. VanBuren and S. W. Owen "A self-refreshing 4K RAM with sub-mW standby power" ISSCC79 Digest, pp. 16-17, February 1979.
59. H. Sunami, T. Kure, N. Hashimoto, K. Ito, T. Toyabe and S. Asai "A corrugated capacitor cell (CCC)" IEEE Trans. on Electron Devices, vol. ED-31, pp. 746-753, June 1984.
60. E. Harari, L. Schmitz, B. Troutman and S. Wang "A 256bit nonvolatile static RAM" ISSCC78 Digest, pp. 108-109, February 1978.
61. R. Cuppens and L. H. M. Sevat "A 256kbit ROM with serial ROM cell structure" IEEE J. Solid-State Circuits, vol. SC-18, pp. 340-344, June 1983.
62. J. T. Koo "Integrated-circuit content-addressable memories" IEEE J. Solid-State Circuits, vol. SC-5, pp. 208-215, October 1970.
63. J. P. Wade and C. G. Sodini "Dynamic cross-coupled bit line content addressable memory cell for high density arrays" IEDM85 Digest, pp. 284-287, December 1985.
64. T. Furuyama, T. Ohsawa, Y. Watanabe, H. Ishiuchi, T. Tanaka, K. Ohuchi, H. Tango, K. Natori and O. Ozawa "An experimental 4Mb CMOS DRAM" ISSCC86 Digest, pp. 272-273, February 1986.
65. J. Miyamoto, S. Saito, H. Momose, H. Shibata, K. Kanzaki and T. Iizuka "A 28ns CMOS SRAM with bipolar sense amplifiers" ISSCC84 Digest, pp. 224-225, February 1984.
66. K. C. Hardee and R. Sud "A fault-tolerant 30ns/375mW 16kx1 NMOS static RAM" IEEE J. Solid-State Circuits, vol. SC-16, pp. 435-443, October 1981.
67. A. H. Shah, C-P Wang, R. H. Womack, J. D. Gallia, H. Shichijo, H. E. Davis, M. Elahy, S. K. Banerjee, G. P. Pollack, W. F. Richardson, D. M. Bordelon, S. D. S. Malhi, C. Pilch, B. Tran and P. K. Chatterjee "A 4Mb DRAM with cross-point trench transistor cell" ISSCC86 Digest, pp. 268-269, February 1986.
68. M. Takada, T. Takeshima, M. Sakamoto, T. Shimizu, H. Abiko, T. Kato, M. Kikuchi, S. Takahashi, Y. Sato and Y. Isobe "A 4Mb DRAM with half internal voltage bit line precharge" ISSCC86 Digest, pp. 270-271, February 1986.
69. S. Asai "Trends in megabit DRAMs" IEDM84 Digest, pp. 6-12, December 1984.
70. L. S. White, Jr., N. H. Hong, D. J. Redwine and G. R. Mohan Rao "A 5V-only 64K dynamic RAM" ISSCC80 Digest, pp. 230-231, February 1980.
71. M. Yoshimoto, K. Anami, H. Shinohara, T. Yoshihara, H. Takagi, S. Nagao, S. Kayano and T. Nakano "A 64Kb full CMOS RAM with divided word line structure" ISSCC83 Digest, pp. 58-59, February 1983 M. Yoshimoto, K. Anami, H. Shinohara, T. Yoshihara, H. Takagi, S. Nagao, S. Kayano and T. Nakano "A divided word line structure in the static RAM and its application to a 64K full CMOS RAM" IEEE J. Solid-State Circuits, vol. SC-

18, pp. 479-485, October 1983.

72. M. Isobe, J. Matsunaga, T. Sakurai, T. Ohtani, K. Sawada, H. Nozawa, T. Iizuka and S. Kohyama "A 46ns 256K CMOS RAM" ISSCC84 Digest, pp. 214-215, February 1984.
73. R. Taylor and M. Johnson "A 1Mb CMOS DRAM with a divided bit line matrix architecture" ISSCC85 Digest, pp. 242-243, February 1985.
74. R. C. Foss "The design of MOS dynamic RAMs" ISSCC79 Digest, pp. 140-141, February 1979.
75. K. Itoh, R. Hori, H. Masuda and Y. Kamigaki "A single 5V 64K dynamic RAM" ISSCC80 Digest, pp. 228-229, February 1980.
76. "Half-V_{DD} bit line sensing scheme in CMOS DRAMs" IEEE J. Solid-State Circuits, vol. SC-19, pp. 451-454, August 1984.
77. O. Minato, T. Masuhara, T. Sasaki and Y. Sakai "Hi-CMOSII 4K static RAM" ISSCC81 Digest, pp. 14-15. February 1981.
78. T. Ozawa, S. Koshimaru, O. Kudo, H. Ito, N. Harashima, N. Yasuoka, H. Asai, T. Yamada and S. Kikuchi "A 25ns 64K SRAM" ISSCC84 Digest, pp. 218-219, February 1984.
79. T. Furuyama, S. Saito and S. Fujii "Threshold voltage compensating sense amplifier for high density dynamic MOS RAMs" Japanese J. Applied Physics, vol. 21, pp. 61-65, 1982.
80. T. Hagiwara, H. Matsuo, M. Fukuda, Y. Matsuno, T. Furuno, K. Kuroda, T. Yasui, H. Katto and A. Shimizu "Page mode programming 1Mb CMOS EPROM" ISSCC85 Digest, pp. 174-175, February 1985.

8

High-Performance CMOS Circuits

8.1 INTRODUCTION

In this chapter two specialized subjects relevant to high-performance CMOS circuit design are discussed in detail. The first subject is fundamental to datapath design: how to propagate the control signals through a tall structure of a multibit arithmetic-logic unit (ALU). The second subject is how to scale FETs in CMOS static logic circuits to achieve minimum delay (dynamic logic gate FET scaling was discussed in Sections 5.13 to 5.15). The two problems show up constantly in CMOS VLSI chip design, and the solutions are quite involved.

8.2 ESTIMATE OF THE CONTROL LINE DELAY

In the following sections, the effects of wiring series resistance on signal propagation delay is analyzed in detail [01] [02]. We first consider what happens if a circuit has wiring series resistance. In the equivalent circuit of Fig. 8.1(a) and (b), the first circuit (a) includes series resistance of wiring R_1, while the second (b) does not. The effect of series resistance R_1, observed from the driver's side, is that capacitance C_1 has less effect. Node N_0 of circuit (a) switches faster than node N_0 of circuit (b). Circuits (a) and (b), however, store the same charge on capacitors C_0 and C_1 in the steady state. The shorter switching transient of node N_0 of circuit (a) is followed by an extended period of gradual drain of the charge stored in capacitance C_1. If circuit (a) is observed from the load's side, the driver becomes less visible and therefore node N_1 takes more time to switch than node N_0 of Fig. 8.1(b). Resistance R_1 makes the switching delays of nodes N_0 and N_1 different.

Suppose that the internal resistance of the CMOS driver shown in Fig. 8.1(c) is R_D (Ω), and the polysilicon line is very long. The polysilicon line has resistance R (Ω/cm) and capacitance C (pF/cm). When the driver pulls down, the driver feels only the capacitance of the polysilicon line of length $L_C = (R_D / R)$ cm from the driving end. To calculate the delay time for driving the output node of the driver down to voltage $V_{DD}/2$, the driver discharges the capacitance CL_C (pF). Therefore, the delay time is given by

$$T_D = R_D C L_C = \left(\frac{{R_D}^2}{R} \right) C \tag{8.1}$$

Equation (8.1) shows that the delay is proportional to the driver resistance squared.

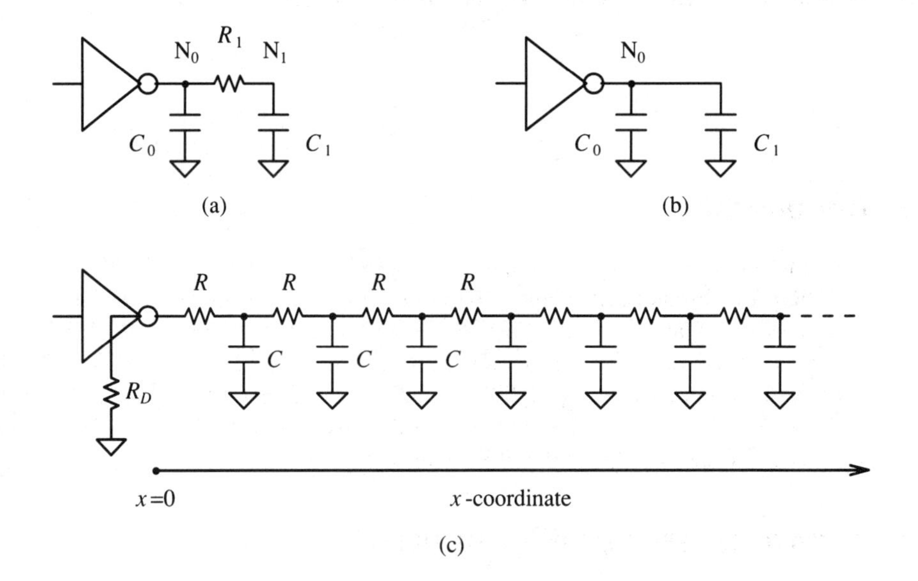

Figure 8.1 Effects of series resistance of interconnects.

If a strong driver is used, the driving point voltage drops very rapidly, but then a trickle current flows for a long time to complete discharge of the entire RC chain. When a polysilicon line is longer than L_C, the switching delay time of the driving point becomes independent of the line length: The driver does not see the portion of the RC chain beyond the point a distance L_C away, at the beginning of switching. The signal delay time to the capacitive load located at the far end of the polysilicon line is no longer determined by the driver's capability [03]. The delay time is determined by the polysilicon line itself.

8.3 DATAPATH CONTROL SIGNAL DISTRIBUTION

The datapath of a microprocessor consists of identical sections stacked schematically as shown in Fig. 8.2(a). The equivalent circuit is shown in Fig. 8.2(b). The number of sections, N, is set by the width of the data the processor handles, and in a state-of-the-art design, $N=32$. The width of each section is typically 60 to 100 μm (W_B). This width is estimated from the observation that in most microprocessor designs, the total height of the datapath is from 25 to 35% of the height of a 1 cm^2 chip. A control signal is typically precharge/discharge clocks, multiplexer select signals, or static or dynamic latch clocks. A control line is a polysilicide or a polysilicon line, and a portion of the line is the gate electrodes of the FETs that the control signals drive. The line is driven from one side of the datapath by a large CMOS static driver.

The signal delay is proportional to the square of N, as was shown in Section 6.8, and the design of a high-performance ALU wider than 16 bits is quite involved. The objective of this section is to analyze the control signal delay and to study several alternative schemes of control signal propagation. Two-level metal interconnects relieve many problems, but for some time, high-conductivity polysilicide interconnects will remain the practical way to control signal distribution. Just because a second-level metal interconnect is available does not resolve the problem: Area-efficient layout techniques to make best use of the second-level interconnect must be developed.

Even if a polysilicide control line is not connected to any FET gates, the parasitic capacitance and the series resistance of the line itself create delay. The intrinsic delay of the line is the ultimate limit for control signal delay. A formula to estimate the delay, T_D, in terms of the series resistance per unit bit width, R (Ω), the parasitic capacitance per unit bit width, C (pF), and the number of bits, N, was given in Section 6.8. The formula is, approximately,

$$T_D = 0.4(RC)N^2 \tag{8.2}$$

In Eq. (8.2) R is determined from the sheet resistance of polysilicide, ρ (Ω), line width W, and the width of one bit section, W_B, as [04]

$$R = \rho \left(\frac{W_B}{W} \right) \ (\Omega)$$

If the capacitance per unit length of the minimum-width polysilicide line (of width $W_{\min}$) is C_p, the capacitance per unit length of a polysilicide line of width W, $C_P(W)$, is very closely approximated by (see Section 1.14)

$$C_p(W) = C_p + \frac{\varepsilon_{OX}(W - W_{\min})}{T_{FO}}$$

where T_{FO} is the thickness of the field oxide. Using $C_p(W)$, C in Eq. (8.2) is given by

$$C = C_p(W)W_B$$

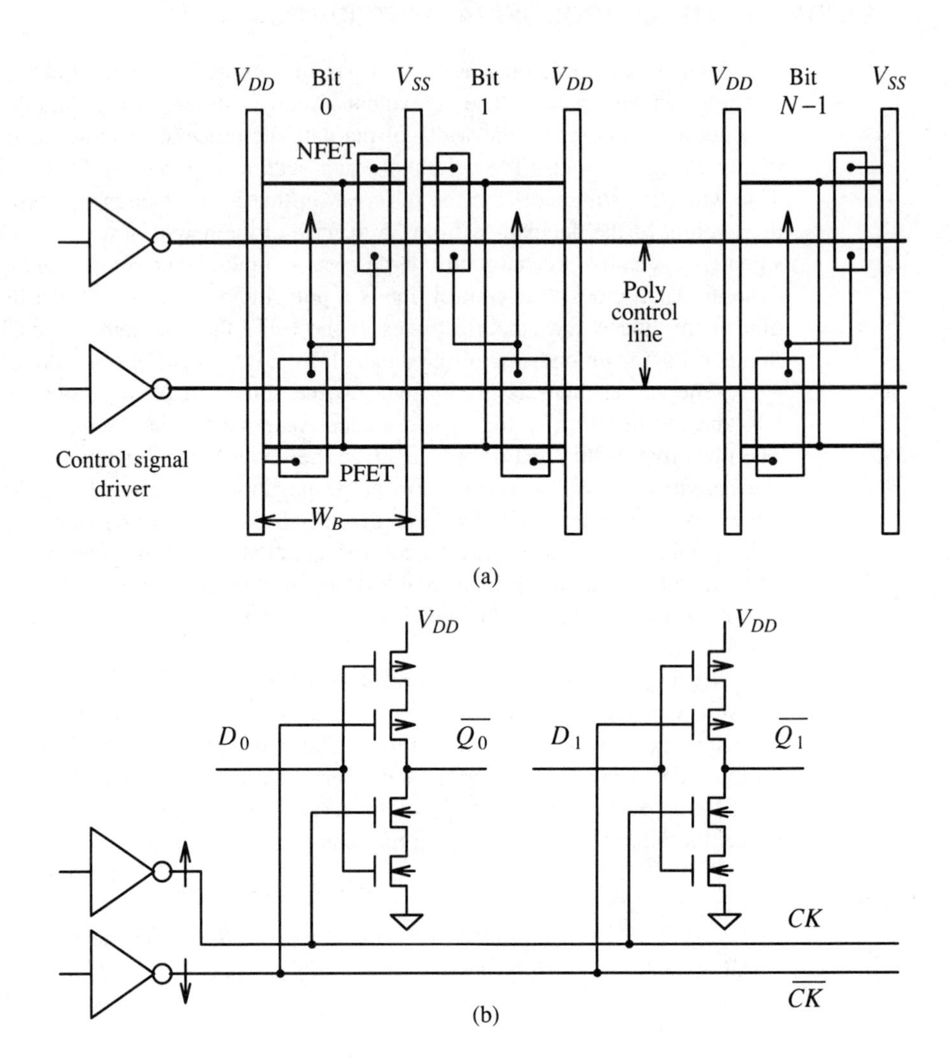

Figure 8.2 Structure of datapath and control lines.

If W is given in μm,

$$RC = (5.056 + 2.76W) \times 10^{-4} \left(\frac{{W_B}^2}{W} \right) \text{(ps)}$$

where typical parameter values for 1.75-μm CMOS are used.

The delay of the entire length of the control line is calculated using Eq. (8.2). If a control line of a 32-bit datapath is designed using 2-μm-wide polysilicide and if

$W_B = 60\,\mu\text{m}$, the intrinsic delay is given by

$$T_D = 0.4 \times 1.903 \times (32)^2 = 779(\text{ps}) = 0.78(\text{ns})$$

and this is the minimum propagation delay. This propagation delay, although considerable, is still tolerable in datapath designs. The problem is how to maintain this small delay in a heavily loaded control line.

The gate capacitance of a FET per unit length is about 13 times that of a polysilicide interconnect. If a minimum-size 1.75-μm CMOS inverter consists of a 4-μm-wide NFET and 7.25-μm-wide PFET, a minimum inverter effectively adds $(4+7.25) \times 13.3 = 150$ μm of equivalent wiring capacitance. Therefore, if one bit datapath width is 60 μm, the delay is increased by a factor of (150 + 60)/60 = 3.5 over the intrinsic delay. The effect of FET gate loading is very significant.

8.4 CONTROL SIGNAL DELAY

Although control signal delay is given by Eq. (8.2) of the preceding section, there is one more problem that must be resolved before using the formula correctly. The problem is to determine the exact capacitance per section, C. The components of capacitance attributable to the metal and to the polysilicide wires are well defined. The gate capacitance can be computed without ambiguity (the gate capacitance depends slightly on the gate voltage, and therefore an average must be taken. This problem was discussed in Section 4.8). The problem is how to account for the Miller capacitance of gates driven by the control line, which depends strongly on voltage.

As was shown in Section 4.8, Miller capacitance is about 20 to 25% of the input capacitance to the gate and is strongly voltage dependent. In the limit of slow input transition, Miller capacitance shows up like an impulse at the input voltage where the output of the gate switches. Miller capacitance depends on the loading of the gate: If the loading is too heavy, the gate does not switch before the input signal completes its transition. Then Miller capacitance does not show up, and therefore does not add up to delay.

Yet another detail is that Miller capacitance may or may not exist, depending on the state of the gate driven by the control line. This problem can be understood from Fig. 8.2(b). For instance, the assertion clock (CK) to a dynamic latch (tristatable inverter) does not have Miller capacitance if the input data (D_0) is high and if the output (storage) node ($\overline{Q_0}$) is already low. If the data (D_1) is low and the output node ($\overline{Q_1}$) is low, however, the anti-assertion clock ($\overline{CK}$) to the dynamic latch (that enables the tristatable inverter) creates a "negative" Miller effect by capacitive coupling, which helps pull up the assertion clock (CK) control line. All these details increase the variability of control line delay. The logic-state-dependent cases are so numerous that not all of them can be analyzed by circuit simulations. Therefore it is reasonable to estimate the range of the control line delay by a closed-form analysis.

Let us write the capacitance per bit of the datapath as a sum of two components,

$$C = C_{wire} + C_{DEV} \tag{8.3}$$

where

$$C_{DEV} = C_{GATE} + C_m \tag{8.4}$$

and where C_{GATE} is the part of the gate input capacitance that is only weakly dependent on the input voltage (defined in Section 4.8) and C_m is the Miller capacitance that shows up when the gate switches. If the "negative" Miller effect due to the simultaneous switching of the other control lines is excluded for simplicity, the minimum value of loading capacitance C, occurs when the Miller effect does not exist at all (when the output of the gate does not change state),

$$C_{wire} + C_{GATE} \le C$$

This condition determines the minimum delay of the control line.

The case of maximum delay is more involved. The capacitance for the maximum delay is close but not exactly $C = C_{wire} + C_{GATE} + C_m$, since the Miller capacitance is voltage dependent. This problem is analyzed in the following, using the simple model shown in Fig. 8.3. Suppose that all the nodes of the RC chain are at ground at time t=0. At that time node N_0 is pulled up to V_{DD} and is held there. Capacitor C is assumed, for the purpose of simplicity, to have the following simplified characteristics: Except at $V_{DD}/2$, the capacitance is negligibly small. At $V = V_{DD}/2$, however, the capacitor absorbs charge Q. This means exactly the following: The terminal voltage increases from a little less than $V_{DD}/2$ to a little more than $V_{DD}/2$ when charge Q is supplied to the capacitor. The capacitor stores the constant charge Q when the terminal voltage is higher than $V_{DD}/2$. Then using the method of measurement of the Miller capacitance discussed in Section 4.8, the capacitance is

$$C = \frac{Q}{V_{DD}} \tag{8.5}$$

A short time after $t = 0$ the voltages of all the nodes reach a little less than $V_{DD}/2$, and currents cease to flow into nodes $N_2, N_3, ..., N_N$. To node N_1, however, current $V_{DD}/2R$ flows from node N_0 and after time $2RQ/V_{DD} = 2RC$ charge Q is supplied to the capacitor and the voltage of node N_1 begins to increase again. From this time on, current $V_{DD}/2(2R)$ flows from node N_0 to node N_2, and after time $2(2RQ)/V_{DD} = 2(2RC)$ the voltage at node N_2 increases. Evolution of the potential profile within the RC chain is shown in Fig. 8.3(b). By repeating the same process, node N_N voltage begins to increase after time T_D given by

$$T_D = 2(RC) + 2(2RC) + \ \dots \ + 2(NRC) = N(N+1)(RC) \tag{8.6}$$

T_D is the delay to the end of the chain. By comparing this formula with Eq. (8.2) of the preceding section, the voltage-dependent capacitance of the Miller effect [capacitance determined from Eq. (8.5)] creates approximately twice the delay of a linear, voltage-independent capacitance. Therefore, the maximum delay of the RC chain occurs approximately when

$$C = C_{wire} + C_{GATE} + 2C_m \tag{8.7}$$

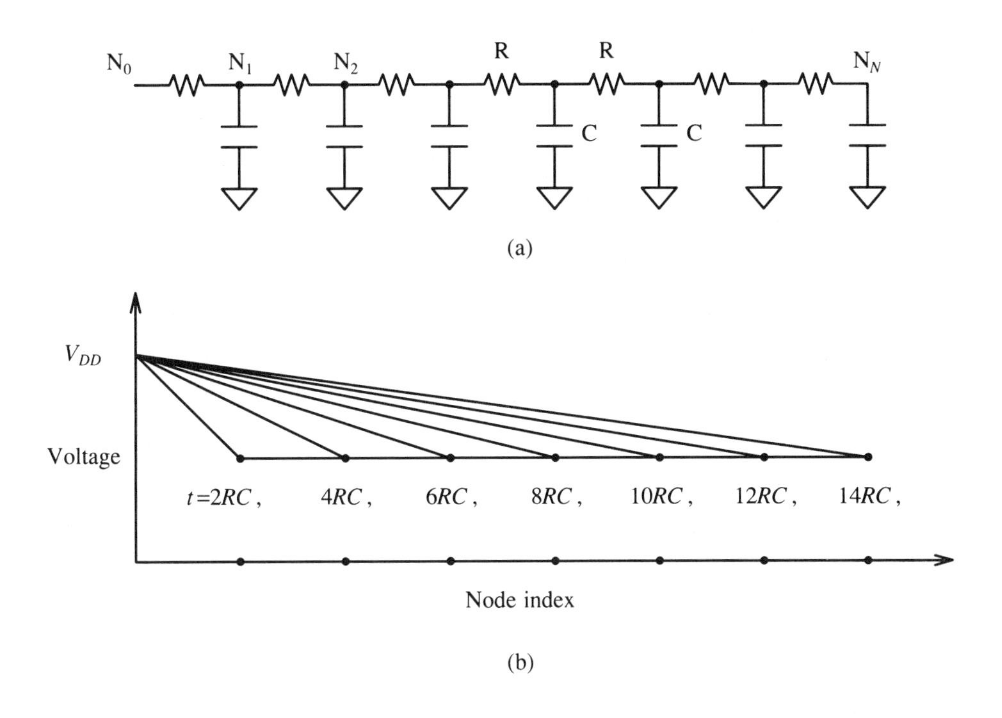

Figure 8.3 Potential profile of a control line loaded by Miller capacitance.

is satisfied, and the extra factor of 2 originates from the reduced voltage ($V_{DD}/2$) available to charge the voltage-dependent capacitor, when the capacitor absorbs charge.

The range of accuracy for the estimate of Eq. (8.2) is checked by the following example. In 1.75-μm CMOS, the minimum-size inverter consists of a 4-μm-wide NFET, whose gate capacitance is 0.0113 pF and a 7.25-μm-wide PFET whose gate capacitance is 0.0223 pF. The total Miller effect capacitance for both FETs is determined by the method of Section 4.8 as 0.0075 pF. The simulation conditions are, $V_{DD} = 5$V, 65 degrees centigrade, and the medium-current process. For simplicity, the wiring capacitance is neglected. The length of the chain is 32 stages, and the input is driven by a voltage source from 0 to 5 V at time $t = 0$, within a negligibly short rise time. At each bit position, an integral number of minimum-size inverters are connected. The circuit delay was determined by simulation, and also by the closed-form theory of this section, using $C = C_{GATE} + C_m$. The results are summarized in Table 8.1. When the loading is light and the series resistance is small, the estimate using $C = C_{GATE} + C_m$ is reasonably accurate. When the loading is heavy and the resistance is high, the switching becomes slow and Eq. (8.7) becomes more accurate. Then the delay is about 18% more than the estimate. The delay agrees better with the theoretical estimate using $C = C_{wire} + C_{GATE} + 2C_m$.

TABLE 8.1 CONTROL LINE DELAY

R (Ω)	Number of Inverters	Delay(theor.)	Delay(simul.)
120	1	2.02	2.10
720	1	12.1	14.50
120	6	12.1	14.50
720	6	72.7	85.90

The theory of this section gives the cumulative effects of Miller capacitance at each node of the control line to the delay of the signal at the end of the control line. When a polysilicide control line such as that shown in Fig. 8.3 is driven by a CMOS driver, the voltages of the node N_0 fluctuate, as the inverters, at nodes $N_1, N_2, \ldots$ switch in succession, and Miller capacitances add to the nodes. Figure 8.4 shows the result of simulation.

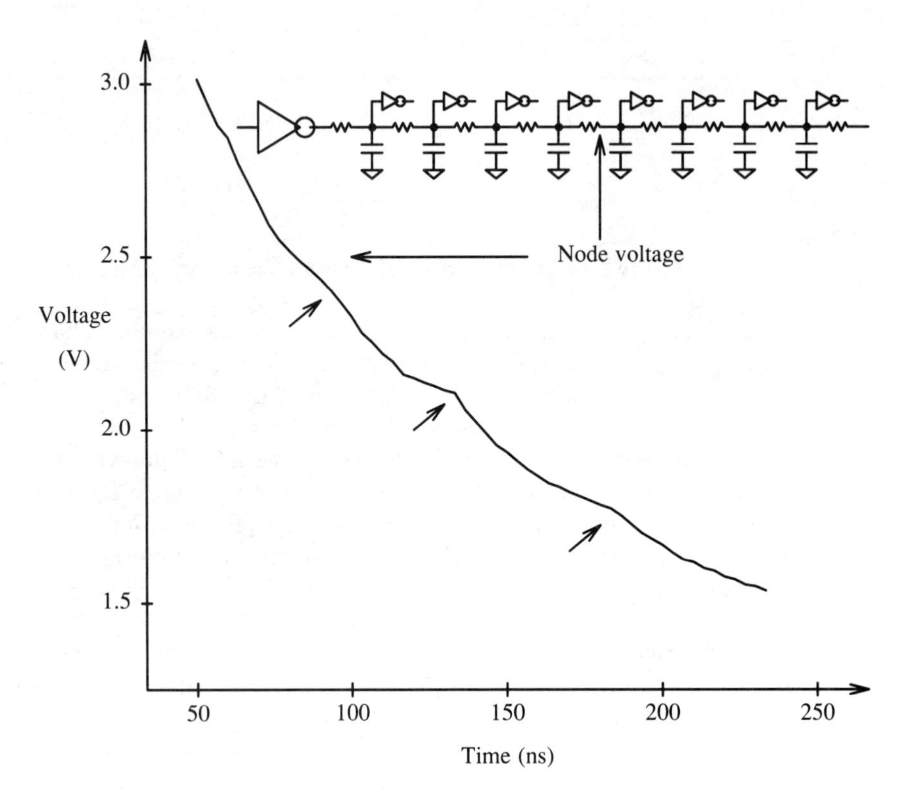

Figure 8.4 Oscillatory Miller effect.

As the node voltage drops, changes in slope, indicated by the arrows, are observed.

8.5 CONTROL LINE BUFFER

The delay of the control line driver is an integral part of the control line delay. The driver sizing problem must be resolved together with the control line delay problem. To minimize the delay to the far end of the control line, the driver size must be increased, but there is an upper limit beyond which further increase is ineffective. This limit is reached when the delay from the driver's input to the output ceases to decrease upon further increase in the driver size. The driver delay depends on the capacitance the driver sees. The capacitance, however, depends on the series resistance of the polysilicide line, and the delay time itself: If the series resistance is large and the delay time is short, the capacitance of the polysilicide line beyond a certain distance from the driving point is invisible to the driver.

To solve this problem, a model similar to that discussed in Section 3.14 can be used. We assume that the control line driver is equivalent to a parallel combination of a resistance R_D and the output capacitance C_D (drain capacitance of FETs), where R_D is inversely proportional to C_D or $R_D C_D =$constant. The polysilicide line presents capacitance C_L to the driver that is determined self consistently as follows. The driving point delay T_D is given by

$$T_D = R_D (C_L + C_D)$$

The polysilicide line capacitance the driver sees, C_L, is determined from T_D, C (pF/cm) and R (Ω/cm) by

$$C_L = \left[\frac{C}{R} T_D \right]^{1/2}$$

because the driver sees only distance $\sqrt{T_D/(CR)}$ of the line from the driver's end within time T_D. By eliminating C_L among the two equations, T_D is solved from

$$\frac{1}{R_D{}^2} T_D{}^2 - \left[2\frac{C_D}{R_D} + \frac{C}{R} \right] T_D + C_D{}^2 = 0 \tag{8.8}$$

as

$$T_D = (R_D C_D) \left\{ 1 + \frac{1}{2} \frac{R_D C}{R C_D} + \left[\frac{C}{C_D} \right] \left[\left[\frac{R_D C_D}{RC} \right] + \frac{1}{4} \left[\frac{R_D}{R} \right]^2 \right]^{1/2} \right\} \tag{8.9}$$

When the driver size is increased, $R_D \to 0$ and $C_D \to \infty$ and $R_D C_D$ remains constant. Then

$$T_D \to (R_D C_D) \left[1 + \frac{R_D}{R_{CR}} \right] \quad (R_D \to 0)$$

where

$$R_{CR} = R \left[\frac{R_D C_D}{RC} \right]^{1/2}$$

This equation shows that when the driver size is increased, $T_D \to R_D C_D$. If $R_D < R_{CR}$ is reached, further increase in driver size does not significantly decrease delay. The maximum driver size is determined from the relationship.

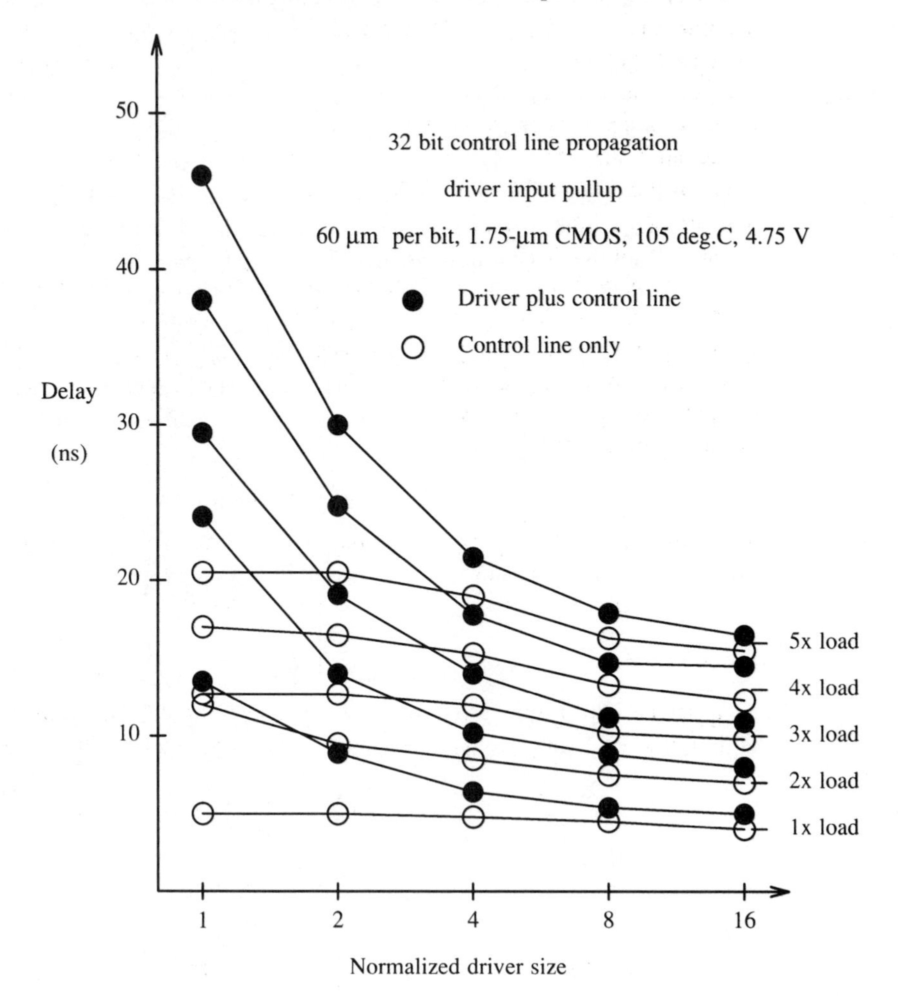

Figure 8.5 Control line driver delay (driver input pullup).

Figure 8.5 shows the delay versus driver size determined by circuit simulations. The simulation conditions are given in the figure. The minimum-size inverter consists of 4-µm-wide NFET and 7.25-µm-wide PFET. The control line driver and the gate loads at every bit position of a 32-bit-wide datapath are measured as multiples of the minimum-size inverter, and the normalized sizes are used in the figures. Width per bit of the datapath (W_B) is 60 µm, and a 2-µm tantalum polysilicide control line (series resistance 4 Ω/square) was assumed. The total delays (including the driver delays) and the polysilicide line delays alone are plotted versus the driver size for a high-to-low transition of the control line. The control line pullup delays are shown in Fig. 8.6.

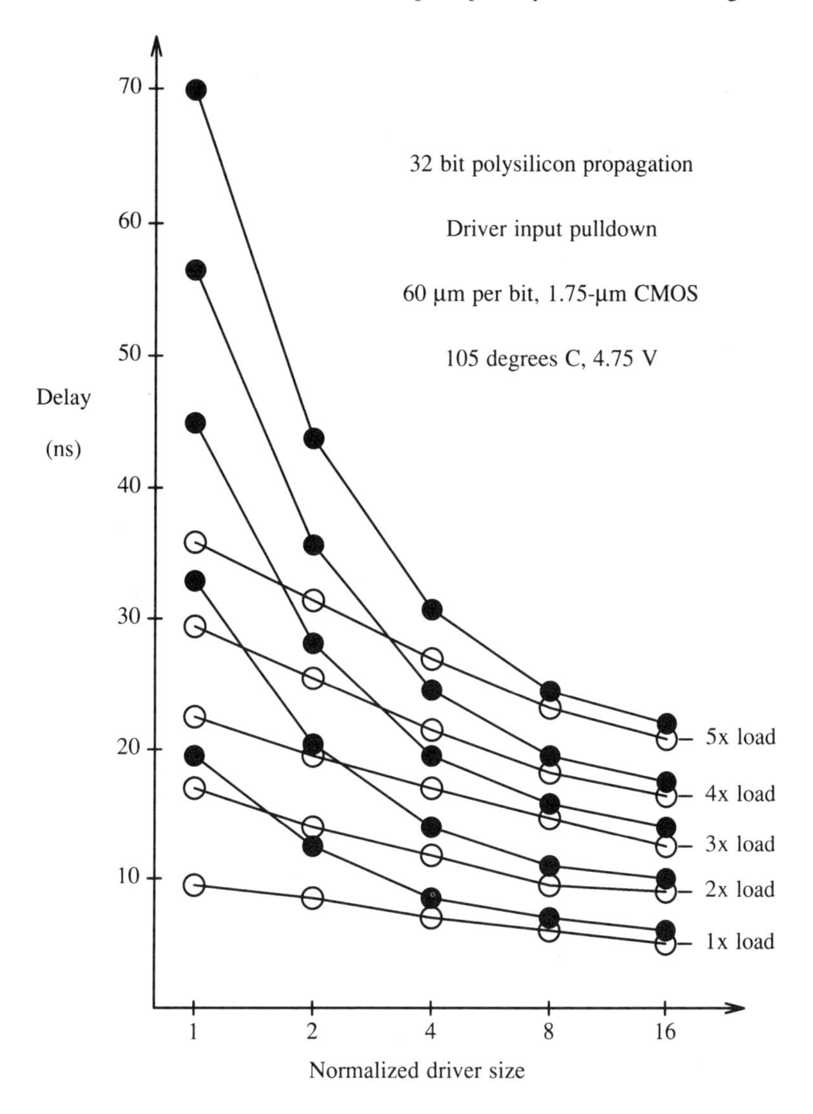

Figure 8.6 Control line driver delay (driver input pulldown).

The driving point delay ceases to decrease when the driver is larger than 4 to 8 times the minimum-size. The delay of the control line loaded with the minimum-size load at each bit position and driven by 8 times the minimum-size driver is about 7 ns.

If the load is five times the minimum-size, the delay is about 25 ns. Although these are the worst-case data, the delay of the resistive control line is much longer than a typical gate delay (1 to 3 ns).

Many techniques are used to reduce the delay [05]. By splitting one control line into two, delay can be reduced. A simple but effective method is to provide a noninverting buffer at the midpoint of the control line [06]. This method is effective when the load is very heavy or the line is long. For light loads, however, the effect of the midpoint buffer is to equalize the control line pullup and the pulldown delays, but not to reduce the smaller one. The most effective method of driving a heavily loaded line is to drive the control line by the complement of the control signal, and to buffer the heavy load at each bit position, or every few bit positions, by a minimum-size buffer as schematically shown in Fig. 8.7. Since the buffering delay is only a few nanoseconds, while the unbuffered control line delay is several tens of nanoseconds, the method significantly reduces the control line delay.

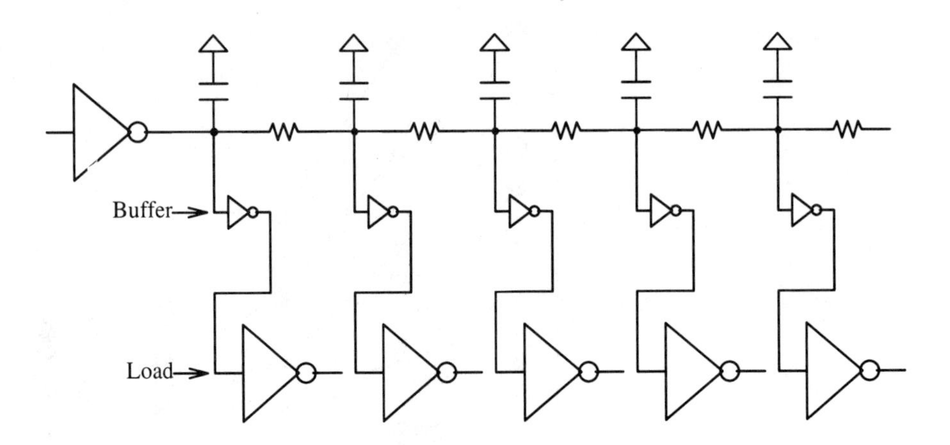

Figure 8.7 Technique of reducing delay of heavily loaded control line.

An interesting circuit technique to reduce the control signal delay is shown in Fig. 8.8. The far end of the control line (node X) is predischarged by NFET MN1 prior to the excitation of the control signal. The switching threshold of the inverter consisting of MP2 and MN2 is set significantly lower than $V_{DD}/2$. The control line is driven high from the left side. As the voltage of node X increases above the inverter threshold large PFET MP1 begins to pull up node X very hard, and the regenerative effect helps speed up the control signal settling [07]. This circuit is not easy to design, and it is noise sensitive. For practical designers the problem of control signal propagation along polysilicon or polysilicide interconnect is so difficult to resolve that techniques of this kind are proposed and considered seriously.

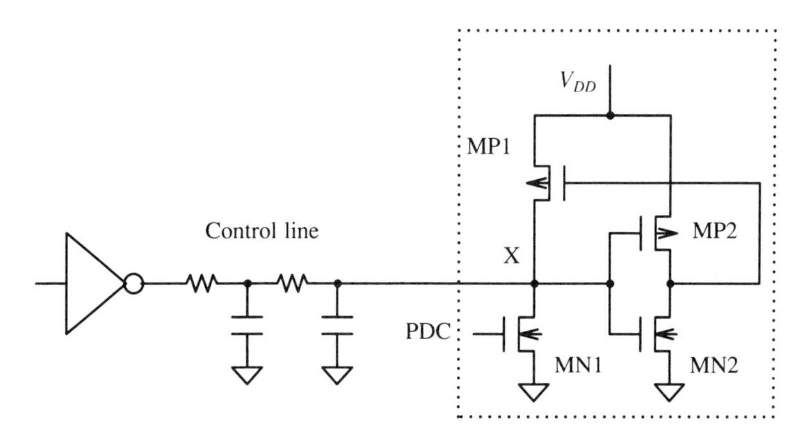

Figure 8.8 A circuit technique to reduce control line delay.

8.6 DELAY OPTIMIZATION BY SCALING FETS

Delay of a CMOS logic gate chain can be minimized by selecting the proper size for the FETs used in the gates. In this section general guidelines for FET sizing are discussed. Practically, FET sizing is carried out by several, significantly different methods. It is convenient to discuss the problem under the following five cases.

1. *FET size determined by the regularity of the structure.* The size of FETs used in a regular, repeated structure like a PLA, a RAM, a ROM, and in many data-path functional blocks is determined by area and layout requirements. In many cases the minimum-size NFETs and the second minimum-size PFETs are used. Logic circuits of this type must satisfy delay specifications using only small FETs because large numbers of FETs are packed in a narrow area.

2. *Output drivers and input buffers.* An output driver must satisfy logic-threshold-voltage-level specification, delay-time (T_D) specification, dT_D/dC_L specification, DC sink current specification, and power-bus noise specification.

An output driver must be able to drive from several tens of picofarads to a few hundred of picofarads of load capacitance. The load can be CMOS devices, TTL devices that require current sinking, or NMOS devices that require TTL logic levels, but that require no current sinking.

A CMOS output driver that drives capacitance load C_L within the allowed delay time T_D must have PFET size, P, and NFET size, N, determined from

$$P \geq \frac{C_L r_P}{T_D} \tag{8.10}$$

and

$$N \geq \frac{C_L r_N}{T_D}$$

where r_P and r_N are the equivalent resistances of unit width PFETs and NFETs, respectively, that are defined in Sections 3.11 and 3.12. In a 1.75-μm CMOS at the medium-current process, $r_P \approx 60\,\mathrm{k\Omega \cdot \mu m}$ and $r_N \approx 30\,\mathrm{k\Omega \cdot \mu m}$. $P/N = r_P/r_N$ is the beta ratio of the FETs. The ratio ranges from 2 to 3 (1 to 2-μm CMOS). A large safety factor of 50% or more is applied to Eq. (8.10), to meet the specifications under the most unfavorable conditions. In addition to process, temperature, and power supply voltage variations, designers must provide some margin for capacitive load variations. Customers always assume that a device works even if the specifications are slightly violated, and "no margin" design is destined to fail. Conservative design and tight specifications often result in large output driver size. If $C_L = 150\,\mathrm{pF}$, $T_D = 10\,\mathrm{ns}$, and a safety factor of 1.5 is assumed, $P = 1350\,\mu\mathrm{m}$ and $N = 675\,\mu\mathrm{m}$.

Output driver design is not just to meet the timing specification of driving the interchip capacitance; delay within the chip is also important. If the logic signal is available from the minimum-size buffer (consisting of 7.25-μm-wide PFET and 4-μm-wide NFET in 1.75-μm CMOS), a scaleup ratio of about 200 requires three to four stages of buffering. If the internal delay by buffering is excessive, either the signal source is designed using larger FETs or an interchip capacitive load C_L must be reduced by negotiating the specifications with the customer.

From Eq. (8.10),

$$T_D = \frac{C_L r_P}{P} = \frac{C_L r_N}{N}$$

and we obtain

$$\frac{\partial T_D}{\partial C_L} = \frac{r_P}{P}(\text{pullup}) = \frac{r_N}{N}(\text{pulldown})\ (\text{ns/pF}) \tag{8.11}$$

$\partial T_D/\partial C_L$ must be less than a certain upper limit to accommodate an unexpected variation of the customer's capacitive load. The range of typical $\partial T_D/\partial C_L$ values is 0.05 to 0.2 ns/pF.

In a TTL driver, logic level low or high is decided at the threshold voltage approximately $2V_{BE} = 1.4$ V above the ground. Then the output driver should drive down from 5 V to 1.4 V in the same time as the driver drives up from the ground potential to 1.4 V. Then the PFET/NFET size ratio P/N is much less than r_P/r_N. In a CMOS process of about 1-μm feature size, $P = N$ is a reasonably close approximation.

In a TTL-compatible output driver the current sinking requirement must be met. Let the minimum resistance r_N' of an NFET having unit width be defined by

$$r_N' = \lim_{V_D \to 0} \frac{V_D}{I_D(V_G = V_{DD}, V_D)}$$

where $r_N' < r_N$. Then the FET size $W(= P = N)$ must satisfy

$$\left(\frac{r_N'}{W}\right) I_S < V_S$$

where I_S is the current that the driver must sink, and V_S is the guaranteed TTL logic low level (usually 400 mV or less). We then have

$$W > \frac{r_N{}'}{(V_S/I_S)} = \frac{r_N{}'}{100} \tag{8.12}$$

if $V_S = 400\,\text{mV}$ and $I_S = 4\,\text{mA}$. A large safety factor of 50% or more is usually applied.

3. *Input buffers.* A chip's input circuit has significant capacitive parasitics. Bonding pads, input protection devices, and the package contribute several picofarads. It is then a rational design to increase the size of the input driver to such a size that the gate capacitance of the first-stage input buffer is a significant fraction of the total parasitic capacitance. The upper limit of the input driver size is set by the series resistance of the input protection circuit (Section 9.9) and the DC current the input buffer draws. Since the input logic level is usually not the CMOS logic level, the first-stage buffer draws DC current. The TTL high logic level can be only 2.4 V above ground. Then both PFET and NFET of the first-stage input buffer conduct, and therefore the first stage carries DC current. Input buffers that interface TTL or ECL logic signals to CMOS logic require an involved design and careful scaling of FETs which is discussed in the next section.

4. *Databus drivers, multiplexers, and latches.* Databus drivers, multiplexers, and CMOS latches share the common feature that a single node is driven by multiple drivers. In databus and in multiplexers, increasing the size of one driver necessitates an increase in the size of all the others, from the requirement that the signal transmission from any source to any destination should be the same. The FET size should be chosen so that the sum of the drain capacitances of all the bus drivers is a few times, but not significantly more than, the capacitance of the bus wire. The bus capacitance is determined by the length of the bus. Further increase in the bus driver size does not reduce the delay significantly.

In CMOS D-latches, two tristatable drivers drive the $\overline{Q}$ node of the latch. The size of the tristatable driver that applies positive feedback to hold the data depends on the requirement of how fast the latch should snap into the final state after clock transition (when the input tristatable driver is unable to complete transition), and that depends on individual requirements.

5. *Other cases.* In the case that does not belong to any of 1 through 4, we need to scale the FETs for optimum delay. There are several different guidelines, and the scaling for each case will be discussed in the following sections.

8.7 TTL LEVEL SIGNAL INPUT CIRCUIT

An ordinary CMOS inverter switches when the input voltage is about $0.5V_{DD}$. In this section we discuss a CMOS inverter design problem whose switching threshold is significantly different from $0.5V_{DD}$. A TTL-to-CMOS logic level converter is an example.

A widely used method of interfacing a TTL signal to CMOS logic gates was to use an open-collector TTL driver with a pullup resistor. Historically, a TTL-to-CMOS converter on a CMOS chip emerged when the feature size of CMOS technology allowed 5 V, single power supply operation. Then, for simplicity and for improved performance gained by eliminating the relatively slow open-collector pullup circuit, TTL-compatible CMOS input circuits were designed.

When the input voltage changes from the absolute-worst-case TTL low input [$V_{TTL}(L)=1$ V] to the absolute-worst-case TTL high input [$V_{TTL}(H)=2$ V], the output of the inverter must switch from 5 V to 0 V. The TTL logic level specifications used here are tighter than conventional specifications. Since failure of the logic level converter is more serious than failure of other circuits, tighter specifications are used in the design. The switching threshold should be at $V_{TTL}=1.4$ V for typical process, temperature, and power supply voltage. When the output node is at 2.5 V, the input node must be at 1.4 V, and the threshold voltage must be maintained well within the range (1 to 2 V) for all the process, temperature, and power supply voltage variations.

Let us consider a CMOS inverter. When the input is at V_G, the excess gate voltage that determines conductivity of the NFET is

$$V_{GNE}=V_G-V_{THN}=1.4-0.75=0.65\,V \quad \text{if } V_G=V_{TTL}$$

where V_{THN} is the threshold voltage of the NFET. The similar excess gate voltage of the PFET is

$$V_{GPE}=V_{DD}-V_G-V_{THP}=5-1.4-0.75=2.85\,V \quad \text{if } V_G=V_{TTL}$$

Let the current of the NFET and the PFET be written as

$$I_N=B_N NJ(V_{GNE})$$

$$I_P=B_P PJ(V_{GPE})$$

where B_N and B_P are the parameters, defined in Section 1.6, and here the $B's$ are defined for unit-width FETs. N and P are the widths of NFET and PFET, respectively, and $J(V_{GE})$ is a function of the excess gate voltage V_{GE}. The drain voltage is set at $V_{DD}/2$. From the steady-state requirement of zero node currents,

$$I_N=I_P \tag{8.13}$$

we have

$$\frac{B_N N}{B_P P}=\frac{J(V_{GPE})}{J(V_{GNE})} \tag{8.14}$$

In the limit of long channel and low field, function $J(V_{GE})$ is given by the FET theory of Section 1.7 as follows:

$$J(V_{GPE})=(V_{GPE}-\frac{1}{4}V_{DD})\frac{V_{DD}}{2} \quad \text{if } V_{GPE}>\frac{1}{2}V_{DD} \tag{8.15}$$

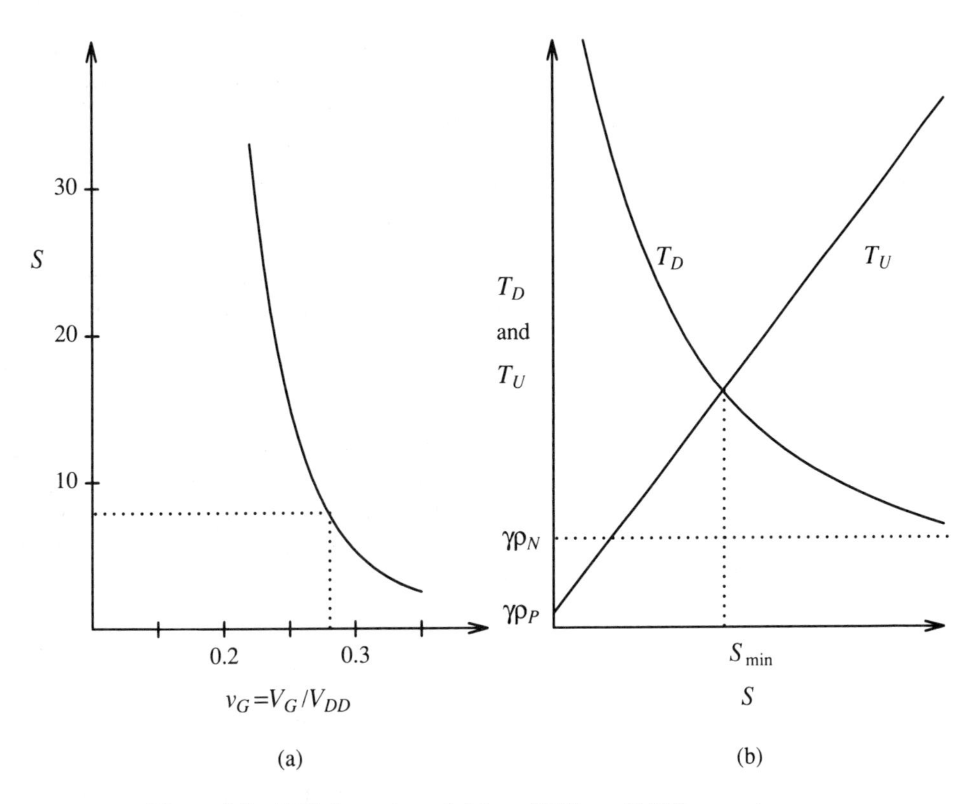

Figure 8.9 FET size ratio and delay of TTL-to-CMOS converter.

$$J(V_{GNE}) = \frac{V_{GNE}^{\ 2}}{2} \quad \text{if} \quad V_{GNE} < \frac{1}{2}V_{DD}$$

Using Eqs. (8.14) and (8.15),

$$\frac{N}{P} = \frac{B_P}{B_N} \frac{V_{DD}((3/4)V_{DD} - V_{THP} - V_G)}{(V_G - V_{THN})^2} \tag{8.16}$$

Assuming that $V_{THP} = V_{THN} = 0.15V_{DD}$ and $B_P/B_N = 0.4$,

$$S = \frac{N}{P} = 0.4\frac{0.6 - v_G}{(v_G - 0.15)^2} \tag{8.17}$$

where $v_G = V_G/V_{DD}$. The limit of validity of Eq. (8.16) is

$$V_{THN} < V_G < (1/2)V_{DD} - V_{THP} \quad \text{or} \quad 0.15 < v_G < 0.35$$

Equation (8.17) is plotted in Fig. 8.9(a). At the TTL threshold voltage $v_G = 0.28$, the ratio of NFET size to PFET size is 7.6 (dotted line). Figure 8.9(a) shows that the lower the threshold voltage, the larger the NFET/PFET size ratio. The ratio diverges like $(V_G - V_{THN})^{-2}$ when $V_G \to V_{THN}$. Since the NFET/PFET size ratio of a conventional CMOS inverter is about 0.3 to 0.5 as discussed in Section 8.9, the size ratio required for a TTL-to-CMOS interface circuit is quite unusual. The unusual size ratio creates a problem of unequal pullup/pulldown delay times.

Parasitic capacitances of the NFET and the PFET drain islands at the output node are given by, respectively,

$$C_P = \gamma P, \quad C_N = \gamma N$$

where γ is a constant (approximately $2 \times 10^{-15} F/\mu m$). The equivalent resistance used to determine the delay, R_P and R_N, of the PFET and the NFET, respectively, is given by

$$R_P = \frac{\rho_P}{P} \quad R_N = \frac{\rho_N}{N}$$

where ρ_P is the equivalent resistance of unit size PFET, whose gate voltage is TTL logic 0 voltage $V_{TTL}(L)$ (400 mV relative to ground), and ρ_N is the equivalent resistance of unit size NFET, whose gate voltage is TTL logic 1 voltage $V_{TTL}(H)$ (2.4 V relative to ground). The pullup and pulldown delays of the inverter with no load (this assumption is practical, since such a heavily scaled inverter is usually loaded very lightly) are given by

$$T_U = R_P(C_P + C_N) = \frac{\gamma \rho_P}{P}(P + N)$$

$$T_D = R_N(C_P + C_N) = \frac{\gamma \rho_N}{N}(P + N)$$

where the delays are defined assuming that input voltage transition occurs in a negligibly short time. If the size ratio $S = N/P$ is used, we obtain

$$T_U = \gamma \rho_P (1 + S), \quad T_D = \gamma \rho_N (1 + \frac{1}{S})$$

The larger of T_U or T_D is the delay of the scaled inverter. T_U and T_D versus size ratio S is schematically shown in Fig. 8.9(b). By equating T_U and T_D and by solving for S, the delay minimum occurs at $S = S_{min} = \rho_N/\rho_P$. If $S < S_{min}$, the pullup delay is smaller than the pulldown delay, and if $S > S_{min}$, the pulldown delay is smaller. Since the size ratio is very large, the level converter delay is determined by the larger pullup delay. A larger pullup delay is the immediate result of setting the threshold voltage at a very low level. Even if level converter is designed to be large, the delay is still significant because of large unbalance of pullup and pulldown delay times.

8.8 FET SCALING OF CASCADED INVERTER STAGES

When a logic signal generated by a minimum-size gate is to drive a heavy capacitive load, several stages of buffering are necessary for minimum delay. The cascaded buffers shown in Fig. 8.10 consist of an individual stage that has pullup delay time T_U

and pulldown delay time T_D. They are given by

$$T_U(i) = \tau_P \frac{P_{i+1} + N_{i+1} + f \cdot (P_i + N_i)}{P_i} \tag{8.18}$$

and

$$T_D(i) = \tau_N \frac{P_{i+1} + N_{i+1} + f \cdot (P_i + N_i)}{N_i} \tag{8.19}$$

where P_i and N_i are the sizes of the PFET and the NFET of the ith buffer stage, respectively, that drive the $i+1$st stage, consisting of a PFET (P_{i+1}) wide and an NFET (N_{i+1}) wide. In Eqs. (8.18) and (8.19), τ_P is the time constant defined by the product of the gate capacitance of a unit size FET (NFET and PFET are assumed to be the same) and the equivalent resistance of the unit size PFET. τ_N is defined in the same way for NFETs. Parameter f is the ratio of drain island capacitance to gate capacitance. The size of first-stage FETs, P_0 and N_0, and of the last stage, P_n and N_n, are given and fixed. FET sizes $P_1, P_2, \ldots, P_{n-1}$ and $N_1, N_2, \ldots, N_{n-1}$ are determined by requiring that the delay is the minimum.

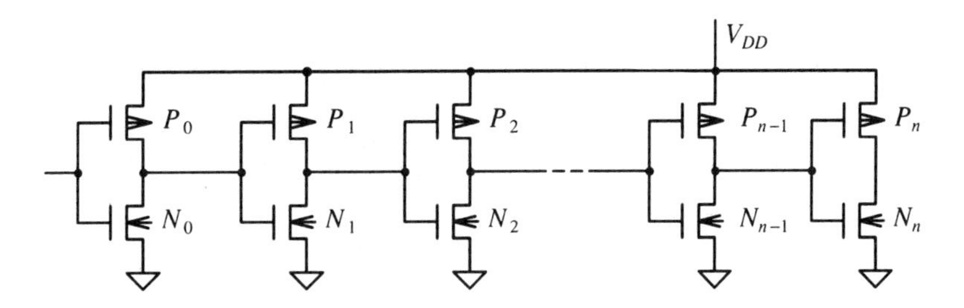

Figure 8.10 Optimum scale-up of a long chain of CMOS inverters.

Let us first assume that each inverter has equal pullup and pulldown delays. From Eqs. (8.18) and (8.19),

$$\frac{\tau_P}{P_i} = \frac{\tau_N}{N_i} = \frac{\tau}{S_i}, \quad i = 0, 1, 2, \ldots, n \tag{8.20}$$

where $S_i = P_i + N_i$ is the size of the gate, $\tau = \tau_P + \tau_N$ and

$$\frac{P_i}{N_i} = \frac{\tau_P}{\tau_N} = \beta$$

The equal pullup and the pulldown delays of the ith stage are T_i. We have

$$T_i = \tau \left(\frac{S_{i+1}}{S_i} + f \right) \tag{8.21}$$

and therefore,

$$T_D = \text{total delay} = \tau \sum_{i=0}^{n-1} \frac{S_{i+1}}{S_i} + \tau f n \tag{8.22}$$

It can be shown that the arithmetic average of n positive numbers $a_0, a_1, \ldots, a_{n-1}$ is always larger than the geometrical average,

$$\frac{1}{n} \sum_{i=0}^{n-1} a_i \geq \left(\prod_{i=0}^{n-1} a_i \right)^{\frac{1}{n}}$$

and the equal sign holds only when

$$a_0 = a_1 = \cdots = a_{n-1}$$

Using this relationship in Eq. (8.22),

$$T_D \geq \tau f n + \tau n \left(\frac{S_n}{S_0} \right)^{\frac{1}{n}} \tag{8.23}$$

and [08],

$$\frac{S_{i+1}}{S_i} (i = 0, 1, \ldots, n-1) = \left(\frac{S_n}{S_0} \right)^{\frac{1}{n}}$$

This equation and Eq. (8.21) show that the delay minimum is attained when the ratio of scale-up of any two consecutive inverters, are the same. The minimum delay, T_{Dmin} is given by

$$T_{Dmin} = \tau n \left\{ f + \left(\frac{S_n}{S_0} \right)^{\frac{1}{n}} \right\} \tag{8.24}$$

Parameter n is the number of stages, and therefore the allowed values are integers. We assume, for the moment, that n is a continuous variable and differentiate T_{Dmin} by n. Then

$$\frac{1}{\tau} \frac{\partial T_{Dmin}}{\partial n} = f + \left(\frac{S_n}{S_0} \right)^{\frac{1}{n}} - \frac{1}{n} \log \left(\frac{S_n}{S_0} \right) \left(\frac{S_n}{S_0} \right)^{\frac{1}{n}} \tag{8.25}$$

When $f = 0$, $\partial T_{Dmin} / \partial n = 0$ if

$$n = \log \left(\frac{S_n}{S_0} \right) \tag{8.26}$$

Equation (8.26) does not always give an integral number; the closest integral number is the optimum number of stages. By substituting Eq. (8.26) into Eq. (8.23) with

$f = 0$, the absolute minimum delay is given by [09] [10],

$$\frac{\left(T_{Dmin}\right)_{min}}{\tau} = \log\left(\frac{S_n}{S_0}\right)\left(\frac{S_n}{S_0}\right)^{\frac{1}{\log\left(\frac{S_n}{S_0}\right)}} = e \log\left(\frac{S_n}{S_0}\right) \tag{8.27}$$

This relationship shows that the minimum delay of a chain of symmetrical inverters is proportional to the logarithm of the ratio of the sizes of the last inverter to the first inverter. Equation (8.26) shows that the optimum scale-up factor between two consecutive stages is 2.7, or approximately, 3 [09].

When parameter f of Eq. (8.25) (the drain/gate capacitance ratio) is not zero, the optimum number of stages is less than that given by Eq. (8.26). If f is small, Eq.(8.26) becomes

$$n = \frac{\log\left(\frac{S_n}{S_0}\right)}{1 + \left(\frac{f}{e}\right)} \tag{8.26a}$$

A computer program to determine the optimum n and S_i has been written [11]. An extensive review of area-delay trade-off techniques using this method has been published [12].

8.9 PFET/NFET SIZE RATIO

When the PFET/NFET size ratio does not satisfy the condition of equal pullup and pulldown delays, delay optimization is more involved. When we consider two cascaded stages of such inverters, however, the delay for input pullup and the delay for input pulldown of two stages can still be made equal. Then the two-stage, noninverting buffer is pullup-pulldown symmetrical. Suppose that many stages of symmetrical noninverting buffers are cascaded, as shown in Fig. 8.11.

The FET sizes satisfy

$$P_j = bN_j \quad \text{(for all } j\text{)} \tag{8.28}$$

and

$$\ldots N_{i+1} = \alpha N_i = \alpha^2 N_{i-1} \ldots = \alpha^{K+1} N_{i-K}$$

The delay of two cascaded stages $i-1$ and i is given by, using Eqs. (8.18) and (8.19),

$$T_D = (\alpha + f)\left(\tau_P + \tau_N + \frac{\tau_P}{b} + \tau_N b\right)$$

and the minimum delay occurs when

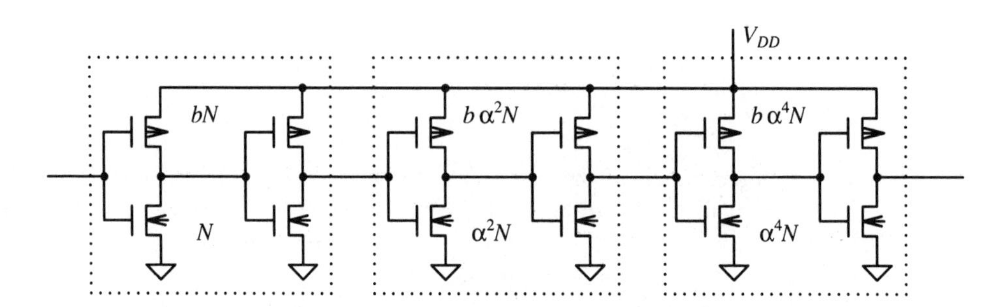

Figure 8.11 Optimum PFET/NFET size ratio.

$$\frac{\tau_P}{b} = \tau_N b \quad \text{or when} \quad b = \left(\frac{\tau_P}{\tau_N} \right)^{1/2} \tag{8.29}$$

and the minimum delay is

$$T_{Dmin} \geq (\alpha + f)(\tau_P + \tau_N + 2\sqrt{\tau_P \tau_N}) \tag{8.30}$$

Equation (8.29) shows that the optimum PFET/NFET size ratio is $\sqrt{\tau_P/\tau_N} = \sqrt{\beta}$ [13] [14]. The symmetrical inverter (with size ratio τ_P/τ_N) of Section 8.8 is not the optimum. If pullup delay and pulldown delay of each stage need not be equal, the total delay of the chain can be reduced by increasing the NFET size and simultaneously decreasing the PFET size.

The delay minimum is, however, very shallow. Figure 8.12 shows averaged delay for pullup and pulldown of the minimum-size inverter of 1.75-μm CMOS for the worst-case conditions (low-current process, 105 degrees centigrade and 4.75 V power supply voltage), determined from simulations. Parameter α is 3. The delay minimum occurs when the PFET/NFET size ratio is 1.6, which is close to the square root of the beta ratio (approximately 2.8).

By generalizing the analysis of Sections 8.8 and 8.9, the following practical conclusions are reached. When a very large fanout ratio is necessary, or when a very complex gate is used, the PFET/NFET size ratio should be chosen so that the gate pulls up and pulls down with approximately equal delay. In a long chain of relatively simple gates the PFET/NFET ratio should be chosen smaller than the ratio for the delay symmetry. In this latter case the pullup-pulldown asymmetry of individual gates are compensated by the asymmetries of neighboring gates, and the delay asymmetry does not accumulate.

8.10 FET SIZING IN GENERAL CASE

The analysis of the preceding two sections is subject to some restrictions on the FET size, to make the complex multivariable extremum problem tractable. The size of each FET is not allowed to vary arbitrarily. In Section 8.8 we assumed that PFET is β

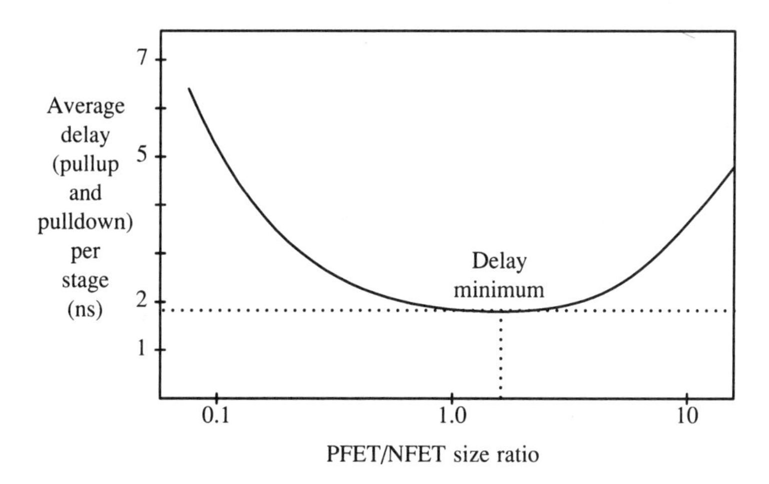

Figure 8.12 Delay of inverter versus PFET/NFET size ratio.

(ratio of conductance of NFET to PFET) times larger than NFET. In Section 8.9 the inverters were uniformly scaled up, and size ratios of PFET and NFET of all the inverters were the same. Complications arise if this restriction is removed. The cause of the complication is that the delays for the two input transition polarities are not the same any more. Then the larger of the two delays must be minimized, and the problem cannot be solved by simple application of differential calculus; many different cases must be analyzed separately.

We discuss here a simple example where either the pullup or the pulldown delay, but not the smaller of the two, of the circuit shown in Fig. 8.13, is optimized by adjusting the size of NFET N_1 or the size of the PFET P_1. If the drain island capacitance is neglected, the input pulldown delay is given by

$$T_D = \tau_P \frac{P_1 + N_1}{P_0} + \tau_N \frac{P_2 + N_2}{\left(\dfrac{N_1}{x} \right)} \tag{8.31}$$

Since the capacitances of the source-drain islands of the FETs are neglected, the analysis is valid when $N_0, P_0 \ll N_1, P_1 \ll N_2, P_2$.

$$= \left(\frac{\tau_P P_1}{P_0} \right) + \left(\frac{\tau_P}{P_0} \right) N_1 + \frac{x \tau_N (P_2 + N_2)}{N_1}$$

The delay minimum occurs when

$$\left(\frac{\tau_P}{P_0} \right) N_1 = \frac{x \tau_N (P_2 + N_2)}{N_1}$$

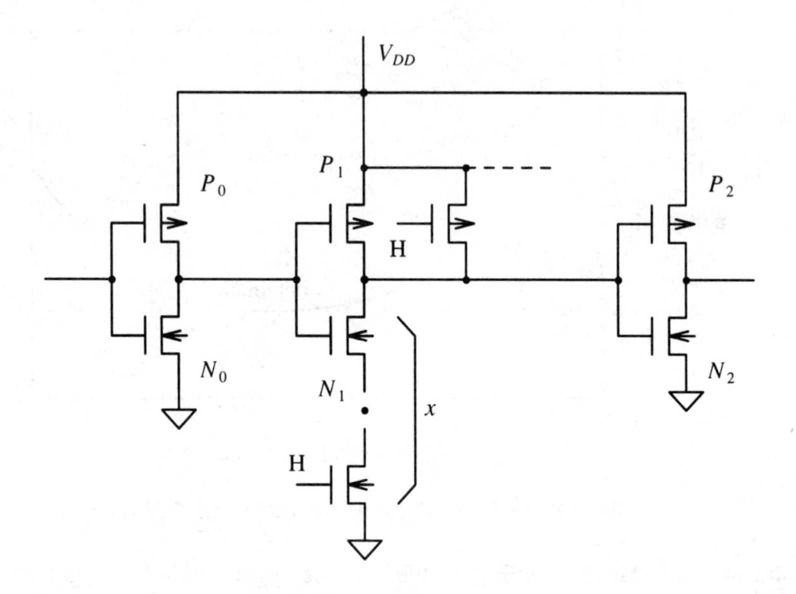

Figure 8.13 General principle of FET scaling.

or

$$N_1 = \left\{ \frac{\tau_N}{\tau_P} x P_0 (P_2 + N_2) \right\}^{1/2} \tag{8.32}$$

In Fig. 8.13, the FETs that are on the signal path of the input pulldown transition of the gate chain are PFET with size P_0 and x series-connected NFETs with size N_1. The size of N_1 is determined by geometrically averaging P_0 and (P_2+N_2) (size of load stage) and multiplying a factor $\sqrt{x}$ and $\sqrt{\tau_N/\tau_P}$ originating from the fact that the N_1 NFETs are connected in series. This rule can be generalized. The input pullup delay of the circuit of Fig. 8.13 is optimized by selecting PFET size P_1 by

$$P_1 = \left\{ \frac{\tau_P}{\tau_N} N_0 (P_2 + N_2) \right\}^{1/2} \tag{8.33}$$

Now we discuss minimization of the larger of the two delays: That is, input pulldown delay (T_D) and input pullup delay (T_U). They are given by [from Eq. (8.31)],

$$T_D = A(P_1 + N_1) + \frac{B}{N_1} \tag{8.34a}$$

$$T_U = D(P_1 + N_1) + \frac{E}{P_1} \tag{8.34b}$$

where

$$A = \frac{\tau_P}{P_0}, \quad B = x\tau_N(P_2+N_2), \quad D = \frac{\tau_N}{N_0}, \quad \text{and} \quad E = \tau_P(P_2+N_2)$$

We must first find which of T_D or T_U is smaller. For this purpose we solve

$$T_D = T_U$$

and answer is as follows:

$$P_1 = p(\xi) = \frac{-\xi + \left\{\xi^2 + 4E(A-D)\right\}^{1/2}}{2(A-D)} \tag{8.35a}$$

where $\xi(N_1) = (A-D)N_1 + (B/N_1)$. This solution is valid when $A > D$. When $D > A$,

$$N_1 = n(\xi) = \frac{-\xi + \left\{\xi^2 + 4B(D-A)\right\}^{1/2}}{2(D-A)} \tag{8.35b}$$

where $\xi(P_1) = (D-A)P_1 + (E/P_1)$. Suppose that T_D and T_U, which are functions of two independent variables N_1 and P_1, are plotted on the coordinate system shown in Fig. 8.14. Figures 8.14(a) and (b) show the coordinate plane for case $A > D$ and $D > A$, respectively, and the curves in Fig. 8.14(a) and (b) represent Eqs. (8.35a) and (8.35b), respectively. The curves have extrema at $N_1 = \sqrt{B/(A-D)}$ and $P_1 = \sqrt{E/(D-A)}$, respectively. Figures 8.14(c) and (d) show cross sections of the plots at vertical line 1 and horizontal line 2 of Fig. 8.14(a) and (b), respectively. T_D is a linear function of argument P_1. As for dependence of T_D on N_1, $T_D \to \infty$ when $N_1 \to 0$ and $N_1 \to \infty$. There is the minimum of T_D when $N_1 = \sqrt{B/A}$. T_U is a linear function of argument N_1 and $T_U \to \infty$ when $P_1 \to 0$ and $P_1 \to \infty$. The minimum occurs when $P_1 \to \sqrt{E/D}$. Delay is the larger of T_D and T_U. Since T_D and T_U are increasing functions of P_1 and N_1, respectively, and decreasing functions of N_1 and P_1, respectively, at their intersection, delay minimum is reached when N_1 and P_1 are on the curves. This minimum (when $A > D$) can be found by requiring that

$$\frac{\partial T_D}{\partial N_1} = A\frac{\partial \xi}{\partial N_1}\frac{\partial P_1(\xi)}{\partial \xi} + A - \frac{B}{N_1^2} = 0$$

where $P_1(\xi)$ is given by Eq. (8.35a). We have

$$X(N_1) = A - \frac{B}{N_1^2}$$

$$+\frac{A\left\{A-D-(B/N_1^2)\right\}}{2(A-D)}\left[-1+\frac{(A-D)N_1+(B/N_1)}{\left\{((A-D)N_1+(B/N_1))^2+4E(A-D)\right\}^{1/2}}\right]=0$$

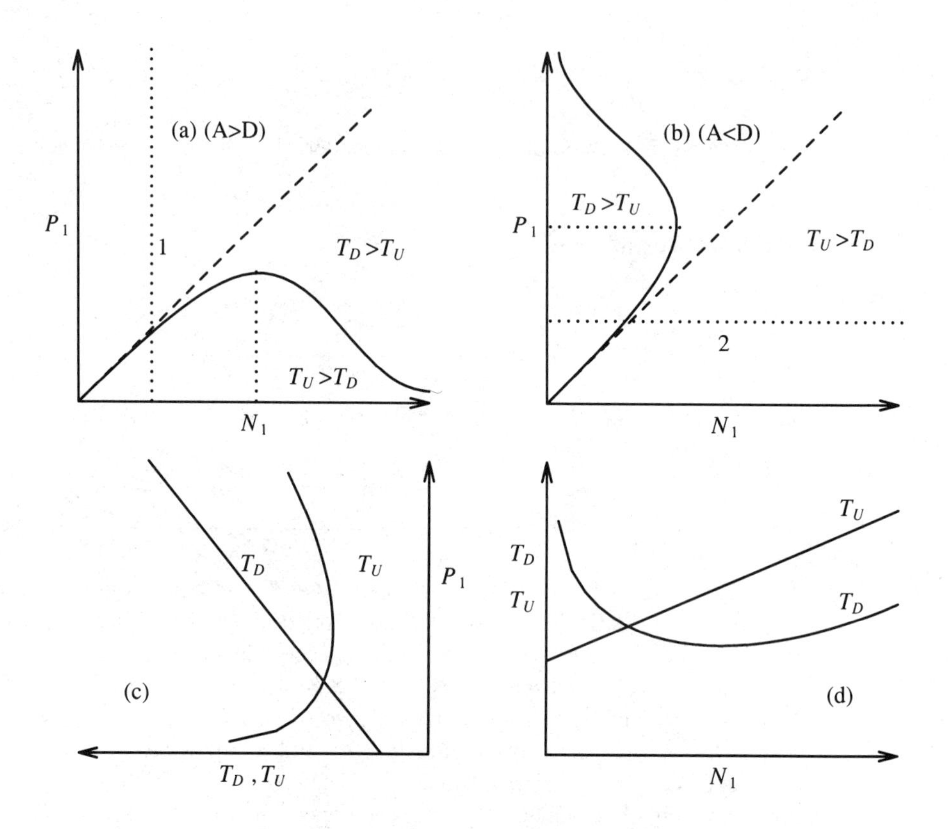

Figure 8.14 T_D and T_U versus P_1 and N_1.

It may be shown that $X(N_1) \to -\infty$ $(N_1 \to 0)$ and $X(N_1) \to A$ $(A > 0)$ when $N_1 \to \infty$. Then there is a positive solution of $X(N_1) = 0$ for N_1 that gives the optimum FET size.

As is observed from this theory, optimization of FET size for both pullup and pulldown, without assuming any constraints an FET sizes, is very complex. Therefore, FET sizing is often carried out by software [15] [16] [17].

8.11 FET SIZE OPTIMIZATION FOR SINGLE-TRANSITION POLARITY

Signals that appear in microprocessor logic are of two kinds: One is a signal that represents data or values and a signal of this type assumes high and low states with equal probability. We call signals of this type "data signals." The other type is a signal that controls the states of the machine. Signals of this type are normally in either high or low quiescent states (control inactive). When a particular control is exercised, the signal takes the value that is complementary to the quiescent value.

When the signal takes this unusual value, the particular control is enabled. We call such signals "control signals."

The difference in the nature of data and control signals is reflected in the timing optimization problem. Any signal that is processed by a chain of logic gates must satisfy the timing condition that the worst-case delay is less than a certain limit. The worst-case delay of a data signal is the longer of the input pulldown and pullup delay. In an MOS circuits, reducing either delay increases the other. Therefore, a logic circuit that processes a data signal is designed such that the input pullup and input pulldown delays are equal. At present, this design practice is generally also followed for control signals. The practice is conservative but is safe, since the circuit may be exercised in a way not immediately obvious to the circuit designer.

There are control signals where both transition polarities are equally important: A signal that controls tristating of the address output drivers of a microprocessor is a signal of this kind. The majority of the control signals of a microprocessor are, however, not of this kind. This is because a control event ends by any of the following:

1. The event completes within a known number of machine cycles. Examples are register select, precharge, and discharge controls. Most of the signals belong to this category.
2. The event has a natural end. Examples are the interrupt, reset, and restart of the processor.
3. The event is terminated by activating another control signal. Examples are READ/WRITE and memory access signals.

In cases 1 to 3 the signal transition edge that initiates the control event must be processed quickly. Once the desired event occurs the signal returns to the quiescent state. The response of this recovery transition need not be very fast. Thus the circuit that processes the control signals need be fast only for one transition polarity.

Figure 8.15(a) shows a circuit that combines two signals from a PLA, C_1 and C_2, to generate a control signal C_3. Signals C_1 and C_2 are supplied by PLA latches 1 and 2, which may belong to different PLAs. The bit lines of the PLAs are precharged during phase 2, and the output latches become transparent during phase 3 (see Section 5.5). If the bit lines both remain high after PLA discharge, nodes C_1 and C_2 are pulled up if they were originally low, and as a result node C_3 is pulled down. If control signal C_3 is required at the beginning of phase 1 of the next cycle, the allowed propagation delay is two phases (phases 3 and 4). If the PLA output latch clock is adjusted so that the latch is transparent during most of phases 2 and 3 the allowed delay can even be three phases.

If any one of the bit lines pulls down, however, the bit line may not reach the logic zero level until the end of phase 3. This is the time required for the bit lines to combine many logic signals from the word lines. In this case the maximum propagation time of the downgoing edge at the output of the PLA latch is only one phase (phase 4). From this observation we conclude that the NAND2 gate G_1 should be

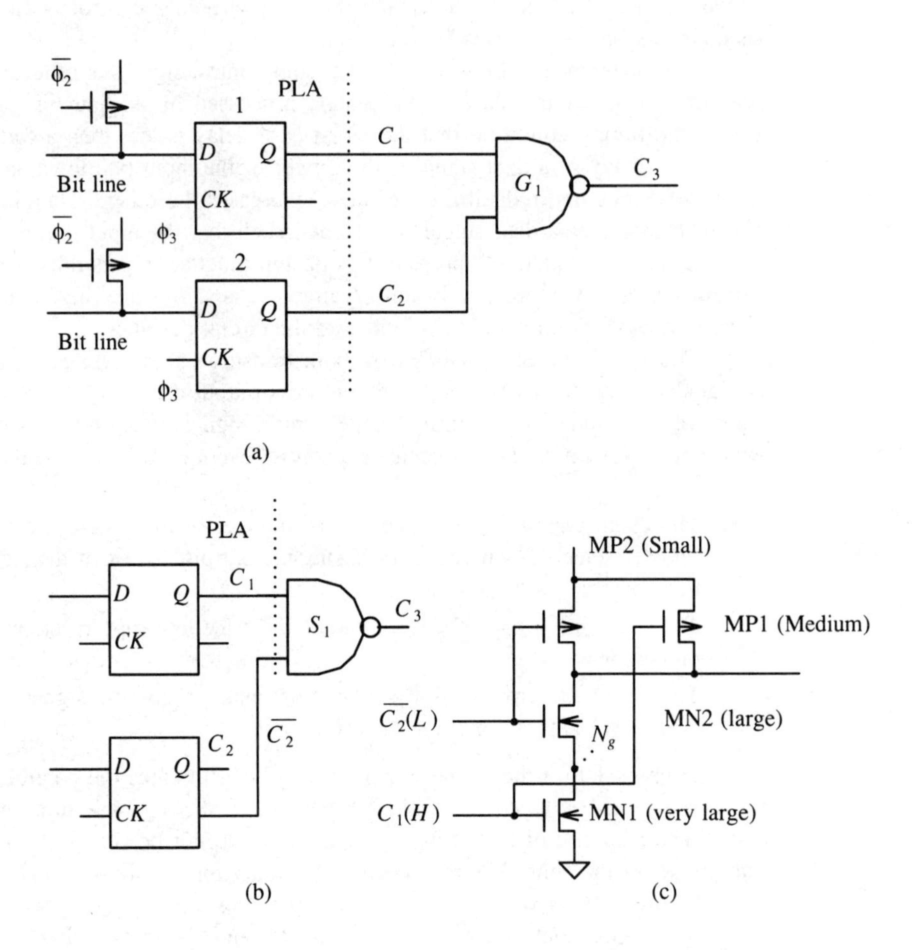

Figure 8.15 Delay optimization for only one polarity of transition.

designed so that the input pulldown delay of gate G_1 is much shorter than the input pullup delay. The size ratio of the PFET and NFET in a typical gate is about 2, but the size ratio to be used in the NAND2 gate G_1 is larger than that. The large size ratio allows gate G_1 to pull up node C_3 rapidly, especially when node C_3 is heavily loaded.

Since signal $\overline{C_2}$ shown in Fig. 8.15(b) is low in phase 3, the optimization of gate S_1 is more difficult. Signal C_3 is high in phase 3. If $\overline{C_2}$ makes a low-to-high transition, the gate must respond very quickly. With reference to Fig. 8.15(c) PFET MP2 can be small, since an extra phase is available to pull node C_3 up. NFET MN2 must be large, since the pulldown of node C_3 must be fast. The pulldown delay, however, is also influenced by the series NFET MN1. By designing MN1 very large, node N_g becomes a virtual ground when MN2 turns on. PFET MP1 is of medium size since the FET takes over the pullup in case $\overline{C_2}$ pulls up and C_1 pulls down simultaneously. This pullup, however, has been accomplished by MP2 by spending the entire phase 3, and therefore the role of MP1 is to prevent a glitch. As is observed from this example, the delay optimization by FET sizing when the initial logic levels are mixed is tedious and ineffective.

The initial logic levels can be adjusted by modifying the PLA encoding. Suppose that the logic circuit was as shown in Fig. 8.15(b). For the purpose of explanation, signal C_2 is generated from the input signals to the PLA, X, Y, and Z by

$$C_2 = XY + Z$$

Since a CMOS dynamic PLA is two-stages of cascaded dynamic NOR gates, the equation above can be synthesized by NOR operations as

$$C_2 = \overline{\overline{X + Z} + \overline{Y + Z}}$$

The complement of C_2, $\overline{C_2}$ can be generated as follows:

$$\overline{C_2} = \overline{XY + Z} = \overline{\overline{\overline{X} + \overline{Y}} + \overline{\overline{\overline{Z}}}} \tag{8.36}$$

If the PLA is encoded this way, signal $\overline{C_2}$ is now available from the Q output of the PLA latch, and therefore the signal is high at the beginning of phase 3. Then the logic circuit of Fig. 8.15(a) can be used instead of the circuit of Fig. 8.15(b), and the optimization of the circuit becomes easier and more effective. As for PLA logic optimization, the reader is referred to reference [18].

In the tandem combination of logic gates shown in Fig. 8.16 signal $\overline{C_3}$ should have a low initial value for optimization, and this polarity can be obtained by a similar PLA encoding change. By a combination of proper PLA encoding and FET scaling of the static gate chain, the timing optimization scheme speeds up the PLA signal combination logic. Two PLA bit lines can be joined together using the same technique. When the Q outputs of two PLA latches are ANDed, the two bit lines are joined together. When $\overline{Q}$ outputs of two PLA latches are ORed the bit lines are also joined. The two operations are logically equivalent, and the second version is faster.

To make a static CMOS logic chain fast in one polarity of transition, complex and simple gates are alternated as shown in Fig. 8.17(a) and (b). The preferred input transition is low to high, as indicated in the figure. Five-input gate symbols represent complex gates, while two-input gate symbols represent simple gates. When NAND and NOR gates are mixed, the logic is as shown in Fig. 8.17(c) and (d). The input pullup and the pulldown delays of logic (e) and (f) are very significantly different if the PFET/NFET size ratio is about 2 (this ratio is used most frequently).

If a long static logic chain is optimized to only one transition polarity, the circuit can be too slow in the unpriviledged polarity of input transition. This problem

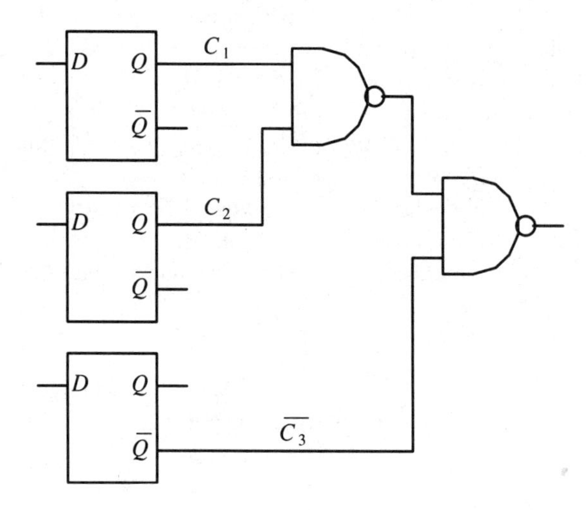

Figure 8.16 Combination of signals having different quiescent levels.

can be alleviated by providing a "precharge" of the node when the nodes return to the control disabled state. By further extending this idea, we arrive at an interesting viewpoint that unifies static and dynamic logic gates.

Static CMOS circuits designed by following the single transition polarity optimization look in many ways like dynamic CMOS circuits. The NORA dynamic circuits discussed in Sections 5.9 to 5.10 have both precharge FETs and ground (or V_{DD}) switches. The ground and V_{DD} switches of some gates can be removed. Figure 8.18 shows cascaded AOI222-OAI222 static gates. The circuit is designed following single-transition-polarity optimization. FETs in the dotted boxes are shrunk to small sizes and precharge/predischarge FETs are added. Then the circuit is very similar to a NORA CMOS dynamic logic, and they are related.

The close relationship gives useful insights. First, if the FETs in the boxes are not removed but are designed very small, the switching delay of the preferred transition polarity would not be much different from that of NORA. Second, the circuit has all the advantages of static CMOS, most notably immunity against noise and charge sharing. Third, if we have a logic design using NORA or using static CMOS, one design can readily be converted to the other. There is a continuous spectrum of implementation between CMOS static logic and NORA dynamic logic. NORA logic is a useful theoretical concept providing a link between CMOS static and CMOS dynamic logic. This point is not obvious from Domino CMOS logic. The concept of single-transition-polarity timing optimization provides not only a powerful timing design technique, but also a unified viewpoint on dynamic and static CMOS gates.

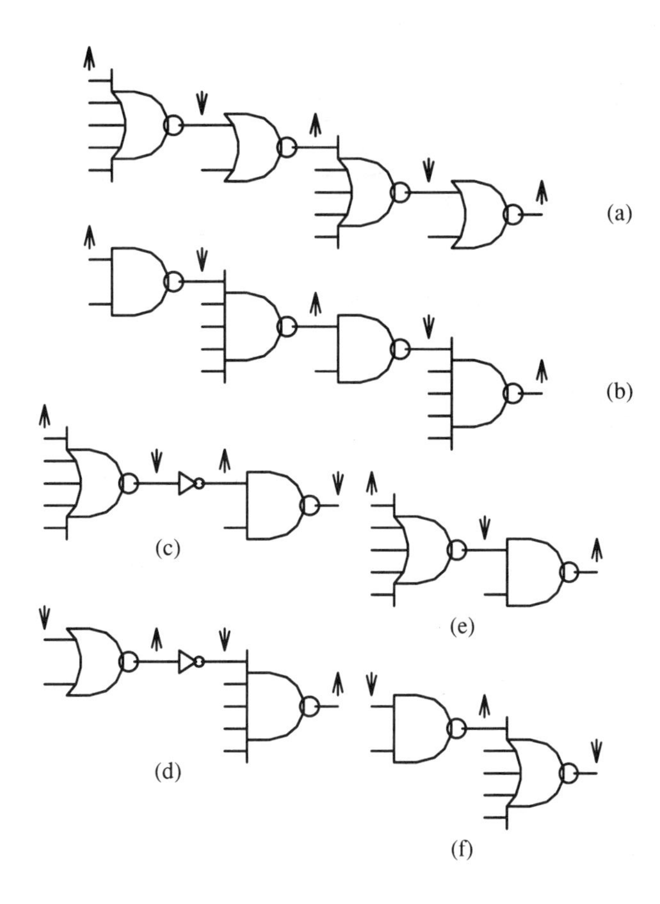

Figure 8.17 Delay optimization of cascaded chain.

8.12 AN ADVANCED FET SCALING TECHNIQUE

In the future CMOS VLSI chips having higher FET counts and higher operating frequencies, clocking schemes based on edge-triggered master-slave latches will be used more and more. Absence of a "dead phase" in edge-triggered timing demands higher precision in clock generation; in a scaled-down CMOS technology a typical gate delay is less than 1 ns. Then 1 ns of clock skew creates data flow-through across master-slave latch pairs (Section 7.7). However, less than 1 ns of clock skew over an entire chip is not an easy design objective to achieve. The problem is to equalize delays of all clock decoders and clock buffers and to maintain their matching under all process, temperature, and voltage conditions. Typical clock decoder and clock buffer circuits are shown in Fig. 8.19, and representative clock waveforms with and without skew are shown in Fig. 8.20.

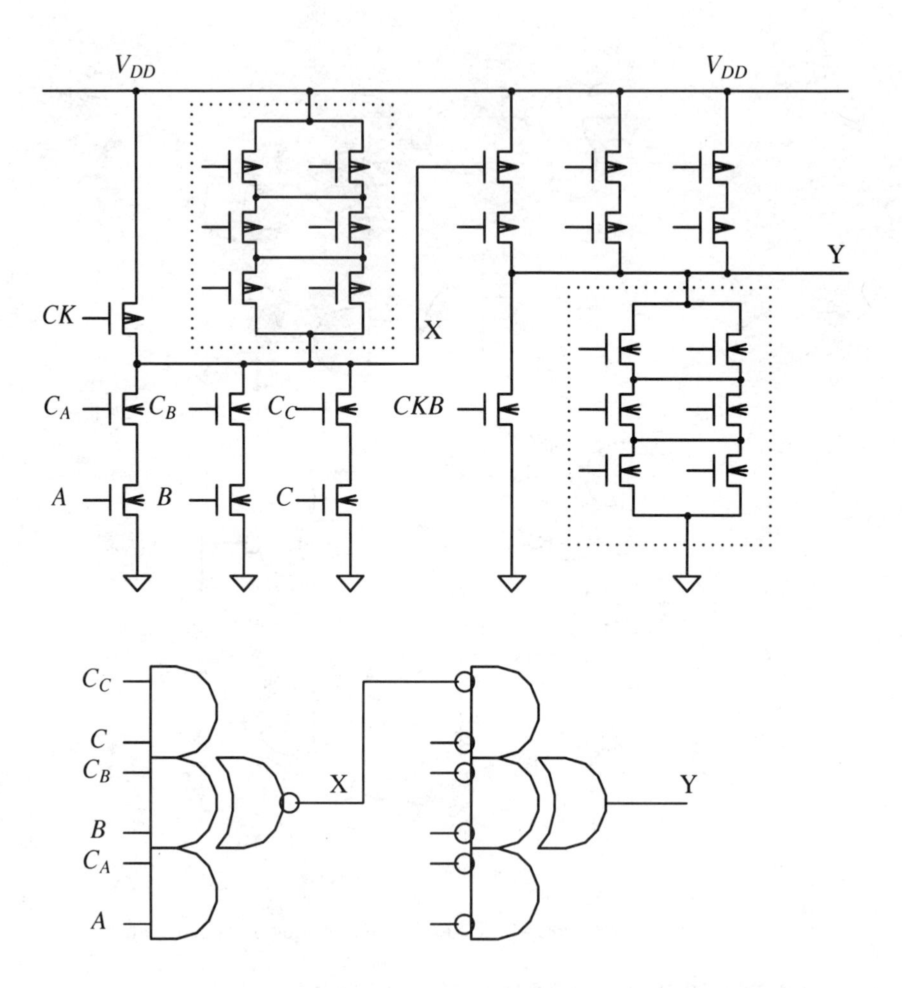

Figure 8.18 NORA dynamic gate and CMOS static gate.

Delay of a CMOS gate is defined as the difference time between the input voltage crossing $V_{DD}/2$ and the output voltage crossing $V_{DD}/2$. In Section 3.16 this definition was shown to be useful when the transition time of the input signals is less than several times (typically, three times) the delay of a gate for the input signal switching instantly. In clock circuits the condition is well satisfied.

Variation in the clock edges caused by processing spread is evaluated by extracting three sets of process parameters of CMOS FETs for a given CMOS process. They are; high-speed (or high-current) process parameters, typical (or medium-current) process parameters, and low-speed (or low-current) process parameters. The high-current process typically gives a half to a third of the gate delay of the low-current

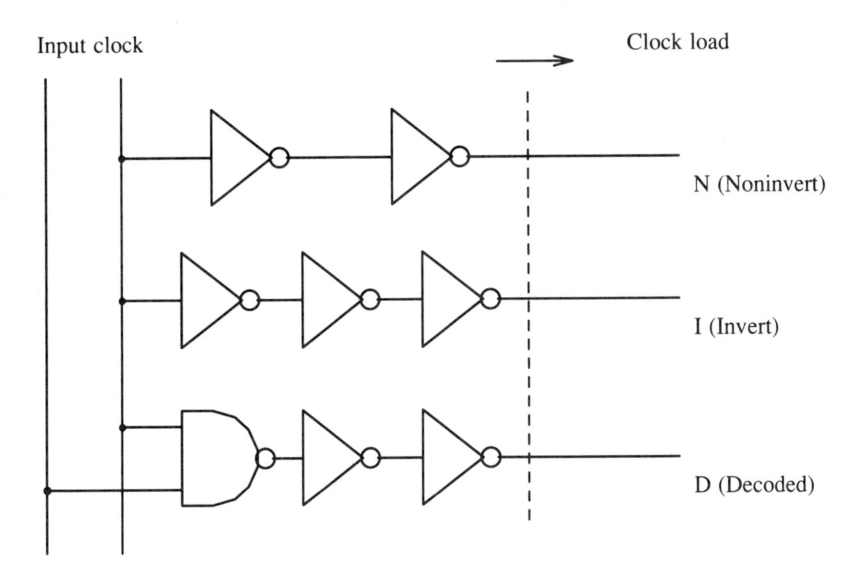

Figure 8.19 Clock signal distribution.

process. Even if all the clock edges are accurately aligned for a typical process, they may not be aligned for deviations from the typical process.

A conventional design verification procedure is to simulate designed circuits by a combined high PFET current/high NFET current process [abbreviated to (PH-NH) process] and low PFET current/low NFET current process [abbreviated to (PL-NL) process]. In this restricted combination the current drive capabilities of a PFET and an NFET vary approximately proportionally, and therefore the delays vary proportionally also. Clock skew generated by this type of variation is relatively small. This is, however, not representative of an actual chip production environment. PFETs and NFETs can vary somewhat independently. Such a process unbalance may occur by failure of control over the threshold voltages of PFETs and NFETs. It is important to note that by unbalancing a process, a larger clock skew is created than by speeding up or slowing down the process in a balanced way. The worst "unbalanced" process is either a combination of high PFET current and low NFET current (PH-NL process) or a combination of low PFET current and high NFET current (PL-NH process). Therefore, the problem is to produce a design that withstands this type of variation, as well as more balanced process variations.

High-speed and low-speed process parameters, are defined such that only 2% of the processed devices are expected to have FET drive capabilities beyond the respective specifications; 96% of the entire population should fall within the high and the low FET speed limits. A convenient parameter to define the drive capability is the time constant of a FET, T_{FET}, defined by

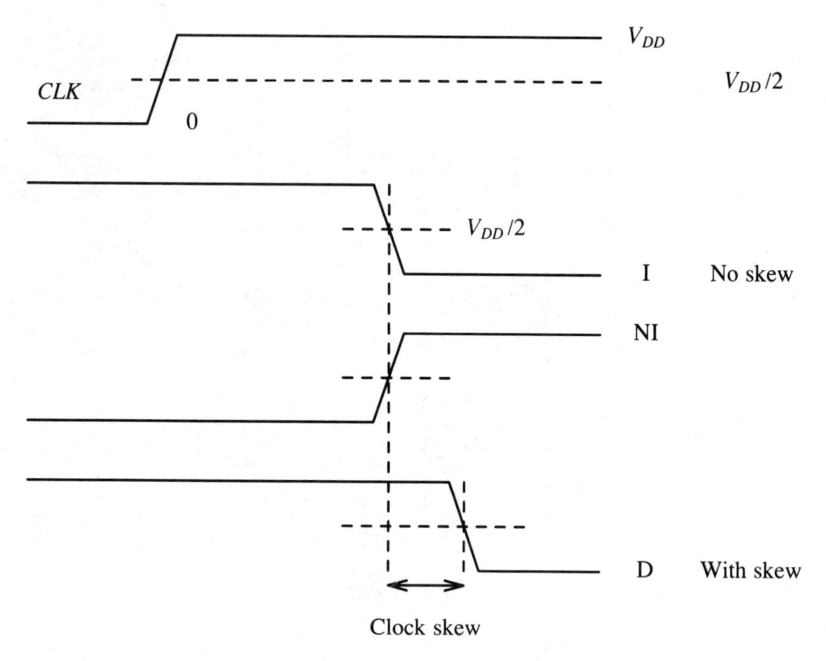

Figure 8.20 Definition of clock skew.

$$T_{FET} = (C_G + C_D)\frac{V_{DD}}{I_{Dmax}} \tag{8.37}$$

where C_G is the gate capacitance, C_D is the drain diffused island capacitance, and I_{Dmax} is the drain current when both gate and the drain voltages are set at V_{DD} (Section 1.12). Table 8.2 lists T_{FET}, for the three process conditions, for a twin-tub 1.75-μm CMOS process [19].

TABLE 8.2 TIME CONSTANT OF FETs

Process	NFET		PFET	
	ps	Ratio	ps	Ratio
High current (H)	88.0	0.556	201.0	0.620
Typical (M)	158.0	1.000	324.0	1.000
Low current (L)	273.0	1.730	530.0	1.630

Drive capability is the inverse of the FET time constant. This result shows that ratio of drive capabilities of NFET to PFET varies from 6.02 for the PL-NH process to 0.73 for the PH-NL process. This variation is the absolute limit, since parameters like

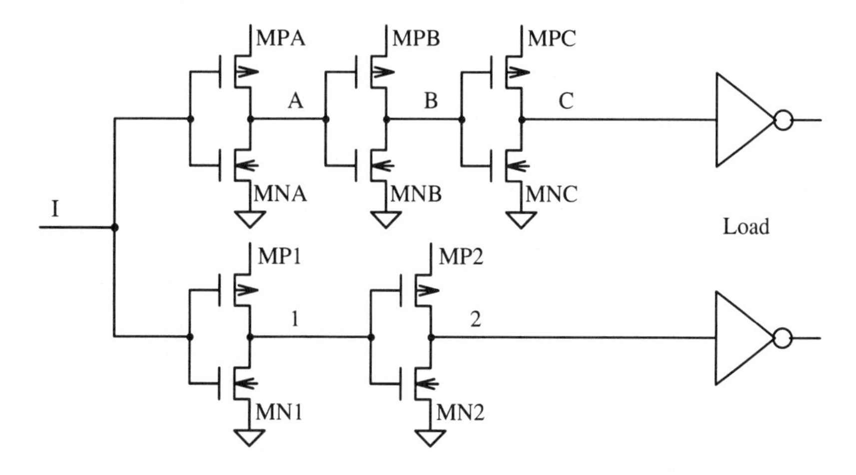

Figure 8.21 A circuit that generates a pair of CMOS clocks.

NFET channel length and thin oxide thickness are likely to track the respective parameters of PFET.

Figure 8.21 shows CMOS inverter chains used to generate an inverting and noninverting clock. When input node I makes a low-to-high transition, the waveforms at the intermediate nodes A, B, C, 1, and 2 are as shown in Fig. 8.22(a) and (b). To make the assertion and antiassertion clocks skew-free, the FETs are scaled to satisfy

$$T_I = T_{NI} = T_0 \tag{8.38}$$

where

$$T_I = T_A + T_B + T_C$$

$$T_{NI} = T_1 + T_2$$

This design is carried out assuming typical process conditions. When the same circuit is processed by the high-current process, delays T_A, T_B, ..., T_2 will be different. Delay T_A depends on the drive capability of an NFET (MNA of Fig. 8.21) which changes by a factor f_N, where $f_N > 1$. Then

$$T_A(H) = T_A(1/f_N) \quad (< T_A)$$

In this equation, $(1/f_N)$ can be found from the (NFET, ratio) column of Table 8.2, as 0.556. Delay T_B depends, however, on the driving capability of a PFET (MPB in Fig. 8.21), which changed by a factor f_P. Then

$$T_B(H) = T_B(1/f_P)$$

and

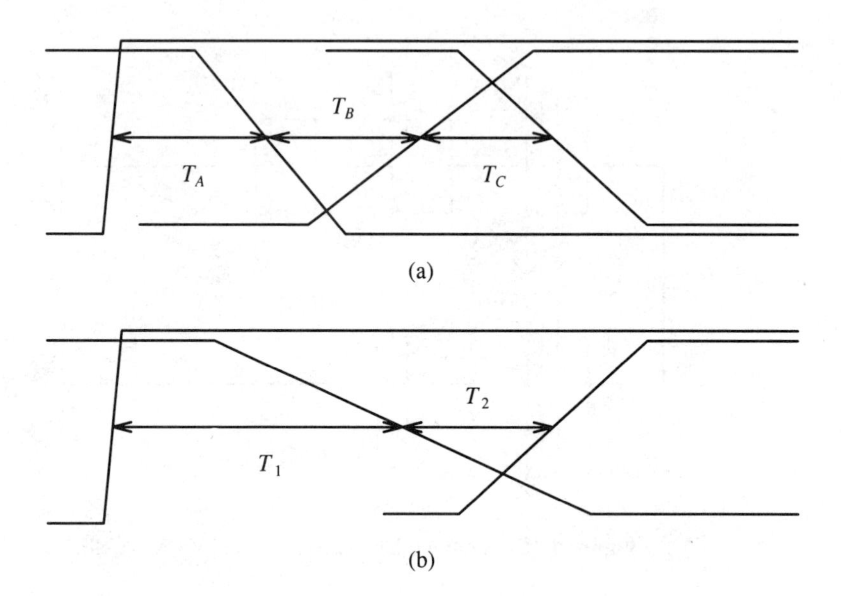

Figure 8.22 Waveforms of nodes.

$$(1/f_P)=0.620$$

from Table 8.2. In this way we obtain

$$T_I(H)=(T_A+T_C)(1/f_N)+T_B(1/f_P)$$
$$T_{NI}(H)=T_1(1/f_N)+T_2(1/f_P) \tag{8.39}$$

and therefore

$$T_I(H)-T_{NI}(H)=(T_B-T_2)\left\{(1/f_P)-(1/f_N)\right\} \tag{8.40}$$

From Eqs. (8.39) and (8.40),

$$T_I(H)=T_{NI}(H)$$

when $f_N=f_P$ or $T_2=T_B$, or when both are satisfied. If the circuit is processed by the PH-NH or PL-NL process, $f_P=f_N$ is approximately satisfied (within 10%, from Table 8.2). Then the clocks remain skew-free against this type of variation. If the process varies from the typical (PM-NM) process to a (PL-NH) process, however, $f_N>1$ and $f_P<1$. Then the circuit will have clocks with large skew.

Equation (8.40) shows, however, that if the circuit is originally designed to satisfy

$$T_B=T_2 \tag{8.41}$$

in addition to the condition of zero skew at the typical process [Eq. (8.38)], the circuit remains skew-free whatever the process may be [20]. From Eqs. (8.38) and (8.41) we obtain

$$T_A + T_C = T_1 \tag{8.42}$$

Equations (8.41) and (8.42) mean the following. When input node I of Fig. 8.21 pulls up, gates A, C, and 1 pull down, by NFET MNA, MNC, and MN1, respectively. Equation (8.42) means that the sum of the pulldown delays by NFET MNA (gate A) and MNC (gate C) of the first logic chain should be equal to the pulldown delay of NFET MN1 (gate 1) of the second logic chain. Further, gate B of the first chain and gate 2 of the second chain pull up. Equation (41) means that the pullup delays of PFET MPB (gate B) and that of MP2 (gate 2) should be equal. By Eqs. (8.41) and (8.42) the delays contributed by NFETs and PFETs are separately and precisely matched. If the precision matching is exercised, the delays of the two chains are robustly matched under any process variation. If only total delays of the two chains are matched, the delay will not be matched with process variation.

This precision design procedure is always possible. Delays are determined by the size of the FETs in Fig. 8.21 (MPA, MNA ,..., MN2). There are 10 FET sizes, and there are only two equations, so there are many solutions to the problem. The way to choose the most desirable solution is discussed later.

Several versions of the circuit shown in Fig. 8.21 were designed and their skew characteristics examined by circuit simulation. Table 8.3 summarizes the results. The inverter chains had fixed 40 times the minimum-size inverter load, and had delays for the typical process ranging from 2 to 3 ns. Wiring parasitics were ignored, since clock generator circuits are usually laid out very compactly.

TABLE 8.3 CLOCK SKEW DATA

Design	$(T_A + T_C)/T_B$	T_1/T_2	Skew (ns)				
—	—	—	PH-NH	PL-NL	PM-NM	PH-NL	PL-NH
I	1.34/0.88	1.22/1.01	0.02	0.02	0.00	0.02	0.00
II	1.50/0.62	0.82/1.31	0.00	0.08	0.00	0.38	0.07
III	1.46/1.38	0.61/2.23	0.04	0.00	0.00	0.46	1.30

Design data (FET sizes):

I. MPA(10) MNA(4) MPB(10) MNB(4) MPC(40) MNC(16)
MP1(7) MN1(3) MP2(50) MN2(20) LOAD PFET(100) NFET(40)

II. MPA(10) MNA(4) MPB(19) MNB(8) MPC(32) MNC(13)
MP1(5) MN1(2) MP2(20) MN2(8) LOAD PFET(100) NFET(40)

III. MPA(10) MNA(4) MPB(5) MNB(2) MPC(40) MNC(16)
MP1(5) MN1(2) MP2(9) MN2(7) LOAD PFET(100) NFET(40)
(FET sizes are multiples of minimum-size FET)

In the Table, FET size was measured in units of the minimum-size FET, which is 3.25 μm in our 1.75-μm CMOS. Design I approximately follows the new design principle, and therefore the skew is quite small: Eqs. (8.41) and (8.42) are satisfied within an accuracy of 13%. This design attained the best skew compensation. The best skew compensation occurs when Eqs. (8.41) and (8.42) are closely satisfied, but not necessarily when they are exactly satisfied. The theory assumes that the delay of an inverter chain is a sum of the delays of the component inverters, which is only an approximation. That there is an FET scaling of the inverter chains that provides the minimum skew remains true independent of the accuracy of approximation.

Design II, which is typical design before the new design principle was found, and design III, which is a very unbalanced design, give increasingly more skew as process variation increases. Unbalanced processes such as PH-NL or PL-NH consistently generate more skew than balanced processes such as PH-NH or PL-NL. The worst-case skew of design I (0.02 ns) is one order of magnitude smaller than the worst-case skew of design II (0.38 ns), showing how effective the design practice is. The input voltage to the chain was a pulse that had a constant rise time of 0.5 ns. This was convenient to allow accurate determination of very small skew by simulation.

Figure 8.23 shows a map of skew versus percentage of the pullup delay in each of the two chains. With reference to Fig. 8.22(a) and (b), the horizontal axis is the quantity h defined by

$$h = (T_2 / T_0) \times 100\,(\%),$$

and the vertical axis is the quantity v defined by

$$v = (T_B / T_0) \times 100\,(\%)$$

where T_0 is the delay of the chains, at the typical process. The number associated with each point in Fig. 8.23 is the percentage skew defined by

$$\left(\frac{|Skew|}{T_0} \right) \times 100\%$$

Figure 8.23 shows that minimum skew occurs approximately when $h = v$ is satisfied, but not exactly. For very accurate compensation, the range of h is practically limited to 35 to 50%. The theory of the preceding section assumes that the delay of the cascaded inverter chain is equal to the sum of the individual inverter delays. In reality the delay depends on the waveforms of the signals (Section 3.16), and the theory does not hold when excessive waveform distortion occurs due to heavy capacitive loading or Miller effect. In spite of this complication the skew compensation does work: The range of h for the best compensation falls within the range where the design of the chains is the easiest. Practical applications never suffer from the limited range of h.

The discussion of this section is based on the input clock of Fig. 8.21 making a low-to-high transition. To design a pair of symmetrical CMOS clock buffers whose

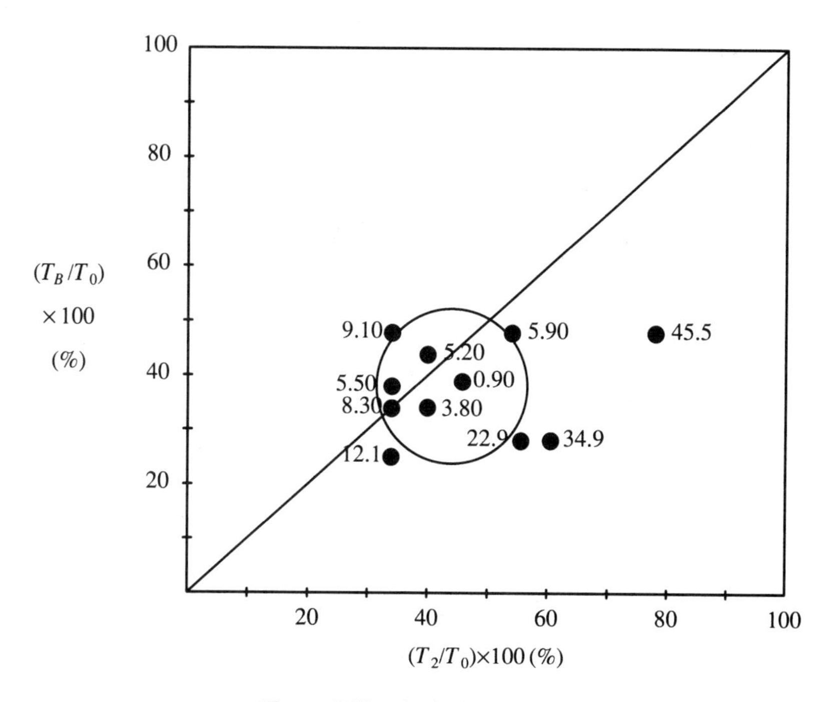

Figure 8.23 Optimization of skew.

outputs are matched for both input transitions is more involved. Our best design is summarized in Table 8.4.

TABLE 8.4 CLOCK SKEW DATA

		Skew (ns)				
$(T_A + T_C)/T_B$	T_1/T_2	PH-NH	PL-NL	PM-NM	PH-NL	PL-NH
1.34/1.03	1.27/1.10	0.00	0.06	0.00	0.00	0.07

Design data:

MPA(15) MNA(6) MPB(13) MNB(6) MPC(66) MNC(27)

MP1(7) MN1(3) MP2(44) MN2(18) LOAD PFET(132) NFET(54)

(FET sizes are multiples of minimum-size FET)

Although the skew is not as good as design I, this inverter chain generates a practically perfect pair of CMOS clocks against all the process variations.

Application of the basic concept of this section is wide. For example, a pair of clock generator logic chains, one of which includes a NAND2 decoder, the other a

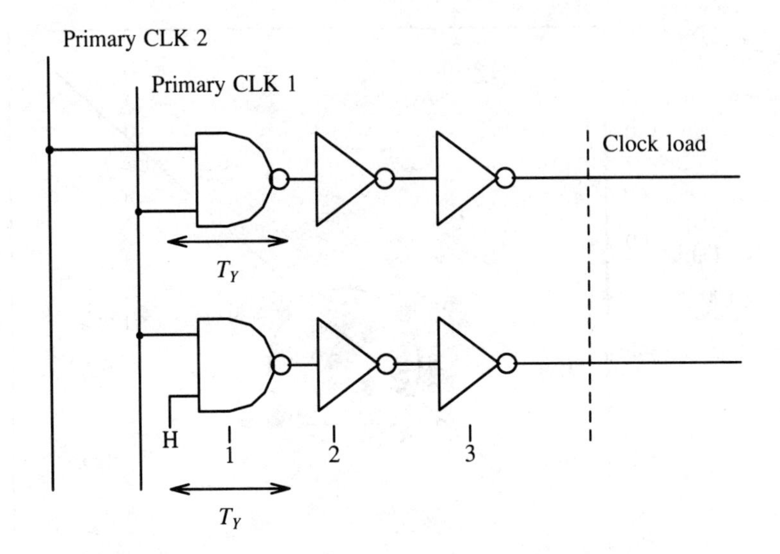

Figure 8.24 Local clock decoder.

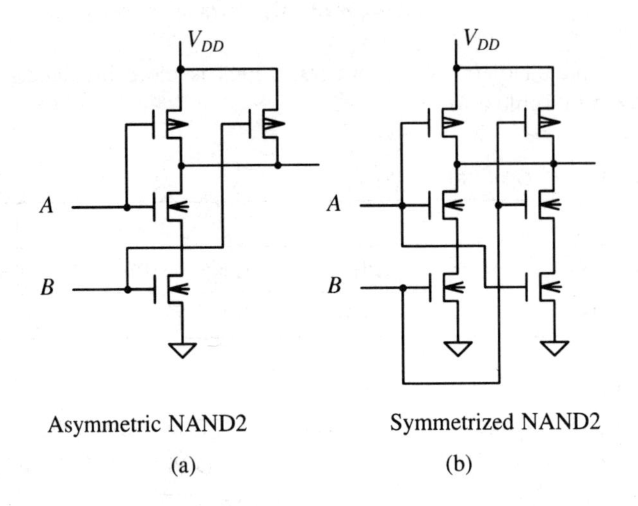

Figure 8.25 Symmetric and asymmetric NAND2 gates.

straight chain of inverters, can be matched by replacing the first stage of the inverter chain by a NAND2 gate, with one input permanently connected to V_{DD}, as shown in Fig. 8.24. This is necessary because process dependence of delay of series-connected NFETs is different from that of a single NFET. In general, if one of the chains includes a gate other than an inverter, the other chain must include the same kind of gate at any one of the equivalent locations. The NAND2 gate of Fig. 8.24 can be either at location 1 or 3, but not at 2. Pulldown delays Tx and Ty (by the series-connected NFETs of NAND gates) must be matched separately. In this precision application, the difference in pulldown delay in response to a transition from input A and delay in response to a transition from input B of the NAND2 gate shown in Fig. 8.25(a) is important. The delay from input A is shorter than that from input B. The delays can be made equal by using the symmetrized NAND2 gate shown in Fig. 8.25(b).

8.13 REFERENCES

01. H. B. Bakoglu and J. D. Meindl "Optimal interconnect circuits for VLSI" ISSCC84 Digest, pp. 164-165, February 1984.
02. H. B. Bakoglu and J. D. Meindl "Optimal interconnect circuits for VLSI" IEEE Trans. on Electron Devices, vol. ED-32, pp. 903-909, May 1985.
03. G. Bilardi, M. Pracchi and F. P. Preparata "A critique of network speed in VLSI-Models of computation" IEEE J. Solid-State Circuits, vol. SC-17, pp. 696-702, August 1982.
04. K. C. Saraswat and F. Mohammadi "Effect of scaling of interconnections on the time delay of VLSI circuits" IEEE Trans. on Electron Devices, vol. ED-29, pp. 645-650, April 1982.
05. W. W. Troutman, P. W. Diodato, A. K. Goksel, M-S Tsay and R. H. Krambeck "Design of a standard floating-point chip" IEEE J. Solid-State Circuits, vol. SC-21, pp. 396-399, June 1986.
06. K. Hardee, M. Griffus and R. Galvas "A 30ns 64K CMOS RAM" ISSCC84 Digest, pp. 216-217, February 1984.
07. W. Ip, T. L. Chiu, T. C. Wu and G. Perlegos "256K CMOS EPROM" ISSCC84 Digest, pp. 138-139, February 1984.
08. H. C. Lin and L. W. Linholm "An optimized output stage for MOS integrated circuits" IEEE J. Solid-State Circuits, vol. SC-10, pp. 106-109, April 1975.
09. C. Mead and M. Rem "Minimum propagation delays in VLSI" IEEE J. Solid-State Circuits, vol. SC-17, pp. 773-775, August 1982.
10. A. M. Moshen and C. A. Mead "Delay-time optimization for driving and sensing of signals on high-capacitance paths of VLSI systems" IEEE J. Solid-State Circuits, vol. SC-14, pp. 462-470, April 1979.
11. C. M. Lee and H. Soukup "An algorithm for CMOS timing and area optimization" IEEE J. Solid-State Circuits, vol. SC-19, pp. 781-787, October 1984.
12. E. T. Lewis "Optimization of device area and overall delay for CMOS VLSI designs" Proc. IEEE, vol. 72, pp. 670-689, June 1984.
13. S. M. Kang "A design of CMOS polycells for LSI circuits" IEEE Trans. on Circuits and Systems, vol. CAS-28, pp. 838-843, August 1981.
14. A. Kanuma "CMOS circuit optimization" Solid-State Electronics, vol. 26, pp. 47-58, January 1983.

15. J. P. Fishburn and A. E. Dunlup "Tilos: A polynomial programming approach to transistor sizing" ICCAD85 Digest, pp. 326-328, 1985.
16. K. S. Hedlund "Models and algorithms for transistor sizing in MOS circuits" ICCAD84 Digest, pp. 12-14, 1984.
17. L. P. J. Hoyte "Automated calculation of device sizes for digital IC designs" MIT VLSI Memo No. 82-116, September 1982.
18. J. D. Ullman "Computational aspects of VLSI" Computer Science Press, Rockville, Md., 1984.
19. L. C. Parrillo, L. K. Wang, R. D. Swenumson, R. L. Field, R. C. Melin and R. A. Levy, "Twin-tub CMOS II-advanced VLSI technology" IEDM82 Digest, p. 706, October 1982.
20. M. Shoji "Elimination of process-dependent clock skew in CMOS VLSI" IEEE J. Solid-State Circuits, vol. SC-21, pp. 875-880, October 1986.

9

Practical Problems of CMOS VLSI

9.1 INTRODUCTION

In this chapter we discuss several practical problems of CMOS VLSI chip design. The discussion covers the following areas:

1. How to estimate the noise amplitude, and how to design a circuit that is resistant to noise
2. Packaging problems including input protection and chip power estimate
3. Testability measure and methods of making a large VLSI chip testable
4. How to improve a chip's reliability in hazardous environments
5. How to estimate the yield of VLSI chip manufacturing, and how to improve the yield
6. Chip layout methods

Digital circuits were originally introduced to process signals reliably. Scale-down of the device size and integration of huge number of devices has reached the point that even a digital circuit is vulnerable to many environmental effects, such as noise. How to control the noise generated by switching a huge number of CMOS gates is now a major concern of designers.

Chip packaging technique influences chip design significantly. Chips must withstand electrostatic discharge (ESD) during handling. In high-speed chips it is often difficult to design an input protection circuit that has an acceptably short delay. Power dissipation of high-speed chips is often so high that accurate chip power estimate and special package development have become necessary. Advanced packaging technology being developed now will solve these problems.

The ever-increasing number of circuits integrated into a single chip, and the relatively small number of terminals that can be attached to the chip's periphery created a problem of observability of the internal circuit operation of a digital VLSI chip which hampered effective testing. It is now taken for granted that a functional test of a VLSI chip is never able to detect all the problems that might exist in the chip. There is a possibility that a faulty chip will be delivered to the user. How to reduce this possibility is a serious concern of designers.

Modern high-performance VLSI chips are often used in a harsh environment, including those of high temperature, high mechanical stress, and strong ionizing radiation generated by nuclear detonation. How to harden a chip to the environment is the fundamental concern in a military VLSI project. As for civilian applications, however, the profit from the product ultimately depends on the yield of the chip, and a designer must predict the yield. How to carry out the layout of a VLSI chip that contains several hundred thousand FETs is a serious project organization problem. These problems are discussed in this chapter.

9.2 NOISE IN CMOS VLSI

The noise phenomena discussed in the following sections are not caused by thermal noise or the discreteness of charge. In a CMOS VLSI chip, insulation or shielding of a logic gate from the rest of the circuits of the chip is not perfect. Switching of a circuit located in the vicinity affects the node voltage within the gate through various coupling mechanisms. Although the noise voltage is predictable in the strict sense, in a VLSI chip the induced voltage has most of the attributes of stochastic noise. Further, the induced voltage is never orders of magnitude less than the signal voltage, and therefore severe noise may cause malfunction of the circuit. Mechanisms of noise voltage coupling, estimates of the coupled noise voltages, and the technique of preventing noise-induced circuit malfunctions are the subjects of the following sections. Wallmark has summarized the noise phenomena in VLSI chips [01].

Two distinct mechanisms causing noise in CMOS digital circuits exist. The first is a local phenomenon, that the noise voltage is induced from one signal node to another; the mechanism is called an induced noise. The other is noise voltages on the ground or V_{DD} power bus that affect all circuits on the chip, and is called power bus noise. We begin with induced noise.

In Fig. 9.1, driver D_1 holds node B at ground potential. Between node B (which is an extended metal connection) and driver D_1 is a large resistance, which can be a polysilicon crossover resistance, resistance of a transmission gate, or the like. The second node, A, is also an extensive metal connection, and node A is in close proximity to node B. Nodes A and B are coupled capacitively. Node A is driven by driver D_2. When driver D_2 drives node A from 0 to V_{DD} V, noise voltage is induced on node B by capacitive coupling, as shown in the figure. Because of the high wiring

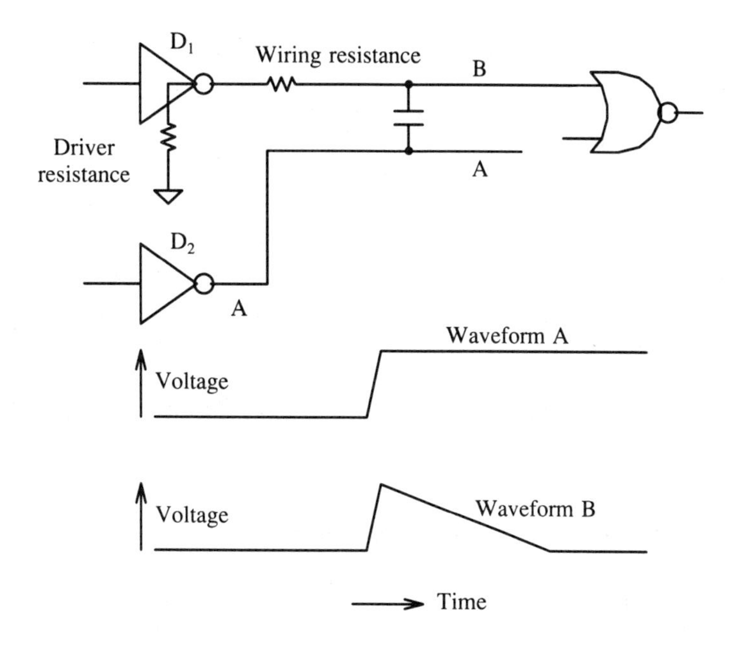

Figure 9.1 Mechanism of induced noise.

resistance and the internal resistance of driver D_1, the induced voltage on node B takes some time to decay. If the induced voltage is high and if the induced voltage spike lasts longer than a certain minimum, NOR2 gate driven by node B generates a glitch, thereby causing logic error.

9.3 INDUCED NOISE

Noise induction phenomena in a CMOS VLSI chip can be classified into four cases, shown in the equivalent circuit of Figs. 9.2(a) to (d). Figure 9.2(a) shows noise induction from a metal wire (top) to another metal wire (bottom). Figure 9.2(b) shows noise induction from a polysilicon wire to a metal wire, Fig. 9.2(c) shows noise induction from a metal wire to a polysilicon wire, and Fig. 9.2(d) shows noise induction from a polysilicon wire to another polysilicon wire. We assume doped polysilicon interconnects, since silicided interconnects have significantly less induced noise.

(A) Metal-Metal Induced Noise

Driver D_2 in Fig. 9.2(a) is assumed to have negligibly small internal resistance, and therefore node A is driven up from 0 to V_{DD} V within a short time. At time $t = +0$

(+ 0 means a little after transition), the voltage of node B is given by

$$V_B(+0) = \frac{C_1}{C_0 + C_1} V_{DD} \tag{9.1}$$

and after that time, node capacitance $(C_0 + C_1)$ discharges through resistor R. $V_B(t)$ is given by

$$V_B(t) = \left[\frac{C_1}{C_0 + C_1} \right] V_{DD} e^{-\frac{t}{R(C_0 + C_1)}} \tag{9.2}$$

The noise waveform is an exponential spike. The height of the noise spike is determined by capacitive coupling coefficients. If we consider the worst case of two metal wires overlapping, $C_1/(C_0 + C_1)$ is 0.5. Then the height is $(1/2)V_{DD}$. The resistance R depends, however, on the circuit. If $C_0 + C_1 = 5\,pF$ (typical for 1 cm-long wire) and $R = 5\,k\Omega$ (typical for resistance of a transmission gate), the spike width is 25 ns. This induced noise is significant when the resistance R, $C_0 + C_1$, and $(C_1/(C_0 + C_1))V_{DD}$ are large, and consequently the spike is high and wide.

(B) Polysilicon-Metal Induced Noise

Figures 9.2(b)–(d) show equivalent circuits of distributed RC chains. The parameters R and C in the figures are the resistance and the capacitance per unit length of the line. Spatial coordinate x is chosen such that the driving point is $x = 0$. Calculation of the coupling capacitance is a problem in electrostatics which is straightforward but usually very tedious. The dependence of coupling capacitance on line width, gap, and oxide thickness is complex [02]. Several papers on the subject have been published [03]. Here we assume that the capacitance values are known.

Induced noise from a polysilicon line to a metal line is analyzed using the equivalent circuit of Fig. 9.2(b). Induced noise vanishes if the length of the coupled lines is infinite. We assume finite length and that the induced noise voltage $V(t)$ is much smaller than V_{DD}. We analyze the case $C_m \gg C_1 L$, where L is the length of the line and C_1 is the coupling capacitance per unit length. The left end of the polysilicon line is driven up from 0 V to V_{DD} V at $t = 0$. After time t, the distributed capacitance of the polysilicon line between the driving point P and point Q are charged to voltages higher than $V_{DD}/2$, where the distance $\overline{PQ}$ is given by (see Section 6.8)

$$x = \overline{PQ} = \left\{ \frac{t}{R(C_0 + C_1)} \right\}^{1/2} \tag{9.3}$$

For simplification of the problem, we assume that the capacitances C_0 and C_1 between points P and Q are charged to V_{DD} and that the rest of the capacitances are not charged at all. Since the boundary of the charged/not-charged region moves with time by Eq. (9.3), current $I(t)$ is injected into the metal line,

$$I(t) = \left\{ V_{DD} - V(t) \right\} \frac{d}{dt}(C_1 x) \approx \frac{C_1 V_{DD}}{2\sqrt{R(C_0 + C_1)t}} \tag{9.4}$$

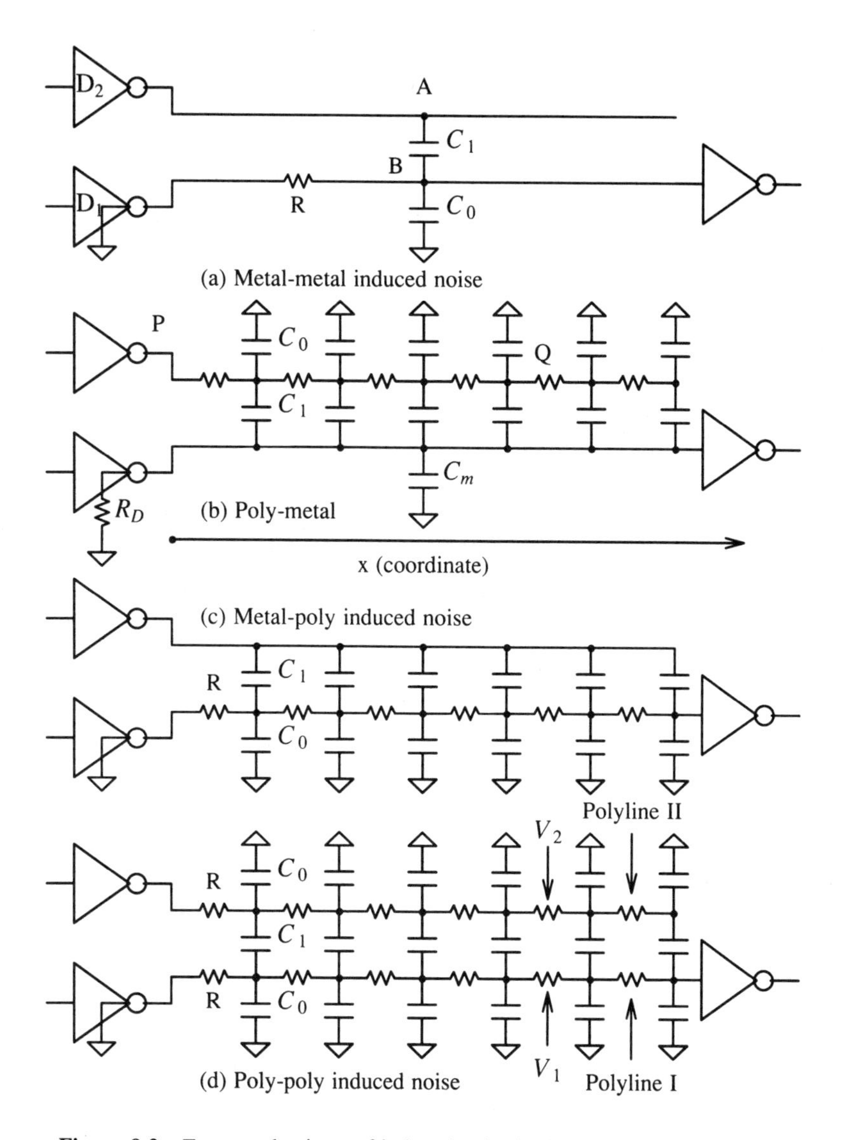

Figure 9.2 Four mechanisms of induced noise in CMOS VLSI chip.

where we assume that $V(t) \ll V_{DD}$. $V(t)$ satisfies the equation

$$\frac{d}{dt}V(t) + \frac{1}{C_m R_D}V(t) = \frac{I(t)}{C_m} = \frac{C_1 V_{DD}}{2C_m \left\{ R(C_0 + C_1)t \right\}^{1/2}}$$

since $C_m \gg C_1 L$.

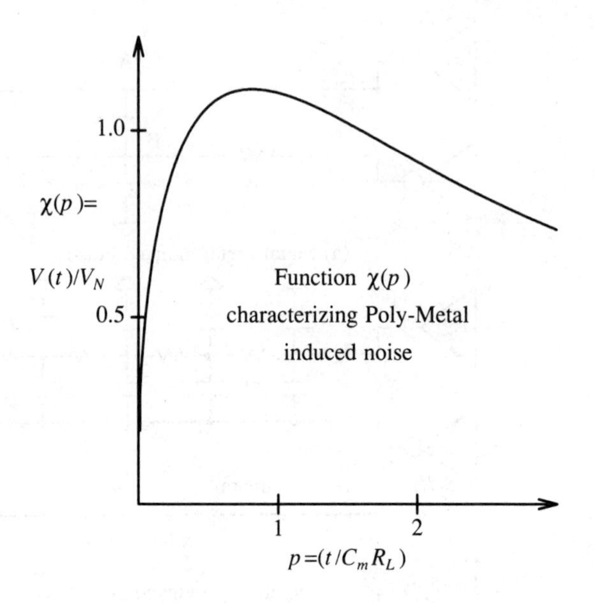

Figure 9.3 Polysilicon-to-metal induced noise.

This equation can be solved in a closed form subject to the initial condition, $V(0) = 0$, as

$$V(t) = V_N \chi \left[\frac{t}{C_m R_D} \right] \tag{9.5}$$

$$V_N = \frac{V_{DD}}{2} \frac{(C_m R_D)^{1/2}}{\left\{ R(C_0 + C_1) \right\}^{1/2} \left[\frac{C_m}{C_1} \right]}$$

where

$$\chi(p) = e^{-p} \int_0^p \frac{e^{\xi}}{\sqrt{\xi}} d\xi$$

Function $\chi(p)$ is plotted in Fig. 9.3. The function has a maximum at $p = 0.9$. Equation (9.5) is valid until the moving point Q reaches the end of the polysilicon line at time t_f given by

$$t_f = R(C_0 + C_1)L^2$$

where L is the length of the line. When $t > t_f$, $V(t)$ is given by

$$V(t) = V(t_f) \exp \left\{ -(t - t_f)/R_D C_m \right\} \tag{9.6}$$

The height of the spike is at most

$$0.54V_{DD}\left[\frac{C_1}{C_0+C_1}\right]\left\{\left[\frac{(C_0+C_1)L}{C_m}\right]\left[\frac{R_D}{R\cdot L}\right]\right\}^{1/2}$$

When $C_1/(C_0+C_1)=0.5$, $(C_0+C_1)L/C_m=1$, and $R_D/R\cdot L=1/7$, the height of the spike is $0.10V_{DD}$. This type of induced noise has smaller amplitude than that of metal-metal induced noise and metal-polysilicon induced noise.

(C) Metal-Polysilicon Induced Noise

At time $t=+0$ (immediately after $t=0$) the noise voltage induced on the polysilicon line is given by Eq. (9.1). Evolution of the potential profile on the polysilicon line is determined from the circuit equation

$$\frac{\partial^2}{\partial x^2}V(x,t)=R(C_0+C_1)\frac{\partial}{\partial t}V(x,t) \tag{9.7}$$

subject to the boundary condition

$$V(0,t)=0$$

where we assumed that the driver has negligible internal resistance. For simplicity, the line is assumed to be infinitely long. The solution is then given by

$$V(x,t)=\frac{C_1}{C_0+C_1}V_{DD}\,\Phi\left\{\left[\frac{R(C_0+C_1)}{t}\right]^{1/2}\frac{x}{2}\right\} \tag{9.8}$$

where

$$\Phi(p)=\frac{2}{\sqrt{\pi}}\int_0^p e^{-u^2}du$$

is the error function. Equation (9.8) is plotted in Figs. 9.4(a) and 9.4(b). Figure 9.4(b) shows that at a point a distance x away from the driver's end, the width of the noise voltage spike, t_{sp}, is given by

$$t_{sp}=1.1R(C_0+C_1)x^2 \tag{9.9}$$

In a 3.5-μm CMOS having a doped polysilicon gate, $R=6.25\times10^4\Omega/cm$, $C_0=3.05$ pF/cm, and $C_1=3.14$ pF/cm, if the metal line completely overlaps the polysilicon line. Since $V_{DD}=5$ V, the pulse height is 2.54 V and the pulse width is given by

$$t_{sp}(\text{ns})=425x^2\,(x\text{ in cm})$$

When $x=0.2$ cm, $t_{sp}=17$ ns. This numerical example is for the old-generation, 3.5-μm CMOS technology that does not have polysilicide interconnects. There is indeed a

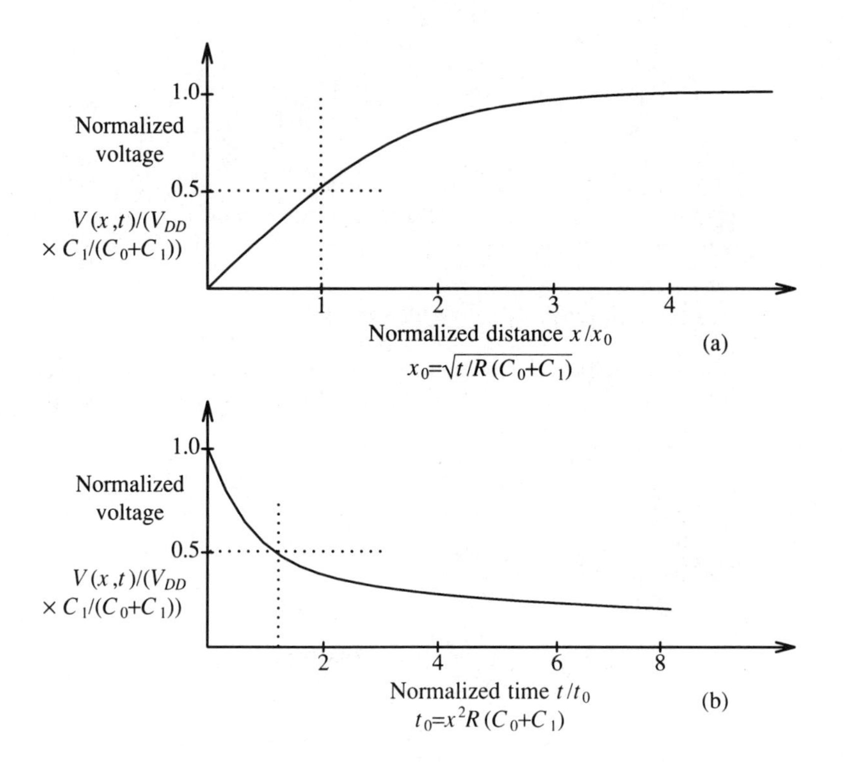

Figure 9.4 Metal to polysilicon induced noise.

significant noise effect. Since the delay of a 3.5-μm CMOS technology gate is typically 3 to 5 ns, a 17 ns wide noise pulse has a significant effect on the circuit operation. The example is pessimistic, since the metal and the polysilicon wires were assumed to overlap entirely, but in spite of that, this estimate warns that induced noise must be controlled in a doped-polysilicon CMOS technology having only one level of metal. To reduce t_{sp} the practical ways are to reduce coupling capacitance C_1 by layout techniques and to increase the conductivity of the polysilicon line by silicided technology. When the resistance is reduced from the 25 Ω/square of doped polysilicon to the 4 Ω/square of polysilicide, $t_{sp} = 68x^2$, or $t_{sp} = 2.7$ ns for $x = 0.2$ cm. This narrow noise pulse is unable to influence gate operations.

When $t \to \infty$ Eq. (9.8) gives,

$$V(x,t) \to 0.564 V_{DD} \frac{C_1}{C_0 + C_1} \left\{ \frac{R(C_0 + C_1)x^2}{t} \right\}^{1/2}$$

This result indicates that the induced noise voltage decays very slowly. This conclusion, however, is subject to the assumption that the polysilicon line is infinitely long. When the line has finite length the problem is as discussed in Section 9.4.

Figure 9.4(b) shows that the noise voltage has a narrow spike at $t = 0$. This spike occurs when the voltage of the metal wire changes from 0 to V_{DD} within a negligibly short time. When the metal wire voltage is not a step function of time but is a continuous waveform described by $V_m(t)$, the noise voltage for small t is given by

$$V(x,t) = \frac{C_1}{C_0 + C_1} V_m(t)$$

If the rise time of the metal line voltage is comparable to the width of the spike, the spike is significantly smoothed out.

(D) Polysilicon-Polysilicon Induced Noise

The problem of induced noise from a polysilicon line to another polysilicon line can be analyzed using the equivalent circuit of two mutually coupled RC chains shown in Fig. 9.2(d). The two chains are assumed to have a symmetrical configuration. Voltages on the two polysilicon lines, $V_1(x,t)$ and $V_2(x,t)$, are determined from a pair of coupled partial differential equations,

$$\frac{\partial^2 V_1}{\partial x^2} = RC_0 \frac{\partial V_1}{\partial t} + RC_1 \frac{\partial}{\partial t}(V_1 - V_2)$$

$$\frac{\partial^2 V_2}{\partial x^2} = RC_0 \frac{\partial V_2}{\partial t} - RC_1 \frac{\partial}{\partial t}(V_1 - V_2)$$

$V_1(x,t)$ and $V_2(x,t)$ are transformed to a set of variables ψ_1 and ψ_2 defined by

$$V_1(x,t) = \frac{1}{2}\left\{\psi_0(x,t) + \psi_1(x,t)\right\} \tag{9.10}$$

$$V_2(x,t) = V_{DD} + \frac{1}{2}\left\{\psi_0(x,t) - \psi_1(x,t)\right\}$$

and $\psi_0(x,t)$ and $\psi_1(x,t)$ satisfy,

$$\frac{\partial^2}{\partial x^2}\psi_0(x,t) = RC_0 \frac{\partial}{\partial t}\psi_0(x,t) \tag{9.11}$$

$$\frac{\partial^2}{\partial x^2}\psi_1(x,t) = R(C_0 + 2C_1)\frac{\partial}{\partial t}\psi_1(x,t)$$

If both polysilicon lines are at zero potential at $t = 0$, and if the end of the second line is pulled up from 0 to V_{DD} volts at that time, the initial conditions are

$$\psi_0(x,0) = -V_{DD} \quad \text{and} \quad \psi_1(x,0) = V_{DD}$$

and the boundary conditions are

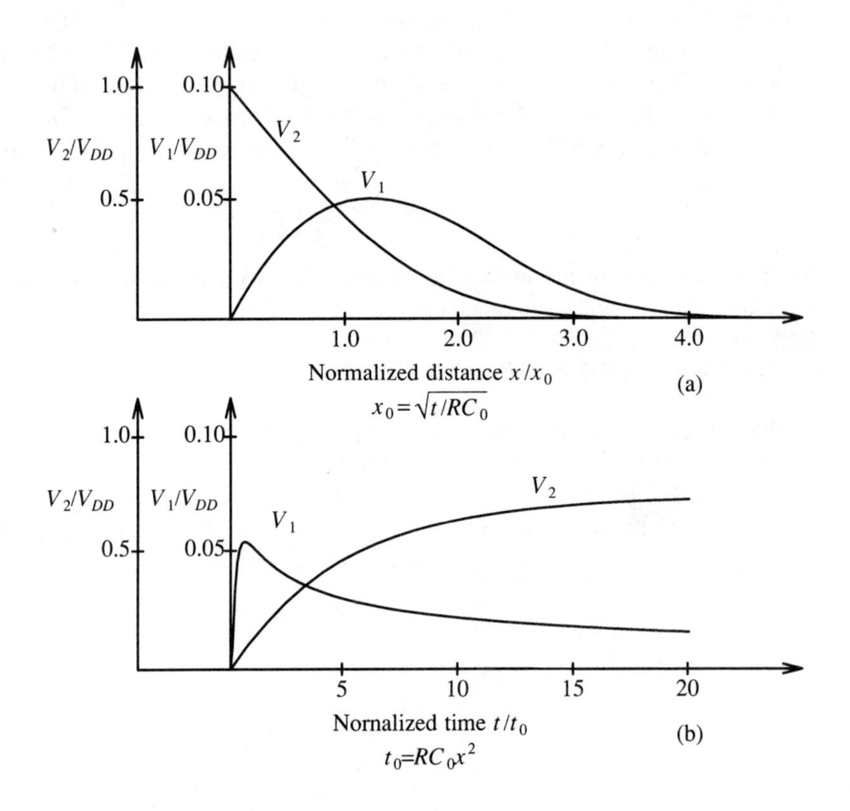

Figure 9.5 Polysilicon-to-polysilicon induced noise.

$$\psi_0(0,t)=0\ ,\ \ \psi_1(0,t)=0$$

The solutions of these equations satisfying the initial and boundary conditions are given by

$$V_1(x,t)=\frac{V_{DD}}{2}\left[\Phi\left\{\left[\frac{R(C_0+2C_1)}{t}\right]^{1/2}\frac{x}{2}\right\}-\Phi\left\{\left[\frac{RC_0}{t}\right]^{1/2}\frac{x}{2}\right\}\right] \tag{9.12}$$

$$V_2(x,t)=V_{DD}-\frac{V_{DD}}{2}\left[\Phi\left\{\left[\frac{R(C_0+2C_1)}{t}\right]^{1/2}\frac{x}{2}\right\}+\Phi\left\{\left[\frac{RC_0}{t}\right]^{1/2}\frac{x}{2}\right\}\right]$$

Voltage profiles on the polysilicon lines and the voltage waveforms are plotted in Fig. 9.5(a) and 9.5(b), respectively, assuming typical values of $\sqrt{(C_0+2C_1)/C_0}=1.25$. The voltage profile on polysilicon line I has a maximum at $x=1.2(t/RC_0)^{1/2}$, which moves with time. This maximum occurs by the following mechanism. The right-hand side of the maximum on the polysilicon line II has not yet been charged up, and the left-hand side of the polysilicon line I has already been discharged through the grounded end.

The voltage spike at position x shown in Fig. 9.5(b) has a maximum when $t = 0.625RC_0x^2$, and decays very slowly with time when $t \to \infty$, like

$$\frac{V_1(x,t)}{V_{DD}} \to \left(\frac{x}{2\sqrt{\pi}} \right) \frac{\sqrt{R(C_0+2C_1)} - \sqrt{RC_0}}{\sqrt{t}}$$

The height of noise spike ($0.05V_{DD}$) of this typical case is about an order of magnitude less than the height of metal-to-metal or metal-to-polysilicon induced noise, and therefore the polysilicon-to-polysilicon noise mechanism is not important.

9.4 SOME DETAILS OF THE INDUCED-NOISE PROBLEM

It was shown in Section 9.3 that the significant induced-noise mechanisms are the induction from metal line to metal line and from metal line to polysilicon line. A metal-metal induced noise problem is mathematically trivial. The noise voltage on a polysilicon line derived in Section 9.3, however, assumes a long polysilicon line that is not loaded by any other capacitance. Practically, a polysilicon line is loaded by gates of FETs, through which the induced noise influences the circuits. The effects of capacitive loading on the height and the pulse width are considered in this section.

(A) Effects of Capacitive Loading in the Middle of a Polysilicon Line.

In Fig. 9.6(a), a polysilicon line that is infinitely long is loaded by a capacitance C_G at point P, a distance x away from the grounded end. The metal line makes a low-to-high transition at time $t = 0$. The entire polysilicon line is driven up to voltage $V_{DD}C_1/(C_0+C_1)$, where C_0 and C_1 are defined in Fig. 9.6(a). At the moment of transition, the voltage of capacitance C_G is zero. Charge induced on the nearby polysilicon line moves into capacitance C_G and the capacitor charges up. After time t the charge on the polysilicon line within distance λ from point P has moved into capacitor C_G to make the voltage within the region of length 2λ centered at point P equal to the voltage of the capacitor C_G. The distance λ is given by

$$R(C_0+C_1)\lambda^2 = t \tag{9.13}$$

and therefore the voltage at point P, $V(t)$, is given by,

$$V(t) = V_{DD}\frac{C_1}{C_0+C_1}\frac{\lambda}{\lambda + \left[C_G/2(C_0+C_1) \right]} \tag{9.14}$$

Since the end of the polysilicon line is grounded, approximately after time $R(C_0+C_1)x^2$ the part of the RC chain near point P begins to discharge. When this happens the charge in the capacitor C_G is drained, and therefore the progress of discharge along the line slows down. The effect is approximately equivalent to adding a polysilicon line of length $(C_G/(C_0+C_1))$ at point P.

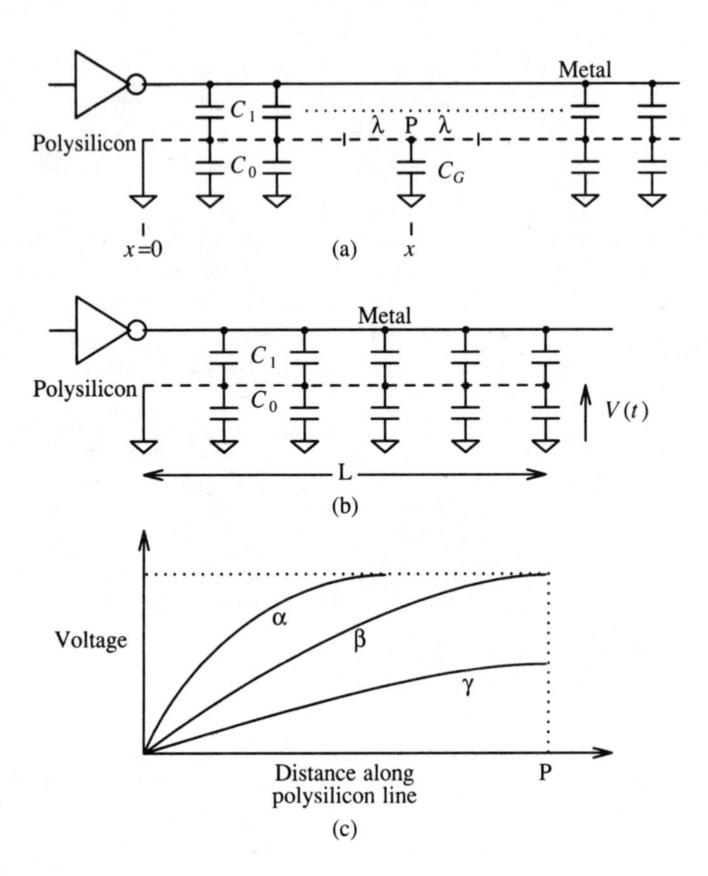

Figure 9.6 Some details of induced-noise phenomena.

(B) Effects of Finite Length of the Polysilicon Line

Potential profiles on the polysilicon line of Fig. 9.6(b) change with time as shown by curves α, β, and γ of Fig. 9.6(c). Until time t_β (profile β) is reached, the voltage at the end of the polysilicon line does not change. After that time the voltage decreases exponentially. This exponential decrease cannot be explained from the model assuming an infinitely long polysilicon line. The problem with a finite-length polysilicon line can be solved by seeking a solution of the form

$$V(x,t)=e^{-kt}\sin\sqrt{k(C_0+C_1)R}\,x \tag{9.15}$$

that approximates the potential profile in the chain. The voltage waveform at the end of the line, $V(t)$, is then given by

$$V(t)=\frac{V_{DD}C_1}{C_0+C_1} \quad 0<t<t_c \tag{9.16}$$

$$V(t) = \frac{V_{DD} C_1}{C_0 + C_1} \exp\left[-\frac{t - t_c}{t_2} \right] \quad t > t_c$$

where

$$t_2 = \frac{4(C_0 + C_1) R L^2}{\pi^2}$$

and where t_c is determined assuming that the polysilicon line is infinitely long, and that at the point distance L away from the grounded end the voltage drops by 3% of V_{DD} when $t = t_c$. Then we have

$$t_c = 0.106 R (C_0 + C_1) L^2$$

The mathematical methods of this section are convenient to determine the induced-noise waveforms and to solve many other problems related to resistive interconnects.

9.5 POWER BUS NOISE

Every time a CMOS gate changes state, a current spike flows through the power bus. The DC current of a chip is the sum of the current spikes from all gates. The power bus should be conductive enough not to develop a significant voltage when the averaged DC current flows, but that is not enough. The current spikes from the gates never average out to constant DC current. At a certain phase within a machine cycle, the chip drives the data or the address output, thereby generating exceptionally large current spikes. The current spikes flowing through the resistance and inductance of the chip power bus, of the bonding wires, and of the package interconnects develop power bus noise V_N as shown in Fig. 9.7 [05].

Let us consider any power-carrying interconnect. Let the inductance per unit length be L, the resistance per unit length be R, the current be I, and the maximum rate of current change be $(dI/dt)_{\max}$. Then we have

$$\frac{\Delta V_{IND}}{\Delta V_{RES}} = \frac{L}{R} \times \frac{1}{I} \left(\frac{dI}{dt} \right)_{\max}$$

where ΔV_{IND} is the noise due to the inductance and ΔV_{RES} is the noise due to the resistance. If we write

$$\left(\frac{dI}{dt} \right)_{\max} = f \frac{I}{T_S}$$

where T_S is the typical switching time of a gate and f is a numerical factor that takes the effects of discharge current waveform. We have

$$\frac{\Delta V_{IND}}{\Delta V_{RES}} = f \frac{L}{R T_S} = f \frac{t_{LR}}{T_S} \tag{9.17}$$

where $t_{LR} = L/R$ is the inductance-resistance time constant of the power bus, the

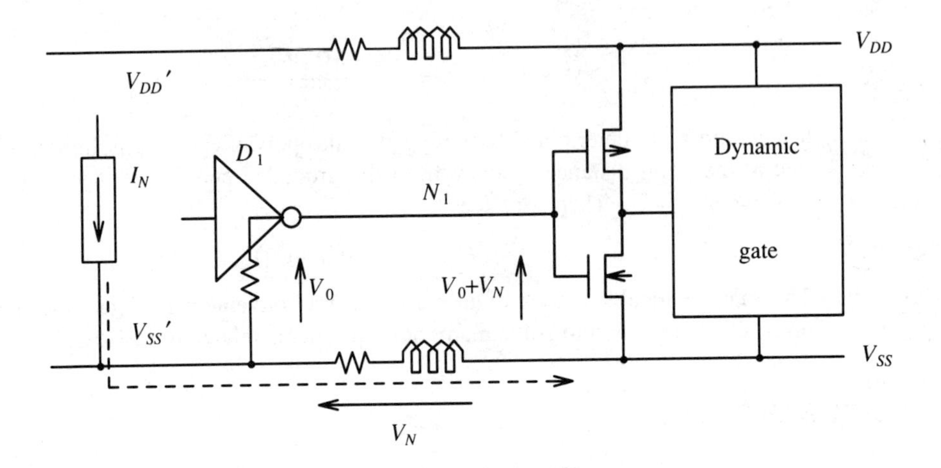

Figure 9.7 Mechanism of power bus noise.

TABLE 9.1 *L/R* TIME CONSTANT OF CONDUCTORS

Conductors	L (nH/cm)	R (Ω/cm)	Time constant (ns)
Bond wire (1 mil)	7.37	0.58	12.80
Bus (100 μm)	7.42	2.82	2.63
Bus (40 μm)	9.21	7.06	1.30
Bus (20 μm)	10.59	14.12	0.75
Bus (5 μm)	13.36	56.4	0.24

bonding wire, or any other current-carrying interconnect. The inductance of a power bus on a silicon chip must be determined by including the skin effect of bulk silicon (see Section 1.14). Table 9.1 shows t_{LR} for some typical interconnects. Assuming that $f = 1$ and $T_S = 1$ ns, the inductance effect predominates in commonly used 1-mil-thick bonding wire and in an on-chip power bus whose width is more than 40 μm.

Power bus noise generated by switching of the output drivers is a difficult design problem in state-of-art 32-bit microprocessors. All 32 output drivers send the data or address out simultaneously, by driving the large capacitance of the external bus. In the worst case, all the drivers may pull down. If the external bus has capacitance C_B and the switching time (10 to 90%) of the driver is T_S, the relationship derived in Section 7.12 gives

$$T_s^{\,2} \left| \frac{dI}{dt} \right|_{\max} \geq 4 C_B V_{DD} \tag{9.18}$$

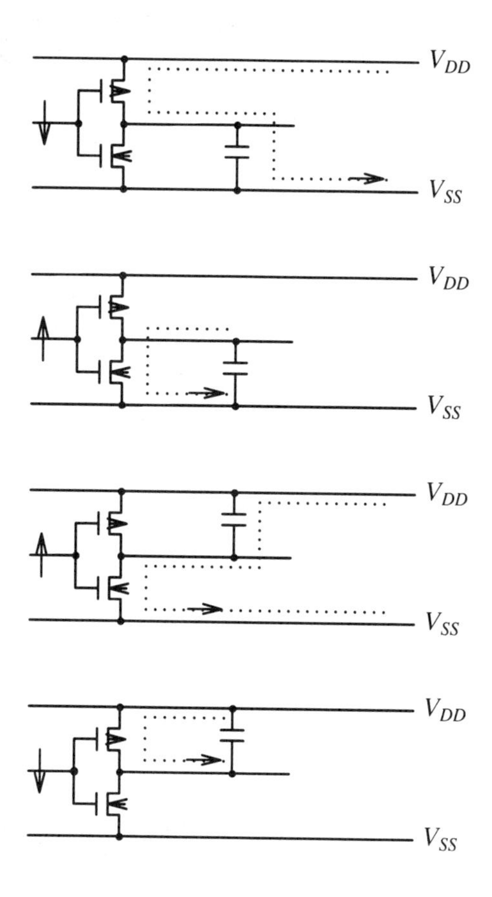

Figure 9.8 Switching current of CMOS inverter.

If a total of N drivers share N_g ground connections, and if each connection has inductance L, the noise generated on the power bus is given by

$$V_N = \left(\frac{N}{N_g}\right) L \frac{4 C_B V_{DD}}{T_s^{\,2}} = \left(\frac{N}{N_g}\right)\left(\frac{2T_{LC}}{T_S}\right)^2 V_{DD} \tag{9.19}$$

where $T_{LC} = \sqrt{LC_B}$ is an inductance-capacitance time constant that depends on the packaging technology. Assuming that $N = 32$, $N_g = 3$, $C_B = 130\,\text{pF}$, $V_{DD} = 5\ \text{V}$, $L = 2\,nH$, $T_{LC} = 0.51\,\text{ns}$ and $T_s = 5\,\text{ns}$, $V_N = 2.2\ \text{V}$. This noise amounts to half the power supply voltage. It is an important design consideration to prevent the noise from affecting the integrity of logic operation of the chip.

The worst-case power bus noise occurs at the open end of the chip's internal power bus. The worst-case noise is proportional to the square of the power bus length, because the resistance and inductance of the bus are proportional to the length of the bus, and the number of gates that charge from or discharge to the bus is also proportional to the length.

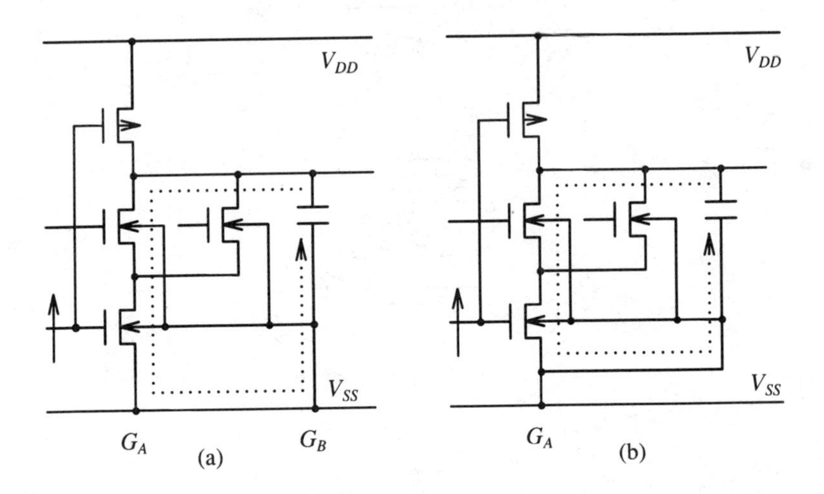

Figure 9.9 One-point ground technique for power bus noise control.

When the driver switches, how a power bus noise is generated depends on the details of the circuit layout. Figure 9.8 shows that if the capacitive load is between the output node of the driver and the ground bus, the current flows from V_{DD} to V_{SS} when the input of the inverter pulls down. When the input pulls up, however, the stored charge in the capacitor discharges within a local loop. When the capacitor is between V_{DD} and the output node, the $V_{DD} - V_{SS}$ current flows when the input of the inverter pulls up. When the input pulls down, however, the stored charge neutralizes through a local loop. Either case may occur. The first case occurs when extensive wiring capacitance is on P-tubs and the second case when the wiring capacitance is on N-tubs.

A layout technique that minimizes dumping the current spike on a power or a ground bus is useful. In Fig. 9.9, a schematic diagram of a NOR gate in a CMOS dynamic PLA is shown. If the ground is connected as shown in Fig. 9.9(a), the discharge current flows on part of the ground bus, and the large current spike creates power bus noise. If the circuit is connected as shown in Fig. 9.9(b) the discharge current loops within the PLA and therefore no noise voltage is induced outside the PLA [06].

A common practice is to separate "noisy" and "quiet" power buses. Figure 7.5(b) shows the power bus structure of the WE32100 CPU. A pair of noisy power buses (DC voltages *can be* noisy within certain limits) that feed data and address drivers and a separate quiet power bus (DC voltages are not allowed to be noisy) are observable. Ground contacts of VLSI chips are often placed at the center of the edges, where the bonding wires are the shortest [07].

Many ideas for reducing the power bus noise have been discussed by Ficchi [08].

9.6 EFFECTS OF NOISE ON CMOS CIRCUITS

In this section noise voltage influence on normal circuit operation is studied. The induced noise and power bus noise can be considered together. It is sufficient to study the effects of a positive noise spike relative to ground. The noise voltage is developed across the input terminal of a static or a dynamic gate and the ground (or V_{DD}) reference of the same gate.

Figure 9.10(a) shows the response of a CMOS inverter to a noise spike. Since the switching threshold of an ordinary CMOS inverter is about at $V_{DD}/2$, a noise spike whose height is less than that is strongly attenuated and the noise never appears at the output. Furthermore, a noise spike whose height is high ($\approx V_{DD}$) but whose width is narrower than a lower limit (typically, the gate delay time defined using a step function input) is blocked by the gate, since the output node of the gate is unable to respond. Therefore, the noise spike must be characterized by using both height and width. When the height of the noise spike is higher and the spike width is wider, a downgoing voltage spike is generated at the output. Figure 9.10(b) shows the response of a CMOS dynamic gate to the noise spike. The dynamic gate is assumed precharged before the noise voltage comes in. Upon reception of the noise spike the output node discharges if the height of the noise spike is more than the threshold voltage of NFET. The effect of noise is permanent. The precharged voltage is never recovered.

A static gate is immune to a narrow noise spike whose height is less than the switching threshold voltage of the gate. A problem is that switching threshold voltages of CMOS gates are not set at the same level. If an inverter whose switching threshold voltage is low is cascaded to a second inverter whose threshold voltage is high and if the input to the first inverter is at low logic level, the cascaded inverter is sensitive to the positive-going noise voltage.

A dynamic gate is vulnerable to a noise spike higher than the threshold voltage of an NFET. Switching threshold voltage of static gates and NFET threshold voltages are different by a factor of 3 to 4. Therefore, a dynamic gate is significantly more vulnerable to noise. One way to make a dynamic gate less susceptible to noise is to buffer the signal input to the dynamic gate by a CMOS static buffer that has common power and the ground buses. In this circuit structure, noise immunity comparable to that of a CMOS static gate can be achieved. The other way is to provide a weak PFET pullup that replenishes the lost charge, as shown in Fig. 9.10(b).

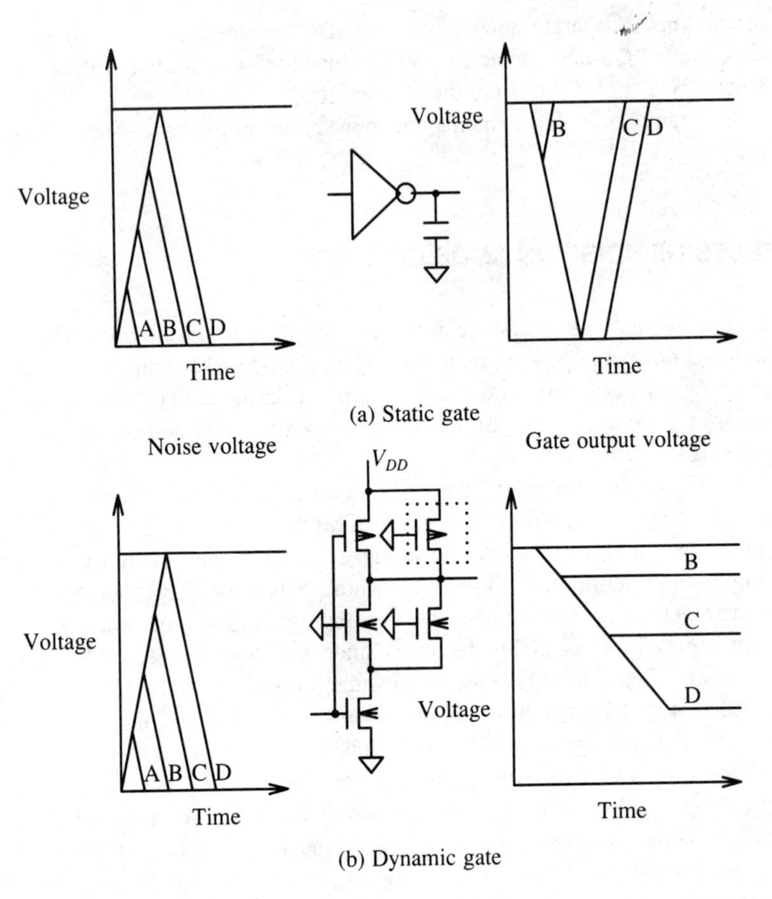

Figure 9.10 Effects of digital noise on CMOS static and dynamic gates.

The mechanisms discussed in this section are relevant to gates that are at noisy locations of a VLSI chip. In the rest of the chip, power supply noise modulates gate delays through power supply voltage, thereby causing logic errors on the critical timing path near the chip's frequency limit [09].

A conventional noise analysis in digital circuits uses the concept of noise margin. Noise margin is the minimum DC noise voltage required to upset circuit operation. Precise definition of noise margin is given by Lohstroh [10]. In CMOS circuits, noise analysis using DC noise voltage level gives unrealistically pessimistic results. Inclusion of noise voltage and noise pulse width simultaneously is required in realistic noise analysis; DC noise analysis is irrelevant in many CMOS circuits, except special circuits like I/O circuits. Recent CMOS VLSI noise analysis is carried out using circuit simulation, including all the significant noise sources and wiring parasitics. Therefore, noise simulation is often the most extensive circuit simulation in a CMOS VLSI project [05].

9.7 ESTIMATING POWER CONSUMPTION OF A CHIP

Estimation of power consumption of a CMOS VLSI chip is a difficult task, especially during the early phase of chip development when there are many uncertainties. Neverthless, estimate of power at an early phase is necessary to gain lead time for package development. In this section a basic method is outlined. To attain reasonable accuracy the parameter values must be carefully chosen.

For simplicity, we assume that the chip drives only other CMOS chips and receives signals only from other CMOS chips: Therefore, no power is spent for sinking and sourcing DC current. An output driver drives off-chip load capacitance C_L. When the driver pulls up, C_L acquires energy $(C_L/2)V_{DD}^2$, and the channel resistance of the PFET of the driver dissipates the same amount of energy (see Section 4.6). When the driver pulls down, the energy stored in capacitor C_L is dissipated by the channel resistance of the NFET of the driver. Pullup and pulldown of the driver therefore consumes energy $C_L V_{DD}^2$. If there are N output drivers, and if they change the state every T_0 (seconds), the power consumed by the output driver is given by

$$C_L V_{DD}^2 N \frac{1}{2T_0}$$

where the factor 1/2 is included because the power consumed by pullup-pulldown operation is $C_L V_{DD}^2$.

In a microprocessor databus, $1/T_0 < f_c$, where f_c is the clock frequency. The databus does not transfer data every cycle. In a microprocessor, however, there are output terminals that send out new signals every clock cycle. Even if the set of signals is sent out every cycle, not all the outputs change level every cycle. In data or address outputs, new data are driven out less often. By taking these correction factors into account, the power consumed by the output drivers, P_0, is written as

$$P_0 = \frac{C_L V_{DD}^2 N}{2} f_c \frac{\Delta_0}{\chi_0} \tag{9.20}$$

for every χ_0 clock cycles the output driver switches, and Δ_0 is the factor representing the fact that different data are sent out. Equation (9.20) was originally derived by Burns [11]. The original formula was adapted to the present problem.

The power consumed within the chip can be estimated by various methods, sometimes with significantly different results. If the percentage area occupied by the FETs is small (as in polycell or in standard cell devices), the power is mostly consumed by charging and discharging the wires. If the total wiring capacitance (the capacitance including the fringing effects) is C_{WT}, and if the wires pull up and pull down every χ_1 cycles, the power consumed is

$$P_1 = \frac{C_{WT} V_{DD}^2}{2} f_c \frac{\Delta_1}{\chi_1} \tag{9.21}$$

where Δ_1 is the probability factor that the data change polarity. If the fringing factor is 2 to 3, C_{WT} is estimated to be

$$C_{WT} \approx \frac{\varepsilon_{OX} S_i}{T_{FCOX}} \quad \text{or} \quad C_{WT}(\mu\text{F}) = 0.575 \times 10^{-2} S_i\,(\text{cm}^2)$$

where T_{FCOX} is the averaged thickness of the field and the composite oxides and S_i is the total area (including the gaps between the wires) occupied by the chip's internal wiring channel. The fringing effect factor is assumed here to compensate for the gaps between the wires.

When the area occupied by FETs is not negligible, C_{WT} in Eq. (9.21) is replaced by

$$C_{WT} \rightarrow C_{WT} + C_D + C_G$$

where C_G and C_D are the gate and the drain capacitances, respectively. Since the field oxide is usually 20 times thicker than the gate oxide and the fringing factor of the wiring capacitance is in the range from 2 to 3, the gates of the FETs have approximately 10 times more capacitance per unit length than polysilicide or metal. The drain island capacitance is, if compared per unit length (in the dimension parallel to the gate), comparable to the capacitance of the gate (Section 1.16). From these crude rules a correction to the capacitance C_{WT} can be added.

Total power of the chip, P_T, is then given by

$$P_T = P_0 + P_1 = \frac{1}{2} C_{eff} V_{DD}{}^2 f_c \tag{9.22}$$

where

$$C_{eff} = C_L N \frac{\Delta_0}{\chi_0} + (C_{WT} + C_D + C_G)\frac{\Delta_1}{\chi_1}$$

From Eq. (9.22) the power is proportional to the power supply voltage squared and is proportional to the frequency. This is the most fundamental relationship for power dissipation in a CMOS VLSI chip. At low clock frequencies the power of a CMOS chip becomes quite small. The minimum power is dependent on the bleeder circuits, if any, and on leakage currents.

By substituting $V_{DD} = 5$ V, we obtain

$$P_T = 12.5 C_{eff}(\mu\text{F}) f_c(\text{MHz})$$

Let us assume that $C_L = 130$ pF, $N = 80, \chi_0 = 1,\ \Delta_0 = 1/2$, and $C_{WT} = 0.00575\ \mu F$ (1 cm^2 chip area). As for C_G and C_D, a reasonably compact design should satisfy

$$C_{WT} \approx C_D + C_G$$

and further if $\Delta_1 = \chi_1 = 1$ is assumed,

$$C_{eff} = 0.0052(\mu\text{F}) + 0.0058(\mu\text{F}) \times 2 = 0.0168(\mu\text{F})$$

then $P_T = 0.21 f_c = 2.1$ W at 10 MHz. This is the typical power consumption of a CMOS VLSI.

9.8 PACKAGE

It has been shown that if a logic circuit consists of N gates and if the circuit has P external connections, N and P are related approximately by,

$$P = kN^{1/2} \tag{9.23}$$

The proof is given in Section 7.3. An advanced 32-bit microprocessor and a peripheral chip contain up to several hundred thousand gates. The number of IO terminals is in the range 100 to 150, excluding the power and the ground terminals.

Packages that are able to accommodate that many IO pins are pin grid arrays (PGA) and chip carriers. Recent 32-bit microprocessors such as the MC68020 and WE32100 use pin grid arrays. The chip is bonded to the back plane by eutectic solder or by epoxy glue, and the bonding pads of the chip and the package pins are connected by wire bonds. A PGA usually has power and ground planes that serve as the internal nodes where all the power and the ground connections from the chip are joined together and then connected to the power and the ground pins of the package. A typical wiring inductance of the ground is 2 to 4 nH per bonding wire. The thermal resistance of the package is about is 20 degrees/W.

Advanced packaging technologies for MOS devices are just about to emerge. An advanced packaging technology that has features of the packaging technology used in ECL-based mainframe computers [12] [13] is a prerequisite to high-performance CMOS VLSI systems whose frequency is above 30 MHz. Lead inductance and parasitic capacitance of the advanced package are typically one order of magnitude less than those of the conventional package on a PC board. The small parasitics open new design possibilities for CMOS VLSI chips.

9.9 INPUT PROTECTION

The input circuit of a CMOS VLSI chip has a very high impedance. The input is usually connected to the FET gates of the input buffer. The capacitance of the high-impedance node is easily charged. Since human operators are likely to touch VLSI chips that are connected to ground, the charge accumulated on the human body (equivalent to a 1 to 200 pF capacitor), as high as 2000 V, could discharge through the VLSI chips to ground. The thin oxide of the FET gates that is on the discharge path is destroyed. Unfortunately, advanced device technology makes FETs very susceptible to electrostatic shocks [14].

The breakdown voltage is determined from the critical breakdown field of oxide, and this field is about $7.6 \times 10^6 V/cm$ [15]. The gate oxide of scaled-down CMOS devices does not withstand more than several tens of volts.

To protect the input circuit of a CMOS VLSI chip, the circuit shown in Fig. 9.11(a) or (b) or their combination is used [16]. The resistance R is usually several hundred ohms. The diodes are made from P+ and N+ diffused islands. Many contacts are applied to make the diode surge current path have small series resistance to divert

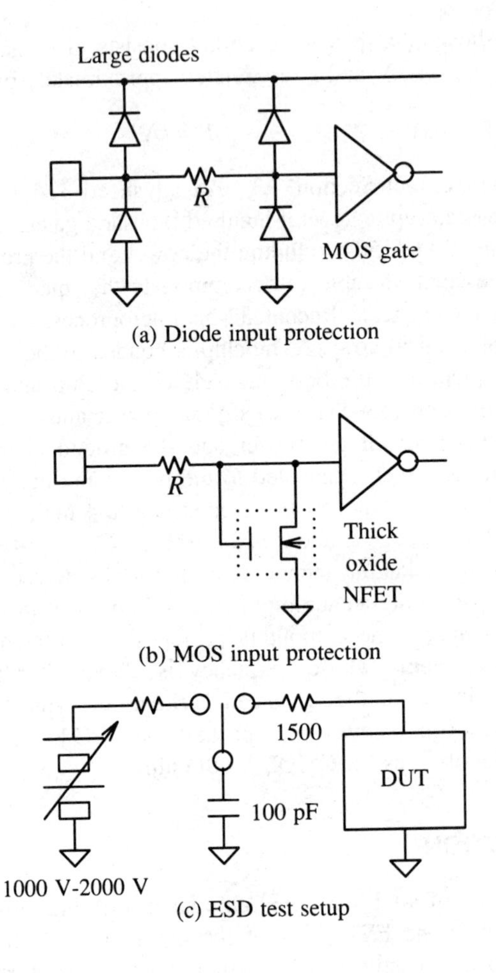

Figure 9.11 Input protection from electrostatic discharge.

the static charge effectively. The resistance, however, cannot be too small lest the protection device itself should be destroyed. The protection circuit may be destroyed by a localized temperature rise [17]. The circuits of Fig. 9.11(a) and (b) have a delay. In a submicron-feature-size CMOS VLSI the delay may cause problems. It can be shown that the degree of protection and the delay are approximately proportional [16].

One way to compromise delay and protection is to connect a large diode directly to the input pad as shown in Fig. 9.11(a). The diode loads the source of the electrostatic charge when it conducts, thereby helping to reduce the surge voltage. Since the driver of the signal source is powerful, the capacitance the protection diodes add (several picofarads) can be tolerated.

Figure 9.11(c) shows a test circuit for the input protection device [17]. A capacitance of 100 pF and 1500 Ω of series resistance simulates a human body. The voltage used is up to 2000 V.

9.10 CMOS GATE FAULT MODEL

It might be thought that all the silicon chips that are processed using a correctly designed photomask would work. Practically, this is not the case. Functional failure of chips occurs when a small dust particle lands on the photoresist and blocks light. Then either circuit nodes that must be connected could be disconnected, or nodes that must not be connected are connected. CMOS gates fail in many different ways. Failures occur from insufficient removal of metal by photolithography, resulting in a short circuit, overremoval of metal resulting in an open circuit, failure in cutting oxide windows, and the like. These process failures create gate failures, which can be identified on the diagram of the NOR2 gate of Fig. 9.12. There are seven basic patterns of failure of a NOR2 gate [18].

TABLE 9.2 FAULTS 1 TO 4

Input logic level		Output logic level				
A	B	(0)	(1)	(2)	(3)	(4)
0	0	1	0	1	1	1
0	1	0	0	1	0	1
1	0	0	0	1	1	0
1	1	0	0	1	0	0

1. Output node O shorted to the ground (stuck at 0 or SA0 fault).
2. Output node O shorted to V_{DD} (stuck at 1 or SA1 fault).
3. Input A is disconnected at point X, and the gates of the FETs are charged to ground potential (input open from A, or IOFA fault). If the gates are charged to V_{DD}, the fault is indistinguishable from the SA0 fault.
4. Same as 3, for input B (input open from B, or IOFB fault).

Failure modes 1 to 4 are simple, in that the failed gate still executes a combinatorial logic. The input-output relationships of a healthy gate 0 and of a gate that suffers from one of faults 1 to 4 are shown in Table 9.2. The following three faults convert the combinatorial NOR2 gate into a sequential circuit.

5. FET MNA is missing (the connections are missing, or the gate of the FET is disconnected and is charged to ground potential). This fault is called an input A stuck open (ASO) fault.

6. Same as 5, but NFET MNB is missing (input B stuck open, or BSO fault).
7. Current path from V_{DD}, PFET MPA, MPB to node O is broken somewhere (V_{DD} stuck open, or $V_{DD}SOP$ fault).

If the state of the gate on the nth clock cycle is $F(n)$, the input-output relationships of the gate at clock cycle $n+1$ with faults 5 to 7, are as shown in Table 9.3.

If a NOR2 gate in a logic circuit fails as shown in Table 9.2 or 9.3, the failed logic levels propagate to the I/O terminals after some clock cycles, and are detected as a discrepancy from the expected signal. The problem is to produce a test pattern that detects all, or as many as possible, of the faults of all the gates in a VLSI chip. Because of the complexity of a VLSI chip, this is a formidable task. A generation of test patterns is usually carried out using a gate-level fault model, where an individual NOR2 gate is replaced by an equivalent faulted gate model that is able to generate the seven failure modes of the gate.

TABLE 9.3 FAULTS 5 TO 7

Input logic level		Output logic level			
A	B	(0)	(5)	(6)	(7)
0	0	1	1	1	F(n)
0	1	0	0	F(n)	0
1	0	0	F(n)	0	0
1	1	0	0	0	0

The gate fault model of a NOR2 gate is shown in Fig. 9.12(b). In this figure the faulted gate (* sign) gives faults 1 to 4, by generating logic levels corresponding to the fault, when the fault type is specified. The rest of the logic circuit generates logic levels when faults 5 to 7 are specified. When any of the three faults 5 to 7 are activated, the transmission gate retains the state of the gate from the preceding cycle.

If this gate model is used in a simulator, sequential operation of a circuit with faulted gates can be correctly tracked down, and it is possible to check whether or not a certain failure is detected by a given set of test vectors. When a logic simulation is run using a faulted gate model, and if the number of faults detected by a given set of test vectors is counted, a fault coverage (in percent) of the given vector set is determined. In a VLSI chip test, 90% fault coverage is considered adequate and is practical. As for the details of chip testability problems, the reader is referred to the textbook by McCluskey [19].

In a digital VLSI chip, it is impossible to test the chip by driving all combinations of the inputs because of the large number of internal states. Tests based on structural knowledge of the tested object can be executed for some combinatorial logic circuits, but if the same approach is used for a sequential circuit, the test becomes very long. For this reason, the only practical choice is a functional test, a test that examines whether or not the implemented functions are correctly executed, using rather

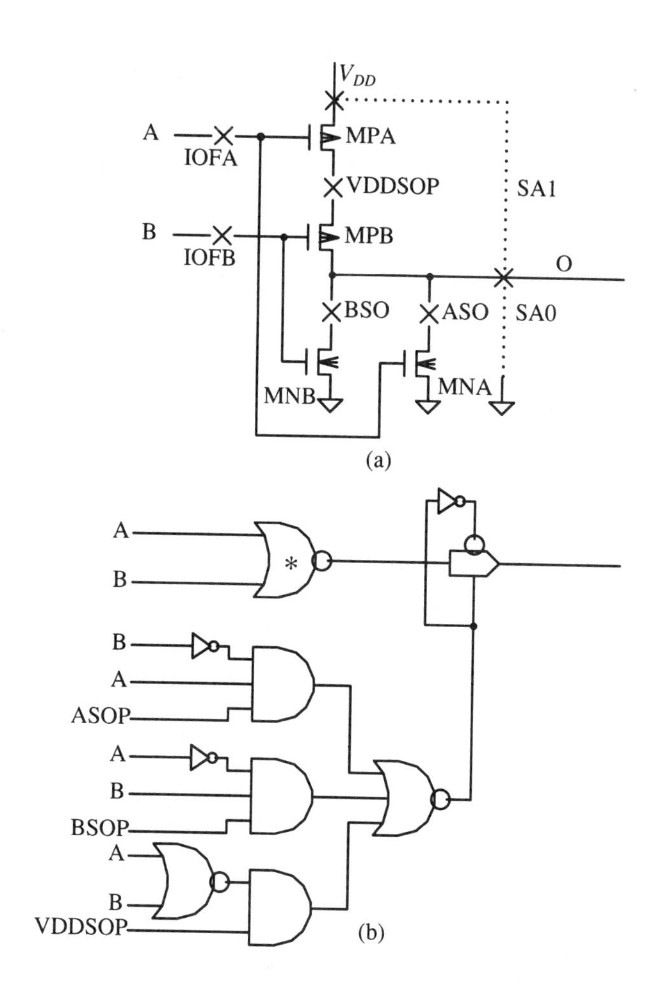

Figure 9.12 CMOS gate fault model.

limited input data [20]. A functional test is never able to detect all the faults with a reasonable number of vectors. Therefore, it is important to know the fault coverage.

Another approach to improving chip testability is to provide hardware testability aids, as discussed in the next section, or even self-test circuits [21] [22]. In a self-test implementation, the test vector generation and the test result evaluation are done internally. Techniques to reduce the test hardware using counters and a signature circuit have been reported [23]. These testability aids often significantly reduce the time required for routine tests in a production environment.

9.11 SCAN DESIGN

In a large-scale sequential circuit having a limited number of terminals, it becomes difficult to exercise all the internal nodes from the access terminals for test purposes. This problem is called the controllability problem. Latches are the roadblocks to propagate signals to the chip's interior. For example, in a simple cascaded N-stage countdown circuit, 2^N vectors are needed to exercise the Nth flip-flop. Latches also prevent observation of the states of internal nodes. This problem is called the observability problem. The difficulties in controllability and observability are called the testability problem [19] [24].

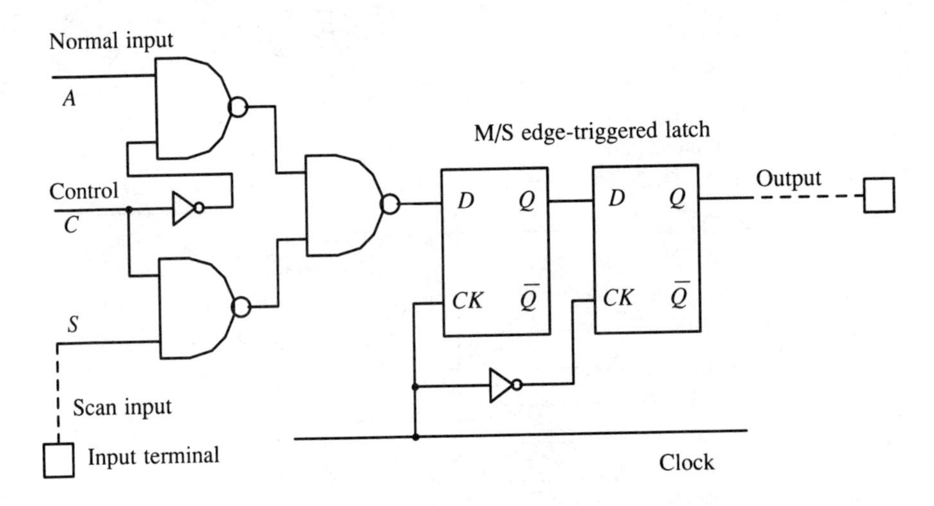

Figure 9.13 Scan design for improved VLSI testability.

Although several different approaches have been proposed to ease the testability problems, the most widely used technique is called scan design. In a system built using edge-triggered M/S latches, each M/S latch pair is fitted with an input multiplexer as shown in Fig. 9.13. If the default input is used (C = low), the system works normally. When C = high, however, the latch accepts its input from the scan line. By connecting the scan input to the output of another latch, all the M/S latches are connected in a long chain. After the chip is clocked normally by the desired number of cycles, control signal C is set high and all the M/S latches are connected in series. Clock signal generators are also reconfigured (not shown in the figure). Further clocking sends out the content of the latches as captured when C went high to the output terminal. The data are examined for the chip's functional integrity.

The shift register arrangement also allows loading of the internal latches from the input terminal. After all the latches are loaded control signal C is set low and the chip is clocked normally, for further tests. The chip's response to the new set of internal latch data is used to further diagnose the chip's integrity. The scan design allows a test engineer to control the most inaccessible area of the chip, with relatively little overhead hardware, and now the idea is used in many designs. [24] [25]. Scan design is also used in gate-array devices [26].

9.12 PRACTICAL ASPECTS OF CMOS VLSI TEST

Whether or not a set of given test vectors are able to detect all the faults of a chip is a fundamental testability question. There are other, more practical testability problems. Often a chip must be designed to provide easy solutions to these problems.

1. It is necessary to provide a way to bring a VLSI chip into a known state, using a small number of vectors. If a chip cannot be *reset* easily, testing is often impossible, or at least much more difficult.

2. State-of-the-art high-speed automatic VLSI test machines are unable to test chips that have more than a certain number of pins: The present limit is in the range 128 to 256. High speed and large pincount are incompatible.

3. The upper limit of the clock frequency of state-of-the-art test machines is 40 to 50MHz. The clock frequency of present-day CMOS chips can be higher than that frequency. If this is the case, the maximum clock frequency must be determined by indirect methods.

4. Ground and power voltages provided by a test machine via a prober do not have low impedance. It is necessary to bypass the power supply by capacitors, but if a chip generates too much switching noise, and if the noise influences the chip's internal circuits, the test may become altogether impossible. Output strobes and input strobes must be separated enough so that the output driver noise decays before the chip strobes in the data. Large and fast output drivers aggravate this problem.

5. A high-speed automatic test machine usually has high pin-card (data analyzer) capacitance. Typically, the capacitance is larger than 50 pF. The output driver must be able to drive the capacitance within the specified time.

6. CMOS chips must be powered up carefully so that the chip does not latch-up. Means of detecting latch-up are required in the test program.

7. Testing cost can be a very significant fraction of the chip cost. Some means of quickly rejecting a faulty chip is desirable. Self-test capability of a VLSI chip is now considered a trend that reduces testing cost problems.

It is obsolete design methodology to detect design errors by chip testing. Design problems must be weeded out before the design is committed on silicon. Not only must the problems of logic design and circuit design be corrected before the photomask is run by the process line, but the chip should run at a reasonable clock frequency.

9.13 LATCH-UP

Latch-up of bulk CMOS IC was once considered the most negative attribute of CMOS technology. Progress in device fabrication technology and design techniques based on understanding the phenomenon has eliminated this problem. In spite of progress we must be cautious, since the physical model predicts that submicron CMOS devices could have resurgent latch-up problems. Here we review the mechanisms and various techniques to prevent latch-up.

Figure 9.14 shows a cross section of a bulk CMOS IC. The circuit shown here is an output driver. Suppose that the output terminal is driven below V_{SS} by the external source. Then electrons are injected from the drain contact of the NFET into the P-tub, and these electrons are collected by the N-type substrate. The electron current flows to the N-substrate contact "b" (dotted). The majority carrier current creates a potential between the PFET source contacts "e" and "b." If the potential is enough to inject holes from "e" the injected holes are captured by the P-tub, and they flow to the P-tub contact "B" (solid). This majority hole current creates a potential between the source contacts "E" and "B" and by the positive potential, electrons are injected into the P-tub from contact "E." The process is obviously regenerative and soon a large current begins to flow from V_{DD} to V_{SS}. The same regenerative phenomenon can be initiated by overbiasing the chip, thereby creating P-tub/substrate avalanche breakdown, or by subjecting the sample to intense ionizing radiation. Latch-up is reset by reducing power supply voltage, but the chip may be destroyed before that.

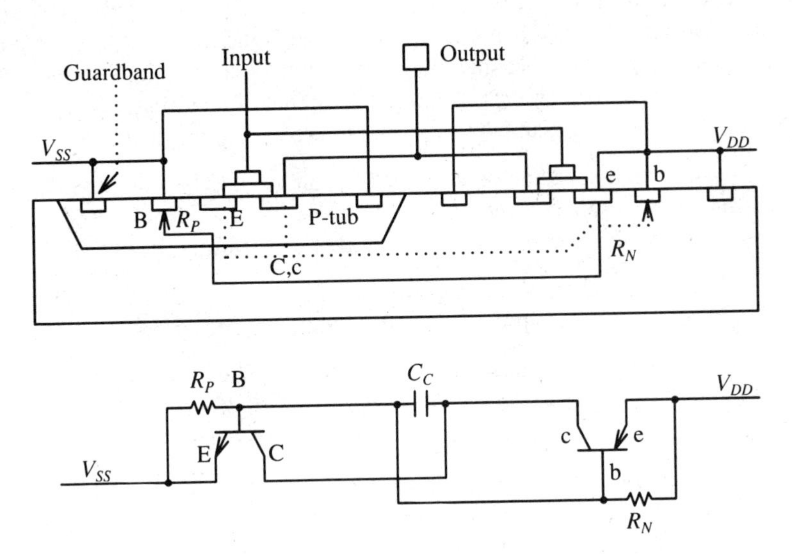

Figure 9.14 Mechanism of latch-up and equivalent circuit.

The equivalent circuit that represents the latch-up model is shown in Fig. 9.14. To study the dynamic characteristics of latch-up, P-tub to substrate capacitance C_C must be added to the equivalent circuit. Troutman and Zappe have analyzed the problem in detail [27]. Here we study only DC characteristics of latch-up. Then this phenomenon can be understood by the equivalent circuit consisting of two bipolar transistors and two resistors, R_P and R_N. If resistors R_P and R_N are very large, the regenerative effects occur when $\beta_N \beta_P > 1$, where β_N and β_P are the current gains of NPN and PNP transistors, respectively. The NPN bipolar transistor has a narrow base (vertical dimension of a P-tub, only a few micrometers), and consequently β_N is high (from 100 to several hundred). The PNP transistor has a wide base region amounting to several tens of micrometers, but there can be an electric field within the base region that helps minority holes to be collected efficiently by the P-tub. Then β_P may not be as small as expected from the base width [28]. P+ guardbands on N-type bulk may not be very effective if such a transverse field exists. Advanced computer simulations revealed detailed mechanisms of latch-up [29].

As CMOS technology scales down, the width of the base of the bipolar transistor decreases and susceptibility to latch-up increases. There is a theoretical reason that if there is no special means to prevent latch-up, latch-up susceptibility is proportional to $1/\Lambda^2 N_A$, where Λ is the feature size and N_A is substrate doping.

Latch-up susceptibility can be reduced by taking measures to reduce β_P and β_N. Neutron irradiation and gold doping reduce substrate minority carrier lifetime to less than 100 ns, and this technique effectively controls latch-up.

When these techniques are not desired, the most effective way is to provide a highly conductive substrate underneath the active surface, using epitaxial N-on-N^+ or P-on-P^+ material as the starting material for CMOS processing [30] or by providing a buried layer underneath the P-tub [28]. A highly conductive substrate effectively reduces resistances R_P and R_N and has the same effect as reducing β of the bipolar transistors. In a quantitative analysis, effects of conductivity modulation of substrate must be included [31]. A buried layer under a P-tub creates a reverse built-in field within the base of the NPN transistor and is effective in reducing β_N. Still another way is to provide a deep N^+ isolation diffusion around P-tub. The highly doped N^+ region effectively stops diffusion of holes to the P-tub.

A widely practiced means for latch-up prevention is to provide a guardband around the sensitive area. In recent designs, NFETs used in the output driver are surrounded by a band of P^+ diffusion that is grounded using contact windows and metal strapping conductors. The PFETs of the driver are similarly protected by N^+ diffusion, contacts, and metal that are all connected to V_{DD}. The guardbands keep the potential of the surrounded area constant, thereby preventing the development of potential difference. A series resistor inserted between the source of the output driver FET and the power bus is also effective, by limiting the minority carrier current that is injected from the source island.

By combining several of these techniques, latch-up has been effectively controlled and CMOS technology produces useful and reliable devices. SOS devices have no latch-up problem. For applications in nuclear radiation environments SOS devices have unique advantages. Bipolar output drivers available in BiMOS technology are also free from latch-up problems [32].

9.14 EFFECTS OF IONIZING RADIATION

In this section the effects of ionizing radiation on CMOS VLSI devices are summarized. A dose of ionizing radiation is measured in rads (silicon). When 1 gram of silicon is irradiated by 1 rad (silicon) of radiation, 100 ergs of energy are transferred from the radiation to the silicon. Another common unit of dose of radiation is the roentgen. Radiation of 1 roentgen generates 1 esu of charge per 1.293×10^{-3} gram of air. Since the reference materials of a rad unit and roentgen unit are different, the measures cannot be correlated without further specifying radiation by its energy. For example, 1 rad (silicon) equals 1.16 roentgen if radiation energy is 1 MeV.

Intensity of radiation is measured in rad/s. The typical range of dose and intensity of radiation studied in a radiation hardening problem is 10^4 to 10^7 rad (silicon) and 10^6 to 10^9 rad/sec, respectively. When a neutron beam generates lattice defects in silicon, the range of dose studied is 10^{14} to 10^{15} neutrons/cm^2.

Radiation is absorbed by silicon bulk, creating electron-hole pairs. To create a pair, energy of 3.6 eV is required. Then 1 rad (silicon) of radiation generates total of

$$\frac{100(\text{erg/gram}) \times 2.33(\text{gram/cm}^3)}{10^7(\text{erg/Joule}) \times 1.6 \times 10^{-19}(\text{Joule/eV}) \times 3.6(\text{eV/pair})} = 4.04 \times 10^{13} \text{pairs/cm}^3$$

electrons and holes. Using this formula, the rad unit is conveniently correlated to the electrical phenomena in the silicon bulk.

Effects of neutron and γ radiation on semiconductor devices are summarized in the report by Gover [33]. Neutron irradiation of semiconductors results in formation of defect clusters of the size of several hundred angstroms, and the defect center region traps majority carriers. The structure attracts minority carriers and works as a recombination center. By neutron irradiation, effective bulk doping decreases, and the threshold voltage of the FETs decreases accordingly. Reduction in minority-carrier lifetime hardens CMOS ICs against latch-up, and this technique is now one method of latchup prevention.

The effects of γ radiation on MOSFETs is complicated. At low doses [$<10^4$ rad (silicon)], electrons and holes generated within the gate oxide drift toward the gate and the silicon-silicon oxide interface. In an NFET with positive gate voltage, electrons are quickly swept out to the gate and they are neutralized there. Holes slowly migrate to the silicon-silicon dioxide interface and some (a few to as much as 50%) of them are trapped there. The trapped positive charge reduces the threshold voltage of the NFET. Many devices fail by leakage current at low krad (silicon) exposure. At very low temperatures holes cannot migrate very far: They are trapped immediately after being generated, and therefore the threshold shift begins at lower doses.

The positive charge trapped at the interface is proportional to T_{OX}. Positive charge that exists at the silicon-silicon dioxide interface is distance T_{OX} away from the gate electrode, and therefore the charge shifts the gate voltage proportional to T_{OX}. Therefore, the observable threshold voltage shift, ΔV_{TH} is given by

$$\Delta V_{TH} = -\text{const}\, T_{OX}{}^2 \qquad (9.24)$$

Equation (9.24) shows that the thinner the oxide, the lower the threshold voltage shift. This relationship is generally confirmed by experiments. The threshold shift is enhanced if the gate of the NFET is biased positively with respect to the source. By the same mechanism, the worst-case γ radiation effect on PFETs is observed when the gate is kept at the same potential as the source. Then the holes are not attracted to the gate, and the threshold voltage increases with increasing γ-ray dose. In a very narrow FET, the threshold shift is enhanced by the effects of the charge in the nearby thick oxide [34].

When the γ ray dose is over 10^4 to 10^5 rad (silicon), negatively charged states begin to form at the silicon-silicon dioxide interface by an as-yet-unknown mechanism. Once the interface state formation begins, the threshold voltage of an NFET begins to increase.

Shifts of the switching threshold voltages of static NAND3 and NOR3 gates of 1.5-μm CMOS were studied by Hatano [35]. By γ irradiation, the threshold voltage of the optimized NAND3 gates shifted from 3.5 V (before irradiation) to 1.5 V (after 10^5 rad irradiation).

The failure mechanism originating from capture of radiation-generated holes has many common aspects with the failure mechanism due to trapping of hot carriers [36]. The hot-carrier effect is most severe when the FET begins to conduct. The drain-source potential at that time is still high, and the capacitive induction from the gate adds to the potential (Section 4.9). For the other switching polarity the effect is not very significant. It is possible to control the hot-carrier effect by reducing the load capacitance and by inserting an extra FET that is always on [37].

9.15 SINGLE EVENT ERROR

When a CMOS device is struck by a single heavy cosmic-ray ion such as an iron or krypton ion, typically 1 to 7 (pC) of charge is deposited on the target node. In scaled-down CMOS ICs, capacitances of many nodes are so small that the deposited charge seriously affects the state of the node. The capacitance of the output node of a 1.75-μm CMOS minimum-size inverter driving three minimum-size fanouts and 250 μm of wiring is estimated to be,

$$30\text{pF/cm} \times 11 \times 10^{-4}\text{cm} \times 3(\text{gate}) + 20\text{pF/cm} \times 11 \times 10^{-4}\text{cm} \times 1(\text{drain})$$
$$+ 0.06\text{pF(wire)} = 0.17\text{pF}$$

If this node is charged to 5 V the node holds 0.85 pC of charge. Therefore, a single strike by a heavy ion is able to deposit more charge. If such an event should occur, the static circuit generates a glitch, and a dynamic circuit loses data. As is understood from this estimate, there is no fundamental solution to the single-event error problem, except using relatively large devices and error-detecting-correcting circuits (Section 7.18).

An α particle penetrates 20 to 30 μm in silicon and generates 1 to 1.5×10^6 electron-hole pairs. The charge is collected at the diffused islands of the memory cell. To reduce α-induced failures, plastic packages are preferred to ceramic packages. Organic coating on the chip often improves reliability [38] [39]. N-substrate CMOS memory cells are less susceptible to α-particle induced single-event errors than P-substrate NMOS cells, because the α-induced charges are drained by the reverse-biased P-tub substrate junction [40]. An estimate of an α-particle induced error rate using a Monte Carlo method has been reported [41].

When an ion or α particle hits a memory cell, the data stored can be lost. This is a problem in satellite-borne CMOS VLSI devices. Susceptibility to a single-event error of a CMOS static memory cell can be significantly reduced by inserting series resistance in the feedback path. The resistance is chosen so that the RC time constant is 3 to 5 times the time of cosmic-ray charge collection (2 ns). Then the state of the node that was struck by the ion recovers with help from the other node and the cell retains correct data [42]. Memory cells are designed by taking the stability considerations for both read/write access and for the single-event upset by α particles [43].

9.16 YIELD OF CMOS VLSI

Yield of an integrated circuit is modeled using the concept of fatal point defects [44]. Fatal defects consist of photolithography defects, oxide pinholes, and leakage centers [45]. When at least one of the fatal defects falls within the area of a VLSI chip, the chip is considered failed.

Suppose that there are D_0 defects per unit area on a silicon wafer. The yield of good chips, Y, depends on D_0 and chip area a. The relationship between Y, D_0, and a is derived as follows. Let the area of a silicon wafer be S, on which there are n fatal defects. The defects are assumed to be scattered randomly within area S. We then have $D_0 = n/S$. Let us divide the wafer into m small cells of equal area S/m. The division is so fine that a cell contains no fatal defect, or at most only one fatal defect. The probability that contiguous k cells contain no fatal defect is given by

$$Y = \frac{m-n}{m} \frac{m-1-n}{m-1} \cdots \frac{m-k+1-n}{m-k+1}$$

$$= \left(1 - \frac{n}{m}\right)\left(1 - \frac{n}{m-1}\right) \cdots \left(1 - \frac{n}{m-k+1}\right)$$

By taking the logarithm, expanding the logarithm, and neglecting the higher-order terms of n/m, we obtain

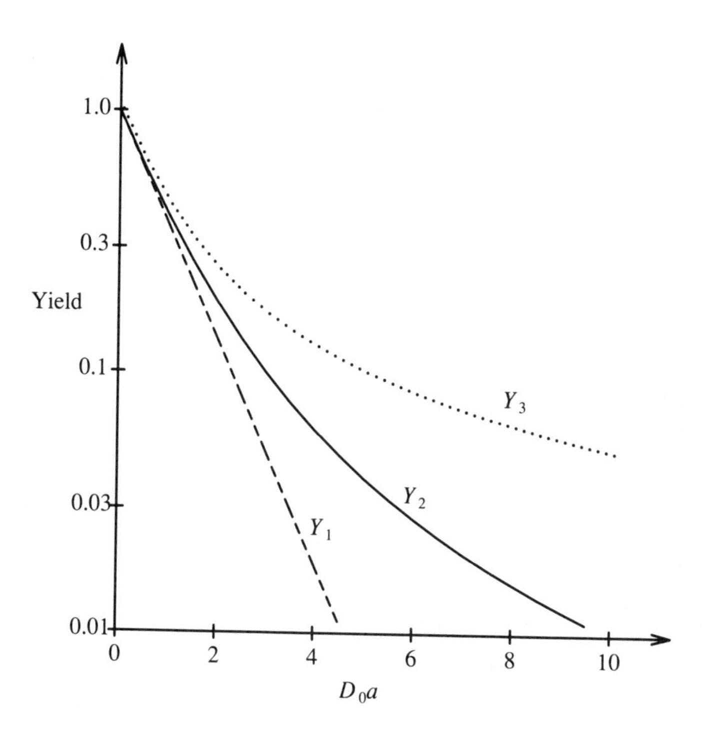

Figure 9.15 Yield of a CMOS VLSI chip.

$$\log Y = -n\left(\frac{1}{m-k+1} + \ldots + \frac{1}{m-1} + \frac{1}{m}\right) = -n\int_{m-k}^{m}\frac{dx}{x} = -n\left(\frac{k}{m}\right)$$

where $k \ll m$. If k contiguous cells have area a,

$$k\frac{S}{m} = a \quad \text{or} \quad \frac{k}{m} = \frac{a}{S}$$

and therefore

$$Y = e^{-D_0 a} \tag{9.25}$$

where $D_0 = n/S$.

In a real chip production environment, the average defect density varies over some range. By taking this variation into account, yield is given by

$$Y = \int_0^\infty e^{-Da} f(D)dD \tag{9.26}$$

where $\int_0^\infty f(D)dD = 1$ is a weighting function for D. Several different functions have been proposed.

1. $f_1(D) = \delta(D - D_0)$. Then $Y_1 = e^{-D_0 a}$.
2. $f_2(D) = D/D_0^2$ $(0<D<D_0)$ and $f_2(D) = 2/D_0 - D/D_0^2$ $(D_0<D<2D_0)$.

   Then $Y_2 = \left\{ \dfrac{1-e^{-D_0 a}}{D_0 a} \right\}^2$.
3. $F_3(D) = 1/2D_0$ $(0<D<2D_0)$. Then $Y_3 = \dfrac{1-e^{-2D_0 a}}{2D_0 a}$.

The three curves are plotted versus $D_0 a$ in Fig. 9.15.

The practical lower limit of the defect density D_0 is about 3. If $D_0 = 3$, the yield of 1 cm square chips estimated from Y_1 is 4.98% and by Y_2 about 10%. The last formula appears to give too optimistic an estimate of yield. Experience shows that the real yield appears to fall between Y_1 and Y_2.

A yield of 5 to 10% is the limit in the competitive integrated-circuit business. Except for a very expensive special-purpose chip, the maximum size of a practical IC is 1 cm times 1 cm. This IC size limit poses significant constraints on VLSI development projects.

9.17 VLSI LAYOUT METHODS

The layout of a chip that contains several hundred thousand FETs is not an easy task. If a designer is totally overwhelmed by the layout work and has no time to do anything else, a VLSI project is destined to fail. The thrust of the effort should be how to ease the pressure of layout.

There is no single effective approach to this problem. Based on the author's own experience, the best way is first to architecture a chip so that the regular structures occupy as large a fraction of the chip area as possible, and to lay out each of the regular structures by dedicated CAD tools written for each of them. As for irregular structures, we use the most efficient layout tools and the most experienced drafting persons. The skill and dedication of drafting workers contribute very significantly to the success of a VLSI project.

Regular structures include PLA-like structures, whose code FETs are at very irregular locations. Irregularity of this kind can be dealt with easily by a computer. A CMOS VLSI chip contains the following regular structures, for each of which dedicated software can be generated. Many of them are already available.

1. Datapath components, such as registers, ROMs, RAMs, adders, shifters, $A = B$ detectors, and multipliers
2. Control components, such as PLAs, microcode ROMs, static/dynamic decoders, local clock decoders, state counters, and datapath interface drivers
3. I/O components, such as output drivers, tristatable drivers, open-drain drivers, input level converters, input/output latches, and clock buffers

It is pleasing to see that the list of regular components is quite long. As for VLSI microprocessor, it is rather difficult to find logic components that are truly irregular. They are mostly random logic circuits interfacing the control section to the datapath section.

The best strategy to design these regular components is to start from past experience. There are always established and reputable designs of these components. They are compact, electrically sound, and fast. There is no need to reinvent these designs. A designer catalogs these designs carefully and provides simple software to modify the design for changing needs. The modification includes stretching the bit width to match the rest of the regular structure (a good layout must always be a compact one), passing extra buses through the structure or removing them, concatenating latch circuits to construct master-slave latches, or increasing control line width for faster control signal propagation. In this way a large part of a datapath can be assembled within a very short time from the stored database.

The most complex datapath component is a multibit adder. A full carry-look-ahead adder is often quite complex. There are, however, other methods of carry generation, as discussed in Sections 7.14 and 7.15, that provide much simpler and regular layout structure.

The design problems of CMOS PLAs have already been studied in considerable detail (Section 5.5), and there is established software that generates layout code from the PLA specification. The PLA can even be folded if performance or compactness is required. A similar approach can be used for microcode ROM, registers, and RAMs, including static and dynamic decoders to drive the word lines. State counters are usually very simple countdown circuits with a few pattern detectors.

Similarly, local clock decoders and datapath control line drivers can be laid out automatically by giving the overall size, location of terminals, and the load. The software to do so is very simple, but must be written individually.

I/O frame design belongs to the same category of jobs. Since I/O buffers surround a VLSI chip as a wooden frame of a painting, the set of I/O buffers and the associated circuits are called I/O frame. The layout of I/O frame is extremely tedious, since many features of I/O frame (e.g., power/ground connections, clock bus and databus connections) must match those of the internal circuit layouts, whose size and coordinates are not precisely known at the early phase of VLSI project. Because of that, design of an I/O frame is considered difficult, and the job is assigned to the most experienced designer. The fact is that the design does not require much experience, if the specifications of the design are well defined and understood. The output drivers do not come with great variety, and neither do the input buffers. In this area, cooperation between a designer and a CAD writer is most desired. Tight input-output critical paths can be dealt with individually once the entire frame is generated by dedicated software. For the time-consuming job of I/O frame design, general-purpose layout software is useless. CAD must be custom-made.

Before any layout work is started, the approximate sizes of all the FETs must be specified by the designer. The designer determines the FET sizes from fanout and from a crude idea about wiring parasitics. It is possible to produce a near-optimum design if this step is carefully executed. The author does not recommend starting layout by using minimum-size FETs. This mode of operation is often supported with the false argument that the most important matter is to generate a "functional" layout as fast as possible. A layout carried out in this way is often unimprovable to meet timing specifications.

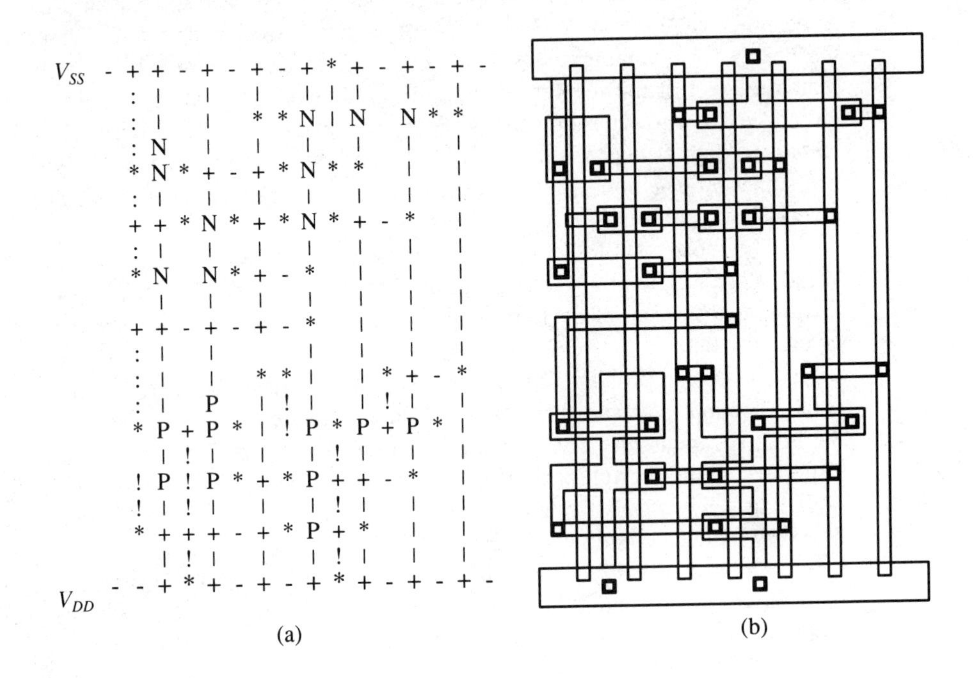

Figure 9.16 Gate-Matrix symbolic layout.

9.18 LAYOUT STYLE

In recent years layout have been carried out more and more symbolically. Creating the final form of mask code by specifying coordinates of all the geometrical features is too inflexible in the environment where designs are updated every time the processing technology is updated, and where the last detail of design rules depends on the accumulated experience of process development.

In a symbolic style of layout, the designer's intentions are stored in one computer file which is readable by a human designer. The Gate-Matrix symbolic code [46] is shown in Fig. 9.16(a). It is easy to identify PFETs, NFETs, polysilicon and metal wires. The information about the design rules is contained in a second file. A

Gate-Matrix compiler reads the two files and produces mask data, shown in Fig. 9.16(b). The symbolic code can be created manually on paper, by computer software [47], or by a CRT terminal. As the symbolic layouts are compiled, some limited compaction is made, and a layout as small as the other specifications permit is generated. Design verification procedures are carried out using the layout data [48].

9.19 FUTURE DIRECTIONS OF CMOS DIGITAL CIRCUIT TECHNOLOGY

Performance of an integrated system measured by operations per second is proportional to the system's clock frequency. Higher frequency means proportionally higher throughput, and from this viewpoint an objective of CMOS digital circuits technology is, simply stated, how to make a circuit work as fast as possible.

A new technology of low-temperature CMOS emerged recently. CMOS logic circuits at liquid-nitrogen temperature are typically two to three times faster than at room temperature. In addition, there are a number of advantages, such as significant reduction in leakage/subthreshold currents, total absence of latch-up, decrease in interconnect resistance, increase in silicon substrate thermal conductivity, tightly controlled circuit parameters, and potentially high reliability [49] [50]. The impressive advantages come with costs. The power necessary for refrigeration is high, the hot carrier degradation mechanism and thermal cycling create new reliability problems, and testing chips and systems is difficult. At present the disadvantages are considered more serious than the advantages, but the evaluation may change. The author believes that a low-temperature CMOS system should be assembled from full-custom, optimized performance CMOS chips. It is wasteful to use slow gate-array or standard-cell chips at low temperatures, since refrigeration cost is high. If we decide to pay the cost, we should not aim at anything less than the highest performance attainable.

Scale-down of the processing technology has been the major driving force to higher speed. Several problems are, however, discernible in this direction [51]. Processing has many new problems [52]. Increased susceptibility of latch-up of submicron devices demands techniques to alleviate this disastrous problem. Increased power dissipation at higher frequencies demands techniques for reducing power while maintaining circuit delay. Increased leakage current of submicron FETs will make it difficult to combine a huge number of variables using PLA-like dynamic gates. An increase in the ratio of wiring capacitance to the device capacitance deteriorates speed improvement by scale-down of technology. These problems present big challenges to CMOS circuit designers of the future. It is necessary to recover the lost speed improvement by better circuit design [53].

Higher performance of an integrated system, however, depends not only on the circuit and device technology that makes the clock frequency higher, but also on the system's architecture. High-performance systems often require chips that contain large amounts of memory. Memory cells have the highest device density in a CMOS VLSI chip. Typical static memory cells have a 400 to 700-μm square area with 1.5 to 2.0-

μm technology, however, do not allow a large memory on a chip. To use high-capacity dynamic memory will be an interesting possibility. Dynamic memory cells will scale down to the smaller size, using trenched capacitor storage cells.

Much room for improvement is left for device density in a CMOS VLSI chip. The average area occupied by a device (around 500 μm square) is several times larger than the minimum area necessary, even if due allowance is made for FET scaling. Multilevel metal interconnects solve the problem only partially. Proper use of multilevel interconnects is not obvious, and the optimum way must be worked out.

The increased number of I/O terminals will require an entirely new technology, Wafer-Scale Interconnects (WSI). The advantage offered by the new technology is quite significant. Reduction of parasitic inductance and capacitance by an order of magnitude appears feasible. The small parasitics make interchip communication easier, to the point that the chips on a single WSI module look like a single large chip. The performance advantage of such a system is great. In a WSI environment, synchronization of clocks of all the chips on the substrate is very important. How to achieve this objective is a new challenge to circuit designers.

The roles that CAD will play in future CMOS VLSI technology will be diverse and essential. What CAD software should do, however, has not been defined satisfactorily. The compatibility of CADs that are used to design a single chip, the logic simulator, the circuit simulator, the layout aids, the design rule checker, and the timing verifier are the most important. The software should work on a set of consistent data that are produced from the logic design specifications and the layout data (these two are the basic design data of CMOS VLSI chip). For CMOS VLSI design, improvement in the gate-level timing simulators is the most important. It is desired that a CMOS VLSI design can be executed without heavy dependence on costly circuit simulation.

There is an unresolved problem of how to conceive good circuits. This essential starting point of design, in the author's opinion, belongs permanently to experienced designers. To conceive good circuits or to choose the best one out of many alternatives requires wide experience and deep understanding of how a CMOS circuit works. This important point has escaped public attention for too long.

Several topics in this book have originated from the author's own work. The ideas appeared unfamiliar to all when they were first presented but neverthless were immediately accepted as resolving the most urgent design problems of the time. This book is intended to show that much more should be learned from seemingly trivial CMOS circuits, and that a deep understanding of CMOS circuits often resolves design problems that once looked formidable.

9.20 REFERENCES

01. J. T. Wallmark "Noise spikes in digital VLSI circuits" IEEE Trans. on Electron Devices, vol. ED-29, pp. 451-458, March 1982

02. R. L. M. Dang and N. Shigyo "Coupling capacitances for two-dimensional wires" IEEE

Electron Device Letters, vol. EDL-2, pp. 196-197, August 1981.
03. E. T. Lewis "An analysis of interconnect line capacitance and coupling for VLSI circuits" Solid-State Electronics, vol. 27, pp. 741-749, August 1984.
04. M. Shoji "Reliable chip design method in high performance CMOS VLSI" ICCD86 Digest, pp. 389-392, October 1986.
05. M. Shoji "Electrical design of BellMac32A microprocessor" ICCC82 Digest, pp. 112-115, October 1982.
06. H-F S. Law and M. Shoji "PLA design for BellMac32A microprocessor" ICCC82 Digest, pp. 161-164, October 1982.
07. S. P. Sacarisen, M. A. Stambaugh, P. W. Lou, A. Khosrovi and K. S. Chang "A VLSI communication processor designed for testability" ISSCC84 Digest, pp. 172-173, February 1984.
08. R. F. Ficchi "Electrical interference" Hayden Book Company, New York, 1964.
09. M. L. Cortes, E. J. McCluskey, K. D. Wagner and D. J. Lu "Modeling power-supply disturbances in digital circuit" ISSCC86 Digest, pp. 164-165, February 1986.
10. J. Lohstroh, E. Seevinck and J. de Groot "Worst-case static noise margin criteria for logic circuits and their mathematical equivalence" IEEE J. Solid-State Circuits, vol. SC-18, pp. 803-807, December 1983.
11. J. R. Burns "Switching response of complementary-symmetry MOS transistor logic circuits" RCA Review, vol. 25, pp. 627-661, December 1964.
12. C. W. Ho, D. A. Chance, C. H. Bajorek and R. E. Acosta "The thin-film module as a high-performance semiconductor package" IBM J. Res. and Develop., vol. 26, pp. 286-296, May 1982.
13. H. J. Levinstein, C. J. Bartlett and W. J. Bertram, Jr. Multi-chip packaging technology for VLSI-based system" ISSCC87 dugest, pp. 224-225, February 1987.
14. K. L. Chen, G. Giles and D. B. Scott "Electrostatic discharge protection for one micron CMOS devices and circuits" IEDM86 Digest, pp. 484-487, December 1986.
15. Y. Mizutani, K. Maeguchi, T. Mochizuki, M. Kimura, M. Isobe, Y. Uchida and H. Tango "High speed $MoSi_2$ -gate CMOS/SOS devices Proceedings of 12th conference on solid state devices" Japanese J. Appl. Phys., vol. 20, pp. 117-122, 1981.
16. E. Fujishin, K. Garrett, M. P. Levis, R. F. Motta and M. Hartranft "Optimized ESD protection circuits for high-speed MOS/VLSI" IEEE J. Solid-State Circuits, vol. 594-596, April 1985.
17. R. N. Rountree and C. L. Hutchins "NMOS protection circuitry" IEEE Trans. on Electron Devices, vol. ED-32, pp. 910-917, May 1985.
18. R. L. Wadsack "Fault modeling and logic simulation of CMOS and MOS integrated circuits" Bell System Technical Journal, vol. 57, pp. 1449-1474, May 1978.
19. E. J. McCluskey "Logic design principles with emphasis on testable semicustom circuits" Prentice Hall, Inc., Englewood Cliffs, N. J., 1986.
20. M. T. M. R. Segres "The impact of testing on VLSI design methods" IEEE J. Solid-State Circuits, vol. SC-17, pp. 481-486, June 1982.
21. S. P. Sacarisen, M. A. Stambaugh, P. W. Lou, A. Khosrovi and K. S. Chang "A VLSI communication processor designed for testability" ISSCC84 Digest, pp. 172-173, February 1984.
22. R. W. Heckelman and D. K. Bhavsar "Self-testing VLSI" ISSCC81 Digest, pp. 174-175, February 1981.
23. U. Theus "A self-testing ROM device" ISSCC81 Digest, pp. 176-177, February 1981.

24. E. J. McCluskey "VLSI design for testability" 1984 Symposium on VLSI technology, pp. 2-5.
25. R. Hoshikawa, H. Kikuchi, S. Baba, S. Sato, K. Kawato, N. Inui and O. Wada "A 10, 000-gate CMOS LSI processor" ISSCC80 Digest, pp. 106-107, February 1980.
26. S. Kuboki, I. Masuda, T. Hayashi and S. Torii "A 4K CMOS gate array with automatically-generated test circuits" ISSCC85 Digest, pp. 128-129, February 1985.
27. R. R. Troutman and H. P. Zappe "A transient analysis of latchup in bulk CMOS" IEEE Trans. on Electron Devices, vol. ED-30, pp. 170-179, February 1983.
28. D. B. Estreich, A. Ochoa, Jr. and R. W. Dutton "An analysis of latch-up prevention in CMOS IC's using an epitaxial-buried layer process" IEDM78 Digest, pp. 230-234, December 1978.
29. R. E. Bank, W. M. Coughran, Jr., W. Fichtner, E. H. Grosse, D. J. Rose and K. Smith "Transient simulation of silicon devices and circuits" IEEE Trans. on Electron Devices, vol. ED-32, pp. 1992-2007, October 1985.
30. R. S. Payne, W. N. Grant and W. J. Bertram "Elimination of latchup in bulk CMOS" IEDM80 Digest, pp. 248-250, December 1980.
31. Y. Niitsu, G. Sasaki, H. Nihira and K. Kanzaki "Resistance modulation effect in n-well CMOS" IEEE Trans. on Electron Devices, vol. ED-32, pp. 2227-2231, November 1985.
32. C. N. Anagnostopoulos, P. M. Zeitzoff, K. Y. Wong and B. Brandt "An isolated vertical npn bipolar transistor in an n-well CMOS process" CICC84 Digest, pp. 588-593, 1984.
33. J. E. Gover "Overview of the basic effects of radiation in semiconductors" Sandia Laboratories Report 82-0815, April 1982.
34. J. Y. Chen, R. C. Henderson, R. Martin and D. O. Patterson "Enhanced radiation effects on submicron narrow-channel NMOS" IEEE Trans. on Nuclear Science, vol. NS-29, pp. 1681-1684, December 1982.
35. H. Hatano and M. Shibuya "CMOS logic circuit optimum design for radiation tolerance" Electronics letters, vol. 19, pp. 977-978, November 1983.
36. J. T. Nelson "Hot carrier and trapping" Wafer level reliability assessment workshop, pp. 33-57, Stanford Sierra Lodge, October 24-27, 1982.
37. T. Sakurai, M. Kakumu and T. Iizuka "Hot-carrier suppressed VLSI with submicron geometry" ISSCC85 Digest, pp. 272-273, February 1985.
38. J. Peeples "Soft errors" Wafer level reliability assessment workshop, pp. 79-87, Stanford Sierra Lodge, October 24-27, 1982.
39. C. Hu "Drift collection of alpha generated carriers and design implications" ISSCC82 Digest, pp. 18-19, February 1982.
40. T. Masuhara, O. Minato, T. Sakai, H. Nakamura and Y. Sakai "2K $\times$ 8b HCMOS static RAMs" ISSCC80 Digest, pp. 224-225, February 1980.
41. G. A. Sai-Halasz "Alpha-particle-induced soft error rate modeling" ISSCC82 Digest, pp. 20-21, February 1982.
42. J. L. Andrews, J. E. Schroeder, B. L. Gingerich, W. A. Kolasinski, R. Koga and S. E. Diehl "Single event error immune CMOS RAM" IEEE Trans. on Nuclear Science, vol. NS-29, pp. 2040-2043, December 1982.
43. B. Chappell, S. Schuster and G. Sai-Halasz "Stability and soft error rates of SRAM cells" ISSCC84 Digest, pp. 162-163, February 1984.
44. B. T. Murphy "Cost-size optima of monolithic integrated circuits" Proc. IEEE vol. 52, pp. 1537-1545, December 1964.
45. C. H. Stapper "Yield model for 256K RAMs and beyond" ISSCC82 Digest, pp. 12-13,

February 1982.
46. A. D. Lopez and H. F. Law "A dense Gate-Matrix layout style for MOS LSI" ISSCC80 Digest, pp. 212-213, February 1980.
47. O. Wing, S. Huang and R. Wang "Gate-Matrix layout" IEEE Trans. on Computer-Aided Design, vol. CAD-4, pp. 220-231, July 1985.
48. S. M. Kang, R. H. Krambeck, H. F. S. Law and A. D. Lopez "Gate-Matrix layout of random control logic in a 32-bit CMOS CPU chip adaptable to evolving logic design" IEEE Trans. om Computer-Aided Design, vol. CAD-2, pp. 18-29, January 1983.
49. F. H. Gaensslen, V. L. Rideout and J. J. Walker "Very small MOSFETs for low-temperature operation" IEEE Trans. on Electron Devices, vol. ED-24, pp. 218-229, 1977.
50. J. D. Plummer "Low temperature CMOS devices and technology" IEDM86 Digest, pp. 378-381, December 1986.
51. J. D. Meindl "Theoretical, practical and analogical limits in ULSI" IEDM83 Digest, pp. 8-13, December 1983.
52. E. Arai "VLSI fine technology and its problems" Japan J. Appl. Phys., vol. 21, pp. 43-49, January 1982.
53. E. Takeda, G. A. Jones and H. Ahmed "Constraints on the application of 0. 5 micrometer MOSFETs" IEEE J. Solid-State Circuits, vol. SC-20, pp. 242-247, February 1985.

Index

E

F

G

H

I

L

M

N

O

P

Q

R

S

T

V

W

Y

Z